Sensors: Devices, Circuits and Systems

Sensors: Devices, Circuits and Systems

Edited by Isabella Edwards

STATES
ACADEMIC PRESS
www.statesacademicpress.com

States Academic Press,
109 South 5th Street,
Brooklyn, NY 11249, USA

ISBN: 978-1-63989-480-2

Cataloging-in-Publication Data

Sensors : devices, circuits and systems / edited by Isabella Edwards.
p. cm.
Includes bibliographical references and index.
ISBN 978-1-63989-480-2
1. Detectors. 2. Detectors--Design and construction. 3. Sensor networks.
4. Detectors--Data processing. I. Edwards, Isabella.
TK7872.D48 S46 2021
681.2--dc23

For information on all States Academic Press publications
visit our website at www.statesacademicpress.com

STATES
ACADEMIC PRESS

Contents

Preface

A sensor is a device, subsystem or module which detects changes or events in its environment. It sends the information to electronic devices such as a computer processor. All living organisms have biological sensors that are sensitive to light, motion, magnetic fields, temperature, humidity or pressure, or that can detect toxins, pheromones, nutrients, or internal metabolic indicators. Sensor devices can be designed to measure different physical parameters, such as geophones, hydrophones and seismometers can measure sounds, while air pollution sensors, actinometers and gas detectors can detect changes in the environment. Sensor accuracy and precision, and sensor deviations are essential attributes of sensors that determine their efficiency. This book elucidates new techniques and applications of sensors in a multidisciplinary manner. It includes some of the vital pieces of work being conducted across the world, on sensor devices, circuits and systems. This book aims to equip students and experts with the advanced topics and upcoming concepts of this technology.

This book is the end result of constructive efforts and intensive research done by experts in this field. The aim of this book is to enlighten the readers with recent information in this area of research. The information provided in this profound book would serve as a valuable reference to students and researchers in this field.

At the end, I would like to thank all the authors for devoting their precious time and providing their valuable contribution to this book. I would also like to express my gratitude to my fellow colleagues who encouraged me throughout the process.

Editor

The Use of Sensors for Monitoring the Feeding Process and Adjusting the Feed Supply Velocity in Fish Farms

Lorena Parra,[1] **Laura García**,[1] **Sandra Sendra**,[2] **and Jaime Lloret**[1]

[1]*Instituto de Investigación para la Gestión Integrada de Zonas Costeras, Universitat Politècnica de València, Spain*
[2]*Dep. de Teoría de la Señal, Telemática y Comunicaciones, ETS Ingenierías Informática y de Telecomunicación, Universidad de Granada, C/Periodista Daniel Saucedo Aranda, s/n E-18071 Granada, Spain*

Correspondence should be addressed to Jaime Lloret; jlloret@dcom.upv.es

Academic Editor: Romeo Bernini

Aquaculture is a growing industry, and its sustainability is crucial. One of its major environmental impacts is the uneaten feed that pollutes the water. To minimize the uneaten feed, many systems have been developed. Nevertheless, current systems can be improved by considering the fish position in the tank and the falling feed. In this paper, we propose a system based on fish presence sensors set at different tank heights and a feed detection sensor located in the drainage tubes. The fish presence sensor is based on light-dependent resistor (LDR). The calibration of these sensors is shown. When the output voltage is higher than 1.467 V, we can consider that fish are present. On the other side, the falling feed sensor is based on a CMOS sensor. The calibration process is performed with 40 pictures. The summation of pixels, with brightness value between 0 and 15 in the blue histogram, is used as an indicator of feed presence. If this value is higher than 520 pixels, we can consider that there is feed in the picture. Moreover, a verification process of both sensors is done. The results of the verification confirm the calibration. Finally, the operation of the system is shown.

1. Introduction

Sustainability of aquaculture is a pressing matter as the demand of fish and seafood is constantly increasing. A production of 59.9 MT/year was obtained in 2010, and in 2030, it is expected to reach 85 MT/year [1]. Fish farms can be implemented both in the sea and inland. Sea facilities are comprised of cages where fish are kept separated from other wildlife. Inland facilities build tanks where the conditions that affect the performance of fish are controlled. Illumination, turbidity, temperature, are salinity are an example of the factors that can influence the behavior of fish and must be monitored [2]. Principally, sustainable aquaculture facilities should not disrupt the ecosystem where they are placed. Business and social factors, such as being profitable and aiding in the well-being of the community surrounding the facilities, are considered as well [3]. Therefore, if only the environmental factors are considered, the key factors to evaluate a sustainable aquaculture facility are land use, water use, energy use, feed use, and freshwater seed. Moreover, diverse technological systems have been developed in order to assess the sustainability of fish farms. Some of these solutions are designed as fixed tanks and filters that measure specific values of water quality [4]; other solutions can be implemented in existing aquaculture facilities incorporating wireless functionalities that allow visualizing the information on a PC or smartphone [5]. The information available on sustainable aquaculture allows implementing environment-conscious fish farms. However, although aquaculture allows providing access to food for a large part of the population, several side effects have been noticed.

Aquaculture can introduce a set of problems to the environment [6]. Particularly, excess feeding is the cause of some of the side effects of fish farming. Eutrophication is an increase in nitrogen and phosphorous in the water as a result of overfeeding fish [7]. Moreover, an increase in inorganic nitrogen can derive in acidification of freshwater ecosystems and survival, growth, and reproduction impairments when toxic levels are reached. This contaminated water affects not only the fish consumed by end users but can also affect the environment where said water is discarded. In addition, consumers of polluted water may develop methemoglobinemia, cancer in the digestive tract, birth defects, mutagenicity, and teratogenicity. Nausea, diarrhea, vomiting, muscular cramps, gastroenteritis, or pneumonia can be caused by algal toxins. Furthermore, the lack of consistent regulation does not contribute to the solution of these problems [8]. Different countries implement different regulations, so even in places with stricter regulations, fish with toxicity problems produced in other areas can be commercialized to other countries spreading the health problems derived from it. Although guidelines have been provided in order to reduce the amount of pollution, there is a great need of implementing solutions to control the feeding process performed in fish farms [9]. These solutions are not only of interest for the environment but also to the producers as reducing the amount of feed results in a reduction of the cost on their commercialization.

For the detection of excess feeding, it is necessary to determine the behavior of fish in the cage during the feeding process. When fish are hungry, their behavior is to rush to the feeding area to ingest the food [10]. However, when excess feeding is provided, a big part of the fish in the cage may continue with the usual behavior they have when feed is not provided, letting it drops to the bottom of the cage. This feed is not consumed later on, resulting in the pollution of the water and the subsequent dangers to the environment and human health. For that reason, detecting when feed is not being eaten is of great importance.

In this paper, we present a system that automatically adjusts the amount of dispensed feed. In order to do so, the system detects when feed reaches the drainage system. The feed detection is done using a CMOS sensor. From the data gathered by the CMOS sensor, we obtain the histograms. After analyzing them, we can find a correlation between the number of pixels with certain brightness value and the presence of feed. Moreover, the height at which fish are swimming is detected by employing light-dependent resistor (LDR) strips deployed from the top to the bottom of the cage. The fish are detected due to the changes in the incident light in the LDR caused by the fish swimming behavior. The fish covered by scales acts as mirrors, reflecting the light and some of the flashes inside in the LDRs placed in the tanks. We show the calibration and verification process of both sensors. In addition, we present the results of the simulated feeding process with our proposed system and how the feed supply velocity changes. Our system allows saving feed during feeding time and ensures that all the fish have time to eat, making aquacultural facilities more economically profitable as well as more sustainable. Although the system has been designed for fish tanks, this paper is the first step of the development of a fish feeding system that can be utilized in both inland fish tanks and sea cages. In future research, the system will be developed for sea cages as well, where the benefits of the presented system will be substantial.

The rest of the paper is organized as follows. In Section 2, the related work is presented. Section 3 details the proposed system, including the scenario description and the hardware employed to gather the data. The results of the calibration and verification process and discursion of the proposed system are presented in Section 4. Finally, Section 5 shows the conclusion and future work.

2. Related Work

In this section, the related work on water quality monitoring systems and fish feeding solutions is presented.

Fish behavior is a crucial factor in determining the performance of the fish in aquacultural facilities. Therefore, various researches have been performed on this matter. A fish behavior monitoring system was implemented by Papadakis et al. in [11]. It employed computer vision in order to monitor 9 fish tanks simultaneously. Stock density was evaluated as a stress factor for fish in fish farms. Real-time images were provided to a remote control application. Only the experiment with the undamaged mesh obtained a significant statistical difference. Saberioon et al. performed in [12] a survey on vision-based fish behavior monitoring systems. Machine vision, thermal imaging, hyperspectral imaging, and X-rays were the technologies for fish monitoring discussed by the authors. They also divided the applications of optical sensors into preharvesting and during cultivation conditions and postharvesting conditions. Moreover, they defined five applications for optical sensors in fish monitoring systems. These applications were physical attributes, chemical attributes, fish sorting, fish quality, and food security. Armstrong et al. presented in [13] a flat passive integrated transponder antenna array that monitored fish behavior by recording the movements performed by salmon shoals. The antenna array did not incite any unusual behavior on fish. Furthermore, a 99% success rate was obtained in the experiments. They were done by varying the number of salmons in each one of the experiments. Fish growth, density, and fish behavior were monitored by Conti et al. in [14]. Fish behavior was detected employing the scattering cross. If an anomaly was detected, an alarm was triggered. Moreover, first-order and second-order polynomial equations were utilized in order to monitor growth. The behavior of sardines, sea bass, and rockfish were studied to perform the experiments. Zhang et al. employed in [15] an imaging sonar called DIDSON (dual-frequency identification sonar) to monitor Chinese sturgeons. Swimming patterns and the length of over 2500 targets were studied in the experiments. Results showed a relation between the body length and swimming pattern and a 35.6% decrease in the length measured by DIDSON in comparison to manual measurements. Fish were found to swim close to the net and in circular motions.

Feeding is another important factor to owners of fish farms as excess feed increases the production cost and contaminates the water. For that reason, the number of researches on fish feeding solutions is constantly increasing. A fish feeding system that considered the behavior of fish in order to determine the best time to provide the food was presented by AlZubi et al. in [10]. The hardware design included an automatic dispenser, a webcam, and an interface circuit. The camera detected the number of fish that went to the feeding area; when the fish learning index (FLI) was higher than the threshold, the system provided food. Results showed the variation of the number of times food is provided for eight days. Atoum et al. introduced in [16] a feeding system for aquaculture tanks that employs visual signal processing. The system included a decision-making process to determine whether fish are eating or not and detection of excess feed. Excess feed was detected employing two different methods, being a support vector machine- (SVM-) based refinement classifier and a correlation filter. Bórquez-Lopez et al. performed in [17] a comparison between mathematical functions and fuzzy logic (FL) feeding techniques for shrimp farming. Authors determined that dissolved oxygen influences the feeding rate the most (74%), and temperature influences it in a 26%. Moreover, they concluded that FL is the better strategy saving up to a 35% of feed without affecting growth or survival. Papandroulakis et al. presented in [18] an automated feeding system for intensive fish farms. The system considered the necessary amount of plankton that had to be distributed to each tank. Experiments were performed by applying the proposed method to four groups of sea bream and comparing it with the results obtained from two groups fed with the standard method. A reduction between 30% and 40% in labor was achieved, and the use of *Artemia nauplii* decreased by 40%. Garcia et al. presented in [19] a fish feeding system that employed sensors that determine when fish need feed. Sensors were deployed both inside and outside the cage. These were temperature, oxygen, displacement speed, biomass distribution placement, and pellet detection sensors for the internal part of the cage and presence and water current sensors for the outside. Covès et al. performed in [20] a set of experiments were an on-demand feeding system was employed to provide feed to 50 sea bass. The two experiments were conducted during 55 and 69 days, respectively. Moreover, a PIT tag was utilized to determine whether fish activated the trigger or not. Results showed that a 67% and 74% of the fish in experiments 1 and 2 activated the trigger at least once. Furthermore, two fish in experiment 1 were responsible for 82% of the triggers and one fish in experiment 2 was responsible of 77% of the triggers. Zhou et al. present in [21] a feeding decision system based on a neuro-fuzzy model and infrared computer vision. Fish feeding behavior was quantized employing an algorithm that obtains an index as a result by employing image texture and Delaunay triangulation. Another algorithm performed the decision of feeding or stop feeding the fish during the feeding process. It employed an adaptive network-based fuzzy inference system (ANFIS). Finally, weight gain, growth and feed conversion rate, and water quality were measured to evaluate the performance of the system. Results show a 98% of ANFIS decision accuracy and a reduction of the feed conversion rate of 10.77%. Lastly, Zhou et al. also present in [22] a near infrared fish feeing behavior system based on imaging techniques that employ the gray-level gradient cooccurrence matrix, the support vector machine, and the Delaunay triangulation. As a result, the flocking index of fish feeding behavior (FIFFB) was obtained. The obtained linear correlation of the FIFFB was 0.945 confirming that the FIFFB can be employed to quantify feeding behavior.

Most feeding management systems rely on fish behavior in order to determine whether fish are eating or not. In this paper, we present a system for automatic detection of excess feeding that determines if the feed is being dropped to the bottom of the cage. Moreover, the height of the cage where fish are swimming can also be detected by employing LED strips on the water tanks.

3. Materials and Methods

In this section, we show the system description. First, we describe the scenario where our system is calibrated and tested. In addition, we detail the hardware part of our system including the employed sensors, node, and connections.

3.1. Scenario Description. In this subsection, we present the scenario where our system is proposed and where the sensors were calibrated. First, we describe the system and its operation. Then, we detail how we test the sensors for the calibration.

The proposed system is able to change the feed supply velocity according to the data gathered by the fish presence sensors and falling feed detector. The system is comprised of an automatic feeder tube, which allows changing the feed supply velocity, see Figure 1. Four different velocities can be selected: 100%, 50%, 25%, and 5% of the usual feed supply velocity. The system always starts to feed with the 100% of the velocity. During the feeding process, the velocity will decrease according to the sensor signals. The system is controlled by the Arduino which is placed in the exterior of the tank.

The fish presence sensors are placed inside a Plexiglas tube as it was presented in [2]. A total of 9 sensors are located along the tank at different depths. The fish presence sensor is comprised of LDRs that are able to detect the changes in the received illumination due to the moving fish. The first LDR, LDR 1, is placed at 5 cm below the water surface. This is the area where the fish used to be during the feeding process. The second LDR, LDR 2, is located at 30 cm from the water surface. The rest of the LDRs, LDR 2 to LDR 9, are spaced 15 cm apart. The Plexiglas tube with the LDRs is fixed to the tank walls in the same side where the feeder tube dispenses the feed. The Plexiglas tube is sealed in both extremes.

Finally, the falling feed detector is located at the bottom of the tank in the drainage tube. This sensor is comprised by a pellet detector sensor, and illumination is provided by a white LED. The first part of the drainage systems, which is usually comprised of an elbow pipe, will be changed to a T-shaped one, and a methacrylate separator will be added in one of the shorter sides. Thus, the camera can have full vision of the drainage system without changing the water flow.

3.2. Node and Sensor Description. In this subsection, we present the employed sensors and the selected node are described.

One of the main elements that forms this system is the control node in charge of monitoring the swimming height of the fish. As Figure 2 shows, to develop our control node, we use an Arduino Mega 2560 Rev. 3 board. This board is based on the ATmega2560 microcontroller. It has 54 digital

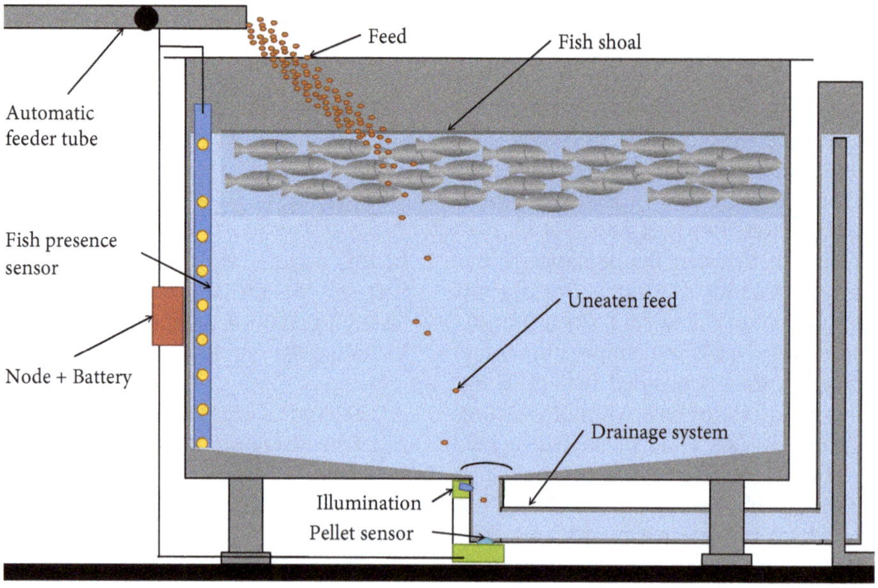

FIGURE 1: Proposed system for feed supply.

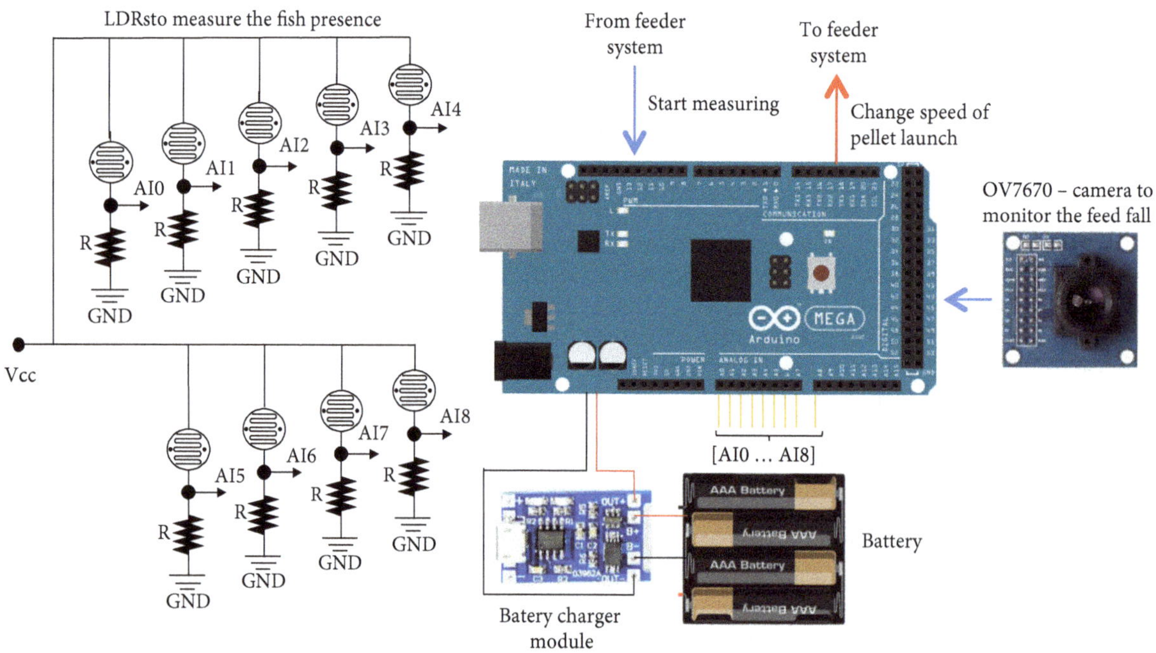

FIGURE 2: Control node description.

input/output pins (where 14 of them can be used as PWM outputs), 16 analog inputs, 4 UARTs (hardware ports), and a 16 MHz crystal oscillator. In addition, the board contains 256 kB of flash memory, 8 kB of SRAM, and 4 kB of EEPROM. The board can be powered by batteries (which is our case) or with an AC-DC adapter.

The operation of the node is as follows. On the one hand, it is not necessary that the set of LDRs is always in operation; it is only required when the process of feeding fish is carried out. Therefore, when the feeder starts pouring food into the water, the node receives a signal to start monitoring. As the fish descends, the node will send different orders to the feeder in order to reduce the speed of food pouring into the water. In this way, the amount of food thrown into the water is adapted to the amount of fish available to eat it. The other important element is the OV7670 camera which controls the point in time the feed starts to reach the floor. The OV7670 camera takes pictures with a resolution of 640 × 480 VGA and presents a high sensitivity for low-light operation. Due to its low energy requirements, the OV7670 camera module is suitable for embedded portable applications. Finally, this implies a lower waste of food which entails important economic improvements in the aquaculture production.

TABLE 1: Code for obtaining the histograms.

```
Function Analyze_image (){
    Read picture ('picture1.png');
    Divide image into RGB components;
    Define brightness levels [1 to 256];
    Count pixels in each brightness level;
    Show RGB histograms:
}
```

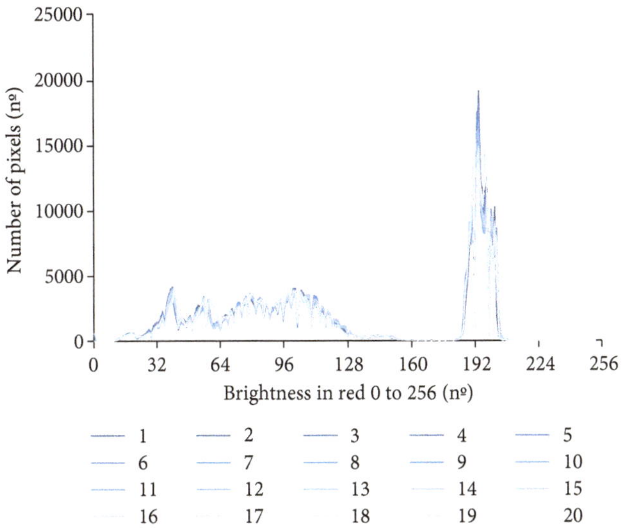

FIGURE 3: Red histogram of pictures without feed.

4. Results and Discussion

In this section, we are going to present the results. First, we show the tests done with the camera for feed detection. Then, the calibration of a system for locating the fish in the tank is presented. Following this, the verification process of both systems is shown. Finally, we present the operation of the system during a feeding process in aquaculture tanks.

4.1. Calibration of the Feed Detection System. In this subsection, the calibration of the feed detection system is presented. We gathered 20 pictures without feed in the water and 20 pictures with feed in controlled conditions. The pictures have a size of 640×480 pixels. In order to obtain the histograms, the following code is used, see Table 1.

In order to find in which part of the histogram it is possible to identify the presence of feed in the water, we represent in Figure 3 the red histogram of the pictures with feed and in Figure 4, the red histogram of pictures without feed. Figure 5 presents the green histogram of the pictures with feed, and in Figure 6, the green histogram of pictures without feed is displayed. Lastly, Figure 7 presents the blue histogram of the pictures with feed and in Figure 8, the blue histogram of pictures without feed is presented. From Figures 3 to 8, we can highlight the following items. First, the histograms of the pictures without feed are much more similar to each other than the pictures with feed. This is due to the pictures without feed being almost the same. However, the pictures with

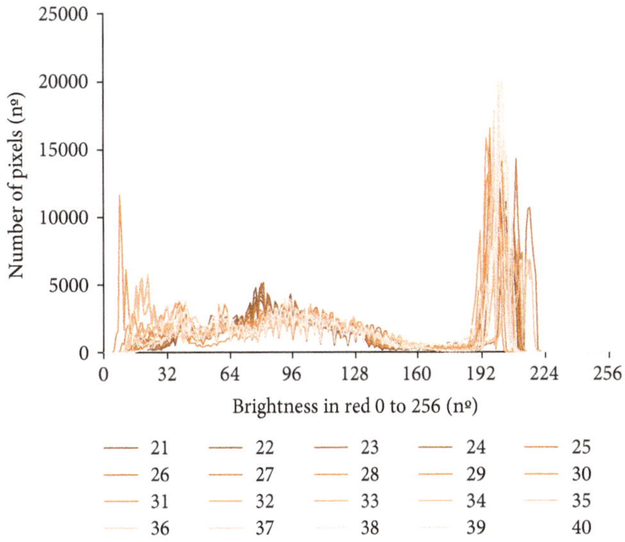

FIGURE 4: Red histogram of pictures with feed.

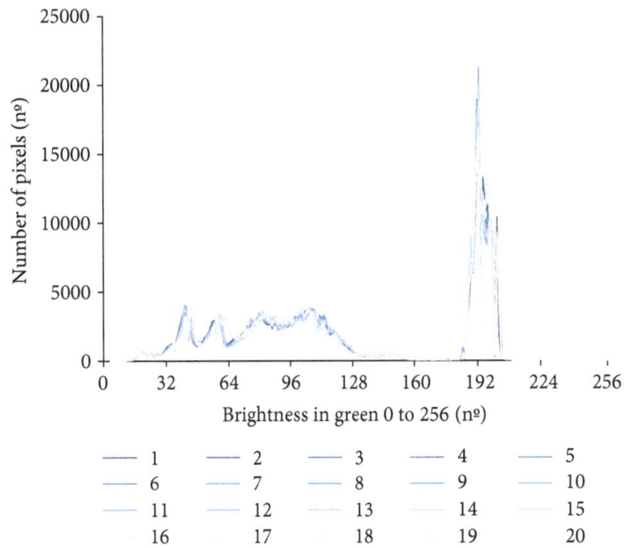

FIGURE 5: Green histogram of pictures without feed.

feed can be very different. In some cases, as in the pictures 28, 31, 33, 34, or 35, the feed pellets cover almost the 20% of the picture. While in other cases as 21, 24, 26, 29, or 38, the pellets cover less than 1% of the picture. Thus, there is a high heterogeneity in the pictures with pellets, which will make the correct detection of pellets in the pictures difficult. The second idea that must be highlighted is that, in general terms, the red, green, and blue histograms are quite similar. This is because the main part of the picture (the PVC tube) is dark grey and the water illuminated by the LED system is light grey. In the histograms of the pictures without pellets, we find two main groups of pixels. The first one is formed by pixels with values between 16 and 157. This part of the histograms represents the PVC tube. The second group of pixels has values between 182 and 205, which represents the water

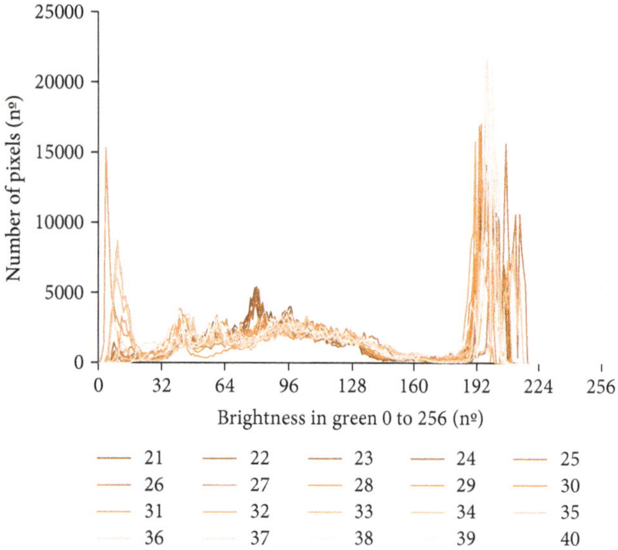

FIGURE 6: Green histogram of pictures with feed.

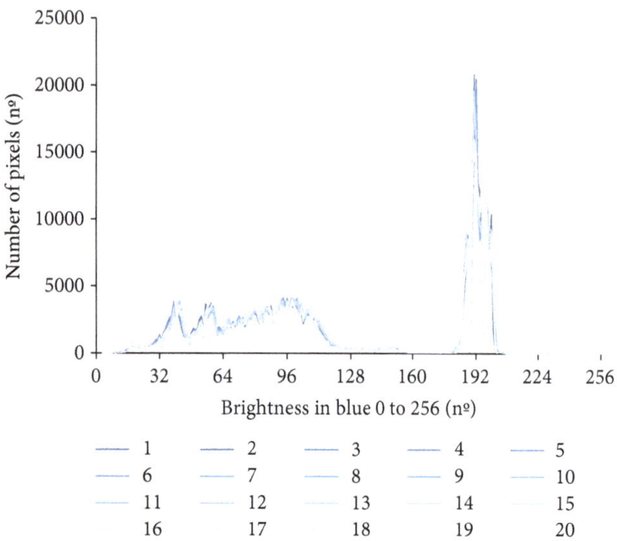

FIGURE 7: Blue histogram of pictures without feed.

illuminated by the flash. Lastly, in Figures 9 to 11, we can observe that in the pictures with feed, there are more pixels in the darkest tones, with low values of brightness (between 0 and 15) than in the pictures without feed. This pattern is found in the three colors. Nevertheless, the difference between the histogram with feed and histogram without feed is greater in the blue histograms.

The next step is to represent the summation of the pixels with lower values of brightness. The summation is done with pixels that present values between 0 and 15. The reason to select those values is due to the pictures without feed which, in general, have a low number of pixels with brightness values lower than 15. The pictures 1 to 20 represent the pictures without pellets that appear in Figures 3, 5, and 7. The pictures 21 to 40 are the pictures with pellets represented in Figures 4, 6, and 8. In Figure 9, we present the summation of pixels with

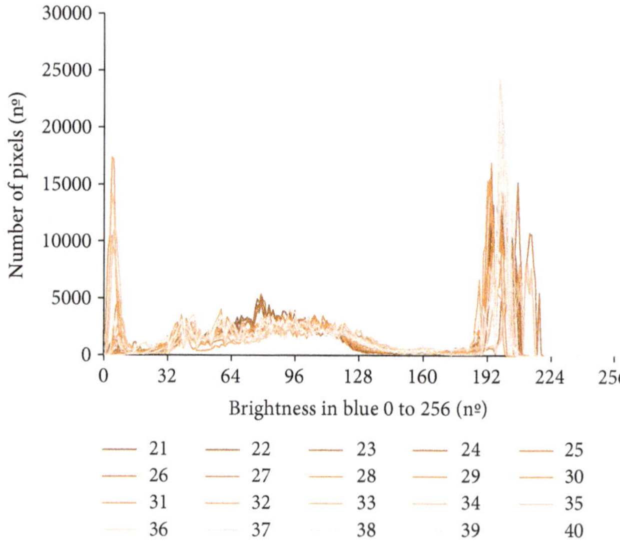

FIGURE 8: Blue histogram of pictures with feed.

low brightness values for red histograms. The summation for green and blue histograms is presented in Figures 10 and 11. In Figures 9 to 11, we present the summation of the pictures with and without feed pellets. In Figure 9, it is possible to see that the pictures 1 to 20 have a mean value of 937 pixels with brightness values between 0 and 15. The minimum and maximum values in pictures 1 to 20 are 672 (picture 6) and 1219 (picture 4), respectively. On the other hand, the mean value in the pictures 21 to 40 is 5038 pixels. Picture 31 presents the minimum value, 42 pixels. The maximum value of pixels, 41134 pixels, is found in picture 40.

Now, we analyze the data from Figure 10. In this case, the mean value of pictures 1 to 20 is 285. The mean value of pixels with brightness values between 0 and 15 in pictures without pellets in the green histogram is lower than in the red histogram. The minimum and maximum values in the summation are 154 (picture 20) and 349 (picture 5), respectively. Secondly, the pictures 21 to 40 present a mean value of 14645 pixels. The minimum value is found in picture 21 with 236 pixels, and the maximum value is in picture 31 with 62594 pixels.

Lastly, we present the summation of pixels with brightness values between 0 and 15 from the blue histogram in Figure 11. The mean value of pictures 1 to 20 is 279 pixels. The maximum and minimum values in the pictures without pellets are 105 (picture 20) and 473 (picture 2), respectively. On the other hand, the mean values of pictures 21 to 40 is 18228 pixels. Picture 40 is the one that presents the lower value, 683 pixels. On the contrary, picture 31 presents the maximum value of pixels with brightness values between 0 and 15, 74005 pixels.

As the maximum differences are found in the blue histogram, we use this histogram for further analysis with statistical software [23]. The first test is a descriptive analysis to confirm or diminish if the data follows a normal distribution. The data is divided into two variables. The variable A represents the data from pictures 1 to 20, and the variable B is the data from pictures 21 to 40. The descriptive analysis of

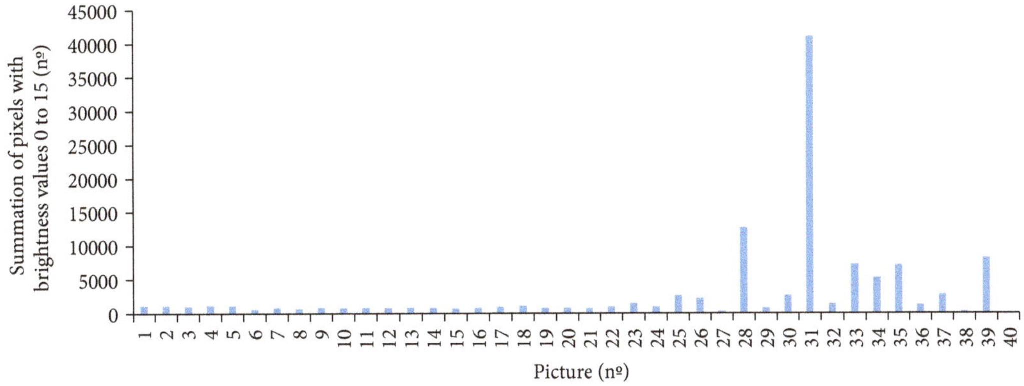

FIGURE 9: Summation of pixels with brightness values between 0 and 15 in the red histogram.

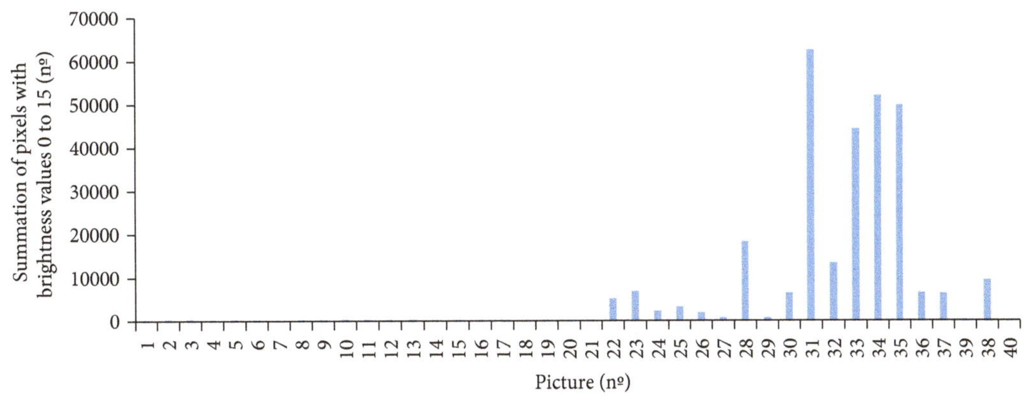

FIGURE 10: Summation of pixels with brightness values between 0 and 15 in the green histogram.

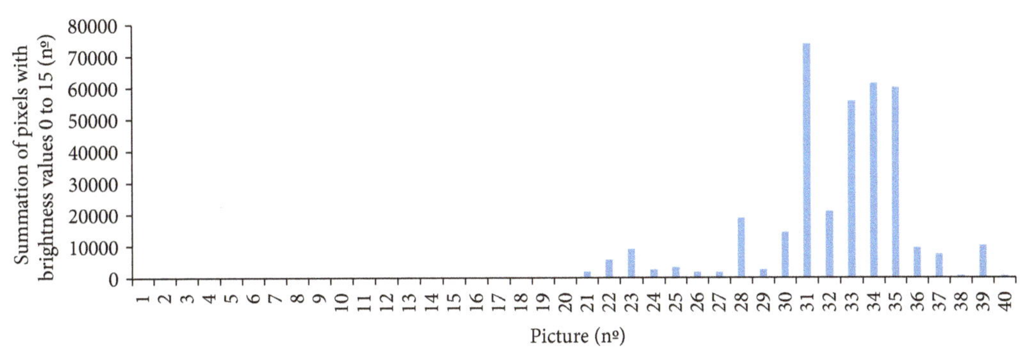

FIGURE 11: Summation of pixels with brightness values between 0 and 15 in the blue histogram.

variable A gives resulting skewness and kurtosis values of 0.876 and 0.417, respectively. Thus, it follows a normal distribution. On the contrary, the results of the test with variable B are 2.706 and 0.647 as skewness and kurtosis values, respectively. Therefore, the variable B does not follow a normal distribution. Consequently, to compare both variables, it is necessary to use nonparametric tests. To compare the medians of both variables, the W of the Mann–Whitney test is employed. The obtained p value of this test is $6.77268E-8$; as it is lower than 0.05, the test concludes that the median of both variables is different. Then, the Kolmogorov-Smirnov test is applied to compare the distribution of both variables. The result of the test is a p value of 0. Thus, it indicates that the distribution of both variables is different. Finally, the Kruskal-Wallis test is done to diminish if the observed differences are due to the randomness of the data or if it is because the data in both situations is statistically different. The result of the Kruskal-Wallis test is a p value of $6.266E-8$. As it is lower than 0.05, it indicates that the observed differences are statistically significant. Thus, the proposed system can be used to identify the presence of feed in the water. Based on the obtained values, we will consider as a threshold value, in order to decide whether there is feed in the water or not, a summation of 520 pixels. This value comes from incrementing 10% of the maximum value of summation in the pictures without feed (picture 2, 473 pixels).

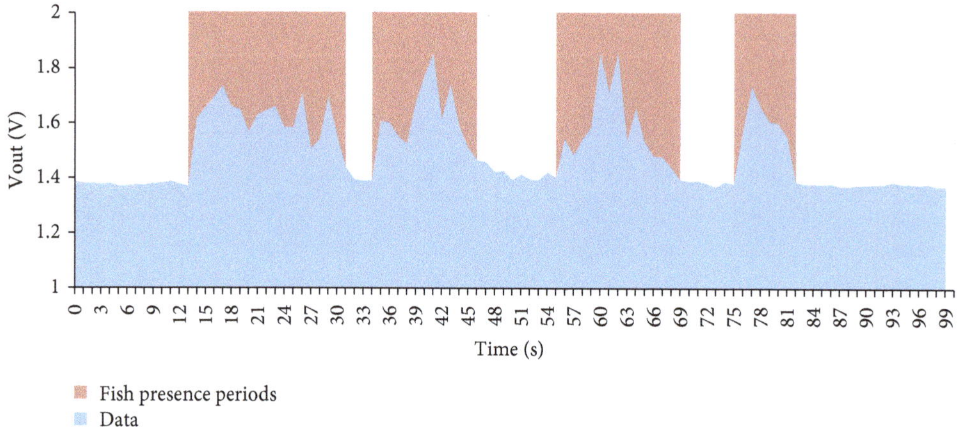

FIGURE 12: Gathered Vout values in the calibration test.

4.2. Calibration of the Fish Presence Sensor. In this subsection, the calibration of the fish presence sensor is shown. This sensor is based on the use of several LDRs placed along the tank. For the calibration of the sensor, a small aquaculture tank is employed. A juvenile of *Sparus aurata* is introduced in the tank, and the values gathered by the LDR and the fish movement are recorded. After processing the video, we can identify the periods where the fish is in the area of the LDR. The data of the LDR during the test is presented in Figure 12. In blue, the voltage of the sensor (Vout) in each second is represented. The periods when the fish are in the area covered by the LDR are marked in red. We can see in Figure 12 that in the periods when the fish is not present, Vout is lower than in the periods when the fish is present. When fish are not present, the values are similar, but when fish are present, the values present high heterogeneity. However, they are always higher than when the fish is not present. The data shown is gathered by the LDR placed in the upper part of the tank. The data when fish are not present have a mean value of 1.390 V and a standard deviation of 0.021. The minimum value is 1.371 V, and the maximum value is 1.467 V. When the fish are present, the mean Vout is 1.624 V, with a standard deviation of 0.096. The minimum value when fish are present is 1.479 V, and the maximum is 1.853 V. As the maximum Vout in the fish presence is lower than the minimum value when the fish is present, it is possible to use this Vout to determine the presence of fish. Now, using the statistical analysis, we are going to evaluate if the values when the fish is present are different from the values when fish are not present. The first step is to assess if the data follows a normal distribution or not. We divide the gathered data into two variables; the first one, variable A, corresponds to the Vout values when the fish is present. On the other hand, the Vout gathered when the fish is not present is considered to be the data of variable B. Variable A has 46 data in total, and variable B has 55. The descriptive analysis of variable A gives resulting skewness and kurtosis values of 2.008 and 0.316, respectively. Thus, it does not follow a normal distribution. The results of the test with variable B for skewness and kurtosis are 6.3094 and 5.696, respectively. In order to evaluate if the observed differences are statistically significant, nonparametric tests must be used. With the *W* of the Mann–Whitney test, we can assess if the medians of both distributions are different. The obtained p value of the test is 0; as it is lower than 0.05, the test concludes that the median of both variables is different. Then, the Kolmogorov-Smirnov test is performed in order to compare the distribution of variables A and B. The result of the test is a p value of 0.037. Thus, it indicates that the distribution of both variables is different. Finally, a Kruskal-Wallis test is done to evaluate the differences in the variance of both variables. The test result gives a p value of 0. Consequently, we can conclude that the differences found in the data when the fish is present and the data when the fish is not present are statistically significant and the sensor can be used to detect the presence of fish. Based on the obtained values, we will consider a threshold value, in order to decide whether or not fish are present, a Vout of 1.467. In this case, it is not possible to increment in a 10% the maximum Vout in the pictures because the maximum Vout without fish presence is similar to the minimum Vout with fish presence.

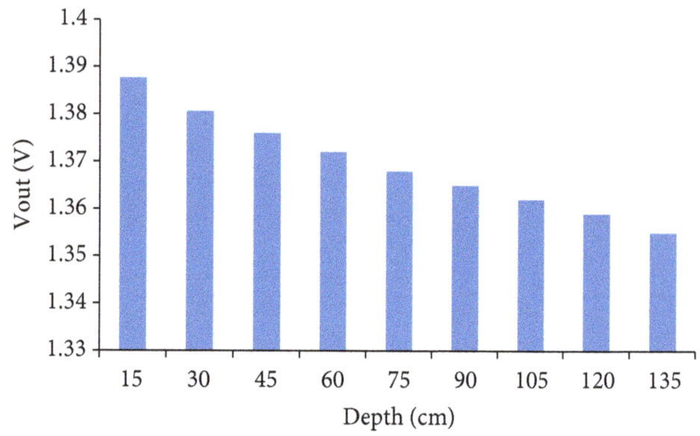

FIGURE 13: Effects of light attenuation in the Vout of the LDR at different depths.

We need to consider the effect of light attenuation caused by the water. Nevertheless, it is important to note that on the surface there is a direct light and the water presents null turbidity. Bearing this in mind, we expect low light attenuation. In addition, in other papers [2], a similar assembly with LDR in other depths, the light attenuation observed is less than

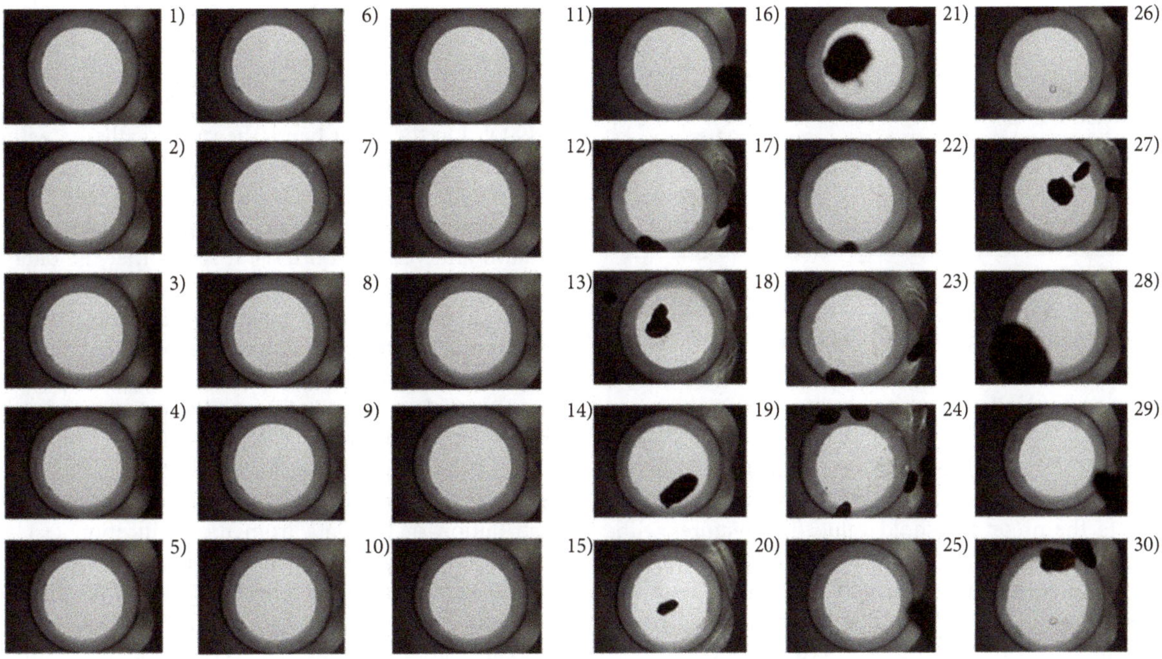

Figure 14: Employed pictures for the verification test.

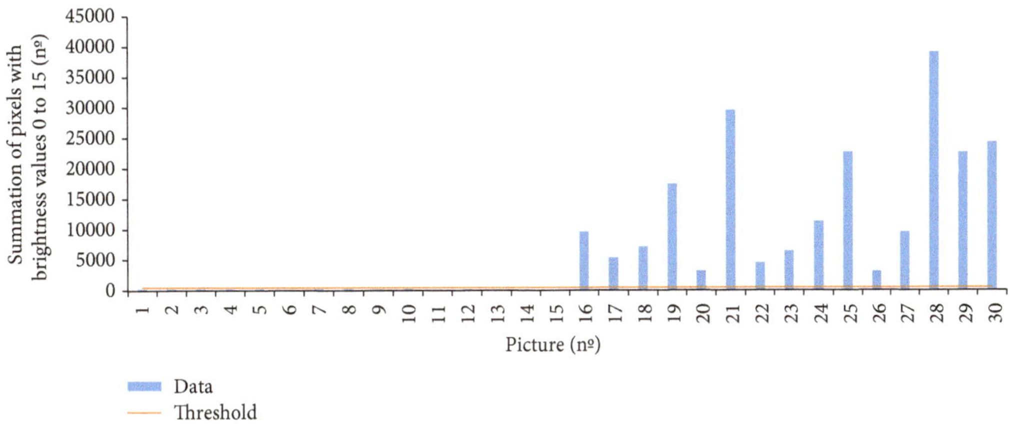

Figure 15: Summation of pixels with brightness values between 0 and 15 in the blue histogram in the verification test.

0.05 V. The values of Vout gathered by the LDR at different depths in a tank without fish are shown in Figure 13. The maximum Vout is 1.388 V at a depth of 15 cm, and the minimum Vout is 1.355 V at 135 cm of depth. On the other side, the light increases when fish presence is higher than 0.2 V. Thus, the set threshold is useful in all the cases, for all the depths. The Vout variation in one LDR (see data in the previous paragraph) due to changes in the light source, the variation in the water surface caused by the water current and fish movement, is higher than the variation caused by the light attenuation at different depths. The effect of light attenuation in tanks can be despised, but this effect will be studied in future applications in cages in the sea.

4.3. Verification Process. In this subsection, we detail the verification process for the two sensors developed in the previous subsections. Firstly, we will show the verification process of the feed sensors and then the verification of the fish presence sensor.

For the verification of the feed sensor, 30 new pictures were used (see Figure 14). From these 30 pictures, 15 of them, from pictures 1 to 15, were taken without feed and the other 15, from pictures 16 to 30, were taken with feed. When the selection of the 15 pictures with feed is performed, pictures with a big area covered by the feed as in 21 and 28 and pictures with a small area covered by feed as in 20, 22, and 26 are sought. After applying the same methodology, obtaining the blue histogram, and doing the summation of pixels with the brightness value between 0 and 15, the data shown in Figure 15 was obtained. In Figure 15, the blue bars indicate the summation of pixels with brightness values between 0 and 15 and the orange line marks indicate the threshold value

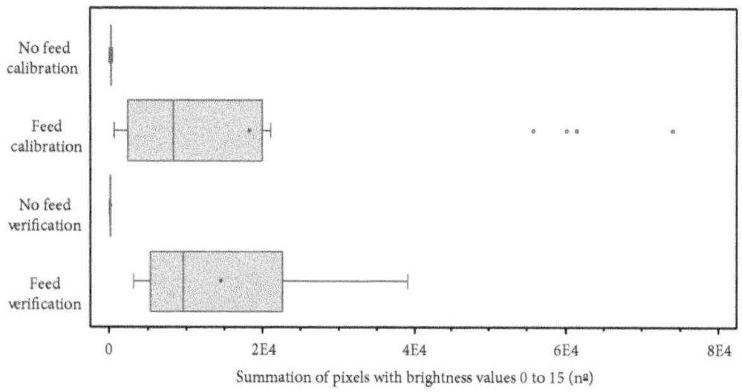

FIGURE 16: Box-whisker graphic of summation of pixels with brightness values between 0 and 15 in calibration and verification test.

established in Section 4.1, being 520. The first idea that we want to highlight is that all the pictures are correctly classified according to the preset threshold value. The pictures without feed, 1 to 15, present a mean value of 205 pixels. The maximum and minimum values are 292 pixels and 108 pixels, respectively. On the other side, the pictures with feed have a mean value of 14427 pixels. Picture 26 presents the minimum value, 3153 pixels, and picture 28 presents the maximum value, 39112 pixels. If we compare the values in the calibration and in the verification test, we can see that, in general terms, the data from the verification test follows the same distribution with the data from the calibration test.

To facilitate the analysis of the distributions of pictures in both the calibration test and verification test, we represent in Figure 16 the box-whisker graphics. Firstly, we see the representation of the data from the calibration test with and without feed and then the data from the verification test with and without feed. In this graph, we can see a summary of the data distribution. Primarily, we want to highlight that, once more, the distribution of the data from the pictures without feed in both tests is almost the same due to the low heterogeneity of these pictures. Moreover, from the distribution of the data of the pictures with feed, we want to foreground that only in the calibration test we found outlier values. The means of both tests are very similar both for the values of pictures with feed, 279 and 205 pixels, and for the pictures without feed, 18228 and 14126 pixels. In addition, the medians are similar, 251 and 218 for pictures without feed and 8393 and 9658 for pictures with feed. We can note that the medians are much more similar than the means due to the outlier values. Finally, a median test of mood is done in order to assess if the medians of the variables from the four samples are identical or not. The result is a p value of 0. Thus, we can conclude that the medians are different. Therefore, we can conclude that the verification test confirms that the proposed sensor can be used to differentiate the presence or absence of feed and the value of 520 pixels is an optimal threshold value.

Following this, we present the verification process of the fish presence sensor. We repeat the same set up as in the calibration process. The data gathered in the verification test can be seen in Figure 17. Vout values are represented in blue, and the periods when fish are present are indicated in red. The set threshold in Section 4.3, 1.467 V, is shown as a black line. In general terms, we can see that the data from the verification test confirms the conclusions of the calibration test. From the 45 gathered values, 10 of them are gathered in the presence of fish. Those 10 values have a mean of 1.615 V. The maximum value is 1.802 V, and the minimum value is 1.490 V. On the other hand, the values gathered without the presence of fish have a mean Vout of 1.368 V. The minimum and maximum gathered values of Vout are 1.388 V and 1.354 V, respectively. After comparing the values of the calibration and verification tests, it seems that the data follows a similar distribution.

With the purpose of facilitating the comparison of the distributions of the gathered Vout in both the calibration test and the verification test, we represent the box-whisker graphic of these data in Figure 18. We represent the data from the calibration test with and without fish presence and then the data from the verification test with and without fish presence. The first idea that we want to highlight is that the distribution of the data from the verification test and the calibration test gathered with fish is much more similar to each other than the data without fish. The means of the gathered Vout without fish are 1.390 V and 1.368 V for the calibration test and the verification test, respectively. On the contrary, the means of Vout with fish presence are 1.624 V for the calibration test and 1.623 V for the verification test. Moreover, the medians are also similar in the verification and calibration tests. The medians of Vout with fish presence are 1.381 V and 1.369 V, and the medians without fish are 1.607 V and 1.602 V for the calibration test and the verification test, respectively. Finally, a median test of mood is done in order to assess if the medians of the variables from the four samples are identical or not. The result is a p value of 0. Thus, we can conclude that the medians are different. Therefore, we can conclude that the verification test confirms that the proposed sensor can be used to differentiate the presence or absence of fish and the value of 1.467 pixels is an optimal threshold value.

4.4. Simulation of the Feeding Process. In this subsection, we show the gathered data during a simulated feeding process. It is known that during the feeding process, the fish shoal rises to the water surface to eat [10]. Thus, when the feed process starts, the fish presence will change, and this change will be detected by the presence sensors. During the first minutes, the fish eats all the supplied feed. However, not all the fish eat at the same time and in the same way. The bigger and more aggressive fish are the first to consume the feed. A few minutes later, when those fish are satiated, the rest of the fish start to eat. At this moment, part of the feed may start to fall. Therefore, it may be necessary to reduce the velocity of the supplied feed to avoid feed waste. Nevertheless, usually the feed is supplied with the same velocity all the time. Our system can detect the falling feed in order to reduce the velocity of feed supply.

For our system, the data from the feed presence and the falling feed is used. The purpose of using both data is due to the data of the fish presence indicating if they are eating or not. However, the data from the falling feed indicates the amount of eaten feed. Therefore, for an optimal monitoring, both data are needed. The data from fish presence acts as a

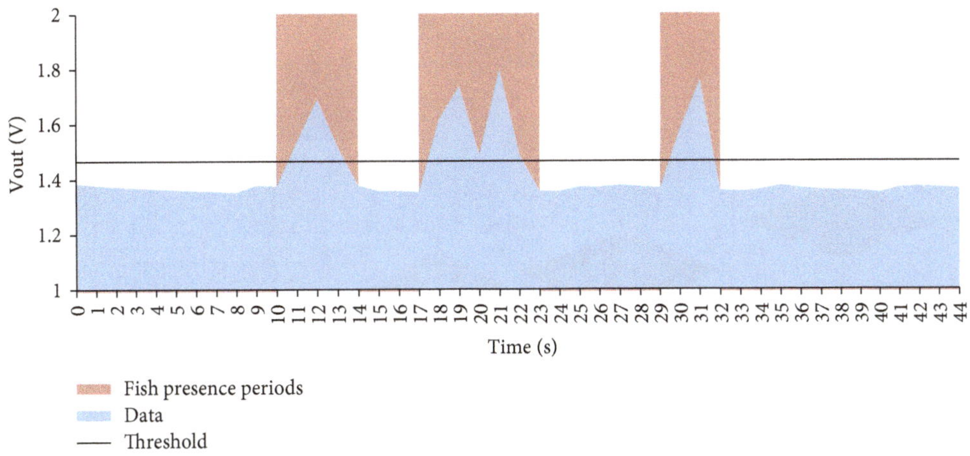

FIGURE 17: Gathered Vout values in the verification test.

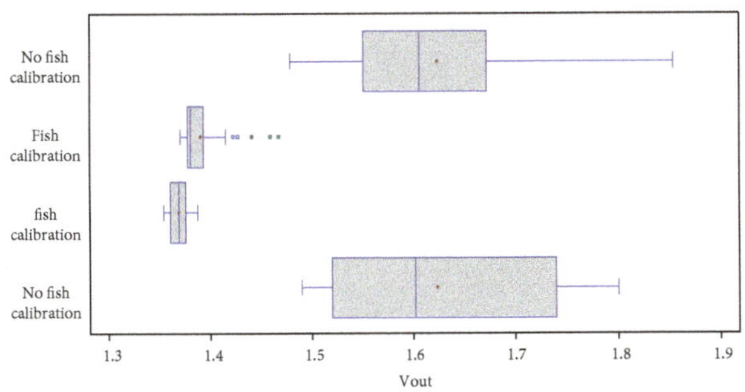

FIGURE 18: Box-whisker graphic of Vout in calibration and verification test.

trigger for turning off the system while the data from falling feed will trigger the changes in feed supply velocity. The operation algorithm can be seen in Figure 19.

When the feeding monitoring system turns on, the first step is to define the order of the analog inputs 1 to 9, which represent from Vout 1 to Vout 9. The Vout 1 is the Vout of the LDR placed at 5 cm above the water surface. Then, the system starts to gather values from the LDRs. If the Vout 2 to 9 are lower than 1.467 V, it means that the whole shoal is in the upper part of the tank and all of them are eating. Thus, the feed velocity must be maintained in 100% and new data must be gathered to continue with the monitoring. Nevertheless, when Vout 2 to 9 are higher than 1.467 V, it means that the whole shoal is not in the upper part of the tank and not all of them are eating. To be sure if all the supplied feed is being consumed, the falling feed system is turned on at 30 s. Then, the pictures are analyzed; if none of the pictures has a summation of pixels greater than 520 pixels, it means that the feed is being consumed, no changes are done, and new data is gathered from the LDR. However, if any of the pictures present a summation of pixels higher than 520 pixels, it indicates that not all the feed is being consumed. Therefore, the feed supply velocity must be reduced to 50% of the initial velocity. Again, the data from the LDR is gathered; if the Vout 1 is higher than 1.467 V, it means that none one fish is in the upper part of tanks and we stop the feeding process. However, the expected situation is that some fish may still be eating in the upper part, resulting in the Vout 1 being higher than 1.467 V. After 1 minute, the camera system is turned on again for 30 seconds. If any of the pictures detect falling feed, the velocity of feed supply will change to a 25% of the initial velocity.

The same system described to change from 50% to 25% of the velocity is used to change from 25% to 5% of the initial velocity. The 5% speed is the slower velocity that can be offered by our system. In order to turn off the system, we use the variable of fish presence. At the moment, when the system detects that there are no fish in the upper part of the tank, the feed suppliers stop feeding. Moreover, the system sends all the gathered data, and then, the system is turned off until the beginning of the new feeding period.

The system uses fish detection as the trigger because we need to ensure that all of the fish have the possibility of eating the feed that they need in order to maximize the fish growth. And it is known that not all the fish eat at the same time. If we do not consider the fish position, our system will stop feeding before all the fish can be satiated. Thus, a decrement on the fish growth will be caused resulting in a reduction in the productivity. On the contrary, if we only use the fish position for the monitoring system, we will not be able to know how to diminish the feed supply velocity and much more feed will not be consumed. Consequently, there will be a percentage of feed waste, which will produce a reduction in the profit of the fish production.

Now, we show the result of applying the algorithm to the data of a simulated feeding period, see Figure 20. We can see the data from the fish presence in Figure 20, feed presence (0 or 1), and feed velocity (100% to 0%). The data from fish presence represents the typical feeding behavior during a normal feeding process. During the first part of the feeding process, all the fish are in the upper part and the Vout of the sensors 2 to 9 give values lower than 1.467 V. Then, the fish presence in the positions 2 to 9 (fish presence 2 to 9) is equal to 0. And the fish presence in the position 2 (fish presence 1) is equal to 1 because the Vout of this LDR is higher than 1.467 V. Nevertheless, at second 450, some of the fish move to the lower part of the tank. Thus, the fish presence 2 to 9 is equal to 1. At this moment, the camera system for

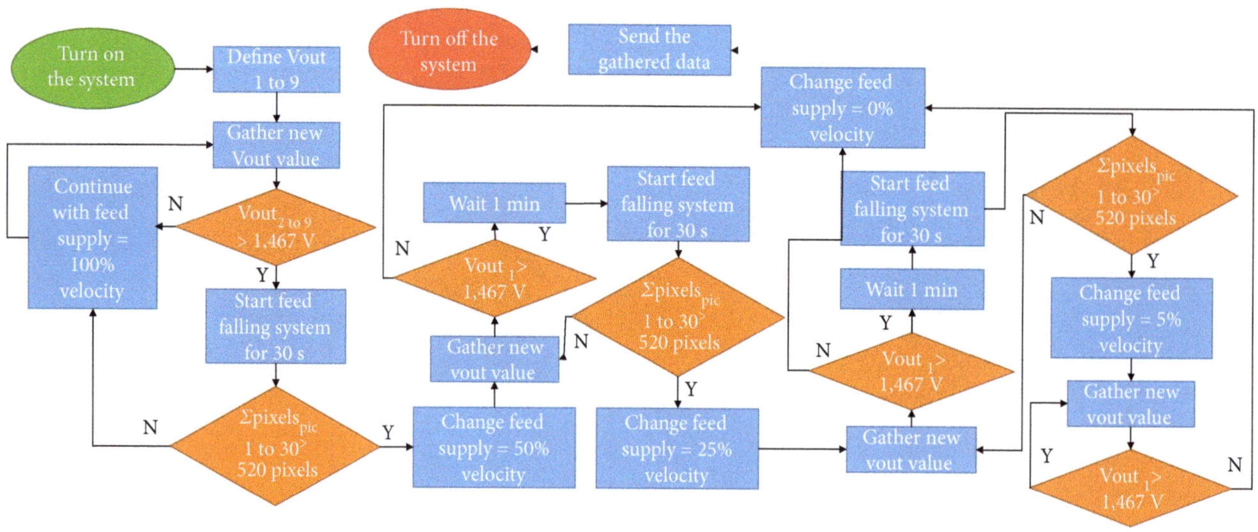

FIGURE 19: Operation algorithm.

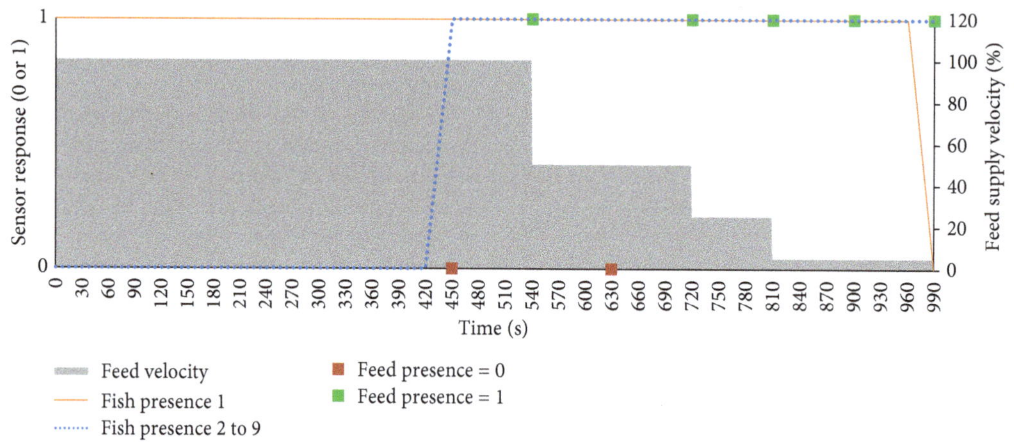

FIGURE 20: Result of applying the algorithm to the data of a simulated feeding period.

feed detection turns on. During the first record, no picture shows a summation of pixels higher than 520; thus, the feed presence is 0. Therefore, the feed velocity remains at 100%. After 1 minute, the camera test is repeated. In this case, the pictures reveal that there is falling feed. Thus, the feed velocity decreases to 50%. At second 720, the camera detects again falling feed and the velocity is reduced to 25%, and at second 810, the feed velocity decreases to 5%. The feed period ends at second 990 when no fish are detected in the upper part of the tank.

4.5. Comparison with Current Systems. Currently, the systems to optimize automatically the feed supply in fish farms are very different from each other. First, we found systems with the purpose of determining the best moment to feed as in [10, 17]. Other systems are for on-demand feeding [20] or are used for feeding with plankton [18]. There is one paper that uses falling feed as the only measure to decrease the feed velocity [19]. Finally, there is other proposed system [16] that uses a combination of fish presence and falling feed for adjusting the feed velocity. Nevertheless, they use an overhead camera to determine the position of the fish. Thus, this system is not able to differentiate when the fish are in the upper part of the tank and are eating feed or when they are in the lower part of the tank and they are not eating.

The sensors and algorithm presented in this paper suppose an improvement of the current methods for adjusting the feed supply to the fish needs. Moreover, the low cost of the employed components facilitates the possibility to implant this system in aquaculture tanks.

5. Conclusions

In this paper, a system for automatic adjustment of feed supply velocity for aquaculture monitoring has been shown. The system is comprised of a fish detector sensor based on multiple LDRs and a falling feed sensor based on a CMOS sensor. The sensors shown in this paper can be used to improve the efficiency of the aquaculture feeding process. The sensors are comprised of simple electronic components and can be connected to an Arduino node. The calibration has been shown, and the threshold values for detecting fish and feed presence have been found. Moreover, a verification test was done in order to ensure that the threshold value set in the calibration are acceptable. Both verification processes show that the calibrations were done correctly. In addition, we show the operation of the proposed system during a simulated regular feeding process.

As future works, we want to test our fish presence sensor with other fish species. Moreover, we want to test the system for larvae fish, considering no systems are developed to adjust the feed due to the small size of fish and the small size of feed. In addition, we plan to use other light sources as color LEDs to illuminate the drainage to differentiate pellets from faeces. The possibility to use different light sources to differentiate substances in the water has already been used for turbidity [24]. In addition, the minimization of the nodes and available sensors is facilitated to monitor the vital sings of fish as it was already done with other animals [25]. The inclusion of similar systems for fish monitoring will help to improve the fish welfare.

Disclosure

The founding sponsors had no role in the design of the study; in the collection, analyses, or interpretation of data; in the writing of the manuscript; and in the decision to publish the results.

Conflicts of Interest

The authors declare no conflict of interest.

Acknowledgments

The authors acknowledged "Ministerio de Educación, Cultura y Deporte," through the "Ayudas para contratos predoctorales de Formación del Profesorado Universitario (FPU) (Convocatoria 2014)" (Grant number FPU14/02953).

References

[1] A. C. Jones, A. Mead, M. J. Kaiser et al., "Prioritization of knowledge needs for sustainable aquaculture: a national and global perspective," *Fish and Fisheries*, vol. 16, no. 4, pp. 668–683, 2015.

[2] L. Parra, S. Sendra, L. García, and J. Lloret, "Design and deployment of low-cost sensors for monitoring the water quality and fish behavior in aquaculture tanks during the feeding process," *Sensors*, vol. 18, no. 3, 2018.

[3] B. A. Costa-Pierce, D. M. Bartley, M. Hasan et al., "Responsible use of resources for sustainable aquaculture," in *Proceedings of the Global Conference on Aquaculture-Farming the Waters for People and Food*, pp. 113–147, Phuket, Thailand, September 2010.

[4] Y. Tal, H. J. Schreier, K. R. Sowers, J. D. Stubblefield, A. R. Place, and Y. Zohar, "Environmentally sustainable land-based marine aquaculture," *Aquaculture*, vol. 286, no. 1-2, pp. 28–35, 2009.

[5] L. Qi, J. Zhang, M. Xu, Z. Fu, W. Chen, and X. Zhang, "Developing WSN-based traceability system for recirculation aquaculture," *Mathematical and Computer Modelling*, vol. 53, no. 11-12, pp. 2162–2172, 2011.

[6] J. H. Primavera, "Overcoming the impacts of aquaculture on the coastal zone," *Ocean & Coastal Management*, vol. 49, no. 9-10, pp. 531–545, 2006.

[7] J. A. Camargo and A. Alonso, "Ecological and toxicological effects of inorganic nitrogen pollution in aquatic ecosystems: a global assessment," *Environment International*, vol. 32, no. 6, pp. 831–849, 2006.

[8] C. Hedley and T. Huntington, "Regulatory and legal constraints for European aquaculture," February 2018, http://www.europarl.europa.eu/RegData/etudes/etudes/join/2009/431568/IPOL-PECH_ET(2009)431568_EN.pdf.

[9] D. Mara, S. Cairncross, and World Health Organization, "Guidelines for the safe use of wastewater and excreta in agriculture and aquaculture: measures for public health protection," 1989, http://apps.who.int/iris/bitstream/10665/41681/1/9241542489.pdf.

[10] H. S. AlZubi, W. Al-Nuaimy, J. Buckley, and I. Young, "An intelligent behavior-based fish feeding system," in *2016 13th International Multi-Conference on Systems, Signals & Devices (SSD)*, vol. 21-24, pp. 22–29, Leipzig, Germany, March 2016.

[11] V. M. Papadakis, I. E. Papadakis, F. Lamprianidou, A. Glaropoulos, and M. Kentouri, "A computer-vision system and methodology for the analysis of fish behavior," *Aquacultural Engineering*, vol. 46, pp. 53–59, 2012.

[12] M. Saberioon, A. Gholizadeh, P. Cisar, A. Pautsina, and J. Urban, "Application of machine vision systems in aquaculture with emphasis on fish: state-of-the-art and key issues," *Reviews in Aquaculture*, vol. 9, no. 4, pp. 369–387, 2017.

[13] J. D. Armstrong, V. A. Braithwaite, and P. Rycroft, "A flat-bed passive integrated transponder antenna array for monitoring behaviour of Atlantic salmon parr and other fish," *Journal of Fish Biology*, vol. 48, no. 3, pp. 539–541, 1996.

[14] S. G. Conti, P. Roux, C. Fauvel, B. D. Maurer, and D. A. Demer, "Acoustical monitoring of fish density, behavior, and growth rate in a tank," *Aquaculture*, vol. 251, no. 2-4, pp. 314–323, 2006.

[15] H. Zhang, Q. Wei, and M. Kang, "Measurement of swimming pattern and body length of cultured Chinese sturgeon by use of imaging sonar," *Aquaculture*, vol. 434, pp. 184–187, 2014.

[16] Y. Atoum, S. Srivastava, and Xiaoming Liu, "Automatic feeding control for dense aquaculture fish tanks," *IEEE Signal Processing Letters*, vol. 22, no. 8, pp. 1089–1093, 2015.

[17] R. A. Bórquez-Lopez, R. Casillas-Hernandez, J. A. Lopez-Elias, R. H. Barraza-Guardado, and L. R. Martinez-Cordova, "Improving feeding strategies for shrimp farming using fuzzy logic, based on water quality parameters," *Aquacultural Engineering*, vol. 81, pp. 38–45, 2018, In Press.

[18] N. Papandroulakis, P. Dimitris, and D. Pascal, "An automated feeding system for intensive hatcheries," *Aquacultural Engineering*, vol. 26, no. 1, pp. 13–26, 2002.

[19] M. Garcia, S. Sendra, G. Lloret, and J. Lloret, "Monitoring and control sensor system for fish feeding in marine fish farms," *IET Communications*, vol. 5, no. 12, pp. 1682–1690, 2011.

[20] D. Covès, M. Beauchaud, J. Attia, G. Dutto, C. Bouchut, and M. L. Bégout, "Long-term monitoring of individual fish triggering activity on a self-feeding system: an example using European sea bass (*Dicentrarchus labrax*)," *Aquaculture*, vol. 253, no. 1-4, pp. 385–392, 2006.

[21] C. Zhou, K. Lin, D. Xu et al., "Near infrared computer vision and neuro-fuzzy model-based feeding decision system for fish in aquaculture," *Computers and Electronics in Agriculture*, vol. 146, pp. 114–124, 2018.

[22] C. Zhou, B. Zhang, K. Lin et al., "Near-infrared imaging to quantify the feeding behavior of fish in aquaculture," *Computers and Electronics in Agriculture*, vol. 135, pp. 233–241, 2017.

[23] STATGRAPHICS, 2017, February 2017, http://www.statgraphics.net/descargas-centurion-xvii/.

[24] L. Parra, J. Rocher, J. Escrivá, and J. Lloret, "Design and development of low cost smart turbidity sensor for water quality monitoring in fish farms," *Aquacultural Engineering*, vol. 81, no. 3, pp. 10–18, 2018.

[25] S. Sendra, F. Llario, L. Parra, and J. Lloret, "Smart wireless sensor network to detect and protect sheep and goats to wolf attacks," *Recent Advances in Communications and Networking Technology*, vol. 2, no. 2, pp. 91–101, 2013.

2

Development and Commissioning of High Temperature FBG Solid Pressure Sensors

Hong-Ying Guo,[1,2] **Zhao-Ba Wang**,[1] **and Hai-yang Li**[1]

[1]*School of Information and Communication Engineering, North University of China, Taiyuan 030051, China*
[2]*Department of Electronics, Xinzhou Teachers University, Xinzhou 034000, China*

Correspondence should be addressed to Zhao-Ba Wang; wangzb@nuc.edu.cn

Academic Editor: Banshi D. Gupta

In this work, a new type of high temperature fiber grating diaphragm pressure sensor is described, including the physical design structure, in-depth analysis of optical response to changes in pressure, and a discussion of the temperature compensation method. Mathematical model of high-temperature compressive shape change and normal-temperature compressive shape change is built, and effective temperature compensation method is proposed, which can both contribute to solve temperature interference of pressure sensor at high temperature. In addition, the simulation computation is conducted for the external encapsulation of the sensor under the high-temperature and pressure according to the design model. Through analysis and comparison of data affecting the changing position of grating signal, the correctness of the simulation result is verified by the development of the prototype. The temperature compensation method is proposed to realize pressure measurement of sensor with a relative error below 4.6%F.S. and range to 50 MPa in the environment of 300, which meets with the safety research requirement of the cartridge system.

1. Introduction

Solid internal pressure in a high temperature environment is an important index in the field of safety and environmental adaptability of the ammunition system. It is the necessary basis for judging whether the whole structure (projectile body) will cause partial or overall instability and failure, which will result in the loss of stability due to the excessive deformation caused by thermal expansion [1], such as solid rocket engine propellant internal pressure, the weapon system inside the warhead internal fire, and explosive pressure, which all require the detection of internal pressure in a high temperature environment. Fiber Bragg grating (FBG) has been proposed and widely used in the optic communication and sensing fields due to their prominent advantages, such as immunity to electromagnetic noise, small profile, lightweight, and high resistance to hostile environments [2–4]. Fiber grating sensor technology exists everywhere in our daily lives, with the advantages of environmental friendliness, effectivity, and sustainability. On this basis, fiber Bragg grating (FBG) sensor has been put forward and widely applied in bridge, landslide safety monitoring [5–8]. The fact that multiple FBGs can be arrayed along a single fiber, various FBG sensors for temperature, pressure, strain, displacement, vibration, and acceleration measurement have been developed and could be applied to the same engineering applications. Pressure sensor based on fiber Bragg grating is feasible for long distance distributed monitoring without on-site power supply. Especially in flammable and explosive, electromagnetic interference, and strong radiation in the harsh environment, it can easily reflect a huge advantage, which can be used in solid pressure testing frequently [9, 10]. The pressure measurement based on FBG is a kind of common sensing technology, and the pressure monitoring can be realized indirectly by checking the parameters of strain and physical measuring object of fiber Bragg grating [11, 12]. The FBG sensor for direct force measurement has also been reported in literature [13, 14]. Many researchers focus on the field of fiber Bragg grating sensing pressure technology. In 2013, Huang et al. [15] reported a diaphragm-type fiber Bragg grating pressure sensor with temperature compensation. The sensor adopts a circular diaphragm with two FBGs adhered on the surface along the radial direction as the pressure sensitive component. The experiment results indicate that the pressure sensitivity of this novel sensor is as high as 1.57 pm/kPa. In 2014, Pachava et al. [16] designed a highly sensitive FBG pressure sensor using a thin metal diaphragm. The experimental pressure sensitivity of the sensor is 2.05×10^{-2} MPa^{-1}. In 2015, Allwood et al. [17] designed a highly sensitive fiber Bragg grating diaphragm pressure transducer. The results show that the transducer has a sensitivity of 0.116 nm/kPa. However, most of these pressure FBG sensor were limited by the pressure-measuring range and operating temperature. The pressure-measuring range of the above-mentioned sensors is within 25 MPa, and the required temperature is 15°C–50°C [15–18].Obviously that cannot meet the demand of pressure-monitoring in the projectile when heated (working temperature: 300°C, pressure range: 50 MPa). In addition, the high sensitivity of these sensors is obtained through the materials with low elastic modulus such as rubber and aluminum [3, 4, 9, 15–17]. These sensors will inevitably be damaged if applied to the safety and environmental adaptability of the ammunition system. For these reasons, it is necessary to research pressure sensors with high temperature and pressure.

Here, we design and present a FBG pressure sensor with high temperature and pressure. In our previous work, the research team has carried out the study of high pressure solid pressure sensor of fiber Bragg grating [19]. In this design, the sensitivity of the sensor is improved completely through the design of the structure. When the pressure perpendicular to the plane of the sensor model is loaded, the structure can convert the pressure into two X direction deformations of the two fixed column. In the Y and Z directions, the two fixed columns have the same deformation and no relative change. When the grating is pasted between two fixed columns, the change in pressure produces only axial strain without radial strain. It can be seen in Figures 1(a)–3(a) of support information (available here). In order to meet the normal working requirements under 300°C, we have realized the ideal pressure-measuring effect, and we used the grating (limit temperature 400°C) which is inscribed on the SMF-28e optical fiber coated with polyimide. The sensor structure which is suitable for high-temperature and high-pressure environment is proposed to make fiber Bragg grating high temperature and pressure-resistant sensors. In a high-temperature environment, the external metal packaging will expand deformation, resulting in grating signal fluctuations and serious pressure measurement errors; therefore, temperature compensation is needed. In this paper, the deformation data of the sensor outer package under different pressure at ambient temperature, the load-free at different temperatures, and the load of 50 MPa pressure are simulated, respectively; 100~300°C of the transducer housing under the above conditions are obtained, and a reasonable temperature compensation method is obtained. By using the method of temperature compensation, the measurement of the sensor's relative error is below 0.836%, and the range 50 MPa is realized under the environment of 300°C, all of which satisfy the requirement of the security research of the ammunition system.

2. Working Principle and Sensor Structure Design

FBG is an optical wavelength modulation component. When a beam of broadband light wave is incident on FBG, the optical wave with the wavelength satisfying Bragg equation after modulation by Bragg grating is coupled by Bragg grating to the inverse transmission mode [20] and form Bragg equation of Bragg reflection wave:

$$\lambda = 2n_{\mathrm{eff}}\Lambda. \tag{1}$$

In (1), Λ is the pitch of grating and n_{eff} is the effective refractive index of optical grating. Both Λ and n_{eff} change owing to the impact of outer world (temperature, strain, pressure, etc.), which causes the reflection wavelength of optical grating moves. The center reflection wavelength of FBG is affected by the temperature and strain. The relationship among wavelength change, axial strain change, and temperature change response [21, 22] is:

$$\frac{\Delta\lambda}{\lambda} = \left(\alpha_f + \xi\right)\Delta T + \left(1 - P_e\right)\Delta\varepsilon. \tag{2}$$

In (2), λ is the initial wavelength of FBG, α_f is the thermal expansion coefficient, ξ is the thermo-optical coefficient, and P_e is the effective photoelastic coefficient. The effective photoelastic coefficient at indoor temperature is 0.22.

The pressure sensitivity of bared grating is low, and it is easy to break off and unpressurized. Thus, it is unsuitable for conventional measurement. We must encapsulate it so as to reach the purpose of pressure sensitivity and compression resistance [23–25]. According to solid mechanics and current research result of matching error of solid sensor [14], if the sensor is designed into cake shape, it can obtain ideal test value. In the design, optical grating pressure sensor with high temperature resistance is designed adopting optical grating with high temperature resistance whose surface is coated with heat-resistant material on the basis of the encapsulation structure of traditional optical grating sensor. The encapsulation structure diagram of sensor is shown in Figure 1.

Solid pressure sensor of optical grating is composed of pressure-bearing membrane, sensor enclosure, and stainless steel top cover. All these three components are made from the same substrate materials. The pressure-bearing membrane is made up of circular elastic membrane and fixed column. The fixed column and elastic membrane as well as the sensor enclosure are formed by wholistic processing so as to reduce the stress concentration or stress nonuniformity problem. The sensor enclosure and SS cover adopt threaded connection. The pressure-bearing part adopts elastic membrane as the elastic component of sensor and two symmetric fixed columns are used for fixing both ends of inductive fiber

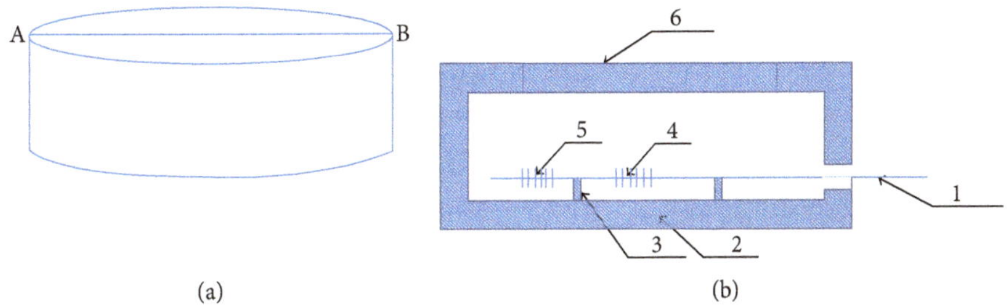

FIGURE 1: Structural diagram of optical grating pressure sensor. (a) Outline of sensor. (b) Structural profile at AB position. (1) Fiber, (2) fixed column, (3) pressure diaphragm, (4) strain grating FBG1, (5) temperature compensation grating FBG2, (6) stainless steel cover.

Bragg grating 1 (FBG1). The elastic shape change of circular elastic membrane transforms the solid pressure into the measurable strain of optical grating. FBG2 is temperature-compensating grating. It is only affected by temperature but without being affected by the enclosure shape change. It is used for realizing temperature compensation. Diameter R of the sensor is selected as 120 mm, the thickness of wall h is 15 mm, the spacing l of two fixed center points l is 40 mm, the diameter r of fixed column is 10 mm, and the height m of fixed column is 10 mm. When external force acts on the pressure-bearing membrane, the membrane behaves small deflection deformation, and the distance between two fixed columns widens. We pull the grating and measure the changing reflection signal and further get applied load information. Therefore, the research of the impact of temperature and pressure to the sensor enclosure focuses on studying the change of the distance between fixed columns due to these factors. There are two factors affecting the distance change of these two fixed columns: (1) the horizontal displacement of the membrane where the fixed columns are (x-axis displacement) and (2) the horizontal projection of fixed columns (directionally proportion to z-axis deformation). The superposition of these two factors decides the distance change of two fixed columns. According to the sensor enclosure structure, it is known: (3)

$$\Delta l = (\Delta x_r - \Delta x_l) + \frac{2m\Delta z}{R}. \quad (3)$$

3. Pressure Test and Temperature Compensation

3.1. Compressive Structure Analysis at Normal Temperature. To obtain the deformation distribution of the sensor enclosure under different pressure, the paper simulates loading using finite element method (ANSYS15). An elastic material made from 304# stainless steel is used for parameter simulation as follows: Young modulus $E = 1.942 \times e^{11}$Pa, Poisson's ratio $\mu = 0.28$, and loading pressure 50 MPa. Simulation results are shown in Figures 2–5, Figure 2 is the total deformation figure of the sensor model in 50 MPa pressure, Figures 3, 4, and 5, respectively, is the directional deformation of the X, Y, Z direction distribution of the sensor model in 50 MPa pressure.

According to the figure above, it can be seen that, after applying pressure, the membrane produces small deflection deformation downward. Two symmetric fixed columns relative to origin are put on the inside wall of the membrane with the diameter y = 0. The two fixed columns are symmetric relative to origin, so the displacement at z-axis is the same, the displacement direction at y-axis is opposite, and the displacement direction at x-axis is opposite. In this way, both ends of optical grating are stuck to the fixed column, the radial shape change of optical grating is eliminated and there is only axial shape change (x-axis direction).

After adding two fixed columns, in the case where the simulated parameter and added load are constant, the internal deformation diagram of the sensor model is shown in Figures 6 and 7. Figure 6 is the overall deformation diagram inside the sensor. Figure 7 shows the deformation distribution of the internal X direction of the sensor.

According to Figure 7, it can be seen that the deformation of two fixed columns at x-axis is the most remarkable and the directions are opposite. Fixing both ends of optical grating on two fixed columns can make the sensor gain maximum sensitivity.

The diameter passing through two fixed columns in Figure 8 is selected as the path. The deformation distribution of each point on this path under the pressure of 50 MPa is shown in Figure 8. The deformation data at two fixed columns under different loads are shown in Table 1.

It can be seen from the above table that the external load is linear between the central points of the two fixed columns, and the two ends of the grating can be bonded to the fixed column to detect the external load. The external pressure is proportional to the distance between the two fixed columns, and the relation is:

$$P = K_d \Delta l. \quad (4)$$

In formula (4), P is the external pressure and Δl is the change of the distance between the two fixed columns. Substitute formula (3) into formula (4), we get:

$$P = K_d\left((\Delta x_r - \Delta x_l) + \frac{2m\Delta z}{R}\right). \quad (5)$$

Substitute the data in Table 1, we get $K_d = 670.48396$ M Pa/mm.

3.2. Impact of Temperature Change on Pressure Test and Temperature Compensation Method

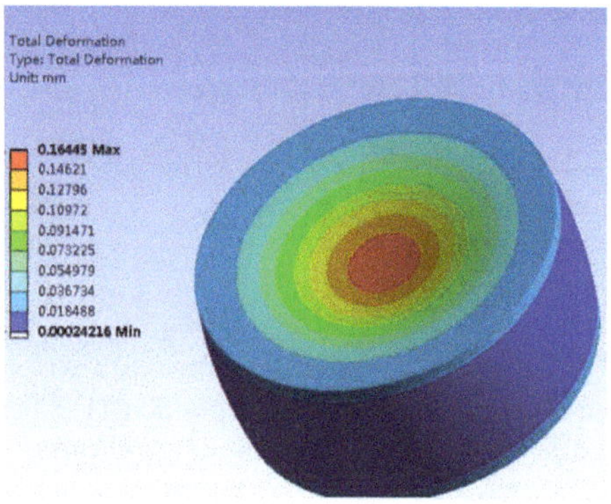

FIGURE 2: Total deformation of sensor outer casing.

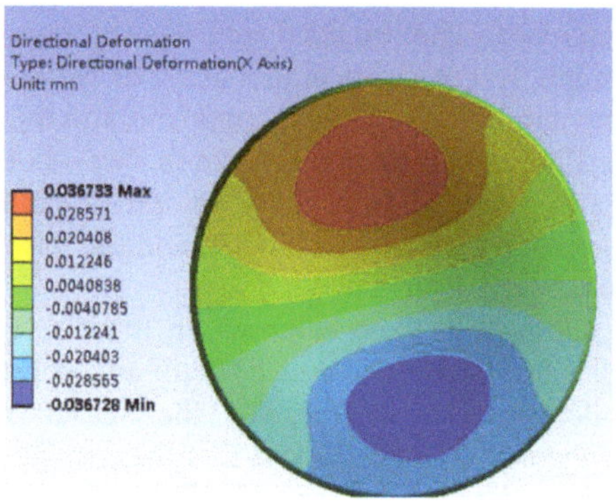

FIGURE 3: x-axis deformation of the sensor outer casing.

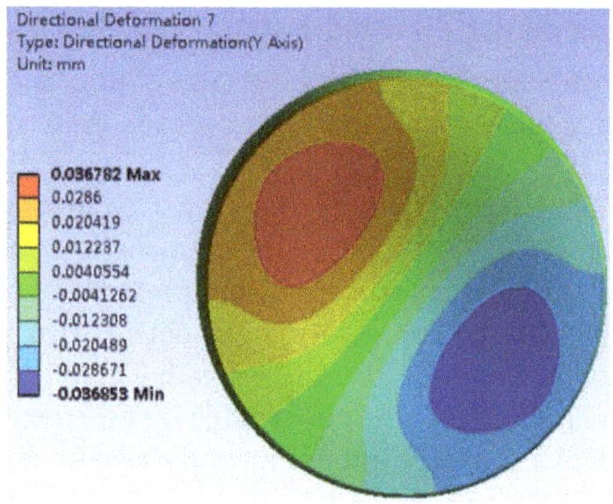

FIGURE 4: y-axis deformation of the sensor outer casing.

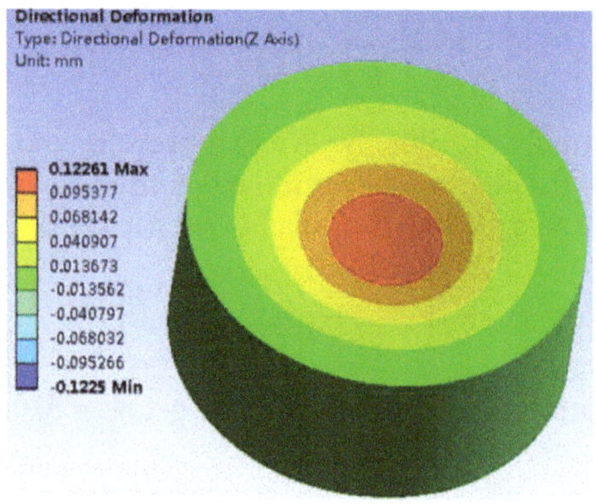

FIGURE 5: z-axis deformation of the sensor outer casing.

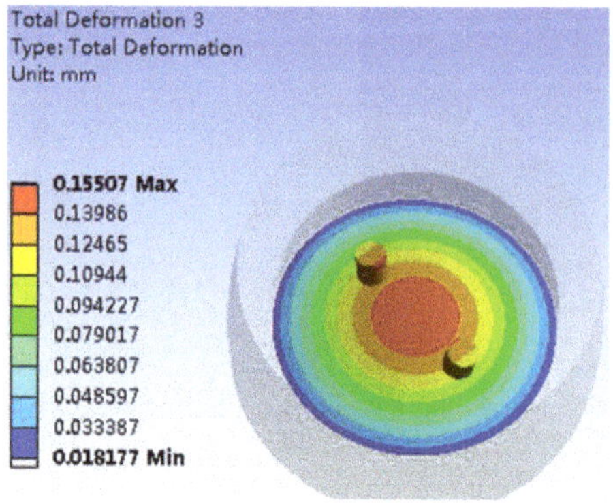

FIGURE 6: Total deformation of inside sensor casing.

3.2.1. Structural Deformation at Different Temperature and Its Impact. The external model of the sensor is simulated at different temperature without any loads applied. The parameter simulated is as below: heat conductivity 16.2 W/m°C, linear expansion coefficient 17.3×10^{-6}/°C, Young modulus $E = 1.942 \times e^{11}$Pa, and Poisson's ratio $\mu = 0.28$. The deformation of the sensor model under the 300°C environment without pressure is shown in Figure 9.

The diameter of compressive membrane passing through two fixed columns is selected as the path. At different temperatures without pressure, the deformation data at x-axis of two fixed columns and z-axis of the center of the membrane is shown in Table 2.

From Table 2, we can see that at high temperature, the sensor shell is heated and expanded, which will cause the grating reflection wavelength to change and result in the error of pressure measurement. Effective temperature compensation must be adopted to counteract the effect of temperature.

3.2.2. Temperature Compensation Method. Taking the above model under the pressure of 50 MPa at 300°C, the material is 304# steel. The following parameters are used to simulate: the thermal conductivity 16.2 and linear expansion coefficient 17.3, Young modulus $E = 1.942 \times e^{11}$Pa, and Poisson's ratio $\mu = 0.28$. The result of simulation is as shown in Figure 10.

The diameter of two fixed columns on the compressive membrane is selected as the path. The deformation data of each point of this path at 300°C before and after applying 50 MPa load are compared as Figures 11 and 12. Figure 11 shows the x-axis deformation data, and Figure 12 is the deformation data of z-axis direction.

In Figures 11 and 12, the red, black, blue, and pink curves are the deformation distribution of each point on this path under different environment. The red is under the 300°C environment without pressure. The black is under the 300°C environment and the pressure of 50 MPa. The blue is shown the difference under the 300°C environment between with 50 MPa pressure and without pressure. The pink curve is at room temperature with 50 MPa pressure. As it can be seen from Figures 11 and 12, the blue curve and the pink curve almost coincide; thus, it indicates that the deformation under the same pressure at different temperature is consistent, and the deformation which is due to temperature and

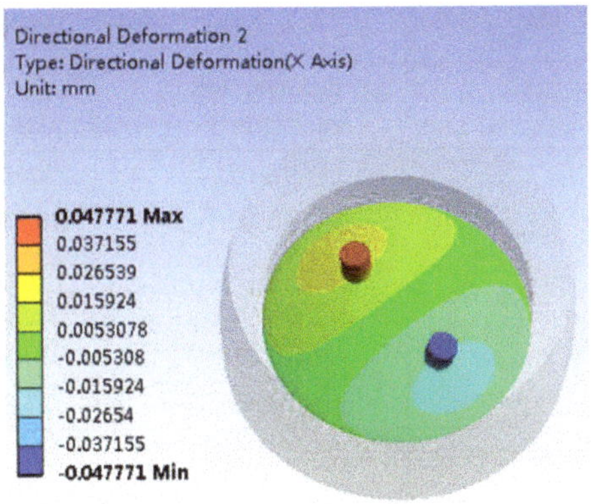

FIGURE 7: x-axis deformation of inside sensor casing.

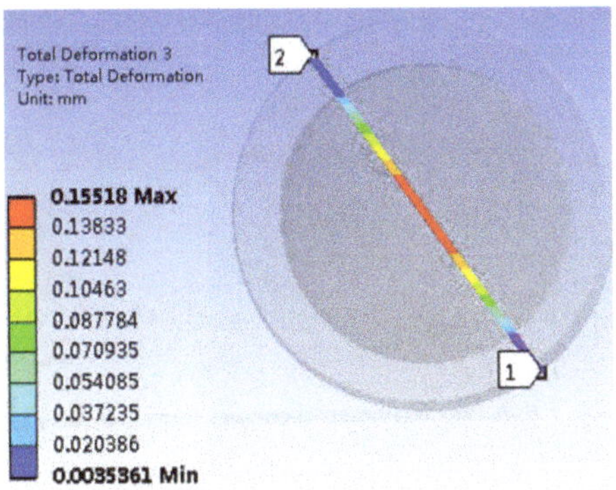

FIGURE 8: Path deformation.

TABLE 1: Deformation under different pressure.

Pressure (MPa)	x-axis deformation of left fixed columns (mm)	x-axis deformation of right fixed columns (mm)	z-axis deformation of center (mm)
10	5.47E-03	−5.48E-03	3.20E-02
20	1.10E-02	−1.10E-02	6.41E-02
30	1.64E-02	−1.65E-02	9.62E-02
40	2.19E-02	−2.19E-02	1.28E-01
50	2.74E-02	−2.75E-02	0.16056

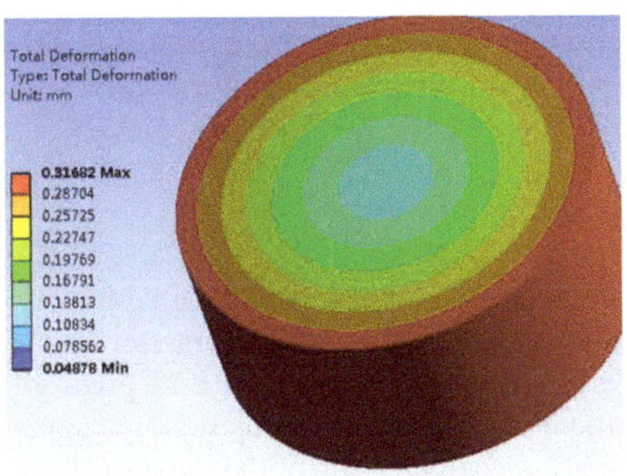

FIGURE 9: Load-free deformation pattern at 300°C.

pressure is linear superposition. We can adopt the signal change caused by thermal deformation under no load after using the measured signal minus the corresponding temperature. That is, the signal change caused by pressure and the deformation caused by pressure at high temperature are nearly totally identical to the deformation caused by the same pressure at normal temperature and the error rate is lower than 0.1%. Thus, the normal temperature formation approximates high temperature deformation.

The diameter of two fixed columns is taken as the path. The deformation data on two fixed columns at different temperature, the deformation data at corresponding temperature after applying 50 MPa load, and that at normal temperature after applying 50 MPa are compared in Table 3.

According to Table 3, it is concluded that the deformation difference at different temperature after applying 50 MPa pressure and applying no load is consistent with the deformation at normal temperature after applying 50 MPa pressure. In the actual test, the optical grating wavelength change at high temperature under stain includes the wavelength change triggered by the impact of temperature on optical grating, thermal deformation, and pressure: (a) A sensor is put in the thermostat to test the wavelength change (b) at different temperature without load. For temperature-compensated optical grating FBG2, owing to the impact of temperature, the wavelength change is $\Delta\lambda_{2T}$. Through wavelength change of temperature-compensated optical grating, it is seen that the temperature of the sensor is ΔT. According to ΔT, we get a corresponding $\Delta\lambda_T$, and the pressure at current temperature:

$$\Delta P = \frac{(\Delta\lambda_1 - K_{1d}{}^{*}\Delta T)}{K} = \frac{(\Delta\lambda_1 - K_{1d}{}^{*}\Delta\lambda_{2T}/K_{2t})}{K}. \quad (6)$$

In formula (6), ΔP is the pressure loaded on the sensor, $\Delta\lambda_1$ is the measured wavelength change of pressure sensitive grating, K is pressure-wavelength coefficient under normal temperature, $\Delta\lambda_{2T}$ is the wavelength change of the temperature compensated optical grating, K_{2t} is the temperature coefficient of temperature-compensated grating, and K_{1d} is the temperature coefficient of pressure sensitive grating. These parameters can be measured by experiment. In this way, the pressure at high temperature is measured.

3.3. Experimental Calibration

3.3.1. Temperature Experiment. The sensor was put into the temperature test box. The preparation part of the experiment is: the sensor is heated from 20°C to 300°C, and then the temperature drops naturally from 300°C to 20°C. The experiment is conducted three times. Then, the output of the sensor's pressure sensitive FBG and temperature compensation FBG was recorded when the temperature went up from 100°C to 300°C, with same intervals of 20°C. During the process, each temperature point should be kept unchanged for 2 hours. Similarly, the output of pressure sensitive FBG and temperature compensation FBG was recorded with 1 h temperature keeping at each temperature point. Taking the average value

Table 2: Deformation data at different temperature.

Temperature (°C)	x-axis deformation of left fixed columns (mm)	x-axis deformation of right fixed columns (mm)	z-axis deformation of center (mm)	Cause deviation (MPa)
100	2.61E-02	−2.79E-02	−3.69E-02	32.070477
140	3.95E-02	−4.22E-02	−5.58E-02	48.5168898
180	5.29E-02	−5.65E-02	−7.47E-02	64.9631909
220	6.63E-02	−7.08E-02	−9.36E-02	81.4096037
260	7.96E-02	−8.51E-02	−0.11249	97.8555695
300	9.30E-02	−9.94E-02	−0.13139	114.302541

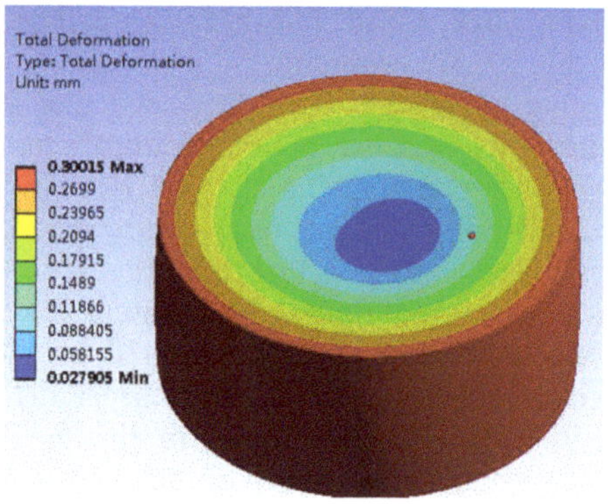

Figure 10: 50 MPa deformation pattern at 300°C.

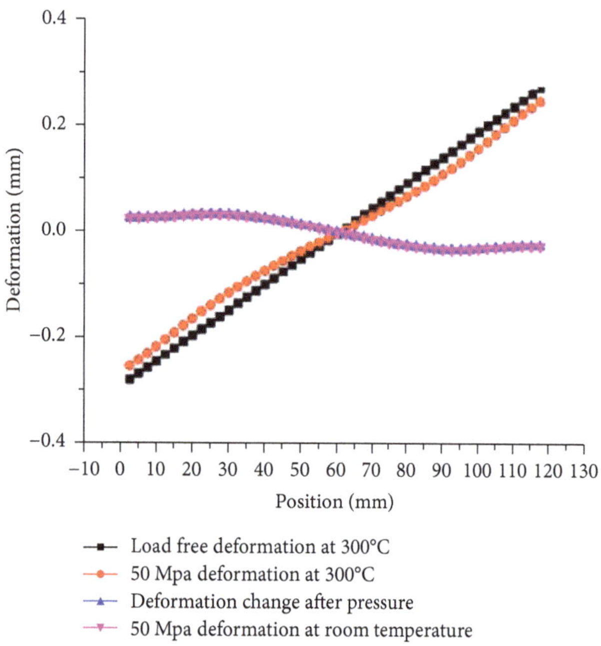

Figure 11: x-axis deformation pattern of point on the path.

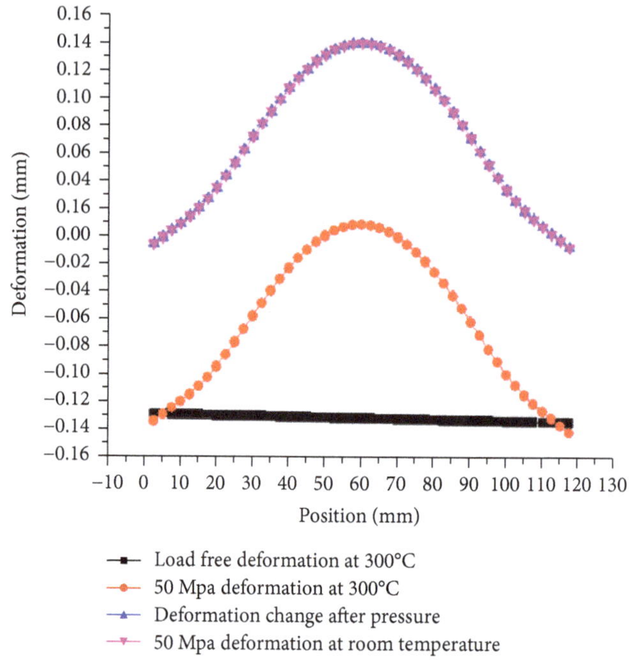

Figure 12: z-axis deformation pattern of point on the path.

of the output caused by the temperature change of the positive and inverse range as the verification value to carry on data analysis and processing. The sensitivity of the sensor's pressure sensitive grating FBG1 and temperature compensation grating FBG2 to temperature response can be obtained. The temperature response curve is shown in Figure 3. Using +20°C as the reference, the drift characteristic curve of the relation between center wavelength of the pressure sensitive grating (FBG1) after temperature compensation under zero pressure load and temperature can be attained by the process of formula (6), as shown in Figure 13.

3.3.2. Temperature and Pressure Cross Experiment. The sensor was placed in a high temperature pressure furnace and tested at 100°C, 150°C, 200°C, 250°C, and 300°C, respectively. The pressure-wavelength data of the sensors at different temperature points were recorded, and the curves were drawn (shown by Figure 14).

4. Conclusion

To meet the requirement of safety and adaptability detection of the cartridge system, the paper proposes a sensor structure applicable to the high temperature and pressure environment. In the mode of ANSYS simulation, the paper discloses the deformation characteristic of the sensor at high temperature and proposes corresponding temperature compensation method, and the correctness of the simulation result is verified by the development of the prototype. In this paper, the corresponding temperature compensation method is put

Table 3: Data comparison before and after applying load at different temperature.

Temperature (°C)	$\Delta x_l - \Delta x_r$(mm)		Δz(mm)		Change after pressure (mm)		Relative error (%)
	No load	50 MPa	No load	50 MPa	$\Delta x_l - \Delta x_r$	Δz	
100	5.40E-02	2.85E-03	−3.69E-02	0.1034	−5.11E-02	1.40E-01	0.065
140	8.17E-02	3.06E-02	−5.58E-02	8.45E-02	−5.11E-02	1.40E-01	0.081
180	1.09E-01	5.83E-02	−7.47E-02	6.55E-02	−5.07E-02	1.40E-01	0.641
220	1.37E-01	8.60E-02	−9.36E-02	4.66E-02	−5.10E-02	1.40E-01	0.258
260	1.65E-01	1.14E-01	−0.11249	2.77E-02	−5.13E-02	1.40E-01	−0.124
300	1.92E-01	1.41E-01	−0.13139	8.77E-03	−5.06E-02	1.40E-01	0.836
Room temperature	0	−5.12E-02	0	0.14031	−5.12E-02	1.40E-01	0.000

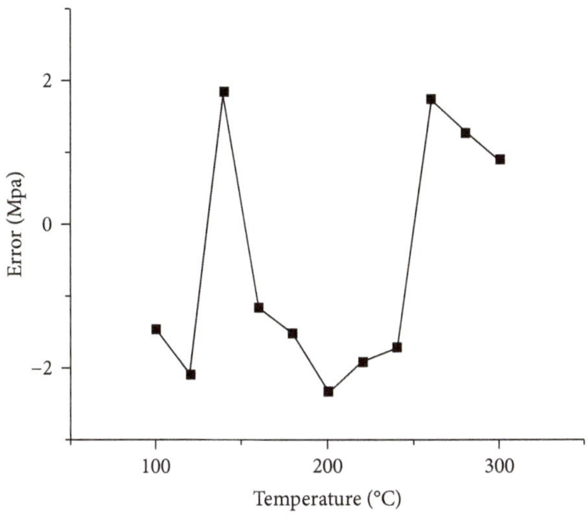

Figure 13: Output error curve.

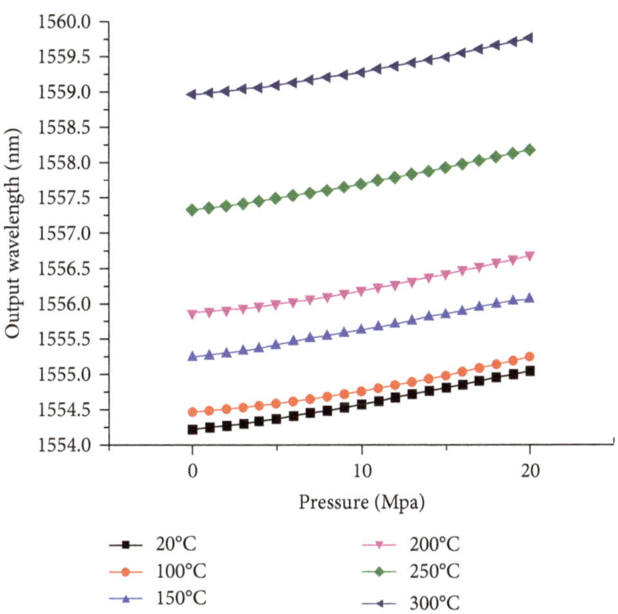

Figure 14: Pressure output curves at different temperatures.

forward, and the correctness of the simulation result is verified by the development of the prototype. Using this method, in 50 MPa pressure range and the environment of 300°C, the relative maximum error is only 4.6%F.S., and that can greatly improve the accuracy of the test data. Therefore, the method is highly worth popularization. It enhances the accuracy of the test data greatly and has high promotion value. The developed sensor can be applied more extensively in such area as various high-temperature and high-pressure measuring occasions of flammable and combustible, electromagnetic interference, and strong radiation-like pressure monitoring of weapon and explosive.

Conflicts of Interest

The authors declare that there are no conflicts of interest regarding the publication of this paper.

Acknowledgments

This work was supported by the Natural Science Foundation of China (no. 11604304), Scientific and Technological Project of Shanxi Province (201603D121006-1), Science and Technology Innovation Project of University in Shanxi Province (no. 201657), Research Funding Project for returned students from Shanxi (2016-084), and Xinzhou Teachers University scientific research fund project (201719). This study was conducted at Science and Technology on Electronic Test & Measurement Laboratory in North University of China, Key Laboratory Instrumentation Science & Dynamic Measurement, Ministry of Education in North University of China, and Metrology Academy of Shanxi Province.

References

[1] A. D. Kersey, M. A. Davis, H. J. Patrick et al., "Fiber grating sensors," *Journal of Lightwave Technology*, vol. 15, no. 8, pp. 1442–1463, 1997.

[2] P. Wang, K. Xie, R. Chen et al., "Test verification and application of a longitudinal temperature force testing method for long seamless rails using FBG strain sensor," *Journal of. Sensors*, vol. 2016, pp. 1–11, 2016.

[3] D. J. Hill and G. A. Cranch, "Gain in hydrostatic pressure sensitivity of coated fibre Bragg grating," *Electronics Letters*, vol. 35, no. 15, pp. 1268-1269, 1999.

[4] A. Sun, "Study of simultaneous measurement of temperature and pressure using double fiber Bragg gratings with polymer package," *Optical Engineering*, vol. 44, no. 3, article 034402, 2005.

[5] J. Qi, "A comparison study of the sensing characteristics of FBG and TFBG," *Sensor Review*, vol. 33, no. 1, pp. 68–79, 2013.

[6] A. Stefani, S. Andresen, W. Yuan, N. Herholdt-Rasmussen, and O. Bang, "High sensitivity polymer optical fiber-Bragg-grating based accelerometer," *IEEE Photonics Technology Letters*, vol. 24, no. 9, pp. 763–765, 2012.

[7] M. H. Yau, T. H. T. Chan, D. P. Thambiratnam, and H. Y. Tam, "Static vertical displacement measurement of bridges using fiber Bragg grating (FBG) sensors," *Advances in Structural Engineering*, vol. 16, no. 1, pp. 165–176, 2016.

[8] Y. Zhan, K. Gu, H. Wu, and J. Luo, "A combined long period fiber grating multi-parameter sensor," *Sensor Review*, vol. 33, no. 3, pp. 220–227, 2013.

[9] Y. Liu, Z. Guo, Y. Zhang, K. S. Chiang, and X. Dong, "Simultaneous pressure and temperature measurement with polymer-coated fibre Bragg grating," *Electronics Letters*, vol. 36, no. 6, pp. 564–562, 2000.

[10] Y. Xiong, J. He, W. Yang, L. Sheng, W. Gao, and Y. Chen, "Research on FBG pressure sensor of flat diaphragm structure," in *Proceedings of 2012 International Conference on Measurement, Information and Control*, Harbin, China, May 2012.

[11] Y. Zhang, D. Feng, Z. Liu et al., "High-sensitivity pressure sensor using a shielded polymer-coated fiber Bragg grating," *IEEE Photonics Technology Letters*, vol. 13, no. 6, pp. 618-619, 2001.

[12] R. Sekar, B. N. Shivananju, K. P. Lakshmi, and S. Asokan, "Dual functional performance of fiber Bragg gratings coated with metals using flash evaporation technique," *Optical Fiber Technology*, vol. 18, no. 4, pp. 183–185, 2012.

[13] Z. G. Jia, L. Ren, D. S. Li, and H. N. Li, "Cable stretching construction monitoring based on FBG sensor," in *Proceedings Volume 7981, Sensors and Smart Structures Technologies for Civil, Mechanical, and Aerospace Systems*, San Diego, California, USA, 2011.

[14] S. Li, "Study and application about differential fiber Bragg grating force-testing ring," in *2010 International Conference on Measuring Technology and Mechatronics Automation*, pp. 519–522, Changsha City, China, March 2010.

[15] J. Huang, Z. Zhou, X. Wen, and D. Zhang, "A diaphragm-type fiber Bragg grating pressure sensor with temperature compensation," *Measurement*, vol. 46, no. 3, pp. 1041–1046, 2012.

[16] V. R. Pachava, S. Kamineni, S. S. Madhuvarasu, and K. Putha, "A high sensitive FBG pressure sensor using thin metal diaphragm," *Journal of Optics*, vol. 43, no. 2, pp. 117–121, 2014.

[17] G. Allwood, G. Wild, A. Lubansky, and S. Hinckley, "A highly sensitive fiber Bragg grating diaphragm pressure transducer," *Optical Fiber Technology*, vol. 25, pp. 25–32, 2015.

[18] F. Li, Y. du, W. Zhang, and F. Li, "Fiber Bragg grating soil-pressure sensor based on dual L-shaped levers," *Optical Engineering*, vol. 52, no. 1, article 014403, 2013.

[19] H. Y. Guo and Z. B. Wang, "Research on solid pressure sensor for high-pressure measurement based on fiber Bragg grating," *Chinese Journal of Analytical Chemistry*, vol. 45, no. 7, pp. 980–986, 2017.

[20] J. Jiang, C. R. Li, M. Guo, and Y. T. Luo, "Development of FBG tension sensor applied to ice monitoring of overhead transmission lines," *High Voltage Engineering*, vol. 36, no. 13, pp. 3028–3034, 2010.

[21] J. Dai, M. Yang, X. Yu, and H. Lu, "Optical hydrogen sensor based on etched fiber Bragg grating sputtered with Pd/Ag composite film," *Optical Fiber Technology*, vol. 19, no. 1, pp. 26–30, 2013.

[22] M. Wang, D. N. Wang, M. Yang, J. Cheng, and J. Li, "In-line Mach-Zehnder interferometer and FBG with Pd film for simultaneous hydrogen and temperature detection," *Sensors and Actuatiors B: Chemical*, vol. 202, pp. 893–896, 2014.

[23] Y. Zhao, Y. Liao, and S. Lai, "Simultaneous measurement of down-hole high pressure and temperature with a bulk-modulus and FBG sensor," *IEEE Photonics Technology Letters*, vol. 14, no. 11, pp. 1584–1586, 2002.

[24] Y. Guo, D. Zhang, Z. Zhou, F. Zhu, and L. Xiong, "Development and commissioning of FBG sensors for impact test of rock fall protective barrier," *Sensor Review*, vol. 34, no. 4, pp. 343–348, 2014.

[25] Q. P. Liu, X. G. Qiao, J. L. Zhao, Z. N. Jia, H. Gao, and M. Shao, "Novel fiber Bragg grating accelerometer based on diaphragm," *IEEE Sensors Journal*, vol. 12, no. 10, pp. 3000–3004, 2012.

3

Research and Application of Heavy-Equipment Parachute Rope Tension Sensor

Min Yao, Yanhua Jin, Min Zhao, and Shaohua Xu

College of Automation Engineering, Nanjing University of Aeronautics and Astronautics, Nanjing, China

Correspondence should be addressed to Min Yao; ym_nuaa@163.com

Academic Editor: Abdellah Touhafi

Heavy-equipment airdrops are mainly used to deliver relief supplies and heavy weapons. Given the heavy weight of the goods, the tension of the extraction and brake ropes of the parachute significantly affects the safety of the aircraft. On the basis of the measurement and installation characteristics of the parachute rope, this study designed the structure of a nondestructive pressure-type parachute rope tension sensor and set the location of the strain gauge patch using the ANSYS simulation software to obtain a high sensor sensitivity. The temperature error of the tension sensor is compensated, and the precision is improved using the LSSVM-PSO (Least Squares Support Vector Machine-Particle Swarm Optimization) algorithm. The developed tension sensor is applied to the extraction parachute test system to measure the traction of 2 and 8 m^2 parachutes. Test results show that the maximum weight of the platform these two parachutes can draw and the effect of parachute opening can be calculated.

1. Introduction

In recent years, large transport aircrafts have rapidly developed, improving the conditions for heavy-equipment airdrops and increasing the demands on the airdrop system [1–3]. The heavy-equipment airdrop system is mainly used by large- and medium-sized transport aircrafts in delivering heavy weapons, equipment, emergency supplies, and other goods to the designated area to support land and naval operations or disaster relief work [4, 5].

In the process of heavy-equipment airdrop, pulling the platform out of the cabin requires a large pulling force of the extraction parachute. Meanwhile, the large tension generated by the parachute affects the safety of the transporting aircraft. Therefore, the tension of the parachute rope during the operation of the parachute is closely related to the platform and the smooth flight of the transport aircraft.

However, such large tension sensors for parachute ropes of heavy-equipment airdrop systems are rarely reported. Zhao et al. from Nanjing University of Aeronautics and Astronautics designed the surface acoustic wave tension sensor for parafoils in 2017 [6]. This sensor has good linearity and repeatability. However, this sensor system needs a reader to send and receive wave signals, which makes the measurement extremely sensitive to interference. Lu et al. from Xidian University designed a compensated SAW sensor to measure the yarn tension [7]. The measurement range is no more than 1 N, and its structure is not suitable for installing in parachute ropes.

The tension sensors for parachute ropes of heavy-equipment airdrop systems may have the same structure as the tension sensors for vehicle safety belts, such as the EL20-S458 sensor produced by U.S. MEAS and LBT-A-20KNSA1 produced by KYOWA. However, this structure has distinct edges and corners. The friction of the sensors may cause damage to the parachute ropes or even burn the ropes. This structure is difficult to install and the rope needs to be cut, all of which are unsuitable for parachutes. Furthermore, the detection range of such sensors for the automotive belt is unsuitable for the heavy-equipment airdrop test.

This study designed a kind of nondestructive tension sensor for the measurement of the tension of extraction and brake parachute ropes. The proposed tension sensor is characterized by wide range, high precision, and easy installation.

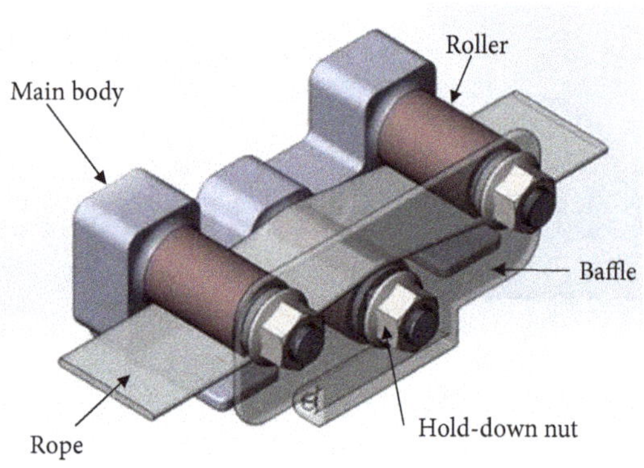

FIGURE 1: Structure of the tension sensor.

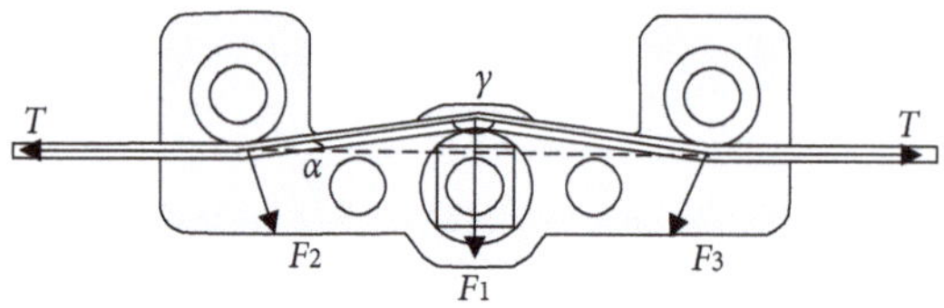

FIGURE 2: Force analysis of the tension sensor.

2. Force Analysis and Structure Design

It is necessary to test the parachutes rigorously before they are put into use. The tests include parachute control performance evaluation, smooth sliding performance evaluation, and stability test. In order to realize these tests, the measurement of the stress of each rope during landing is very important. It reflects the force of the ropes and the degree of stress equalization. It is also an important indicator of parachute control performance.

The overall structure of the tension sensor is shown in Figure 1. The sensor is designed as a three-axis compression type, which consists of four parts, namely, the main body, roller, baffle, and nut. A cylinder roller is used in this sensor to ensure that it is nondestructive and it can help to reduce weight. The roller can effectively reduce the contact area with the parachute ropes and the friction of the sensors.

As shown in Figure 1, the parachute rope is installed and fixed on the side of the sensor. After the parachute rope is stressed, the sensor is tightened and pressed. Then, the parachute rope transmits the force to the sensor and undergoes elastic deformation. The stress on the sensor body is uneven, and the location of the strain gauge is an important part of the design. Sufficient strain should be provided to ensure the sensitivity of the sensor and meet the measurement range. Therefore, force analysis of the sensor structure is required to determine the locations of the strain gauges.

The mechanical structure model of the sensor structure is shown in Figure 2. Assuming that the ropes are subjected to tensile force T, the forces acting on the two rollers of the sensor due to the straightening of the ropes are shown in Figure 2. F_1, F_2, and F_3 are the pressures applied to the sensor structure by the belt tension, and α is the initial belt tension angle.

According to the sensor force diagram shown in Figure 2, the following equations are derived:

$$|F_1| = 2|T| \sin \alpha, \tag{1}$$

$$|F_2| = |F_3| = 2|T| \sin \left(\frac{\alpha}{2}\right). \tag{2}$$

According to Equations (1) and (2), the sensor sensitivity and range are related to the angle α. The smaller the angle α is, the lower the sensitivity of the sensor is and the larger the measurement range is. In finding the sensor to meet the range of the premise, increasing the angle as much as possible under the condition of meeting the range of the sensor is necessary.

The material of the sensor is 42CrMo. Figure 3(a) shows the ANSYS grid meshing, and Figure 3(b) shows the strain analysis. The stress distribution diagram shows the maximum stress areas in red. These areas are the combination of the middle guide cylinder and the force beam, which is consistent with the actual structure. The structure of the beam arm suffers from a high degree of stress, such that the location of the beam arm is selected as the installation position of the strain gauge. The glue patch location is shown in Figure 4.

Strain gauges are widely used in strain measurement [8, 9]. The strain gauge used in this sensor is BE650-4BB (11), the structure of which is shown in Figure 5. The two gantries perpendicular to each other in the "T" gauge are used for the axial and transverse direction train measurements, also known as Poisson strain.

As shown in Figure 4, in position 1, a strain gauge is attached to each force arm. The tension sensor detects four resistances on each force arm, i.e., axial resistance changes ΔR_{x1} and $\Delta R'_{x1}$ and transverse resistance changes ΔR_{y1} and $\Delta R'_{y1}$. Position 1 is opposite to position 2 in terms of strain direction. ΔR_{x1} and ΔR_{y1} are the resistance changes of the left force arm, and $\Delta R'_{x1}$ and $\Delta R'_{y1}$ are the resistance changes of the right force arm, as shown in Figure 4. Overall, these eight resistances can produce two four-arm bridges, as shown in Figure 6.

According to the four-arm bridge principle, the output is derived as follows:

$$U_{\text{out}} = \frac{1}{8} UK \left[\varepsilon_{x_1} + \varepsilon_{y_1} - \left(\varepsilon_{x_2} + \varepsilon_{y_2}\right)\right]. \tag{3}$$

In Equation (3), U_{out} is the output voltage of the bridge, K is the sensitivity coefficient of the strain gauge, and ε is the strain.

From the sensor force shown in Figure 2 and the analysis expressed in Equations (1) and (2), when the sensor receives a certain amount of tension T, ε can be related to α in this sensor structure.

3. Temperature Error Compensation

The sensor is designed for testing the heavy-equipment airdrop extraction parachute and will measure the pulling

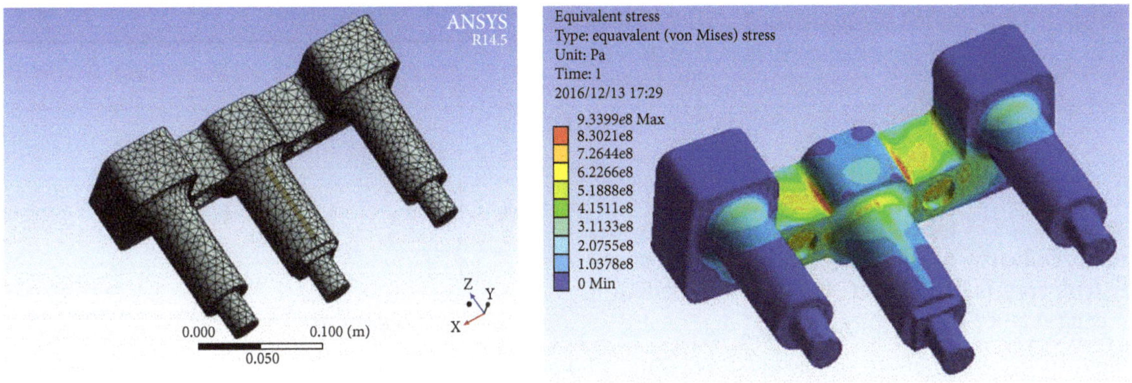

(a) The mesh of the sensor in ANSYS

(b) The stress analysis of the sensor

FIGURE 3: Stress analysis of the sensor by the ANSYS software.

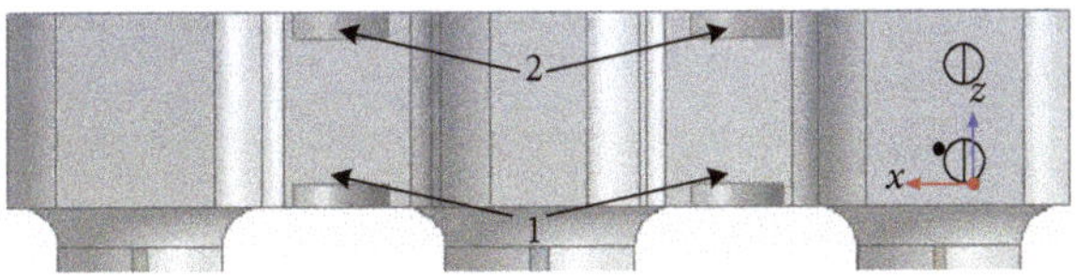

FIGURE 4: Strain gauge patch diagram.

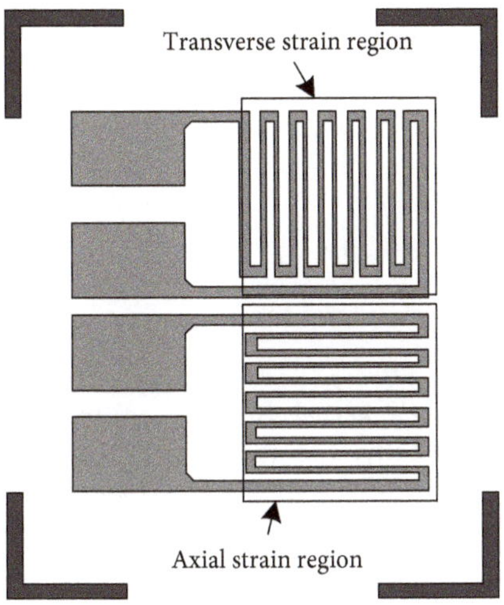

FIGURE 5: Diagram of the strain gauge.

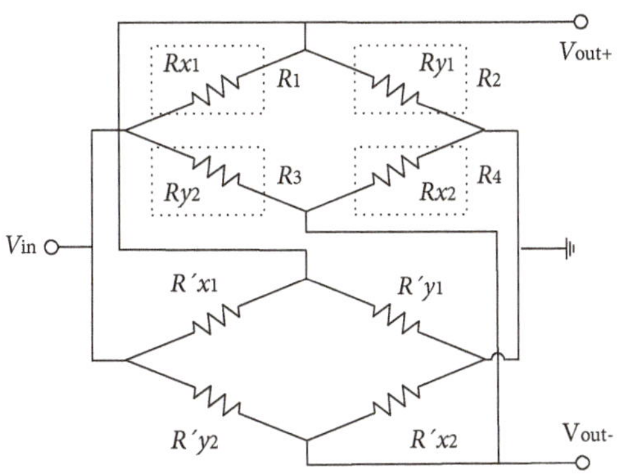

FIGURE 6: Measurement circuit.

FIGURE 7: The sensor system.

force of the platform on the rail out of the plane. The range of traction is 0–100 kN. The sensor system is shown in Figure 7.

Temperature error is one of the main errors of the strain sensor [10–12]. The airdrop height will vary from hundreds of meters to thousands of meters, so that the temperature will change in a wide range. Therefore, the temperature error of the parachute tension sensor cannot be ignored.

The measurement of the strain gauge may be affected by many factors, such as temperature, vibration, and transverse effect. These factors cause a nonlinear error and the temperature is one of the key factors. A traditional temperature compensator is usually used to decrease the temperature error. However, other nonlinear errors caused by other factors are not fully considered [13]. A BP neural network is also used in nonlinear compensation [14]. This method has higher precision, but it does not fully consider the influence of temperature. Nagi et al. used support vector machines and a genetic algorithm to compensate for the nonlinear error [15], but the process of coding, crossover, selection, and variation increases the complexity of the algorithm. In this paper, LSSVM and particle swarm optimization are combined to compensate for the nonlinear error of the tension sensor, especially the temperature error. LSSVM-PSO algorithm maps the input vector to a high-dimensional feature space via nonlinear mapping and constructs the optimized decision function [16–18]. The PSO algorithm featured with memory functions is used to search for optimal value. In this way, it can efficiently reduce calculation complexity and solve nonlinear problems of small samples.

The resistance variation of the strain gauge is the combination of the resistance changes ΔR_E, which is caused by load deformation, and ΔR_T, which is caused by temperature change, and is expressed as follows:

$$\Delta R = \Delta R_E + \Delta R_T. \quad (4)$$

The strain caused by the temperature is influenced by the temperature coefficient of resistance of the strain gauge. When the temperature changes ΔT, the relative variation of the strain gauge can be expressed as follows:

$$\frac{\Delta R_T}{R} = \alpha_R \Delta T. \quad (5)$$

In Equation (5), α_R is the temperature coefficient of resistance of the strain material. In Figure 6, according to the bridge principle, ΔR_{x1} and $\Delta R'_{x1}$ are the axial resistance variations and ΔR_{y1} and $\Delta R'_{y1}$ are the lateral variations. Assuming that the strain errors of each strain gauge are the same, the lateral resistance variation error is δ and the longitudinal variation error is $-\delta$. According to Equations (3), (4), and (5), the result is derived as follows:

$$U_{\text{out}} = \frac{1}{2} U\delta \left[(1+2\mu)\frac{dL}{L} + \alpha_R \Delta T \right]. \quad (6)$$

In building an LSSVM model with good adaptability, the test takes the tension sensor data from different environments such as training samples, with sensor output U and environmental temperature T as inputs. The sample set Q is built as follows:

$$Q = \{(x_1, y_1), (x_2, y_2), \cdots, (x_i, y_i),\ i = 1, \ldots, n\}. \quad (7)$$

In Equation (7), x_i is the ith input, i.e., $x_i = [T_i, U_i]$, and y_i is the algorithm output. The LSSVM model of data approximation can be expressed as follows:

$$\begin{cases} \min & J(\omega, \xi) = \frac{1}{2} w^T w + \frac{1}{2}\gamma \sum_{i=1}^{n} \xi_i^2 \\ \text{s.t.} & \beta_i = w^T \varphi(x_i) + b + \xi_i, \\ & i = 1, \ldots, n. \end{cases} \quad (8)$$

In Equation (8), J is the pending optimized function, ω is the weight vector, $\varphi(x)$ is the nonlinear mapping function, γ is the regularization parameter, and b is the deviation value. The following Lagrangian function is introduced to derive the constrained optimal solution:

$$L(w, b, \xi, a) = \frac{1}{2} w^T w + \frac{1}{2}\gamma \sum_{i=1}^{n} \xi_i^2 - \sum_{i=1}^{n} a_i \left[w^T \varphi(x_i) + b + \xi_i - \beta_i \right]. \quad (9)$$

The parameter a_i is introduced in Equation (9) as the Lagrangian multiplier. The partial derivatives of ω, b, ξ, and a should be equal to 0 to obtain the extreme value using Equation (9). The linear equation set is generated as follows:

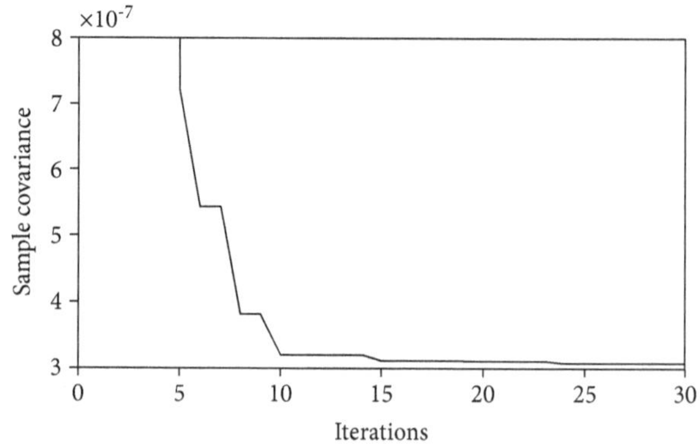

Figure 8: PSO optimal curve.

$$\begin{bmatrix} 0 & 1 & \cdots & 1 \\ 1 & K(x_1, x_1) + \gamma^{-1} & \cdots & K(x_n, x_1) \\ \vdots & \vdots & \ddots & \vdots \\ 1 & K(x_1, x_n) & \cdots & K(x_n, x_n) + \gamma^{-1} \end{bmatrix} \begin{bmatrix} b \\ a_1 \\ \vdots \\ a_n \end{bmatrix} = \begin{bmatrix} 0 \\ \beta_1 \\ \vdots \\ \beta_n \end{bmatrix}. \quad (10)$$

$K(x, x_i)$ is the kernel function of the LSSVM algorithm. In this study, the radial basis kernel function of strong generalization is selected by mapping the original feature into the indefinite dimensions. The expression is as follows:

$$K(x, x_i) = \exp\left(-\frac{\|x - x_i\|^2}{\sigma^2} \right). \quad (11)$$

The values of b and the Lagrangian multiplier a_i can be derived from the Lagrangian function. The tension sensor compensation model, i.e., the nonlinear approximation function of the LSSVM model, is expressed as follows:

$$f(x_i) = \sum_{i=1}^{n} a_i K(x, x_i) + b. \quad (12)$$

In the LSSVM model, parameters $[\sigma, \gamma]$ require optimization. Sample covariance e_{RMSE} is the objective function in the PSO algorithm. When the sample covariance reaches the set precision, the corresponding parameters

TABLE 1: Temperature error of the sensor without compensation.

Tension (kN)	V_{out} (V) −20°C	−10°C	0°C	+10°C	+20°C	+30°C	+40°C	Err_{temp} (%FS)
0	0.012	0.015	0.017	0.017	0.019	0.025	0.044	0.50%
20	0.969	0.972	0.991	1.003	1.011	1.020	1.034	0.84%
40	1.930	1.942	1.961	1.985	2.006	2.198	2.041	1.52%
60	2.892	2.920	2.953	2.962	2.975	3.021	3.033	**1.66%**
80	3.859	3.870	3.894	3.908	3.925	3.940	3.956	1.32%
100	4.817	4.833	4.845	4.857	4.862	4.882	4.896	0.90%

TABLE 2: Temperature error of the sensor with compensation.

Tension (kN)	V_{out} (V) −20°C	−10°C	0°C	+10°C	+20°C	+30°C	+40°C	Err_{temp} (%FS)
0	0.041	0.032	0.028	0.022	0.019	0.021	0.043	0.48%
20	0.998	1.005	1.007	1.010	1.011	1.014	1.019	0.26%
40	1.964	1.973	2.002	1.997	2.006	2.004	2.000	0.84%
60	2.932	2.961	2.958	2.971	2.975	2.982	2.973	**0.86%**
80	3.905	3.913	3.912	3.924	3.925	3.928	3.930	0.40%
100	4.875	4.863	4.857	4.867	4.862	4.870	4.881	0.38%

$[\sigma, \gamma]$ can be regarded as the global optimal solution. The sample covariance e_{RMSE} is expressed as follows:

$$e_{RMSE} = \left[\frac{1}{n}\sum_{i=1}^{n}(f(x_i) - y_i)^2\right]^{1/2} = \left[\frac{1}{n}\sum_{i=1}^{n}\left(\sum_{i=1}^{n} a_i K(x, x_i) + b - y_i\right)^2\right]^{1/2}. \quad (13)$$

We assume that the size of the particle in the D-dimensional search space is M, the position of the ith particle is x_i, and the flight speed is v_i. In the nth iteration, the historical optimal position of the particle is P_{best} and the optimal position of the global particles is g_{best}. The iterative formula is expressed as follows:

$$\begin{cases} V(t+1) = \omega V(t) + c_1 r_1 [P(t) - X(t)] + c_2 r_2 [P(t) - X(t)], \\ X(t+1) = X(t) + \lambda V(t+1). \end{cases} \quad (14)$$

In Equation (12), t is the current evolution algebra, c_1 and c_2 are the nonnegative factors, ω is the inertia weight, λ is the speed factor, and r_1 and r_2 are the random numbers within [0,1]. We set $D = 2$, $M = 25$, $[c_1, c_2] = [2,0.8]$, and $\lambda = 1$. 25 stacks of measured data of the tension sensor tested in different temperatures were brought into the PSO algorithms as samples to search for the optimal results. The optimization process is shown in Figure 8. When the PSO algorithm reaches 24 iterations, the sample covariance e_{RMSE} is 3.084×10^{-7} and the pending optimal parameters are $[\sigma, \gamma] = [2.181,1.395]$.

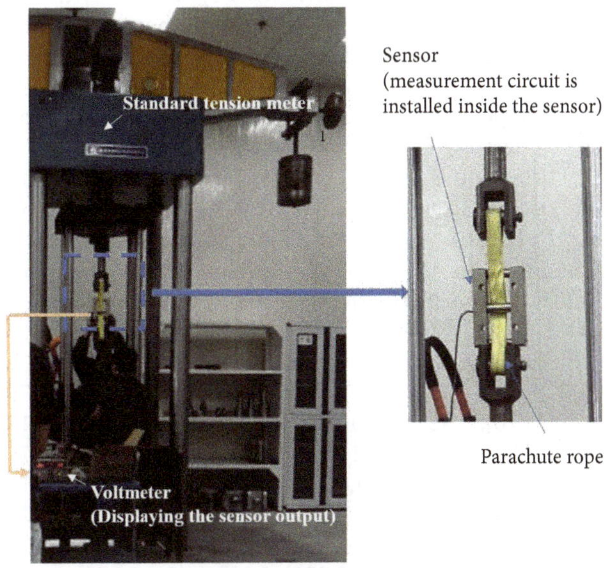

FIGURE 9: The picture of the test.

The tension sensor with the full range of 100 kN was kept in a high- and low-temperature environment for 30 minutes before the test. As the standard tension meter has a large volume, it is difficult to put the whole testing system into a constant temperature environment. Therefore, the tension sensor is processed separately to get different working temperatures and we get its exact working temperature with a temperature sensor stuck to the tension sensor during the test.

Table 3: The parameters of the standard tension meter and the voltmeter used in the test.

Equipment name	Range	Precision
Standard tension meter	1–100 kN	0.01%
Voltmeter	DC: (0–1000 V), AC: (0–750 V)	DC: 0.01%, AC: 0.1%

Table 4: The results of the test in room temperature (temperature: 20°C, humidity is 62% RH).

Standard force value (kN)	Uptravel output (V)				Repeatability (%FS)	Theoretical value (V)	Nonlinearity (%FS)	Downtravel output (V)			Precision (%FS)
	1	2	3	Average				1	2	3	
0	0.039	0.008	0.011	0.019	0.64	0.019	0.00	0.007	0.012	0.017	0.41
20	1.037	0.991	1.005	1.011	0.95	0.988	0.47	1.056	1.020	1.037	1.40
40	2.040	1.978	2.001	2.006	1.28	1.956	**1.03**	2.085	2.030	2.068	2.65
60	3.006	2.942	2.977	2.975	1.32	2.925	1.03	3.072	3.130	3.065	2.88
80	3.942	3.893	3.939	3.925	1.01	3.893	0.66	3.975	4.006	4.036	**2.94**
100	4.860	4.831	4.896	4.862	**1.34**	4.862	0.00				0.70

Temperature error calculation is as follows:

$$\mathrm{Err}_{\mathrm{temp}} = \frac{V_{\mathrm{out}} - V_{20^{\circ}\mathrm{C}}}{V_{\mathrm{FS}}}, \tag{15}$$

where V_{out} is the output of the sensor system in any temperature, $V_{20^{\circ}\mathrm{C}}$ is the output of the sensor in room temperature (20°C), and V_{FS} is the full scale of the output.

Table 1 shows the test data in the range of 0–100 kN at room temperature (+20°C), high temperature (+40°C), and low temperature (−20°C) without compensation.

Three rounds of the tests were carried out. All the data were used to form the training data (x_i, y_i), as fomula (7). $x_i = [T_i, U_i]$ and y_i is the expected output. For example, the data in Table 1 form the training data as follows: ([−20,0.012], 0.019), ([−20,0.969], 1.011), ([−20,1.930], 2.006), ([−20,2.892], 2.975), ([−20,3.859], 3.925), ([−20,4.817], 4.862), ([−10,0.015], 0.019), ([−10,0.972], 1.011), ([−10,1.942], 2.006), ([−10,2.920], 2.975), ([−10,3.870], 3.925), ([−10,4.833], 4.862), ([0,0.017], 0.019), ([0,0.991], 1.011), ([0,1.961], 2.006), ([0,2.953], 2.975), ([0,3.894], 3.925), ([0,4.845], 4.862), ([10,0.017], 0.019), ([10,1.003], 1.011), ([10,1.985], 2.006), ([10,2.962], 2.975), ([10,3.908], 3.925), ([10,4.857], 4.862), ([20,0.019], 0.019), ([20, 1.011], 1.011), ([20, 2.006], 2.006), ([20, 2.975], 2.975), ([20, 3.925], 3.925), ([20, 4.862], 4.862), ([30,0.025], 0.019), ([30, 1.020], 1.011), ([30, 2.198], 2.006), ([30, 3.021], 2.975), ([30, 3.940], 3.925), ([30, 4.882], 4.862), ([40,0.044], 0.019), ([40, 1.034], 1.011), ([40, 2.041], 2.006), ([40, 3.033], 2.975), ([40, 3.956], 3.925), and ([40, 4.896], 4.862).

After training with the LSSVM-PSO algorithm, another round of test was carried out. The output after temperature compensation was shown in Table 2. The test results show that the sensor temperature errors decrease from 1.66% to 0.86% after temperature compensation.

Finally, the sensor was tested and calibrated in Wuxi Institute of Metrology and Testing (WXMTC). The sensor was tested in high temperature (+40°C), room temperature (+20°C), and low temperature (−20°C) separately. Figure 9 is the testing picture. The parameters of the testing machine are shown in Table 3.

The sensor was tested for three rounds in room temperature. The room temperature is 20°C, and the relative humidity was 62% RH. The results of the testing are shown in Table 4.

As shown in Table 4, the repeatability of the sensor is 1.34% and the nonlinearity is 1.03%. The precision of the sensor is 2.94%.

Then, the sensor was tested in high temperature (+40°C) and low temperature (−20°C) separately. After the sensor was kept in high temperature/low temperature for 30 minutes, the sensor was tested from 0 to 100 kN. The test results for high/low temperature are shown in Table 5.

As shown in Table 5, the precision of this sensor in high temperature and low temperature is 2.46%. Temperature has a very small influence on this sensor.

4. Traction Test for Heavy-Equipment Airdrop Extraction Parachute

The proposed sensor was applied to the extraction parachute traction test of large transport aircrafts. The test measured the pulling force of the parachute when the cargo was dragged out of the aircraft cabin in the heavy-weaponry system. The traction test platform is shown in Figures 10 and 11. The heavy-weaponry system requires a tremendous force to drag the cargo out of the cabin for airdrop, which usually uses an extraction parachute. Different extraction parachutes generate varied pulling forces. Each type of transport aircraft equipped with the corresponding type of extraction parachute according to cargo weight must pass rigorous tests. On one hand, the appropriate traction force ensures that the cargo is dragged out of the cabin successfully. On the other hand, the tremendous pulling force, particularly at the moment of opening the parachute, cannot influence the safety of flight [19, 20].

TABLE 5: The results of the test in high/low temperature (high temperature: +40°C, low temperature: −20°C).

Standard force value (kN)	Output (V) −20°C Uptravel	Output (V) −20°C Downtravel	Output (V) +40°C Uptravel	Output (V) +40°C Downtravel	Theoretical value (V)	Precision (%FS)
0	0.029	0.020	0.012	0.032	0.019	0.27
20	1.022	1.034	1.043	1.056	0.988	1.40
40	2.023	2.044	2.004	2.034	1.956	1.80
60	3.001	3.023	3.009	3.045	2.925	**2.46**
80	3.931	3.956	3.907	3.942	3.893	1.29
100	4.848		4.817		4.862	0.96

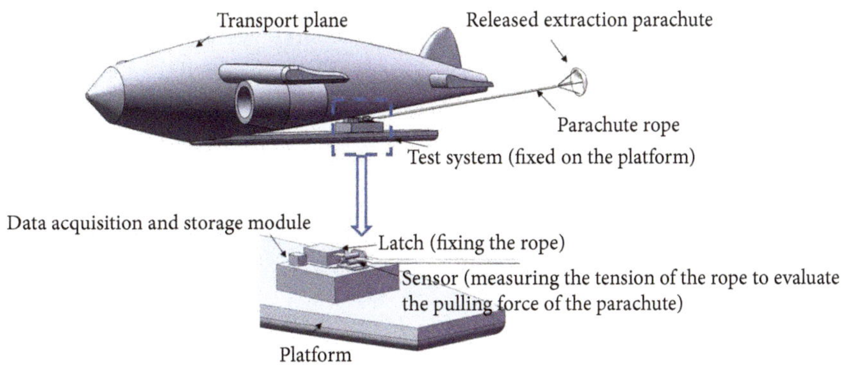

FIGURE 10: Traction test diagram.

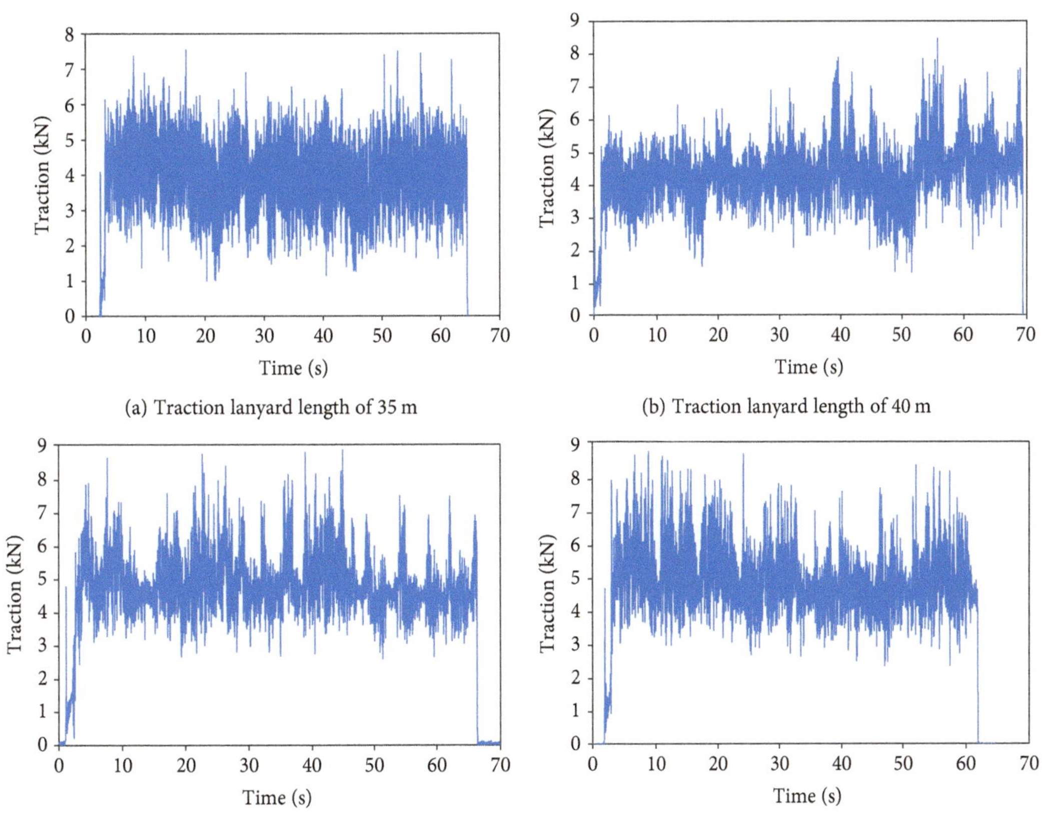

(a) Traction lanyard length of 35 m

(b) Traction lanyard length of 40 m

(c) Traction lanyard length of 45 m

(d) Traction lanyard length of 50 m

FIGURE 11: Parachute test data at an area of 2 m^2.

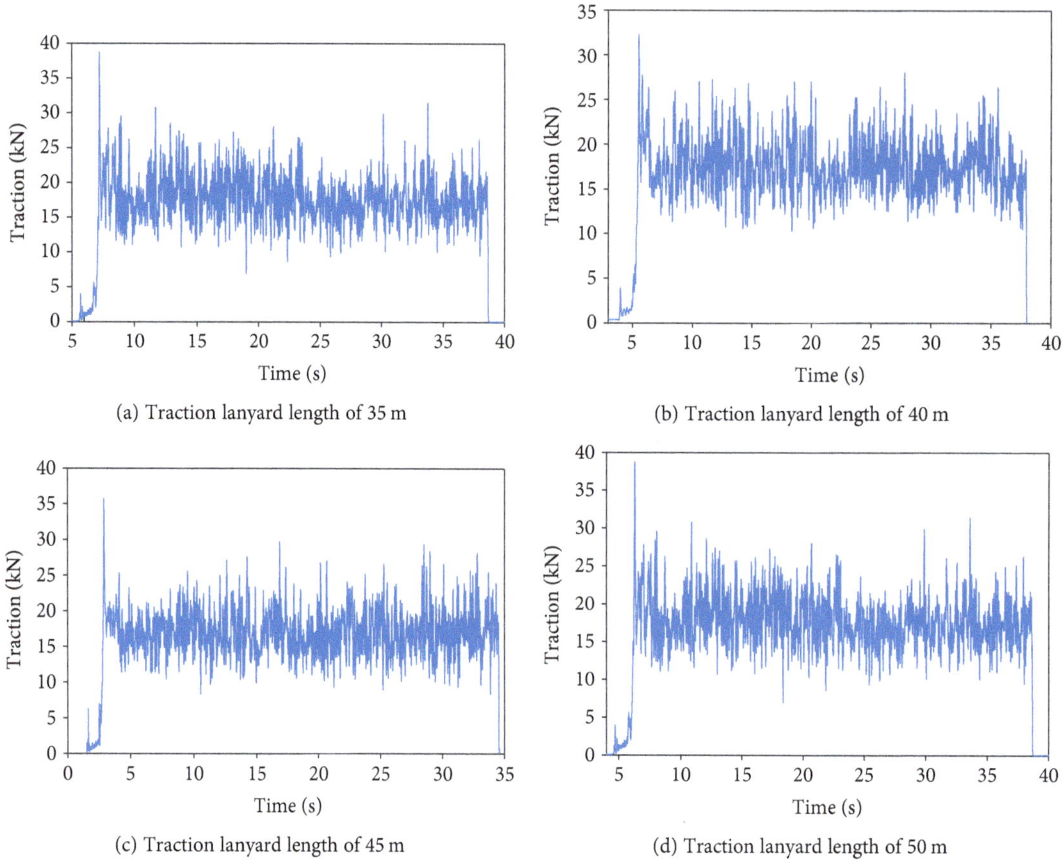

(a) Traction lanyard length of 35 m

(b) Traction lanyard length of 40 m

(c) Traction lanyard length of 45 m

(d) Traction lanyard length of 50 m

FIGURE 12: Parachute test data at an area of 8 m^2.

The test aims at evaluating the pulling force to a transport aircraft using different dimensions of extraction parachutes. The test diagram is shown in Figure 10. The steps of the test are as follows:

(1) The parachute rope was fixed on the cargo table at the rail of the transport aircraft by a fastening bolt. The tension sensor and traction parachute latch were fixed on the heavy-weaponry cargo table and the parachute rope passed through the tension sensor.

(2) The parachute pack for airdrop was hanged at the rear of the cabin. The transport aircraft dragged the parachute and the tension sensor measured the parachute rope tension and saved the data in the storage module.

(3) After 30–60 s, the parachute opened and was released. The sensor system saved the detail test data in the process of the whole test.

In this test, the sampling frequency of the tension sensor was 200 Hz, the areas of the extraction parachutes were 2 and 8 m^2, and the length of the parachute lanyard was 35–50 m.

Figure 11 shows the four test results for tension areas of 2 m^2. They give the tension curves of the ropes from the parachute opening prior to being released. It is obvious that the tension increases rapidly when the parachute opens at the beginning of the curve and decreases rapidly at the end of the curve. During the test, the tension fluctuates violently, which is caused by severe turbulence.

TABLE 6: Traction of different parachutes.

No.	2 m^2 Maximum (kg)	2 m^2 Average (kg)	8 m^2 Maximum (kg)	8 m^2 Average (kg)
1	788	398	3988	1697
2	883	445	3748	1645
3	872	428	3977	1788
4	853	456	4058	1689
Average	849	432	3943	1705

Figure 12 shows the four test results for tension areas of 8 m^2.

According to the parachute test data, the maximum and average tractions generated by two parachutes, i.e., 2 and 8 m^2, are shown in Table 6. The traction generated by an 8 m^2 parachute is much bigger than that generated by 2 m^2 parachute, especially the maximum traction. The traction generated by an 8 m^2 parachute is so big that the tension in the rope increases dramatically at the moment of opening the parachute, which is shown in Figure 12.

There is a relation between the weight of the cargo load and the traction needed to pull the cargo out of the cabin,

Table 7: Weights of cargo loading.

$2\,m^2$		$8\,m^2$	
Traction (kg)	Weight (kg)	Traction (kg)	Weight (kg)
432	900–2100	1705	3400–8500

which is expressed as traction ratio formula λ of the heavy-equipment airdrop:

$$\lambda = \frac{D_s}{G}, \tag{16}$$

where D_s is the traction resistance and G is the total cargo load. In actual engineering, λ is 0.2–0.5. Using formula (16), we can calculate the weights of the cargo loads that these two parachutes with areas of 2 and $8\,m^2$ can pull. The weights of the cargo loads are shown in Table 7.

Table 7 shows that the traction capabilities of 2 and $8\,m^2$ extraction parachutes for airdrop are 900–2100 kg and 3400–8500 kg. At the moment of opening, $8\,m^2$ parachutes generate the maximum pulling forces of 46 kN.

5. Conclusion

This study designs a triaxial side-pressure tension sensor with a wide range that can be easily installed to measure the parachute tension of airdrop equipment in real time. Temperature error is one of the main errors of the strain sensor, which reaches 1.66% from −20°C to +40°C in the range of 0–100 kN. The LSSVM-PSO algorithm is used to reduce the temperature error of the tension sensor to 0.86%. The sensor was tested and calibrated in Wuxi Institute of Metrology and Testing (WXMTC). The results show that the repeatability of the sensor is 1.34%, the nonlinearity is 1.03%, and the precision is 3%.

The tension sensor was used in the heavy-equipment airdrop test to measure the average and maximum tension values of extraction parachutes with areas of 2 and $8\,m^2$. According to the test, the traction capabilities of two kinds of extraction parachutes for airdrop are 900–2100 kg and 3400–8500 kg. The maximum pulling forces generated by $8\,m^2$ parachutes are 10 and 46 kN, respectively. The results of the test can provide a reference to ensure the safe flight of transport aircraft.

Conflicts of Interest

The authors declare that they have no conflicts of interest.

Acknowledgments

This research has been funded by the Natural Science Foundation of China (Nos. 51875289 and 61873124), the Aeronautical Science Foundation of China (No. 2016ZD52036), and Jiangsu Province Youth Funding (No. BK20150746).

References

[1] K. J. Desabrais, J. Riley, J. Sadeck, and C. Lee, "Low-cost high-altitude low-opening cargo airdrop systems," *Journal of Aircraft*, vol. 49, no. 1, pp. 349–354, 2012.

[2] J. Chen, C.-B. Ma, and D. Song, "Kinetic characteristics analysis of aircraft during heavy cargo airdrop," *International Journal of Automation and Computing*, vol. 11, no. 3, pp. 313–319, 2014.

[3] J. Qiu, X. Y. Wanng, Y. K. Gao, Y. H. Han, and W. J. Hei, "Analysis for heavy cargo airdrop on low altitude based on natural airplane," *Journal of System Simulation*, vol. 24, no. 4, pp. 933–937, 2012.

[4] D. C. Meier and S. T. Fiorino, "Application of satellite- and NWP-derived wind profiles to military airdrop operations," *Journal of Applied Meteorology and Climatology*, vol. 55, no. 10, pp. 2197–2209, 2016.

[5] D. M. Ştefănescu and M. A. Anghel, "Electrical methods for force measurement—a brief survey," *Measurement*, vol. 46, no. 2, pp. 949–959, 2013.

[6] M. Zhao, H. Zhang, R. Zhang, M. Yao, and M. Ma, "Design of surface acoustic wave parafoil riser tension sensor," *IEEE Sensors Journal*, vol. 17, no. 10, pp. 3022–3029, 2017.

[7] X. Lu, W. Lu, and C. Zhu, "Compensated SAW yarn tension sensor," *IEEE Transactions on Instrumentation and Measurement*, vol. 63, no. 12, pp. 3162–3168, 2014.

[8] V. R. Bagheri, P. H. Tehrani, and D. Younesian, "Optimal strain gauge placement in instrumented wheelset for measuring wheel-rail contact forces," *International Journal of Precision Engineering and Manufacturing*, vol. 18, no. 11, pp. 1519–1527, 2017.

[9] Y. Zuo, "Strain gauge measuring system depending on high temperature of aero-engine," *Computer Measurement & Control*, vol. 16, no. 11, pp. 1528-1529, 2008.

[10] A. Maskay and M. Pereira da Cunha, "High-temperature static strain langasite SAWR sensor: temperature compensation and numerical calibration for direct strain reading," *Sensors and Actuators A: Physical*, vol. 259, pp. 34–43, 2017.

[11] W. Lu, Y. Feng, C. Zhu, and J. Zheng, "Temperature compensation of the SAW yarn tension sensor," *Ultrasonics*, vol. 76, pp. 87–91, 2017.

[12] T. C. Haber, S. Ferguson, D. Guthrie, T. W. Graver, B. J. Soller, and A. Mendez, "Analysis, compensation and correction of temperature effects on FBG strain sensors," in *Fiber Optic Sensors and Applications X*, vol. 8722 of *Proceedings of SPIE*, p. 872206, Baltimore, MD, USA, May 2013.

[13] F. M. Haran, J. K. Rew, and P. D. Foote, "A strain-isolated fibre Bragg grating sensor for temperature compensation of fibre Bragg grating strain sensors," *Measurement Science and Technology*, vol. 9, no. 8, pp. 1163–1166, 1998.

[14] X. G. Wang, T. Cai, and F. Deng, "Error compensation of photoelectric encoder based on improved BP neural network," *Chinese Journal of Scientific Instrument*, vol. 30, no. 7, pp. 1404–1409, 2009.

[15] J. Nagi, K. S. Yap, and S. K. Tiong, "Detection of abnormalities and electricity theft using genetic support vector machines," in *TENCON 2008 - 2008 IEEE Region 10 Conference*, pp. 1–6, Hyderabad, India, November 2008.

[16] R. G. Gorjaei, R. Songolzadeh, M. Torkaman, M. Safari, and G. Zargar, "A novel PSO-LSSVM model for predicting liquid rate of two phase flow through wellhead chokes," *Journal of*

Natural Gas Science and Engineering, vol. 24, pp. 228–237, 2015.

[17] B. Sun and H. T. Yao, "The short-term wind speed forecast analysis based on the PSO-LSSVM predict model," *Power System Protection and Control*, vol. 40, no. 5, pp. 85–89, 2012.

[18] X. Xue, "Prediction of slope stability based on hybrid PSO and LSSVM," *Journal of Computing in Civil Engineering*, vol. 31, no. 1, article 04016041, 2017.

[19] Y. H. Han and Y. P. Lu, "Dynamics for heavy equipment airdrop and control design," *Flight Dynamics*, vol. 29, no. 3, pp. 28–31, 2011.

[20] T. Jann, S. Geisbauer, N. Bier, W. Krüger, and H. Schmidt, "Multi-fidelity simulation of cargo airdrop: from the payload bay to the ground," in *AIAA Modeling and Simulation Technologies Conference*, pp. 1–20, Dallas, TX, USA, 2015.

4

Wearable Lumbar-Motion Monitoring Device with Stretchable Strain Sensors

Hiroyuki Nakamoto ,[1] Tokiya Yamaji,[1] Akio Yamamoto,[2] Hideo Ootaka,[3] Yusuke Bessho,[3] Futoshi Kobayashi ,[1] and Rei Ono[2]

[1]*Graduate School of System Informatics, Kobe University, 1-1 Rokkodai-cho, Nada, Kobe 657-8501, Japan*
[2]*Graduate School of Health Science, Kobe University, 7-10-2 Tomogaoka, Suma, Kobe 654-0142, Japan*
[3]*Bando Chemical Industries Ltd, 4-6-6 Minatojima Minamimachi, Chuo, Kobe 650-0047, Japan*

Correspondence should be addressed to Hiroyuki Nakamoto; nakamoto@panda.kobe-u.ac.jp

Guest Editor: Aftab M. Hussain

Low-back pain is a common affliction. Epidemiological analyses have reported that periodic cycles of lumbar flexion and rotation are major risk factors for low-back pain. To prevent low-back pain, a lumbar-motion monitoring device could help diagnosticians assess patients' risk for low-back pain. This study proposes such a device that uses lightweight stretchable strain sensors. Six of these strain sensors form a parallel-sensor mechanism that measures rotation angles of lumbar motion in three axes. The parallel-sensor mechanism calculates rotation angles from the lengths of the strain sensors iteratively. Experimental results reveal that the prototype device is effective for lumbar-motion measurement and significantly improves in terms of wearability over comparable devices.

1. Introduction

Low-back pain (LBP) is a common affliction in contemporary life. Globally, two-thirds of adults experience LBP to the extent that treatment is required [1, 2]. Hoy et al. reported that the point prevalence of LBP is as much as 10%, and the 1-month prevalence is over 20% [3]. LBP sufferers often have difficulty maintaining healthy living habits and steady work.

Although the prevalence of LBP is not limited to a particular job category, industrial workers, nurses, and healthcare workers tend to experience a relatively high prevalence of LBP [4–11]. These workers frequently use their back muscles and spines in lumbar motions of flexion, rotation, and side bending. With epidemiological analysis, Hoogendoorn et al. revealed that periodic lumbar flexion and rotation is a major risk factor for LBP [12]. Overall, LBP risk is assessed in terms of accumulated lumbar motions and loads, and patients can benefit from alerts of excess lumbar motions to prevent the onset or worsening of LBP. Waters et al. estimated the risk for LBP in manual lifting tasks with calculations [13]. Norman et al. assessed the risk for LBP in automotive industries using peak motion and cumulative lumbar motions as risk factors [14]. These methods are difficult to compare with each other, as they use different metrics for understanding lumbar risk. A measurement system that can continuously record time-series data during lumbar motions of flexion, rotation, and side bending could provide comparable lumbar data for a range of industries and subject groups.

To obtain time-series measurements of the pressure between lumbar bones, Lisi et al. obtained in vivo measurements with a load sensor inserted between lumbar vertebrae [15]. Wilke et al. measured the pressure between lumbar vertebrae during various motions such as lying, sitting, walking, and lifting and verified the results of Nachemson's study using EMG signals [16, 17]. Although these sensors directly provide the loads on lumbar vertebrae for various motions, such an invasive procedure cannot be used to continuously monitor movements for workers in practical environments. For less-invasive applications, camera-based motion-capture systems are generally used to record human motions. Motion-capture systems require expensive cameras and markers on the bodies to be measured, and the measured subjects must remain within a small area. Wearable sensors could dramatically reduce costs and allow workers to move unimpeded while monitoring their movements.

Milea measured joint motion with lightweight and inexpensive bend sensors [18]. Though the materials used were simple and inexpensive, the three-axis motions possible at the lumbar joints cannot be measured with bend sensors alone.

Inertial measurement unit (IMU) systems have also been used to measure body motions. Since IMUs can estimate three-axis angles using accelerometers and gyroscopes, the measured values will include errors [19]. Marras et al. developed a lumbar-motion monitoring device that uses a three-axis electrogoniometer [20]. With this apparatus, Marras et al. monitored lumbar motion for industrial workers [21] and health care workers [22]. This lumbar-motion monitor effectively recorded time-series data about lumbar motions. However, the device covered the subjects' low back and may have been heavy for the subjects to work with over the course of a full shift. The weight of such a device must be reduced significantly if workers' low backs are to be monitored continuously.

We therefore propose a novel lightweight lumbar-motion measurement device (LMMD). The key component of the device is a flexible and stretchable strain sensor. This strain sensor is lightweight and features low elasticity, high durability, and good repeatability [23]. The strain sensor we use has been applied to measurements of wrist and elbow motions [24] and rapid prototyping human interfaces [25]. These joints rotate in one or two axes and can be modeled with a simple equation. Since lumbar motion includes flexion, rotation, and side bending, lumbar motion must be modeled with three degrees of freedom. To solve this problem, we designed a parallel-sensor device with six strain sensors. This device is lighter than comparable devices, and tests reported in this paper show that it can effectively measure lumbar motion over time.

2. Methods

The stretchable strain sensor we chose is flexible and thin. The main material of the strain sensor is urethane elastomer. Three elastomer sheets sandwich two carbon-nanotube membranes. These carbon-nanotube membranes work as electrodes in the strain sensor and are flexible enough to expand or contract along with the elastomer sheets. The sandwiched carbon-nanotube membranes are effectively a parallel-plate capacitor, the capacitance of which changes as the strain sensor expands and contracts. Hence, the strain sensor works as a variable capacitor, accurate to within 5% of its range [23]. The elastomer sheets are 70 mm long and 15 mm wide. The electrodes are 50 mm long and 10 mm wide. The whole strain sensor is approx. 150 μm thick [23]. The strain sensors we used have protective fabric on both sides, as shown in Figure 1. Snap buttons on both terminals are used to anchor terminals of the strain sensor. The strain

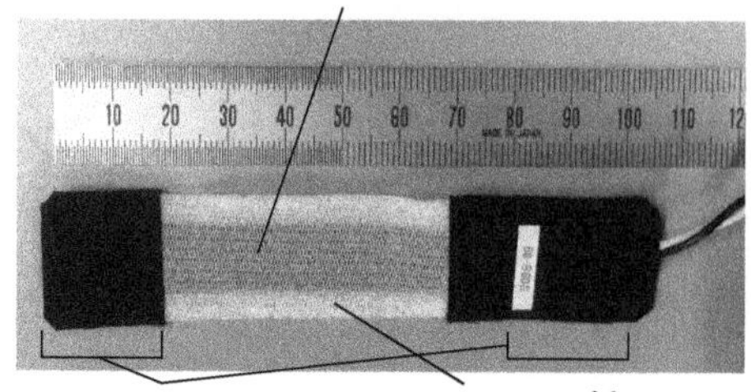

FIGURE 1: Stretchable strain sensor with protective fabric and snap buttons.

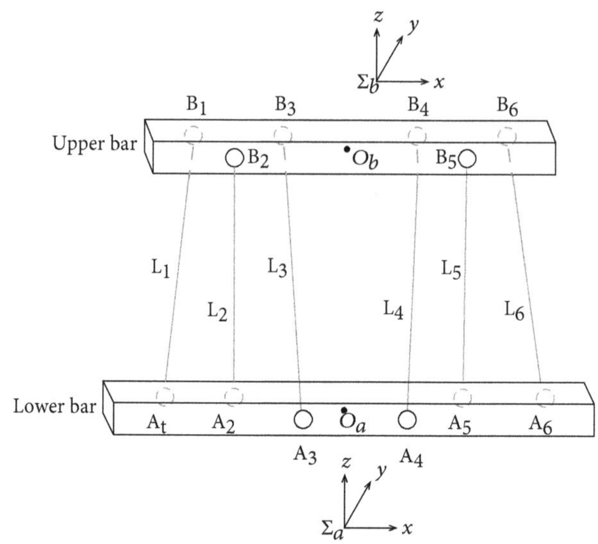

FIGURE 2: Model of the parallel-sensor mechanism. The origins of the coordinate systems of the lower and upper bars, Σ_a and Σ_b, are coincident with the center points of the bars, O_a and O_b, respectively.

sensor shown in Figure 1 is 110 mm long and 20 mm wide and weighs 4.1 g.

The proposed measurement device is built around a parallel-sensor mechanism. The parallel-sensor mechanism is based on the Stewart platform [26]. A diagram of the parallel-sensor mechanism is shown in Figure 2. Two bars are connected by six links. The terminals of each link are attached to the lower and upper bars. When the links are actuators, the mechanism is a positional control device with six degrees of freedom [27]. If the links are linear encoders, the mechanism can function as a measurement system [28]. In this study, as in our previous work, the strain sensors serve as links and the device measures their lengths. When the upper bar moves or tilts relative to the lower bar, the strain sensors expand or contract. In that case, the relative three-dimensional position and rotation of the upper bar can be calculated based on the lengths of the strain sensor.

The terminals of the links in the parallel-sensor model are labeled from A_1 to A_6 and from B_1 to B_6. The coordinate systems of the lower bar and upper bar are Σ_a and Σ_b,

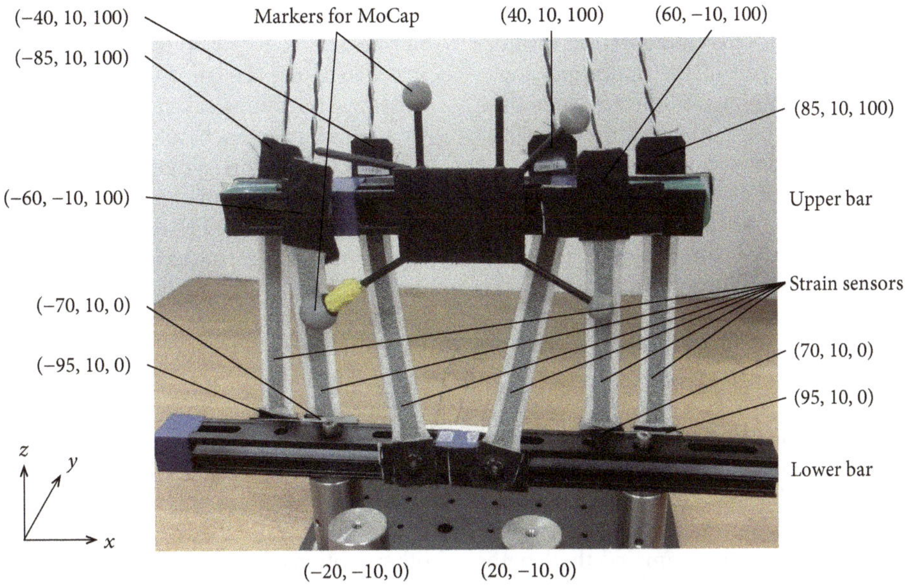

FIGURE 3: Parallel-sensor mechanism prototype. The origin of the coordinate system of the lower bar is at the center of the bar. The position of each terminal of the strain sensor in terms of the coordinate system is marked in units of mm.

respectively. The origin O_b of Σ_b is represented as ${}^aO_b = [{}^aO_{bx}{}^aO_{by}{}^aO_{bz}]^T$ in the coordinate system Σ_a. In coordination system Σ_a, the rotation angles of the upper bar around the x, y, and z axes are represented as θ_x, θ_y, and θ_z, respectively. The position and rotation of the origin of the upper bar are written as $P = \left[{}^aO_{bx} \quad {}^aO_{by} \quad {}^aO_{bz} \quad \theta_x \quad \theta_y \quad \theta_z\right]^T$. Coordinates from B_1 to B_6 in Σ_b are represented as ${}^bB_i = \left[{}^bB_{ix} \quad {}^bB_{iy} \quad {}^bB_{iz}\right]^T$, where i is the index number of the sensor and ranges from 1 to 6. Similarly, the coordinates from B_1 to B_6 in Σ_a are represented as ${}^aB_i = \left[{}^aB_{ix} \quad {}^aB_{iy} \quad {}^aB_{iz}\right]^T$, which is calculated from P. The coordinates from A_1 to A_6 in Σ_a are ${}^aA_i = \left[{}^aA_{ix} \quad {}^aA_{iy} \quad {}^aA_{iz}\right]^T$. ${}^bB_i$ and ${}^aA_i$ are predefined constants. The lengths of the strain sensors are represented as L_{si}. L_i is obtained by the following equation.

$$L_i = {}^aB_i - {}^aA_i = \left[L_{ix} \quad L_{iy} \quad L_{iz}\right]^T. \tag{1}$$

The equation expresses the relation between L_i and P. If we consider the difference between L_{si} and L_i as the length $\dot{L}_i$ changes

$$\dot{L}_i = L_{si} - L_i, \tag{2}$$

the following equation can be derived using the Jacobian determinant J.

$$\dot{P} = J^{-1}\dot{L}_i. \tag{3}$$

From the difference between L_i and L_{si}, the change in position and rotation of the upper bar $\dot{P}$ is determined iteratively by Newton's method. The position and rotational orientation of the upper bar P is calculated by adding $\dot{P}$ to the previous P.

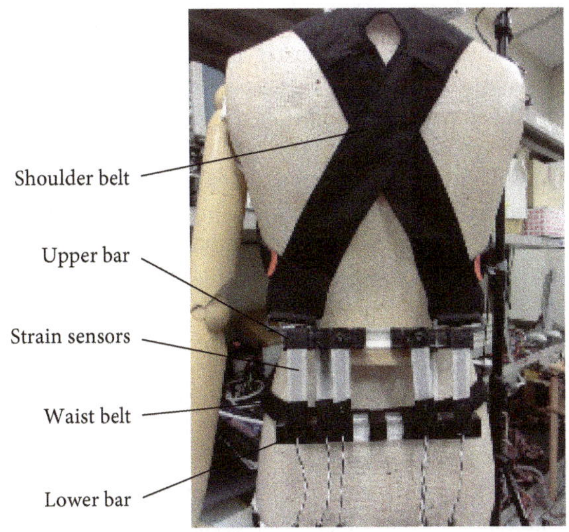

FIGURE 4: Prototype of LMMD on the low back of mannequin.

The prototype parallel-sensor mechanism we constructed is pictured in Figure 3. The parallel-sensor mechanism is built from two rectangular bars and six strain sensors, as shown in Figure 3. The dimensions of the upper and lower rectangular bars are 20 × 200 × 20 mm and 20 × 260 × 20 mm, respectively. One terminal of each strain sensor is fixed on the upper bar, and the other terminal is fixed on the lower bar. Although the length between the two bars is about 100 mm in this picture, the default length is 80 mm. The prototype weighs 235 g.

The upper and lower bars are worn on the back over the first lumbar vertebra and the fifth lumbar vertebra with shoulder straps and a waist belt, as pictured in Figure 4. The overall weight of the sensor mechanism with straps is

480 g. The weight of the LMMD is lighter than that of the lumbar-motion monitor in [20]. This configuration assumes that the relative position and twist of the upper and lower bars change along with the motions of the first and fifth lumbar spines. As the relative position of the two bars changes, the capacitances of the strain sensors change. A capacitance-to-voltage converter processes the signals into voltages that are recorded by an analog-to-digital converter running on a desktop computer. The computer translates voltages into lengths of the strain sensors and iteratively calculates the relative position and rotation of the upper bar.

3. Results and Discussion

3.1. Parallel-Sensor Mechanism. First, a preliminary experiment was conducted to confirm the accuracy of the parallel-sensor mechanism. The lower bar was fixed on a table. The bend angle and rotation of the upper bar were manipulated by a hand. The computer sampled the lengths of the strain sensors at 20 Hz and calculated the displacement and rotation in three dimensions. An optical motion-capture system (Optitrack, NaturalPoint, USA) simultaneously measured the displacement and rotation of the upper bar from markers attached to the bar, as pictured in Figure 3. The motion-capture data were used as reference data. Overall, ten trials were performed to compare data from the parallel-sensor mechanism and motion-capture system. Each trial lasted for about 25 s. Typical results from these tests are shown in Figure 5. We calculated the cross-correlation values between the motion-capture data and the parallel-sensor mechanism data. The cross-correlation values of x, y, z, flexion-extension, side bending, and rotation were 0.84, 0.93, 0.83, 0.84, 0.94, and 0.93, respectively. The parallel-sensor mechanism data agreed well with those from the motion-capture system, with some error that we assume to be caused by errors in the individual length measurements.

The accuracy of the parallel-sensor mechanism was quantified using the residual errors between the parallel-sensor mechanism and the motion-capture data. The mean and standard deviation (SD) of each dataset are listed in Table 1. Table 1 also lists the ranges of motion (ROM). The means for side bending and rotation were within 2° and were small. Compared to the data for these motions, the difference in the mean of the flexion-extension data was large. The means of the displacement data were nearly a function of the measured ROM. The ROM in the x-axis was limited by the maximum length of the sensors and was smaller than that of the other axes.

3.2. Lumbar-Motion Monitoring Device. We also tested LMMD with human subjects. Our study protocol complied with the Declaration of Helsinki. All participants were fully informed and provided written informed consent before participating in the study. The study and all procedures were approved by the Ethical Committee of Graduate School of System Informatics, Kobe University (Permission number: 28-03). The human subjects were ten volunteers of average dimensions: female : male = 1 : 1, mean age: 19.8 ± 1.0 years, height: 164.9 ± 11.9 cm, and weight: 58.1 ± 18.2 kg. The human subjects wore the LMMD and adjusted the tension of the straps to fit it to their low backs. Subjects were instructed to perform a motion typically encountered by nursing professionals, i.e., the transfer of a patient from bed to a wheelchair. Each subject performed the motion twice. The motion-capture system was used to obtain reference data as it was in the preliminary test. A selection of typical results is shown in Figure 6. The motion-capture data for flexion and extension had missing values due to a marker that was not detected by the motion-capture system. Excluding the missing data, the cross-correlation values of flexion-extension, side bending, and rotation were 0.78, 0.89, and 0.79, respectively. Although the results differ somewhat, the LMMD data still agree reasonably with the reference data. Table 2 lists the mean and SD of the residual errors between the LMMD and the motion capture and the ROM of the LMMD. The mean for the side bending angle was relatively small, and flexion-extension motions had more than double the error of the other motions.

3.3. Discussion. The comparison of data from the parallel-sensor mechanism and motion-capture system demonstrates the validity of the proposed measurement device. The cross-correlation values showed strong positive correlations. The difference in the means of the measurement and reference angles was less than 4.5°, which is less than 8.3% of the ROM of the device. These results show that the parallel-sensor mechanism with strain sensors can measure rotation in three dimensions with reasonable accuracy.

In the tests with human subjects, the cross-correlation values also showed strong positive correlations. The ROMs of the side bending and rotation motion were small compared to that of flexion-extension and were less than 35°. The ROM of the flexion-extension was greater than 50°. Although these ROMs are expected for the single motion we tested, the device must measure rotation around all three axes to be effective. The differences in the means of the side bending and rotation measurements were less than 5°. In contrast, measurements of the flexion-extension motion returned errors greater than 10°. Two explanations of these errors seem likely. First, we note that arrangement of the sensors will lead to some error due to geometry. The Stewart platform has circular plates and places links at dispersed positions on those plates; however, the LMMD terminals are attached to sensors linked to a smaller range of positions on the upper and lower bars. In this arrangement, the sensor lengths change less as the upper bar moves than they would in the Stewart platform. The other cause is that the sensors are stretched along the curve of the low back so they do not remain straight as they move. The Jacobian matrix calculation then generates errors in the angle calculations; therefore, the length of the sensor is not accurately recorded as it bends. Improved bars and belts are required to prevent these errors. Sensors could be distributed evenly around the waist, and they could be kept straight. These limitations decrease the latitude of the design of bars and belts. For example, extremely downsizing the device is difficult. The errors yielded by this LMMD are acceptable for practical purposes, though, because the device is more wearable than similar

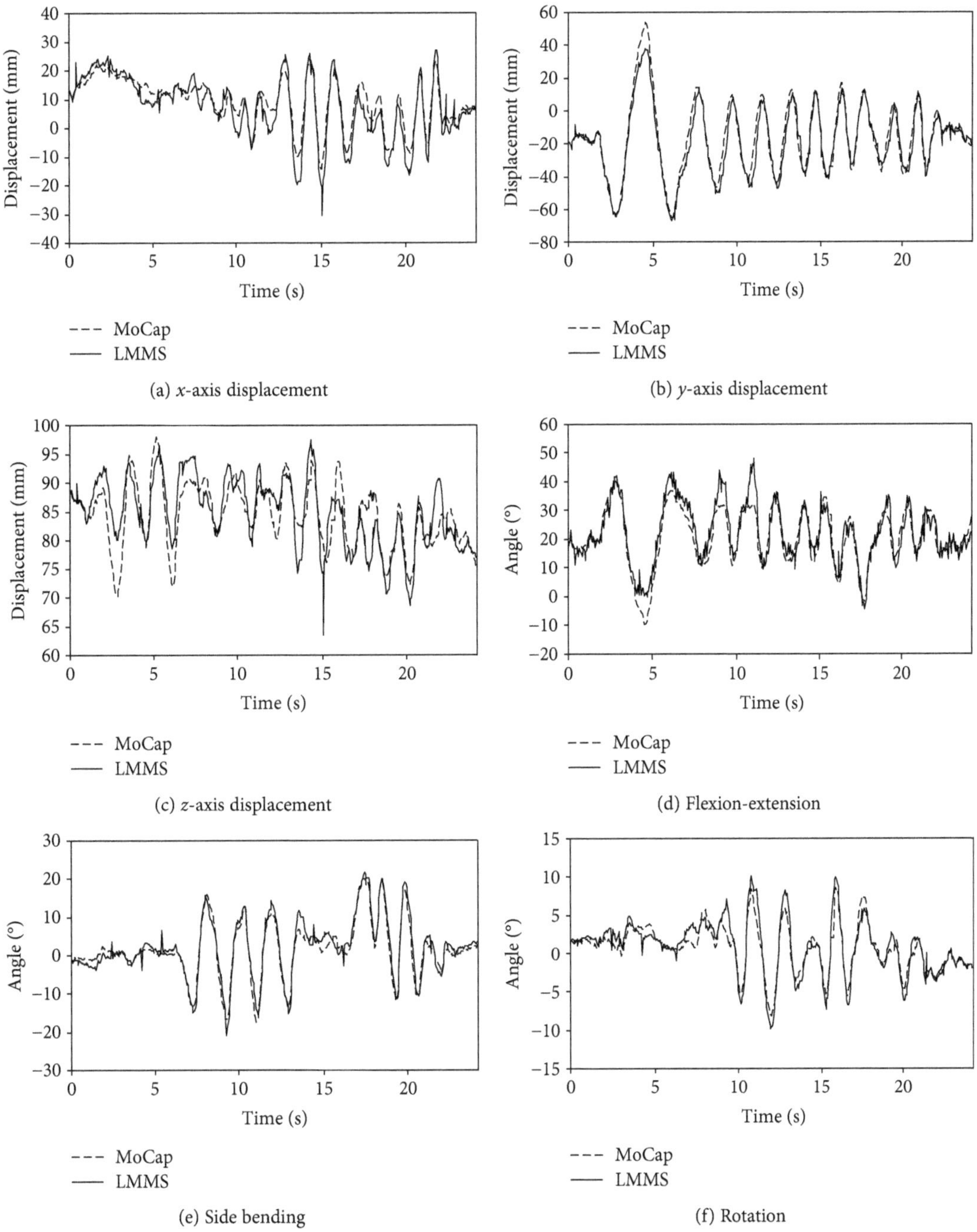

(a) x-axis displacement (b) y-axis displacement (c) z-axis displacement (d) Flexion-extension (e) Side bending (f) Rotation

Figure 5: Typical time-series data for displacements and rotation angles when moving the parallel-sensor mechanism alone.

Table 1: Range of motion values and means and SDs of residual errors between measured and reference displacements and rotation angles in preliminary tests with the parallel-sensor mechanism on a bench.

Axis	Displacement (mm)			Angle (°)		
	x	y	z	Flexion-extension	Side bending	Rotation
ROM	56.69	131.97	84.56	51.24	22.13	67.35
Mean	3.23	5.42	2.79	4.22	1.04	1.52
SD	3.24	5.43	2.36	3.43	1.02	2.17

devices that have been proposed. The device could be effective for rough measurements and monitoring of multiaxis lumbar motion such as the transfer of a patient.

4. Conclusion

We developed a lumbar-motion monitoring device that uses flexible and stretchable strain sensors. The LMMD measured three-axis lumbar motion using a parallel-sensor mechanism with six strain sensors. The accuracy of the parallel-sensor mechanism was evaluated as the residual errors within 4.5°, and this may be sufficient for a wearable lumbar-motion

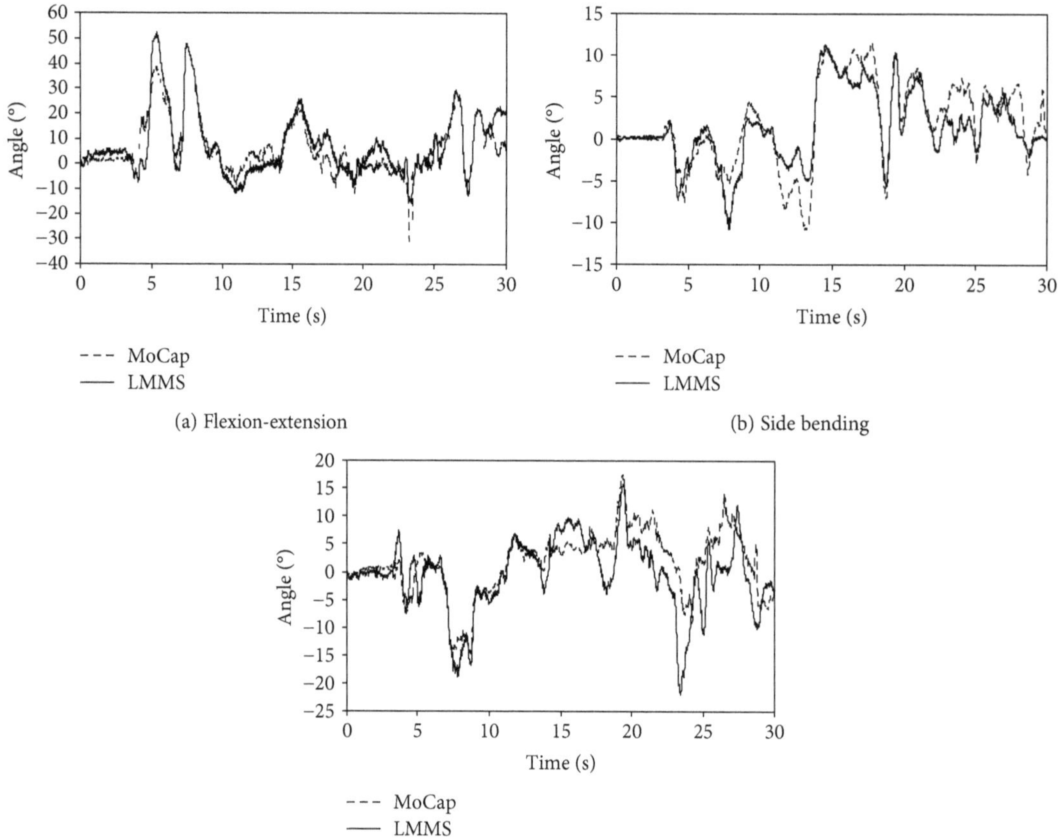

(a) Flexion-extension

(b) Side bending

(c) Rotation

FIGURE 6: Typical time-series data of rotation angles for the device worn by human subjects.

TABLE 2: ROM values and means and SDs of residual errors between measured and reference rotation angles in human-subject experiments.

Axis	Flexion-extension	Side bending	Rotation
ROM°	54.64	31.78	34.01
Mean°	11.11	3.26	4.64
SD°	7.47	3.09	3.79

monitoring device. The accuracy of the LMMD for tracking low-back movements of human subjects was evaluated against reference data from a motion-capture system. The accuracy of the LMMD was slightly inferior when worn by human subjects.

In future work, we plan to implement an arc-shaped bar that fits the user's waist better. If the sensors are evenly distributed to the arc-shaped bar, the LMMD could have high accuracy. We also plan to improve the device as a wireless device and to adapt the measurement device for continuous monitoring to make it practical for preventing LBP in the workplace.

Disclosure

The funders provided part of the experimental materials used for data collection, but this does not alter our adherence to the publication ethics of the journal on sharing data and materials.

Conflicts of Interest

YB and HO are employees of Bando Chemical Industries, Ltd. All other authors declare that they have no conflicts of interest.

Acknowledgments

This project was funded by a grant from Bando Chemical Industries, Ltd.

References

[1] C. Leboeuf-Yde, N. Klougart, and T. Lauritzen, "How common is low back pain in the Nordic population? Data from a recent study on a middle-aged general Danish population and four surveys previously conducted in the Nordic countries," *Spine*, vol. 21, no. 13, pp. 1518–1525, 1996.

[2] R. A. Deyo and J. N. Weinstein, "Low back pain," *The New England Journal of Medicine*, vol. 344, no. 5, pp. 363–370, 2001.

[3] D. Hoy, C. Bain, G. Williams et al., "A systematic review of the global prevalence of low back pain," *Arthritis and Rheumatism*, vol. 64, no. 6, pp. 2028–2037, 2012.

[4] A. Leclerc, F. Tubach, M. F. Landre, and A. Ozguler, "Personal and occupational predictors of sciatica in the GAZEL cohort," *Occupational Medicine*, vol. 53, no. 6, pp. 384–391, 2003.

[5] L. Punnett, L. J. Fine, W. M. Keyserling, G. D. Herrin, and D. B. Chaffin, "Back disorders and nonneutral trunk postures of automobile assembly workers," *Scandinavian Journal of Work, Environment and Health*, vol. 17, no. 5, pp. 337–346, 1991.

[6] H. Miranda, E. Viikari-Juntura, L. Punnett, and H. Riihimäki, "Occupational loading, health behavior and sleep disturbance as predictors of low-back pain," *Scandinavian Journal of Work, Environment and Health*, vol. 34, no. 6, pp. 411–419, 2008.

[7] I. Niedhammer, F. Lert, and M. J. Marne, "Back pain and associated factors in French nurses," *International Archives of Occupational and Environmental Health*, vol. 66, no. 5, pp. 349–357, 1994.

[8] J. Smedley, P. Egger, C. Cooper, and D. Coggon, "Manual handling activities and risk of low back pain in nurses," *Occupational and Environmental Medicine*, vol. 52, no. 3, pp. 160–163, 1995.

[9] A. M. Trinkoff, J. A. Lipscomb, J. Geiger-Brown, C. L. Storr, and B. A. Brady, "Perceived physical demands and reported musculoskeletal problems in registered nurses," *American Journal of Preventive Medicine*, vol. 24, no. 3, pp. 270–275, 2003.

[10] T. Videman, A. Ojajärvi, H. Riihimäki, and J. D. G. Troup, "Low back pain among nurses: a follow-up beginning at entry to the nursing school," *Spine*, vol. 30, no. 20, pp. 2334–2341, 2005.

[11] R. Ono, S. Yamazaki, M. Takegami et al., "Patient-reported disability in the general Japanese population was associated with medical care visits for low back pain, regardless of pain intensity," *Journal of Orthopaedic Science*, vol. 20, no. 4, pp. 742–749, 2015.

[12] W. E. Hoogendoorn, P. M. Bongers, H. C. W. de Vet et al., "Flexion and rotation of the trunk and lifting at work are risk factors for low back pain: results of a prospective cohort study," *Spine*, vol. 25, no. 23, pp. 3087–3092, 2000.

[13] T. R. Waters, V. Putz-Anderson, A. Garg, and L. J. Fine, "Revised NIOSH equation for the design and evaluation of manual lifting tasks," *Ergonomics*, vol. 36, no. 7, pp. 749–776, 1993.

[14] R. Norman, R. Wells, P. Neumann et al., "A comparison of peak vs cumulative physical work exposure risk factors for the reporting of low back pain in the automotive industry," *Clinical Biomechanics*, vol. 13, no. 8, pp. 561–573, 1998.

[15] A. J. Lisi, C. W. O'Neill, D. P. Lindsey, R. Cooperstein, E. Cooperstein, and J. F. Zucherman, "Measurement of in vivo lumbar intervertebral disc pressure during spinal manipulation: a feasibility study," *Journal of Applied Biomechanics*, vol. 22, no. 3, pp. 234–239, 2006.

[16] H.-J. Wilke, P. Neef, M. Caimi, T. Hoogland, and L. E. Claes, "New in vivo measurements of pressures in the intervertebral disc in daily life," *Spine*, vol. 24, no. 8, pp. 755–762, 1999.

[17] A. L. Nachemson, "The lumbar spine an orthopaedic challenge," *Spine*, vol. 1, no. 1, pp. 59–71, 1976.

[18] L. Milea, "Detection and tele-replication of human hand motions by a robotic hand," *American Journal of Aerospace Engineering*, vol. 2, no. 4, p. 30, 2015.

[19] M. C. Schall Jr, N. B. Fethke, H. Chen, and F. Gerr, "A comparison of instrumentation methods to estimate thoracolumbar motion in field-based occupational studies," *Applied Ergonomics*, vol. 48, pp. 224–231, 2015.

[20] W. S. Marras, F. A. Fathallah, R. J. Miller, S. W. Davis, and G. A. Mirka, "Accuracy of a three-dimensional lumbar motion monitor for recording dynamic trunk motion characteristics," *International Journal of Industrial Ergonomics*, vol. 9, no. 1, pp. 75–87, 1992.

[21] W. S. Marras, W. G. Allread, D. L. Burr, and F. A. Fathallah, "Prospective validation of a low-back disorder risk model and assessment of ergonomic interventions associated with manual materials handling tasks," *Ergonomics*, vol. 43, no. 11, pp. 1866–1886, 2000.

[22] W. S. Marras, K. G. Davis, B. C. Kirking, and P. K. Bertsche, "A comprehensive analysis of low-back disorder risk and spinal loading during the transferring and repositioning of patients using different techniques," *Ergonomics*, vol. 42, no. 7, pp. 904–926, 1999.

[23] H. Nakamoto, H. Ootaka, M. Tada, I. Hirata, F. Kobayashi, and F. Kojima, "Stretchable strain sensor based on areal change of carbon nanotube electrode," *IEEE Sensors Journal*, vol. 15, no. 4, pp. 2212–2218, 2015.

[24] H. Nakamoto, H. Ootaka, M. Tada, I. Hirata, F. Kobayashi, and F. Kojima, "Stretchable strain sensor with anisotropy and application for joint angle measurement," *IEEE Sensors Journal*, vol. 16, no. 10, pp. 3572–3579, 2016.

[25] T. Yamaji, H. Nakamoto, H. Ootaka, I. Hirata, and F. Kobayashi, "Rapid prototyping human interfaces using stretchable strain sensor," *Journal of Sensors*, vol. 2017, Article ID 9893758, 9 pages, 2017.

[26] D. Stewart, "A platform with six degrees of freedom," *Proceedings of the Institution of Mechanical Engineers*, vol. 180, no. 1, pp. 371–386, 1965.

[27] R. C. Merkle, "A new family of six degrees of freedom positional devices," *Nanotechnology*, vol. 8, no. 2, pp. 47–52, 1997.

[28] Y. Kitano and K. Yokota, "Method for measuring position and posture of the shoulder skeleton using parallel links," *Transactions of the Japan Society of Mechanical Engeneers, Series C*, vol. 79, no. 802, pp. 2004–2012, 2013.

5

Design of a Smart Sensor Network System for Real-Time Air Quality Monitoring on Green Roof

Zhihe Zhao, Jiaheng Wang, Chenxu Fu, Zhenbang Liu, Dawei Liu, and Bailiang Li

Xi'an Jiaotong-Liverpool University, Suzhou, China

Correspondence should be addressed to Zhihe Zhao; zzh_journal@163.com

Academic Editor: Grigore Stamatescu

The research on the green roof is of great importance in the field of urban beautification and improving ecological effect. According to the previous research, plants have shown a significant impact on the absorption of particulate matter (PM2.5) in the air. Therefore, it is justified that the appropriate planting design or some particular combinations of plants can be considered as a solution, dealing with the urban PM2.5. This paper presented a work in progress on developing wireless sensor networks (WSN) system based on a prototype wind tunnel, which is used for the simulation of the green roof. Several data collection processes are handled by this system, where the concentration of PM2.5, wind speed, temperature, and relative humidity are obtained and stored in the database simultaneously. Additionally, users are able to real-timely define their commands in detail, controlling the sensor's height through a GUI on the website. Experimental and simulation results and measurements have verified the validity of the wind tunnel module as well as the reliability of the sensor network. The system can be operated on thousands of devices when the packet delay maintained in a low level.

1. Introduction

In 2014, 92% of the world population were living in places where the WHO air quality guideline levels were not met, especially for the particulate matter with diameter equal or smaller than 2.5 micrometers, which acts as a major environmental risk to health [1]. Therefore, it is urgent to look for ways to mitigate the PM2.5 pollution [1]. Green roofs become popular in urbanized area to fight air pollution. Green roofs, referred as roofs with planted vegetation on top of buildings, not only have aesthetic and economic benefits but also have been proved to be an effective way to mitigate air pollution [2]. Particularly, green roofs can be used to reduce the particulate matter in the air as shown in one study that one square meter green roof can remove the particulate matter generated by a car [3].

However, until now, there are not many field measurements on the green roof filtering effect on air-borne particles. Most of measurements were conducted in the wind tunnel with artificial aerosols, whose structure, density, and thus dynamics will be significantly different from natural particulate matter. Therefore, one of the main purposes of this study is to design a portable wind tunnel system to monitor the roof vegetation performance on air cleaning under different meteorological conditions outdoors.

Most related research projects are generally based on the theoretical mathematical models [4], whose simulation accuracy relies on the verification from field or lab experiments. However, they cannot be well controlled to obtain the results under the desired environment. On the other hand, the lab work often uses the artificial aerosols, whose shape and dynamics are different from natural aerosols. In addition, other wireless sensor system solutions based on the similar projects such as in [5, 6] did not take the duplex control into consideration, which sometimes is essential when the settings of the sensors need changing from time to time to adapt the change to some circumstances of the environment. The duplex control designed in the system enhances the flexibility and facility of the structure. Moreover, extensibility of the whole system is considered as well. Several solutions presented in the literature such as in [7] provided a limited design concept of the extensible structure of the system. When the growth in the variety of devices is connected to the server, further needs of throughput of the whole system would be driven, which reflects the extensibility of the system especially the choice of the data transmission protocol [8]. In the light of these, a wind tunnel system with the functionality of real-time sensor measuring as well as the duplex control was designed. It is suitable for outdoor use especially the scenario of the roof in urban to study on the actual process of pollutants' movement.

The paper demonstrates the development of a robust data acquisition as well as remote-control system for a wind tunnel design that monitors in-air PM2.5 concentration, temperature, relative humidity, and wind speed at both terminals of working section in the wind tunnel. Wireless sensor networks (WSN) are selected as the solution to deal with works including integrating sensing, data processing, packet formation, and wireless transmission, which are attributed to the combination of a cluster of low costs but easily deployed nodes [9]. Each unit of the monitoring system can be freely joined or separated from the network and adjusted independently, which is designed in parts to meet various demands.

In this project, a real-time environment measurement monitoring, duplex communication, and control system are built, which are connected to a prototype outdoor wind tunnel so as to simulate complex environmental changes that occurred in the urban outdoor space. The system itself can be applied to urban environmental monitoring. Additionally, the prototype wind tunnel can also be regarded as a microsystem utilized in the research process, creating a convenient research environment on how the effectiveness of the green roof is influenced by the different plant species combination.

The primary contribution of this paper are as follows:

(i) A remotely controllable sensor network system used to monitor the environment of a green roof is deployed in a wind tunnel. The sensors equipped in the system include a temperature sensor, a relative humidity sensor, inhalable particle sensor, and the anemometer. All of these sensors can be remotely controlled to adjust their positions, which enables us to monitor a green roof environment actively and quantitatively.

(ii) An extensible server and a user-friendly front for data analysis are used. The server interacts with the sensor network via an Asynchronous JavaScript and XML- (AJAX-) based Web application. The data will be collected and stored in MongoDB every second. A user-friendly graphical user interface (GUI) is developed to facilitate data analysis and remote sensor control.

The remainder of this paper is structured as follows: Section 2 describes the hardware architecture of the whole system including the wind tunnel as well as developed wireless sensor system, listing specific components and their specifications. Section 3 describes details of the software design and comprehensive comparison, and analysis of choices in each par is also included. Section 4 presents

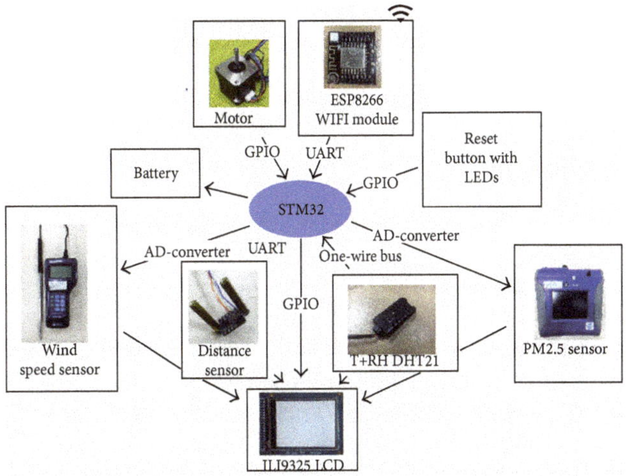

FIGURE 1: Wireless sensor hardware architecture.

experiment, and Section 4.2 presents the results. Finally, the conclusion is brought in Section 5.

2. Hardware Architecture

2.1. General Overview. The module of the proposed system is briefly described in this section as presented in Figure 1. The ambient sensors act as a stand-alone device, which measures the level of the fine particulate matter (PM2.5) in the air. As another influencing factor, wind speed is measured by a portable intelligent wind speed measuring instrument. The humidity coupled with temperature data is recorded by AM2301. The acquired information is sent to the Wi-Fi module via the UART after the formation of the data packet. Simultaneously, acquired data is available to be displayed on a local LCD screen unit, where 8 rows data text representing for data collected from both terminals of the wind tunnel are shown. When the device is power on, users will receive the real-time data with visualizing numerical and graphical values through the website at a remote location. As shown in Figure 1, all sensors together with other peripheral devices such as the LCD are attached to a core microcontroller where the data are concentrated and are communicated to the Wi-Fi module via USART. The specifications are listed in Table 1.

2.2. Hardware Infrastructure. The hardware infrastructure is decomposed into five parts, which is demonstrated below.

2.2.1. The Minimum System of Single-Chip Microcontroller. The minimum system of STM32 ship (MCU) is the smallest hardware design to support the single-chip microcomputer work regularly. The block diagram of the main routines is shown in the next section to demonstrate the design of data collector part.

Considering that outdoor environment may exacerbate aging of the device, the STM32F103VET6 is selected, equipping on the advanced Cortex-M3 kernel in ARM, which has a right balance between high operating speed

Table 1: The measured data.

Measured data	Unit
Concentration of PM2.5	$\mu g/m^3$
Wind speed	m/s
Temperature	°C
Relative humidity	%
Distance	mm

and robustness [10, 11]. Moreover, the cost of minimum system design is represented as another essential criterion implemented in this project. When STM32F103 controller is used, the cost leads this system design to be suitable for mass production. Furthermore, the program occupies 72% of flash memory and 4.2% of SRAM. All related sensors are attached to the MCU, operating simultaneously. All sensors are chosen to respect a reasonable range and requirement of accuracy.

In test environments, sensors for environmental data measurement work continuously, passing information to the MCU, where the packet is formed and transmitted to the host computer once per second.

2.2.2. PM2.5: Wind Speed Sensors. The TSI DUSTTRAK™ II 8530 aerosol monitor was selected for the measurement of the in-air PM2.5 concentration. It can monitor PM2.5 concentration between 0 and 500 $\mu g/m^3$, where real data is translated to a voltage level between 0 and 5 V proportionately (analog signal). It is essential that the level of PM2.5 is possibly affected by the change in temperature and humidity of the outdoor environment. Furthermore, the wind speed may also have an indirect effect on temperature and humidity derived from the agency of plant. The Kanomax Anemomaster™ 6036 was chosen to do the wind speed measurement (±0.1 m/s). Corresponding data is recorded by the core chip through a group of ADC channels. After filtering, signals experience the conversion from analog to digital. Due to the fact that stm32F103Vx is equipped with three 12-bit ADCs which share up to 16 external channels, they perform conversions in a relative fast speed, which aids stability when the core is dealing with an abundant supply of sensor data. Because of the 1-wire bus digital transmission, cable wiring work of the sensor node becomes more flexible [12]. Meanwhile, the antinoise performance of the whole system is further enhanced with the help of insulating and damage-protecting cover.

2.2.3. Temperature and Humidity. The AM2301 digital sensor is a product integrated with 1-wire digital bus and temperature and relative humidity sensor, which has industrial reliability.

Its measuring range is 0~100% for relative humidity and 0~50 degree Celsius for temperature, showing the positive agreement with the typical situation of the atmospheric environment. Therefore, collected data is returned more effectively.

The check bit is applied on the final inspection of a read, where the read error can be avoided. In addition, because AM2301 acts as a typical representative of the device based on 1-wire bus communication pattern, another technical property of AM2301 minimizes its impact on the working system. Its function is merely triggered by receiving a special signal from the microprocessor based on the established telecommunication protocol. Therefore, the stability of the system has been further guaranteed.

2.2.4. Wireless Module. The wireless module performs all the data which exchanges between different nodes. A stand-alone wireless 802.11 b/g/n networking module, the ESP8266EX Wi-Fi chip, was chosen for sending the measurement data to the broker configured in the cloud via the Message Queuing Telemetry Transport protocol [13, 14]. The slave computer, MCU of STM32f103, uses Universal Asynchronous Receiver Transmitter (UART) to reach the connection to the wireless transmission module. It talks in baud-rate 115200, to further realize wireless transceivers of data collecting and motor controlling. According to the requirements of low power consumption of the devices which work outdoors for extended time, the performance of energy consumption of the Wi-Fi chip should be considered. The ESP8266 provides three working modes, which, respectively, correspond to different scenarios: Deep sleep forces all of the components except for the RTC (real-time clock) module of the chip to pause working. The chip can wake up from the deep-sleep state at each set interval of time to upload data. In modem sleep mode, the chip will be automatically woken up each time before the next Beacon arrives. The sleep time is determined by the DTIM Beacon interval time of the router. In this way, ESP8266 can stay connected to the Wi-Fi during sleep mode and receive the interactive information from the MQTT server. Light sleep is similar to the modem sleep mode. The only distinction between them is that the chip in light-sleep mode cannot respond to the signals and interrupts from peripheral hardware ports or the server. Since real time is one of the main requirements of the WSN system, apparently, the deep-sleep mode cannot reach it. ESP8266 will not be in keep-alive connection under this mode, which means that the packet loss rate will be high [5]. With regard to the light-sleep mode, if the mode is applied in, the chip would suspend the internal CPU (central process unit), which results in much less consumption of power than the modem sleep mode together with not receiving any packets from the MQTT server. Therefore, the proper way to apply a low-power resolution in the system is to utilize its modem sleep mode in the aspect of Wi-Fi chip. The system can respond to the external interruption as the occasion demands including the UART, bridge between ESP8266 and STM32, and the MQTT processing unit. It would not be affected with over average 30 mA compared to the normal mode. The transmission mechanism and the details about the MQTT protocol along with the message format will be presented in Section 3.

2.2.5. Mechanical Structure and PCB. Concerning that the distribution of particulate matter has a different pattern at a different height, sensors are transported to different heights, usually fixed at the average height of the plants, to measure data.

A pinion and rack mechanism is supported by a pair of type-42 step motors located at both the terminals of the wind tunnel to support the functionality of duplex control. Figure 2 shows the overview of core functional circuit board which was designed in Altium Designer. The PCB (Figure 3) was double-sided. All components including the MCU, Wi-Fi module, power circuit, and pins for LCD were on the top layer, and the bottom layer was used for traces. The wind tunnel design and the established mechanical structure are demonstrated in Figure 4. The precision of pinion and rack mechanism through minimizing the spacing between each measurement was considered to increase the accuracy of the data. The rack size is maintained to the minimum, in order to have a better control through the step motor. On the other hand, to ensure enough air-tightness, the structure of the elevator screw was abandoned, which may cause a large hole on the side of the tunnel. Detailed explanation will be specified in the experiment part.

Figure 2: Wireless sensor circuit board.

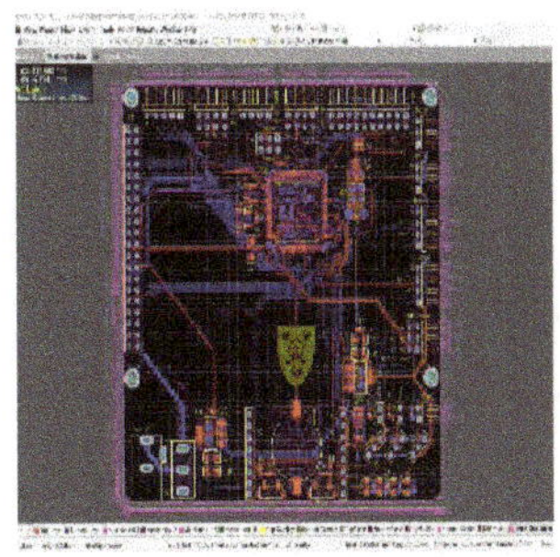

Figure 3: PCB of the wireless sensor circuit board.

3. Software Design

3.1. General Overview. The flow diagram of the main routines is presented in Figure 5. As one of the most common applications employing a large number of wireless sensors, WSN model is the archetype of this system. All environmental sensors together with the LCD work as input, and they are connected to STM32 via different protocols such as SPI, 1-wire bus, and UART. The MCU sends the monitoring packet, messages consist of header, order, data contents, direction, CRC (cyclic redundancy check), and the end of frame observing the designed data transmission protocol, via USART at PA9 (TX) to ESP8266, the Wi-Fi module, according to an established period (1 second by default) which can be set on the website. A timer interrupt is implemented in the STM32, which is responsible for the function of this part. When the wireless module is first used, it should be initialized and configured, including connecting to the gateway (router) together with setting the key information such as host IP, PORT, user name, and password concerning to the MQTT connection.

These actions can be triggered off while resetting or powering up the device via pressing the button on the board. Once the ESP8266 receives messages sent from STM32 via USART, the MQTT client configured in the Wi-Fi module will send out the message momentarily. Another important mechanism applied in the system is the MQTT (Message Queuing Telemetry Transport). Numerous applications utilize the MQTT such as health care, monitoring, and energy meter. The MQTT protocol represents an ideal messaging protocol for the M2M (machine to machine) communication in WSN since it is able to provide routines for small, low power, and low memory devices in unstable and low bandwidth networks [15].

The MQTT server, also commonly called "broker," is configured and bound on the cloud, which provides publishing/subscribing message transportation among (or between) devices that configure an MQTT client. The website backend implemented with node-JS and ESP8266 configures an MQTT client in their internal system.

3.2. Information Processing. In this section, the main information processing as well as details about data transmission protocol parts are described.

The collected ambient sensor data are sometimes not of accuracy or even get corrupted results due to inherent errors and random errors in the system. These unreliable readings should be well coped with. Data packet formation and processing take place at each ambient data transmission cycle. The core data transmission protocol used in the sensor network is MQTT.

3.2.1. Data Smoothing and Reduction of Noise. Unstable sensor hardware, internal inaccuracy of hardware units such as ADC progress, and clutter during collecting lead to the data noise to some extent. The system explores a method from the perspective of software, specifically, the moving average filter method which can alter the significance level.

The MCU collects the sensor data and combines them into the message packet every second (by default). By applying the moving average filter method, the continuous data would be balanced so that the measurement noise can be reduced.

Take the temperature data as an example. The system takes the continuous 6 values where the length of the queue is fixed. Each time a new sample of data is added to the queue, the origin data of the queue would be thrown out as shown in Figure 6. The MCU would compute the citation error at each cycle, if the result is over than the preset threshold, it would obtain and update the calibrated value by calculating the weighted average value of the 6 values (including the latest one) in the queue.

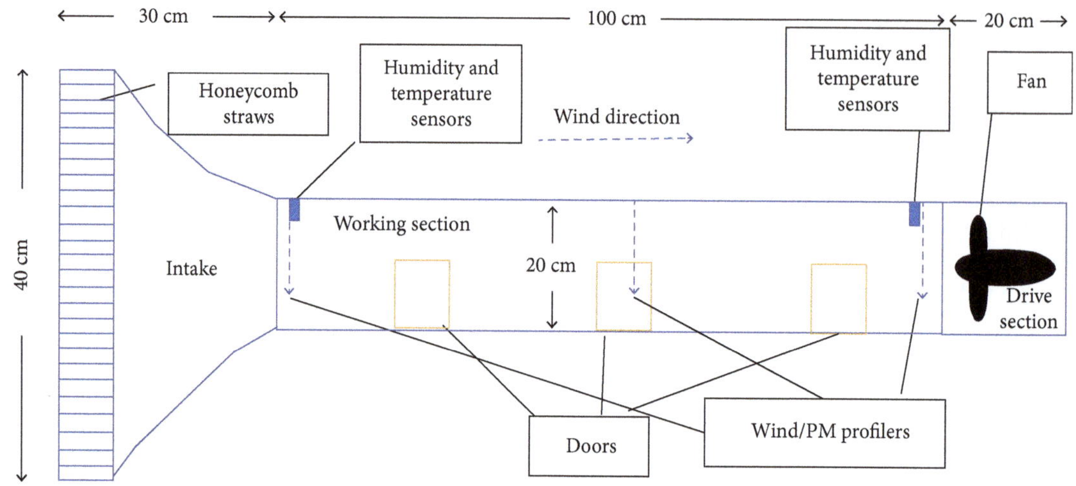

FIGURE 4: Architecture of the wind tunnel.

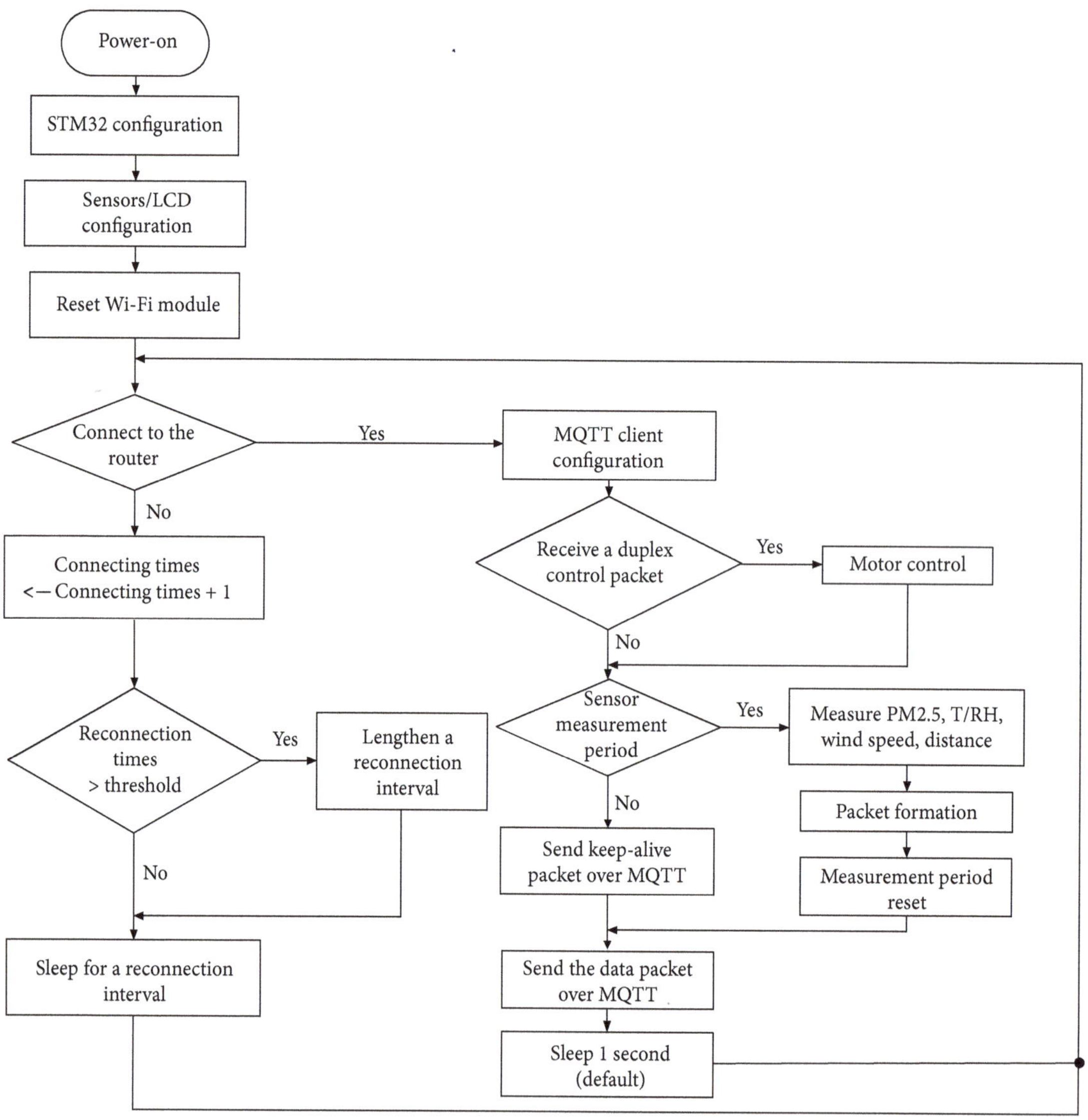

FIGURE 5: Software flow diagram.

V1 (latest)	V2	V3	V4	V5	V6 (oldest)

Figure 6: Sketch of data queue using smooth algorithm.

The random error can be handled to some extent which improves the real-time performance.

3.2.2. Packet Formation. The main data packet formation takes place at every one second (by default) in the MCU, which means the packet is created at the same rate as the MCU sends it out. Also, it is created on the cloud (web back end) whenever the control order is triggered. There is another kind of special data packet, which is created in the MCU whenever an error happens during the data transmission. The main packet consists of six parts, and the whole content is in string format. These describe the settings of the measurement results or action commands.

The specific fields of each part in a main data frame are shown below.

(i) Frame header: fixed as "AA."

(ii) Function: the function field is filled according to the specific action such as setting "00" to represent the function of controlling the device.

(iii) Frame data: if the function is sensor collecting, the frame data is passed from all the sensors including temperature and humidity sensors AM2301, PM2.5 sensors. The field stores the specific numerical values. If the function is to control the device, the field will store the crucial information concerning to the specific action commands.

(iv) Identifier: the identifier comes from the MCU, which represents the whole collecting and duplex-control system. Each stand-alone system is unique as it would be put on the roof, and for each ID of the entire system, there is a corresponding location stored in the database.

(v) End of frame: fixed as "ZZ."

(vi) Cyclic redundancy check: the CRC is used to detect error such as accidental changes to raw data [16]. The CRC is combined in the frame, and the receiver can judge whether the CRC matches the data. If the data does not match the CRC, then the frame would be discarded which ensures the stability of data transmissions to some extent.

(vii) Figure 7 shows the frame packet format.

3.2.3. Transmission Protocol. The Wi-Fi standard was used to accomplish the part of wireless sensor data transmission since the application scenario, roof of buildings in city, is not large where the wireless node can associate and send messages directly via the access to the router in the building. The success of Wi-Fi technology has even catalyzed many city administrations to build city-wide Wi-Fi coverage. Therefore, some more complex routing protocols such as ZIGBEE were not taken into consideration.

The protocol chosen for data transmission was MQTT, which is a lightweight TCP/IP-based publish-subscribe protocol. Compared with other transmission protocols such as UDP, MQTT utilizes the publish/subscribe pattern to provide transition flexibility and simplicity of implementation [15]. Also, MQTT is suitable for resource-constrained devices since each message header can be as short as 2 bytes to minimize the network traffic. To improve the stability and reliability of the whole system, three levels of quality of service (QoS) are used since conditions of wireless connection differ from each roof of the buildings. In accordance with the various connection requirements, the corresponding QoS level can be configured which costs the system low computational resource usage in a weak network condition.

According to the requirement of real-time integration approach to data as well as the potential needs of big data, the extensibility of the system is also of significance. The mode of Sub/Pub realizes a flexible interface, which means each time a client publishes a message on a particular topic, all the clients' devices (including the client configured in the cloud) that have subscribed to the same topic can receive this message.

The broker, regarded as the key component that completes the transfer process by which one-to-many connections are enabled, is portable to be configured on different platforms such as a server with more capacity.

3.3. Web Front and Back End. Considering the accessibility of this monitoring system for the green roof, a website was particularly designed and established to visualize the data acquired from the multiple sensors on which real-time data is displayed and an up-to-date dynamic line chart is rendered accordingly as shown in Figures 8 and 9. The data collected is also stored in the database for further retrieval and analysis. Moreover, for the purpose of collecting adequate data simulating various scenarios and better analyzing while piloting the system, the website added support for manipulating the mechanical parts of the sensors in wind tunnel model through the control panel, controlling the sensors to alter their positions and collect the corresponding data.

3.3.1. Front End. Despite the fact that the front end did not act as an indispensable role in the implementation of the overall system, it defines how the experiments' data is visualized. In this system, the design of the web's front end enables users to monitor the data of all sensors remotely and most importantly in real time. Evaluating the principles as human-computer interaction design approaches described [17], we elaborately designed the user interface as well as the operation logic of the web to make it more intuitive and user-friendly. For instance, particular snippets of JavaScript code enable users to manually specify/set the frequency of the dynamic line chart updating or even temporarily freeze the chart to inspect the current trend of the sensor data. Due to the utilization of AJAX technology, the latest data transferred

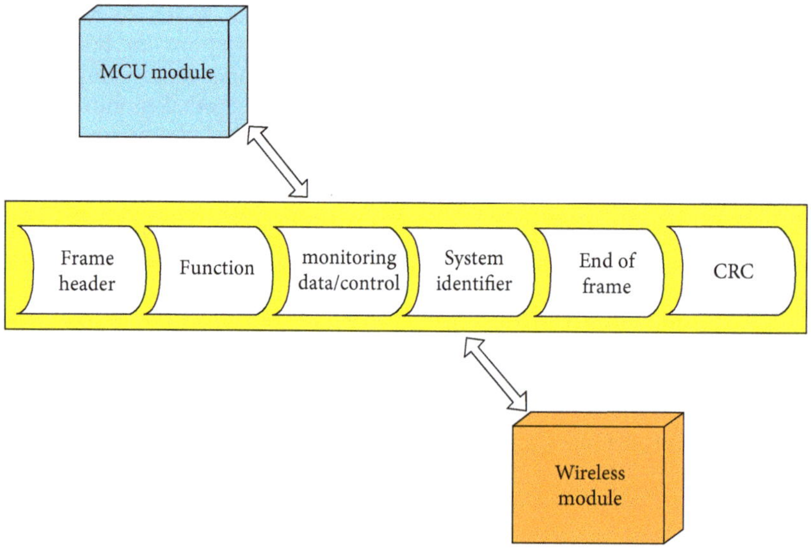

FIGURE 7: Data packet formation.

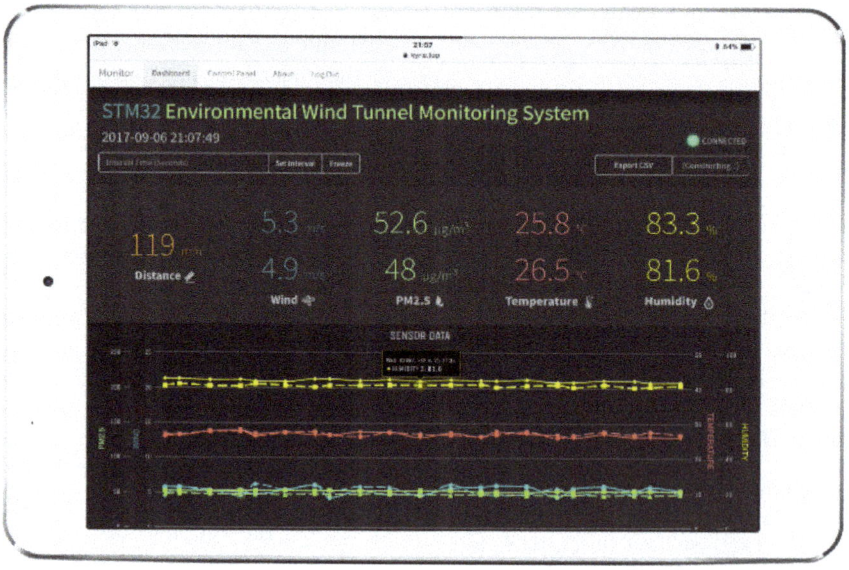

FIGURE 8: Front panel of the web application.

from the back end can be displayed dynamically without the need of refreshing the web page which guarantees the instantaneity and accuracy of the data as further analysis requires.

3.3.2. Back End. The last main component is the back end of the website which undertakes the concrete core tasks including receiving the messages from the sensors while storing the preprocessed data in the database simultaneously and feeding them to the front end when invoked. Furthermore, it also acts as a crucial part in initiating a request for hardware manipulation in wind tunnel model.

We were confronted with several choices when deciding the appropriate structure for the back end of the website. Ones using PHP, JSP, and Node.JS are all examples of the current trending structures, and each of them has its own superiority and drawbacks. At the very beginning, PHP was picked as the back end solution for the primary model of the website due to its convenience of deploying. However, keeping subscribing MQTT messages consumed much memory and server resources, giving rise to latency to be reckoned with which compromised the accuracy, reliability, and promptness of the data.

Therefore, we reconsidered the overall structure of the website and the choice of framework for the back end. According to experimental results acquired from a benchmark test, PHP and Node.JS performed at virtually the same level when "Hello World" module, which was a fundamental module to build an adequate web server, was requested by a total of no more than 100 users at the same time, both handling about 3000 requests. However, when the number of multiuser access reached 200, Node.JS still can handle about 3500 requests while PHP could only handle 500 requests per

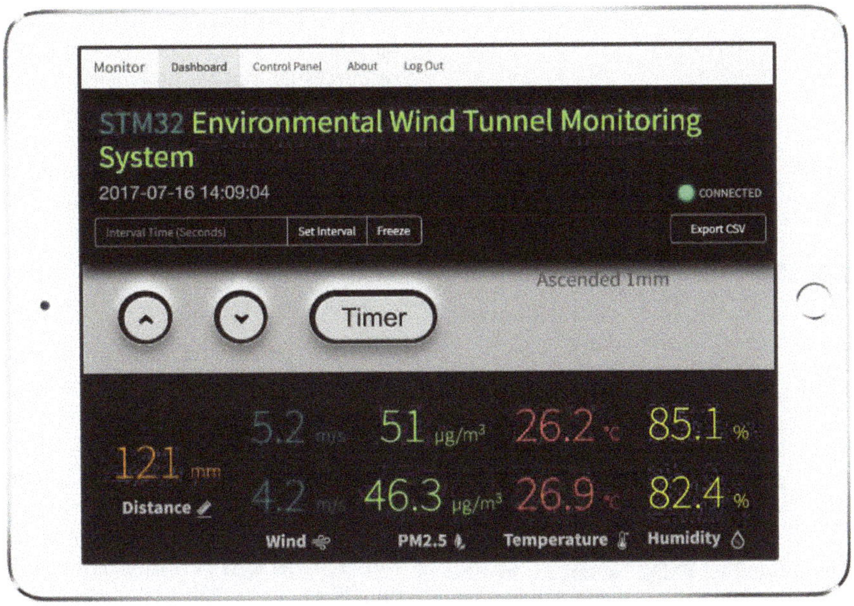

FIGURE 9: User interface to enter control commands.

second, leading to the case that Node.JS was virtually 7 times faster than PHP, and the difference was even more dramatic with the increase of users. The result gave rise to a conclusion drawn by the experiment researcher that PHP was only suitable for applications of small scale and middle scale while Node.JS was an ideal choice for I/O intensive website applications which to some extent matches this project's requirements considering the extensibility and real-time performance of the entire system, and further demonstration is clarified in the experiment part.

4. Experiment and Simulation

Elaborated experiments were carried out in a particularly designed wind tunnel model to simulate the complicated situations that might occur in real applications and to justify the overall effectiveness of the prototype, the monitoring system for the green roof, spotting the potential need for optimization in the process.

4.1. Experimental Setup. A wooden wind tunnel which has three parts including an air intake part in shape of a big trumpet with convergence from the inlet entrance to the head of working section, a cylindrical part with 20 centimeters of diameter and 100 centimeters of length, and an adjustable suction auxiliary system was designed and produced as shown in Figure 4. The suction system consists of a velocity-regulating motor-driven fan which can control the range of simulated wind speed in the tunnel to keep constant between 0 and 15 m/s. The PM2.5 sensor is installed under the lever. The wind tunnel was set on the balcony in the Environmental Science Department at the Xi'an Jiaotong-Liverpool University.

The underlying surface part in the cylinder is the main experimental space in the apparatus. The PM2.5 concentration and other sensor values are measured at the beginning and ending of the working section.

There was a total of three sets of comparison experiments to justify the feasibility of the system by measuring the PM2.5 absorption capacity of plants under separate circumstances, using the same kind of plants with different leaf area indexes (LAI) [18], under different wind speeds, and under different weather conditions which represent different conditions of temperature and humidity. Due to the fact that the distribution of PM2.5 in the air is nonlinear and presents an irregular density, we define the following formula as a metric to describe the flux of PM2.5 flowing through the cross section of the two sides of the working section and to further track the quantity of PM2.5 absorbed by plants.

$$Q = C \cdot V \cdot S, \tag{1}$$

where Q (μg/s) represents the mass flow rate of PM2.5, which passes through a wind tunnel cross section S (m^2) with PM2.5 concentration of C (μg/m^3) and wind speed of V (m/s). PM2.5 concentration is collected at the average height of the leaves of the plants, where the position of a sensor implemented by the duplex control and S has the value of 0.3 square meters. After obtaining the value of Q, it would be practical to acquire the quantity of PM2.5 absorbed by plants by calculating the difference between the amount of PM2.5 at the head and end to justify the absorption capacity. Considering the complex influence factors in the test, *Ophiopogon japonicus*, with a less complex contour shape to be calculated for LAI, was chosen to be the experimental plant [19].

Before the start of the experiment, we carried out a pretest, in which PM2.5 measurement is exempted from the influence of the apparatus and the wind tunnel walls itself by not placing any plants in the working section to minimize the error. The comparison of Q values obtained at head and

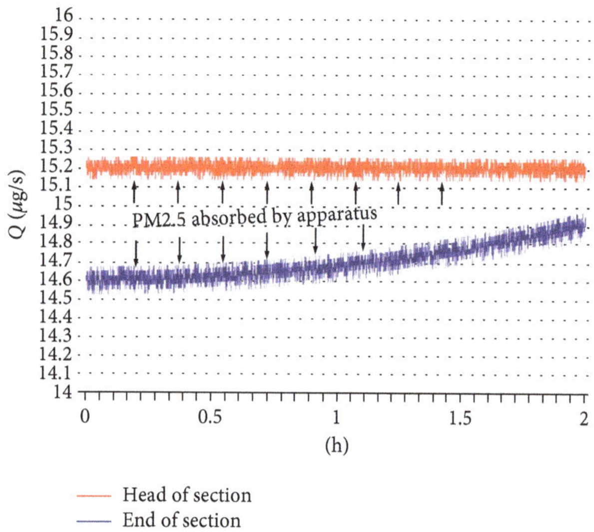

FIGURE 10: Pretest to measure the PM2.5 absorbed by the equipment itself.

end of the working section is shown in Figure 10. The area inside the curves represents the PM2.5 absorbed by apparatus during 2 hours with 10 m/s of wind speed. The cause of this effect is considered in follow-up experiments.

The position of a PM2.5 sensor in each of the test, flexibly controlled by web page due to the functionality of duplex control, should be fixed to minimize the effects of different measuring angles on the entrance. Since the experimental subjects are plants and most PM2.5 was absorbed by the area of dense leaves, the standard position of the measuring sensor we regulated was at the plants' average height. We compared the PM2.5 values measured at different positions under similar environmental conditions to testify the theory and the functionality of duplex control as Figure 11 displays. The pretest was carried out continuously to compare the effects of different measuring positions of the sensor. It is apparent that data collected at the average height of the plants have shown a more dramatic change over the position at the top of tunnel, which can reflect the conclusion of the numerical simulation more intuitive.

To justify the capability of the overall system, we also simulated the situation in which 5000 devices sent and received sensor data simultaneously.

The sampling rate of data transmitted in the whole system is configured to one sample per second by default, which can be customized to any interval value greater than the default value.

For the purpose of collecting more accurate data, the system has added support for timer configuration, where sensors can be set to automatically patrol at multiple heights for certain different periods of time.

Each experiment dedicated to testing and verifying one influencing factor lasted for 2 hours in consideration of the changing environment affecting elements such as different kinds of flowing particles in the air especially under the unstable weather conditions in Suzhou, the city where we processed the experiment.

4.2. Result Analysis. The leaf area index (LAI) is an important parameter to describe how plant populations use the resources as well as the efficiency of capturing the particles, and it can be calculated by the formula below [18]. The leaf area measurement is based on the LI-3000C portable meter which utilizes an electronic method of rectangular approximation to provide 1 mm^2 resolution.

$$\mathrm{LAI} = \frac{S_{\text{leafarea}}}{S_{\text{ground}}}. \tag{2}$$

The experimental result for the effects of LAI is shown in Figure 12. The scale of left y-axis represents the Q condition at the head of working section while the right one represents the real-time value of delta Q which can be obtained by adding all the differences of Q values (measured at head and end of the working section) at each second. Under the same conditions of environmental factors, working section with a higher LAI performs a better ability of PM2.5 absorption. Two curves below the diagram have indicated that the absorption rate of PM2.5 gradually reaches a local minimum point over time, then tends towards stability and shows a modest growth. The experiment was operated under the condition of 10 m/s wind. Additionally, leaves will experience procedures of washing and drying in the clean room during the time gap between experiments. We wiped and dried the surface of the wind tunnel in order to minimize the interrelationship between each group of experiment.

As the demonstration of Figure 13, the displayed real-time data in the situation of LAI = 2 indicates the extent of impact on the absorption rate caused by different wind speed conditions (5 m/s/10 m/s). This is evident: the experiment showed the same trend of delta Q as its counterpart in the LAI experiments. With other equal conditions, the PM2.5, however, is better absorbed under the high-speed wind condition, which agrees well with the real measurement revolving around wind speed and PM2.5 absorption in the ambient environment.

It is complicated to analyze the relationship between the PM2.5 absorption behavior by plants and the weather. However, temperature and relative humidity are main effect elements which can be the dependent variables in tests. Three typical groups of data are chosen for the illustration of PM2.5 absorption circumstance under different weather conditions as shown in Figure 14. Data was recorded in a three-month experiment time span when the weather condition of the high T&RH difference has been experienced. Abnormal measurements were eliminated by moving average algorithm on the software and hardware level.

In the light of the application of WSN, it is essential for the system to satisfy the real-time requirement [20]. Therefore, corresponding packet delay tests were conducted in situations where the levels of QoS in MQTT data transmission are different. The subject in the test is the MQTT publisher whose packet delay performance can be simulated via JMeter-MQTT, a third-party open source that helps do the load testing [21]. We created a JMeter plugin which provides the configuration element interfaces that allow virtual clients to register the server. The result of the test is

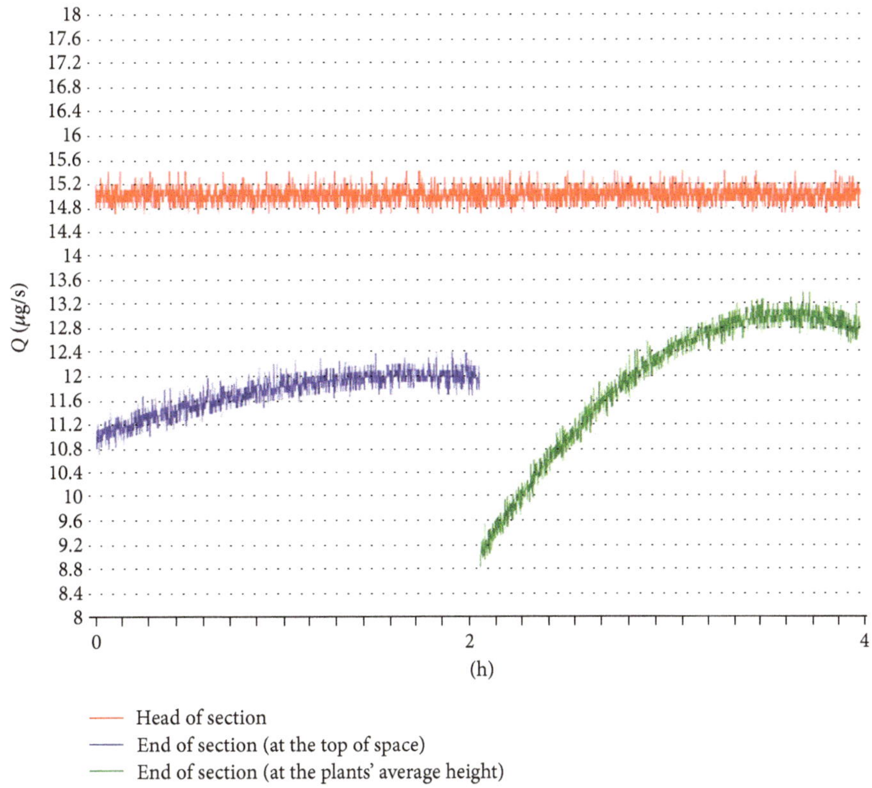

FIGURE 11: Pretest to verify the effects of different measuring positions of the sensor.

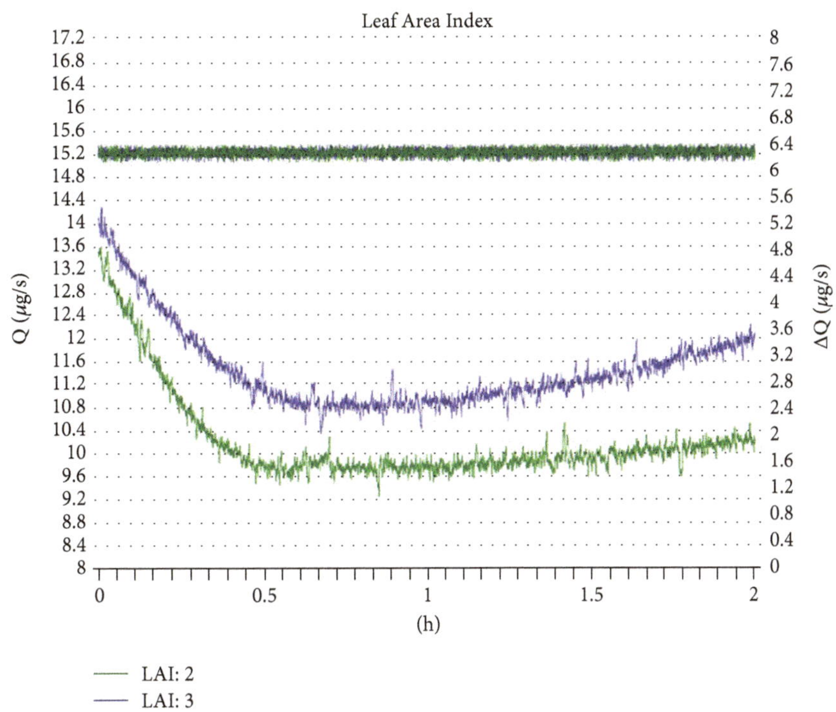

FIGURE 12: Time series representation of PM2.5 absorption measurement under the condition of LAI 2/3.

shown in Figure 15. Obviously, the latency of packets has a positive correlation with the level of QoS [22]. Nevertheless, in the situation of experiments and simulation, a higher level of QoS set in transmission will exactly result in a more serious latency. Simultaneously, the value of QoS, however, has the positive correlation with the stability of communication. Thus, the QoS level for the green roof scenario is level 1 for devices which can ensure the quality of communication as

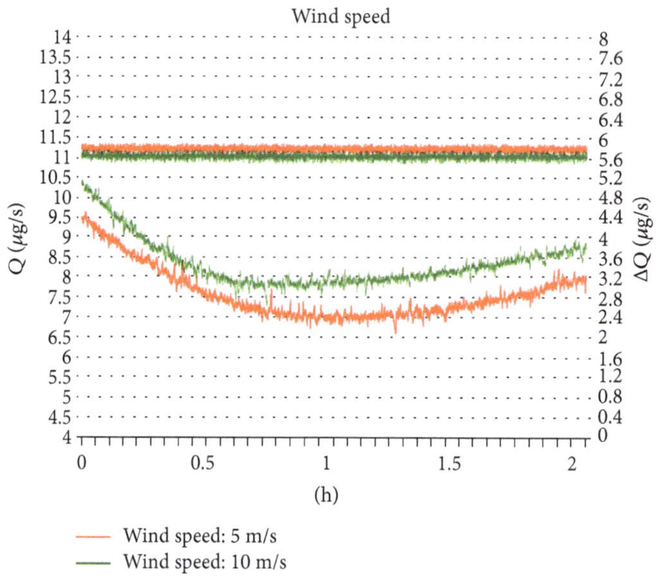

FIGURE 13: Time series representation of PM2.5 absorption measurement under the condition of wind speed: 5 m/s and 10 m/s.

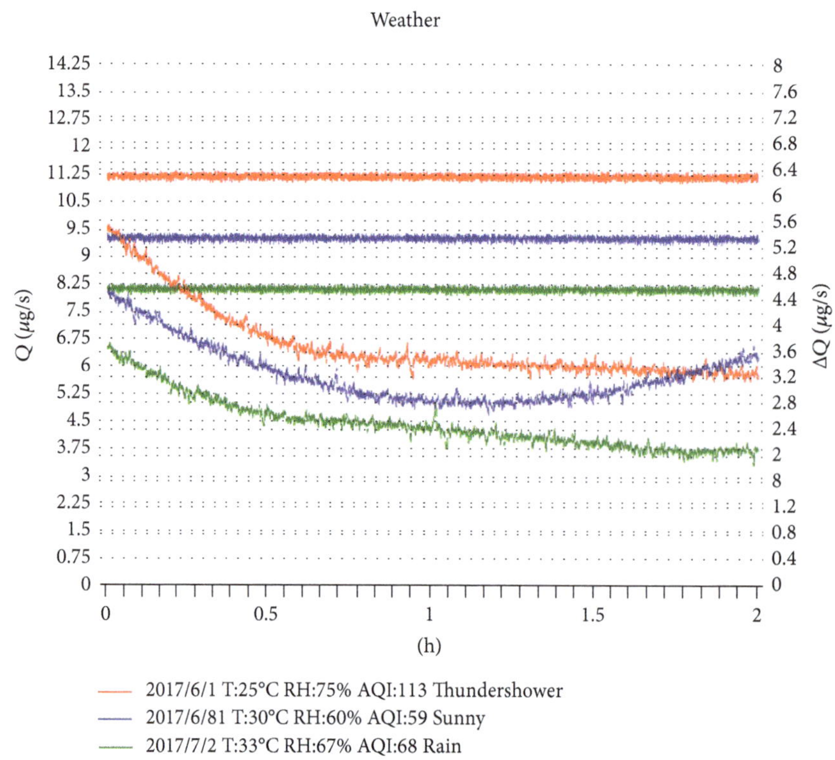

FIGURE 14: Time series representation of PM2.5 absorption measurement under different weather conditions.

well as achieve a relative high concurrency. In general, it has been approved by the results of the tests that the system can ensure a relative low packet delay and low packet loss rate according to the data analyzed while the number of devices is under 50,000. Because of the high extensibility of the MQTT broker, corresponding server configuration operations can be flexibly achieved according to the number of devices. Due to the fact that strategies and methods to reduce errors on both levels of software and hardware are considered at the very beginning, this proposed system is expected to achieve a higher stability and real-time performance. In addition to this, the selection of elements also owns a high-cost performance.

5. Conclusion

Aimed at studying on the problem of urban air pollution which is highly correlated with the excessive particular

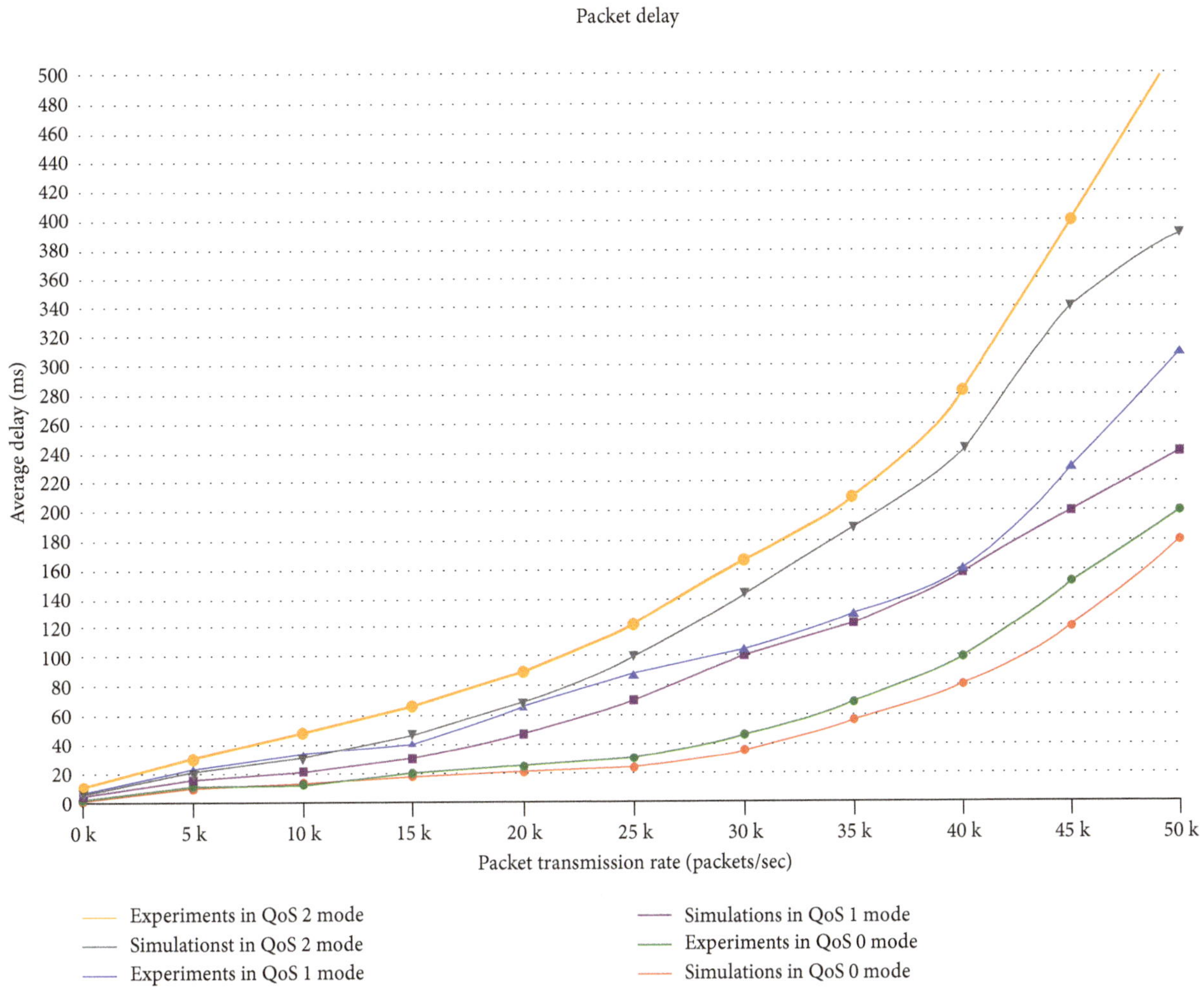

FIGURE 15: Comparison of experimental and simulated packet delay in data transmission.

matter in the air, the work develops a real-time wireless sensor system simulating the scenario of a green roof that monitors the PM2.5 and other related sensor parameters such as relative humidity, temperature, and wind speed which are based on the IEEE 802.11b/g/n standards. The main goal of the approach is to justify the stability, extensibility, and data accuracy of the prototype which was designed and tested. Considering the fact that real applications of green roofs in the city require multiple concurrent devices online, the performance in the monitoring system was elaborately evaluated and optimized for concurrent connections. And we carried out multiple experiments to examine and evaluate the system in terms of key points overall performance, latency, and accuracy of data. Also, by carefully selecting the hardware components as well as software configuration, the solution is highly effective and has a wide potential application.

Conflicts of Interest

The authors declare that they have no conflicts of interest.

Acknowledgments

This project is financially supported by a Summer Undergraduate Research Fellowship (SURF 201762) from Xi'an Jiaotong-Liverpool University. The authors also thank Dr. Bingjian Liu, Mingyang Wang, Chunyu Chen, and many other fellow students for their support on experimental design and wind tunnel construction.

References

[1] "Ambient (outdoor) air quality and health," September 2016, http://www.who.int/mediacentre/factsheets/fs313/en/.

[2] K. Vijayaraghavan, "Green roofs : a critical review on the role of components, benefits, limitations and trends," *Renewable and Sustainable Energy Reviews*, vol. 57, pp. 740–752, 2016.

[3] D. B. Rowe, "Green roofs as a means of pollution abatement," *Environmental Pollution*, vol. 159, no. 8-9, pp. 2100–2110, 2011.

[4] T. Kim, A. Saini, J. Kim et al., "Piezoelectric floating element shear stress sensor for the wind tunnel flow measurement," *IEEE Transactions on Industrial Electronics*, vol. 64, no. 9, pp. 7304–7312, 2017.

[5] P. Spachos and D. Hatzinakos, "Real-time indoor carbon dioxide monitoring through cognitive wireless sensor networks," *IEEE Sensors Journal*, vol. 16, no. 2, pp. 506–514, 2016.

[6] S. C. Folea and G. Mois, "A low-power wireless sensor for online ambient monitoring," *IEEE Sensors Journal*, vol. 15, no. 2, pp. 742–749, 2015.

[7] H. Yang, Y. Qin, G. Feng, and H. Ci, "Online monitoring of geological CO2 storage and leakage based on wireless sensor networks," *IEEE Sensors Journal*, vol. 13, no. 2, pp. 556–562, 2013.

[8] C. Lu, "Improving system extensibility via an IoT-interoperable platform for dynamic smart homes," in *2017 International Conference on Applied System Innovation (ICASI)*, pp. 13–17, Sapporo, Japan, May 2017.

[9] V. Jelicic, M. Magno, D. Brunelli, G. Paci, and L. Benini, "Context-adaptive multimodal wireless sensor network for energy-efficient gas monitoring," *IEEE Sensors Journal*, vol. 13, no. 1, pp. 328–338, 2013.

[10] T. Bai, "A control method to prevent falling from a treadmill based on STM32 microcontroller and ultrasonic transducer," in *2017 29th Chinese Control And Decision Conference (CCDC)*, pp. 28–30, Chongqing, China, May 2017.

[11] STMicroelectronics Corporation, "STM32F103XC/STM32F103XD/STN32F103XE datasheet [EB/OL]. (2014-9-30)[2015-4-17]," http://www.st.com/content/ccc/resource/technical/document/datasheet/59/f6/fa/84/20/4e/4c/59/CD00191185.pdf/files/CD00191185.pdf/jcr:content/translations/en.CD00191185.pdf.

[12] J. Sun and Z. Hu, "The application of 1-wire bus technology in the temperature monitoring on the of rolling mill," in *2009 International Conference on E-Learning, E-Business, Enterprise Information Systems, and E-Government*, pp. 5-6, Hong Kong, December 2009.

[13] R. K. Kodali and K. S. Mahesh, "Low cost ambient monitoring using ESP8266," in *2016 2nd International Conference on Contemporary Computing and Informatics (IC3I)*, pp. 14–17, Noida, India, December 2016.

[14] Espressif Corporation, "ESP8266EX/ datasheet. Version 5, 2018," https://www.espressif.com/en/support/download/documents.

[15] A. Al-Fuqaha, M. Guizani, M. Mohammadi, M. Aledhari, and M. Ayyash, "Internet of things: a survey on enabling technologies, protocols, and applications," *IEEE Communications Surveys & Tutorials*, vol. 17, no. 4, pp. 2347–2376, 2015.

[16] E. Tsimbalo, X. Fafoutis, and R. J. Piechocki, "CRC error correction in IoT applications," *IEEE Transactions on Industrial Informatics*, vol. 13, no. 1, pp. 361–369, 2017.

[17] J. A. Jacko, *Human-Computer Interaction. Design and Development Approaches*, Springer, Berlin Heidelberg, 2011.

[18] J. Tian, L. Wang, X. Li, C. Shi, and H. Gong, "Differentiating tree and shrub LAI in a mixed forest with ICESat/GLAS spaceborne LiDAR," *IEEE Journal of Selected Topics in Applied Earth Observations and Remote Sensing*, vol. 10, no. 1, pp. 87–94, 2017.

[19] Z. Zhu, U. Ge, and Y. Wang, "Development of species-specific SCAR (sequence-characterized amplified region) markers for identification of famous-region drug *Ophiopogon japonicus* (L.f.) Ker-Gawl," in *2011 IEEE International Symposium on IT in Medicine and Education*, pp. 9–11, Cuangzhou, China, December 2011.

[20] N. Lu, Y. Sun, H. Liu, and S. Li, "Intrusion detection system based on evolving rules for wireless sensor networks," *Journal of Sensors*, vol. 2018, Article ID 5948146, 8 pages, 2018.

[21] S. Víctor, "Performance evaluation of scalable and distributed IOT platforms for smart regions. 2017," https://ltu.diva-portal.org/smash/get/diva2:1136272/FULLTEXT01.pdf.

[22] J.-Y. Kim, C.-H. Chu, and S.-M. Shin, "ISSAQ : an integrated sensing systems for real-time indoor air quality monitoring," *IEEE Sensors Journal*, vol. 14, no. 12, pp. 4230–4244, 2014.

6

Evolutionary Multilabel Feature Selection using Promising Feature Subset Generation

Jaesung Lee, Wangduk Seo, Ho Han, and Dae-Won Kim

School of Computer Science and Engineering, Chung-Ang University, 221, Heukseok-Dong, Dongjak-Gu, Seoul 06974, Republic of Korea

Correspondence should be addressed to Dae-Won Kim; dwkim@cau.ac.kr

Academic Editor: Grigore Stamatescu

Recent progress in the development of sensor devices improves information harvesting and allows complex but intelligent applications based on learning hidden relations between collected sensor data and objectives. In this scenario, multilabel feature selection can play an important role in achieving better learning accuracy when constrained with limited resources. However, existing multilabel feature selection methods are search-ineffective because generated feature subsets frequently include unimportant features. In addition, only a few feature subsets compared to the search space are considered, yielding feature subsets with low multilabel learning accuracy. In this study, we propose an effective multilabel feature selection method based on a novel feature subset generation procedure. Experimental results demonstrate that the proposed method can identify better feature subsets than conventional methods.

1. Introduction

Recent progress in the development of sensor networks improves the precision of continuous data sensing [1], which increases the coverage of ambient applications such as activity monitoring in daily routines that may involve the concurrent prediction of the activity level and caloric expenditure [2, 3]. Owing to limitations in computational and storage capability [4, 5] and redundant data sensing for denoising [6, 7], composing a strategy that would produce the best accuracy under given data collection conditions is considered one of the most important issues in this field [8]. Consequently, multilabel learning is considered to be a promising approach because it allows for improvements in accuracy by exploiting the dependency among labels [9, 10].

Let $W \subset \mathbb{R}^{|F|}$ denote the set of patterns described by a set of features $F = \{f_1, \cdots, f_d\}$. Then, each pattern $w_i \in W$, where $1 \leq i \leq |W|$, is assigned to a certain label subset $\lambda_i \subseteq L$ in which $L = \{l_1, l_2, l_3, \cdots, l_{|L|}\}$ and represents a finite set of labels. To attain additional improvements in accuracy, the algorithm has to exploit useful dependencies among labels based on input feature values [11]. For this purpose, the multilabel feature selection that identifies a subset $S \subset F$ with maximum $n << d$ features that provide the largest dependency on L can be used as a promising preprocessing step because it remedies the complicated relation among features and labels by selecting important features and discarding unnecessary ones [12, 13].

Basically, multilabel feature selection is a search problem [14]; it can be achieved by identifying the optimal feature subset that gives the best prediction accuracy from

$$\sum_{k=1}^{n} \binom{d}{k} \tag{1}$$

candidate feature subsets [15]. Because the examination of all feature subsets is impractical, conventional methods employ a heuristic search method that identifies a feasible solution within limited computational costs by sacrificing optimality [16]. Of the many search methods, the evolutionary search method is considered a promising approach because it effectively narrows down the search space by examining neighbor solutions or feature subsets of the best solutions created from past generations [17, 18].

In the evolutionary search method, the best solution is replaced if a newly created neighbor solution yields a better fitness value. Therefore, generating promising solutions determines the effectiveness of the search. Owing to the extensively wide search space and limited computational cost, a conventional strategy tackling this difficulty is to employ a cheap evaluation method that measures the potential of possible solutions, filtering out unpromising solutions and then validating the exact fitness value of the remaining solutions [19]. However, to the best of our knowledge, there is no serious investigation on this direction from the literatures related to intelligent sensor applications and multilabel feature selection.

In this study, we propose a novel effective evolutionary search method for multilabel datasets. Previous studies considering the intelligent sensor applications incurring multilabel feature selection did not tackle the issue related to the generation of promising feature subsets, resulting in a degeneration of search effectiveness. Our contribution can be summarized as follows:

(i) The proposed method improves search effectiveness by producing a large number of feature subsets with important features and then filters out unpromising feature subsets using a cheap evaluation method.

(ii) A cheap feature subset evaluation method is employed to filter out unpromising feature subsets without checking the fitness value which demands expensive computational cost.

(iii) We compared the performance of conventional multilabel feature wrapper methods and the proposed method on 14 multilabel datasets and conducted 53 standard statistical tests to validate the superiority of the proposed method

2. Related Work

Because multilabel feature selection can improve the learning accuracy as well as the efficiency of a later algorithm by highlighting important features such as multilabel classifier for the concurrent prediction, it gained significant attention from diverse fields [20, 21].

Feature selection methods come in two categories: filters and wrappers. Filter methods rank features based on their own criterion by evaluating the importance of each feature. For multilabel feature selection on multilabel datasets, a simple strategy that changes the label sets to a single label set was often considered, such as a label powerset [22]. This method is advantageous because it enables conventional feature selection methods for single-label datasets. Several conventional filter methods have been reported [23]; however, filter methods commonly suffer from low multilabel classification accuracy, owing to noninteraction with multilabel classifiers or subsequent problems such as imbalance in transformed single-label data. By contrast, wrapper methods evaluate created feature subsets and improve them. In detail, they locate promising feature subsets using a search method employed and then evaluate them using a later learning algorithm [17]. Although the learning algorithm can be different according to the application, recent review indicated that the most frequent choice for the search method is the evolutionary search [24] because it is effective at searching for feasible solution in global perspective. Zhang et al. [14] proposed a multilabel feature selection method based on genetic algorithms. However, a major drawback of the genetic algorithm is their premature convergence to unrefined solutions [17]. On the other hand, a genetic algorithm-based nondominated sorting genetic algorithm-II [25] and multiobjective particle swarm optimization [26] have been used for multilabel feature selection.

Although most studies consider single-label sensory datasets, there are several studies on feature selection methods because of the promising potential. To apply automatic view generation, a semisupervised feature selection method for features extracted from very high-resolution remote sensing images was proposed [27]. Specifically, features are categorized into a series of disjoint groups, and then important features in each group are selected by solving the $l_{11,2}$-norm-based minimization problem. Similarly, a refined feature subset from discrete wavelet transform coefficient features, extracted from artificial tongue sensor signals, was selected by using a dispersion ratio computation [12]. Activity recognition using accelerometers was also shown to be improved by feature selection [28]. There are several studies related to the identification of a set of important features based on the fitness or classification accuracy derived from the learning algorithm. For example, a feature subset can be obtained by iteratively including the best feature at each step, which is referred to as the sequential forward selection algorithm [29]. This technique is applied to the application of chiller fault detection [30], which is an instantiation of an automatic fault detection problem in a smart factory [31]. The genetic algorithm which is one of the most famous evolutionary search methods in the machine learning community was also considered for selecting discriminative features for online bearing fault diagnosis [32]. In addition, the particle swarm optimization technique, which is another popular evolutionary search method, was also used to find the optimal feature subset for intrusion detection [13]. Support vector machine recursive feature elimination has been used for the analysis of correlated gas sensor data [7]. Energy consumption was minimized and the classification accuracy was improved by feature selection from sensor data [5].

3. Proposed Method

3.1. Preliminary. Of the various evolutionary search methods, estimation of distribution algorithm (EDA) has proven effective for solving various problems [24, 33]. Unlike typical evolutionary search methods, to generate new feature subsets, EDAs do not use genetic operators [19]. Instead, conventional EDAs generate new solutions or candidates using a probability model and update the probability model based on a statistical distribution estimated from the representation of solutions. Thus, it provides an opportunity to generate promising feature subsets by manipulating the probability model. The probability model can be implemented as follows [33, 34]:

$$P^{t+1}(i) = P^t \times (1 - \text{LR}) + F^t(i) \times \text{LR}, \quad (2)$$

where $P^t(i)$ is the selection probability of the i-th feature in the t-th generation, $F^t(i)$ is the probability associated with the i-th feature in the top 50% feature subsets in the t-th generation that are ranked in terms of their fitness values, and LR is the learning rate, which is a user-defined parameter that controls the influence of $F(i)$ to the probability model in the next generation. Through (2), the probability of selecting a feature in the $(t+1)$-th generation, P^{t+1}, is calculated, and in the $(t+1)$-th generation, feature subsets are built. This process is repeated until the maximum allowed computational cost is exhausted. Although there are many stopping criteria, we set the number of spent fitness function calls (FFCs) as the termination condition for all evolutionary search methods employed in this study for a fair comparison against diversified settings and implementations [35].

In the feature selection problem, the algorithm should be capable of searching a huge parametric space; thus, significant computational cost is associated with finding a promising solution. Although simple probability models are easy to implement, it can be insufficient for solving complicated problems, such as pinpointing promising feature subsets in a large search space [36]. For example, in the conventional EDA-based feature selection method, all features are initially assigned the selection probability of 0.5. This means that nonpromising features can be also present in feature subsets. To overcome this drawback, we devise a process for generation of a promising feature subset. Specifically, when creating a feature subset, the algorithm will consider important features more frequently by setting the priority to such features given by an individual feature filter.

After creating the feature subsets, the next step amounts to selecting promising feature subsets. Although good feature subsets can be created using filter methods, there can be nonpromising feature subsets because the creation process is probabilistic and there can be efficient interaction among features. Nonpromising feature subsets consume FFCs and negatively affect the search efficiency. To overcome this problem, we propose a feature subset evaluation method consuming a cheap computational cost. Using the methods of information theory, the proposed method calculates, for each subset, the relevance and redundancy of the subset features. Then, the proposed method selects feature subsets with maximal relevance and minimal redundancy. Because there is a possibility that the proposed solution will be only locally promising, the proposed method uses roulette wheel selection as the selection algorithm [37]. Thus, nonpromising feature subsets are filtered out from the neighbor set, without exact evaluation.

In the proposed method, there are two key functions for the feature subset generation. *create* function makes candidate feature subsets that is composed of relevant features. *select* function selects promising feature subsets among created ones by using roulette wheel selection based on their potential given by a feature subset evaluation method. Figure 1 schematically shows the proposed method. In the first stage, the probability model is initialized, indicating feature subsets containing randomly chosen features will be created frequently. The probability model is represented as a vector where each element encodes the presence of each feature. In the next step, feature subsets are created using *create* function. All feature subsets are assigned random integers, ranging from one to n. If one feature subset is determined to choose two features, the proposed method ranks the features in terms of their importance, using a filter method. In the first iteration, the most important feature is f_4. Then, the proposed method chooses a random number r between 0 and 1 and compares r to the selection probability of f_4 in the probability model, p_4. Since p_4 is greater than r in the example, f_4 is selected and added to the feature subset. In the second iteration, features are again ranked using the filter method. In this case, the features' importance is measured again in terms of relevance and redundancy under the selection of f_4. Thus, the features' ranks can change. In this example, f_2 is the most important feature. Then, another random number r is drawn and compared to the selection probability of f_2, p_2. However, p_2 is lower than r; thus, f_2 is not selected. Then, the second most important feature can be selected. In this example, f_5 is added to the subset of features, and iteration is terminated. Through this process, the proposed method creates a series of new feature subsets including important features. The next step amounts to selecting promising feature subsets among created feature subsets using *select* function. The proposed method chooses m promising feature subsets by roulette wheel selection biased by the proposed feature subset evaluation. Finally, the probability model is updated using m promising feature subsets and (2), to reflect the presence of features in the best half of new feature subsets ranked by fitness value.

3.2. Proposed Search Procedure. The proposed algorithm creates feature subsets to large searching spaces and filters nonpromising subsets using the proposed subset evaluation method that does not incur exact evaluation. Algorithm 1 shows the proposed method. For the population size m and maximal number of FFCs v, the method initializes feature subsets $O(t)$ and the probability model P (line 3). The method generates a set of m feature subsets $O(t)$ through a random assignment of maximum $|F|$ binary bits. The probability model P is an $|F|$ length vector, and each entry in the vector refers to the probability of choosing coordinated features. Each entry is initialized for the distribution of the features of $O(t)$. Then, the created set $O(t)$ is evaluated (line 4). The method set consumed FFCs u to 0

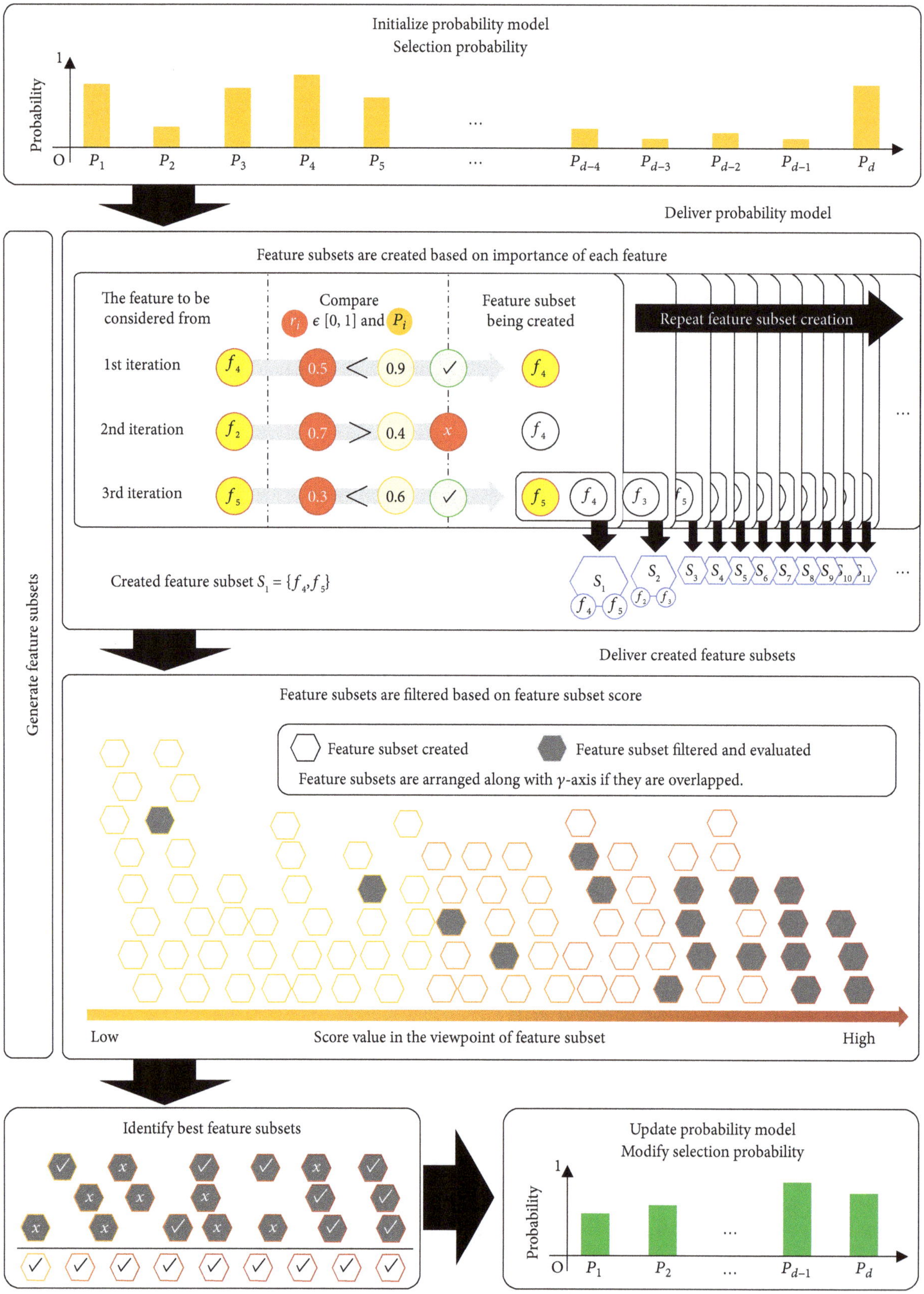

FIGURE 1: Schematic overview of the proposed method.

```
1:  Input: population size m, max FFCs v
2:  Output: best feature subset S_g
3:  initializing O(t) and probability model p
4:  evaluating O(t)
5:  u ← 0                                              ▷ Set consumed FFCs to 0
6:  store the global best feature subset to S_g
7:  while u ≤ v do
8:     update P by Eq. (2)
9:     E(t) ← create neighbor set
10:     O(t + 1) ← select feature subsets in E(t)
11:     evaluate O(t + 1)
12:     u ← u+m                                        ▷ update consumed FFCs
13:     update global best feature subset S_g
14:  end while
```

ALGORITHM 1: Proposed method.

```
1:  Input: neighbor population size e, probability model P
2:  Output: neighbor set E (t)
3:  E(t) ← ∅
4:  for k = 1 to e do
5:     n ← random integer value in [0, |F|] ▷ set a feature subset size randomly
6:     S_k ← {∅}
7:     for i = 1 to n do
8:        R ← {f_1, f_2, f_3, ···}                     ▷ ranked features by Eq. (6)
9:        for j = 1 to |F| do
10:            if P^t(f_j)>random value in [0,1] then
11:              S_k ← S_k ∪ j
12:              break
13:            end if
14:          end for
15:        end for
16:     E(t) ← E(t) ∪ S_k
17:  end for
```

ALGORITHM 2: *create* function.

(line 5) and stores the global best feature subset to S_g (line 6). P is updated by (2) (line 8). The method creates a set of neighbor feature subsets E, which is based on a filter method by *create* function (line 9). Then, m feature subsets are selected by roulette wheel selection weighted by *select* function in the set $E(t)$ and yield the new generation $O(t+1)$ (line 6). The feature subset $O(t+1)$ is evaluated (line 11), and sets consumed FFCs u (line 12). The feature subset S_g, which offers globally optimal performance, is stored and replaced in the procedure (line 13). After all allowed FFCs are consumed, the algorithm returns the feature subset S_g.

Algorithm 2 is a *create* function that shows the process of creating feature subsets. Each feature subset selects random size n features (line 5). To introduce important features more frequently, first, each feature should be ranked by their importance value. To achieve this, we evaluate the importance of each feature using the relevance criterion [20]:

$$I(f_i) = \mathrm{Rel}(f_i) - \mathrm{Red}(f_i), \tag{3}$$

where $\mathrm{Rel}(f_i)$ and $\mathrm{Red}(f_i)$ denote the relevance and redundancy of the i-th feature and $I(f_i)$ denotes the importance of the i-th feature. Although both functions can be implemented differently according to the subject of each study, we use a recent filter method for measuring the importance of features. In the work of [15], we proposed a filter method for multilabel dataset, and was shown to outperform conventional filter methods. Because of this reason, we use this method for measuring the importance of features. Accordingly, $\mathrm{Rel}(f_i)$ can be implemented as

$$\mathrm{Rel}(f_i) = \sum_{l \in L} M(f_i\,; l), \tag{4}$$

where $M(x;y) = H(x) - H(x,y) + H(y)$ indicates the mutual information between variables x and y and $H(x) = -\sum P(x)\log P(x)$ is the joint entropy obtained from the probability $P(x)$, $P(y)$, and $P(x,y)$. Next, $\text{Red}(f_i)$ can be implemented as

$$\text{Red}(f_i) = \sum_{f\in S}\sum_{l\in L}\frac{M(f_i;l)}{H(f_i)}M(f_i;f). \tag{5}$$

Thus, the feature f_i's importance is measured by

$$I(f_i) = \sum_{l\in L}M(f_i;l) - \sum_{f\in S}\sum_{l\in L}\frac{M(f_i;l)}{H(f_i)}M(f_i;f). \tag{6}$$

Then the rank of each feature can be determined by using (6) and remembered (line 8). After then, the function decides whether to choose a feature from the most important subsets by P (lines 9 to 13). If a feature is chosen, it is added to subset S_k (line 11). In addition, after a subset is created, it is added to the set of neighbor feature subsets E (line 16).

It is well-known fact from the feature selection community that a set of individually good features is not necessarily a good feature subset due to the interaction among features. This means that the created feature subset can be unpromising even though (6) only included important features. To achieve this, *select* function described in Algorithm 3 that shows the process for selecting promising feature subsets in a neighbor set is necessary. In *select* function, a new feature subset filter method is employed [38]. Specifically, it evaluates the fitness of the feature subset as

$$E(S) = \sum_{f_i\in S}\sum_{l\in L}M(f_i;l) - \sum_{f_i\in S}\sum_{f_j\in S}M\left(f_i;f_j\right). \tag{7}$$

By using (7), *select* function ranks feature subsets in the neighbor set E (line 3). Next, the algorithm selects m feature subsets $G(t)$ using roulette wheel selection [37], which is a biased selection weight by (7) (line 4).

In summary, in the generation of a feature subset, the algorithm ranks the importance of features using the filter method and selects the most important feature i based on the probability $P^t(i)$ considering subset S selected at this point. If the i-th feature is not chosen, the next most important feature j can be selected with the probability $P^t(j)$, and the process repeats until a feature is selected. Then, for each neighbor feature subset, (7) ranks the importance of feature subsets, and feature subsets with highest $E(\cdot)$ values are likely to selected.

```
1: Input: neighbor set E, population size m
2: Output: filtered set G(t)
3: rank feature subsets in set E by Eq. (7)
4: select m feature subsets by roulette wheel selection
5: G(t) ← selected feature subsets
```

ALGORITHM 3: *select* function.

4. Experimental Results

We conducted experiments on 14 datasets from various domains. The Birds dataset is audio data containing samples of multiple bird calls. The Enron and Language Log (Llog) datasets are generated from text mining applications, where each feature corresponds to the presence of a word and each label represents the relevance of each text pattern to a specific subject. The Mediamill dataset contains video data from an automatic detection system. The Medical dataset is sampled from a large corpus of suicide letters obtained from the natural language processing of clinical free texts. The TMC2007 dataset contains safety reports of a complex space system. The remaining eight datasets came from the Yahoo dataset collection. We performed unsupervised dimensionality reduction on datasets, including the TMC2007 and Yahoo collections, consisting of more than 10,000 features. Because our algorithm uses information theory, for numeric features, we performed discretization using the supervised discretization method [39]. Table 1 shows the standard characteristics of the multilabel datasets used in our experiments, including the number of patterns in the datasets $|W|$, number of features $|F|$, type of features, and number of labels $|L|$. The label cardinality measure C ard represents the average number of labels for each instance. The label density measure Den is the label cardinality over the total number of labels. The number of distinct label sets Distinct indicates the number of unique label subsets in L. *Domain* represents the applications associated with the extracted datasets.

We compared the proposed method with conventional methods, including the genetic algorithm (GA) [14], non-dominated sorting genetic algorithm-II (NSGA-II) [25], and multiobjective particle swarm optimization feature selection (MPSOFS) [26]. We considered a conventional multilabel classifier, namely, the multilabel naïve Bayes (MLNB) classifier [14]. We used conventional hold-out cross-validation for each dataset. Of the patterns, 80% were randomly chosen as a training set and the remaining 20% were chosen as a test set. We set the size of the population to 20, and the maximal number of FFCs was limited to 100. In our proposed method, we created 500 feature subsets using the probability model and set the learning rate (LR) to 0.4. The GA and NSGA-II created two offspring feature subsets and one feature subset from mutation operators in each generation. The MPSOFS preserved the global best particle solutions and each particle's best solutions. Thereafter, the MPSOFS updated the velocity values. All experiments were repeated 10 times, and the average measured values were used to compare the performances of the methods.

To measure the methods' performances, we employed the following four evaluation metrics: multilabel accuracy,

Table 1: Standard characteristics of employed datasets.

Dataset	$\|W\|$	$\|F\|$	Type	$\|L\|$	Card	Den	Distinct	Domain
Birds	645	260	Mixed	19	1.014	0.053	133	Audio
Enron	1702	1001	Nominal	53	3.378	0.064	753	Text
Llog	1460	1004	Nominal	75	1.180	0.016	304	Text
Mediamill	43,907	120	Numeric	101	4.376	0.043	6555	Video
Medical	978	1449	Nominal	45	1.245	0.028	94	Text
TMC2007	28,596	49,060	Numeric	22	2.158	0.098	1341	Text
Business	11,214	1096	Numeric	30	1.599	0.053	233	Text
Education	12,030	1377	Numeric	33	1.463	0.044	511	Text
Entertainment	12,730	1600	Numeric	21	1.414	0.067	337	Text
Health	9205	1530	Numeric	32	1.644	0.051	335	Text
Reference	8027	1984	Numeric	33	1.174	0.036	275	Text
Science	6428	1859	Numeric	40	1.450	0.036	457	Text
Social	12,111	2618	Numeric	39	1.279	0.033	361	Text
Society	14,512	1590	Numeric	27	1.670	0.062	1054	Text

hamming loss, ranking loss, and normalized coverage. Multilabel accuracy is defined as

$$\mathrm{mlacc}(T) = \frac{1}{|T|}\sum_{i=1}^{|T|}\frac{|\lambda_i \cap Y_i|}{|\lambda_i \cup Y_i|}, \quad (8)$$

where T is a given test set. Hamming loss is defined by

$$\mathrm{hloss}(T) = \frac{1}{|T|}\sum_{i=1}^{|T|}\frac{1}{|L|}|\lambda_i \Delta Y_i|, \quad (9)$$

where λ denotes the correct label subset and Δ denotes the symmetric difference between the two sets. Ranking loss is defined by

$$\mathrm{rloss}(T) = \frac{1}{|T|}\sum_{i=1}^{|T|}\frac{\left|\left\{(a,b)\middle| a \in \overline{\lambda}_i, \psi_{i,a} \le \psi_{i,b}\right\}\right|}{|\lambda_i|\left|\overline{\lambda}_i\right|}, \quad (10)$$

where $\overline{\lambda}_i$ is a complementary set of λ_i. Ranking loss measures the average fraction of (a, b) pairs with $\psi_{i,a} \le \psi_{i,b}$ over all possible relevant and irrelevant label pairs. Finally, normalized coverage is defined as:

$$\mathrm{ncov}(T) = \frac{1}{|L|}\left(\frac{1}{|T|}\sum_{i=1}^{|T|}\max_{l\in\lambda_i} \mathrm{rank}\ (l) - 1\right), \quad (11)$$

where rank $(\cdot)$ returns the rank of the corresponding relevant label $l \in \lambda_i$ according to $\psi_{i,l}$ in nonincreasing order. Therefore, normalized coverage measures how many labels must be marked as positive for all relevant labels to be positive. Higher values of multilabel accuracy and lower values of hamming loss, ranking loss, and normalized coverage indicate good classification performance.

Tables 2, 3, 4, and 5 list the experimental results for the different performance measures as averages over the experiments on the employed datasets. The best performance of each dataset is indicated by a bold font. In each table, the last column shows the average rank (Avg. rank) of each comparison method over all the multilabel datasets. In terms of the multilabel accuracy and ranking loss measures, the proposed method outperformed the GA, NSGA-II, and MPSOFS, on all datasets. In terms of the hamming loss, the proposed method outperformed conventional methods on all datasets except TMC2007. In terms of the normalized coverage, the proposed method outperformed the conventional methods on all datasets except Llog.

After measuring the performance of the methods on all datasets, we analyzed the performance using statistical tools. We employed the Friedman test, a widely used statistical test, for comparing multiple methods over a number of datasets [40]. Supposing there are k methods and N datasets, and let R_j denote the average rank for the j-th method under the null hypothesis (i.e., when all of the methods perform equally well). Then, the following Friedman statistic F_F is distributed according to the F-distribution with $k - 1$ numerator degrees of freedom and $(k - 1)\ (N - 1)$ denominator degrees of freedom as parameters:

$$F_F \frac{(N-1)\chi_F^2}{N(k-1)-\chi_F^2}, \quad (12)$$

where χ_F^2 is defined as

$$\chi_F^2 = \frac{12N}{k(k+1)}\left[\sum_{j=1}^{k} R_j^2 - \frac{k(k+1)^2}{4}\right]. \quad (13)$$

If F_F is larger than the critical value at a significance level α, the null hypothesis is rejected, implying that the compared methods have different performances. After the

TABLE 2: Comparison results in terms of multilabel accuracy.

Method	Birds	Enron	Llog	Mediamill	Medical
Proposed	**0.527 ± 0.044**	**0.383 ± 0.012**	**0.249 ± 0.014**	**0.362 ± 0.004**	**0.427 ± 0.030**
GA	0.491 ± 0.055	0.284 ± 0.022	0.209 ± 0.018	0.336 ± 0.031	0.303 ± 0.058
NSGA	0.480 ± 0.040	0.282 ± 0.022	0.208 ± 0.014	0.347 ± 0.013	0.297 ± 0.018
MPSOFS	0.453 ± 0.031	0.206 ± 0.012	0.042 ± 0.002	0.163 ± 0.007	0.286 ± 0.033
Method	TMC2007	Business	Education	Entertainment	Health
Proposed	**0.441 ± 0.004**	**0.672 ± 0.011**	**0.318 ± 0.009**	**0.396 ± 0.004**	**0.537 ± 0.011**
GA	0.435 ± 0.010	0.657 ± 0.021	0.316 ± 0.010	0.361 ± 0.011	0.499 ± 0.011
NSGA	0.434 ± 0.005	0.662 ± 0.013	0.318 ± 0.010	0.362 ± 0.010	0.495 ± 0.009
MPSOFS	0.420 ± 0.005	0.634 ± 0.008	0.283 ± 0.005	0.365 ± 0.010	0.496 ± 0.013
Method	Reference	Science	Social	Society	Avg. rank
Proposed	**0.436 ± 0.009**	**0.288 ± 0.010**	**0.546 ± 0.008**	**0.371 ± 0.008**	**1.00**
GA	0.422 ± 0.012	0.231 ± 0.009	0.517 ± 0.012	0.258 ± 0.011	2.79
NSGA	0.429 ± 0.009	0.237 ± 0.010	0.526 ± 0.009	0.267 ± 0.010	2.64
MPSOFS	0.414 ± 0.017	0.234 ± 0.013	0.527 ± 0.012	0.239 ± 0.007	3.57

TABLE 3: Comparison results in terms of hamming loss.

Method	Birds	Enron	Llog	Mediamill	Medical
Proposed	**0.061 ± 0.008**	**0.060 ± 0.003**	**0.016 ± 0.001**	**0.034 ± 0.000**	**0.020 ± 0.001**
GA	0.072 ± 0.015	0.100 ± 0.033	0.075 ± 0.073	0.048 ± 0.022	0.023 ± 0.001
NSGA	0.064 ± 0.008	0.104 ± 0.026	0.072 ± 0.071	0.054 ± 0.032	0.022 ± 0.001
MPSOFS	0.135 ± 0.018	0.198 ± 0.007	0.292 ± 0.008	0.174 ± 0.005	0.023 ± 0.001
Method	TMC2007	Business	Education	Entertainment	Health
Proposed	**0.088 ± 0.002**	**0.029 ± 0.001**	**0.042 ± 0.001**	**0.055 ± 0.001**	**0.039 ± 0.001**
GA	0.088 ± 0.004	0.035 ± 0.005	0.046 ± 0.003	0.070 ± 0.007	0.051 ± 0.004
NSGA	0.086 ± 0.004	0.037 ± 0.010	0.048 ± 0.004	0.068 ± 0.004	0.054 ± 0.005
MPSOFS	0.117 ± 0.002	0.079 ± 0.003	0.061 ± 0.002	0.105 ± 0.003	0.067 ± 0.002
Method	Reference	Science	Social	Society	Avg. rank
Proposed	**0.034 ± 0.006**	**0.035 ± 0.003**	**0.025 ± 0.003**	**0.054 ± 0.002**	**1.07**
GA	0.055 ± 0.013	0.051 ± 0.012	0.042 ± 0.007	0.062 ± 0.005	2.71
NSGA	0.047 ± 0.008	0.045 ± 0.008	0.040 ± 0.010	0.060 ± 0.002	2.29
MPSOFS	0.086 ± 0.005	0.110 ± 0.005	0.070 ± 0.002	0.144 ± 0.007	3.93

null hypothesis is rejected, we perform a post hoc test to analyze whether the proposed method performs significantly better than other methods. The Bonferroni–Dunn test is employed [41]. Critical difference (CD) is used to compare the proposed method and one comparison method. CD is defined as

$$\mathrm{CD} = q_{\alpha}\sqrt{\frac{k(k+1)}{6N}}, \tag{14}$$

where the critical value q_{α} is constant and is determined by the number of methods and the significance level. If the difference between the two compared methods' average ranks is greater than CD, the better-ranking method is concluded to perform significantly better than the other method. Because our experiment used four methods, including the proposed method, and 14 datasets, we set $k = 4$ and $N = 14$. We employed the Friedman test when the significance level α was 0.05. Table 6 shows the summary of the employed Friedman test. The critical value for 3 and 39 degrees of freedom was 2.845. The Friedman statistic F_F for all performance measures was above the critical value. Thus, the null hypothesis that the compared methods perform equally well was rejected.

To employ the Bonferroni–Dunn test, the calculated CD with $\alpha = 0.05$ was 1.168 since $q_{\alpha} = 2.394$ at the significance level α of 0.05. Figure 2 shows the CD diagrams for all evaluation measures, where the average rank of each method is on the top of each figure. Our proposed method significantly outperforms other, conventional, methods on all evaluation measures.

TABLE 4: Comparison results in terms of ranking loss.

Method	Birds	Enron	Llog	Mediamill	Medical
Proposed	**0.115 ± 0.017**	**0.100 ± 0.008**	**0.155 ± 0.023**	**0.060 ± 0.001**	**0.115 ± 0.026**
GA	0.129 ± 0.015	0.133 ± 0.024	0.164 ± 0.023	0.066 ± 0.007	0.145 ± 0.026
NSGA	0.125 ± 0.017	0.149 ± 0.031	0.163 ± 0.023	0.067 ± 0.011	0.139 ± 0.026
MPSOFS	0.132 ± 0.017	0.194 ± 0.011	0.180 ± 0.021	0.159 ± 0.004	0.140 ± 0.024
Method	TMC2007	Business	Education	Entertainment	Health
Proposed	**0.073 ± 0.001**	**0.062 ± 0.027**	**0.089 ± 0.003**	**0.111 ± 0.002**	**0.085 ± 0.028**
GA	0.075 ± 0.002	0.070 ± 0.029	0.100 ± 0.004	0.140 ± 0.007	0.098 ± 0.027
NSGA	0.076 ± 0.002	0.067 ± 0.027	0.100 ± 0.003	0.137 ± 0.009	0.097 ± 0.028
MPSOFS	0.078 ± 0.002	0.096 ± 0.027	0.101 ± 0.003	0.153 ± 0.005	0.098 ± 0.028
Method	Reference	Science	Social	Society	Avg. rank
Proposed	**0.111 ± 0.023**	**0.118 ± 0.003**	**0.075 ± 0.010**	**0.135 ± 0.003**	**1.00**
GA	0.130 ± 0.021	0.154 ± 0.007	0.088 ± 0.012	0.151 ± 0.008	2.86
NSGA	0.128 ± 0.025	0.150 ± 0.006	0.085 ± 0.011	0.153 ± 0.011	2.29
MPSOFS	0.140 ± 0.023	0.157 ± 0.004	0.097 ± 0.012	0.212 ± 0.005	3.86

TABLE 5: Comparison results in terms of normalized coverage.

Method	Birds	Enron	Llog	Mediamill	Medical
Proposed	**0.194 ± 0.015**	**0.277 ± 0.015**	**0.202 ± 0.025**	**0.196 ± 0.003**	**0.155 ± 0.029**
GA	0.223 ± 0.025	0.337 ± 0.038	0.201 ± 0.021	0.205 ± 0.005	0.180 ± 0.027
NSGA	0.212 ± 0.020	0.336 ± 0.038	0.199 ± 0.025	0.208 ± 0.019	0.178 ± 0.029
MPSOFS	0.222 ± 0.023	0.413 ± 0.014	0.201 ± 0.023	0.330 ± 0.006	0.179 ± 0.027
Method	TMC2007	Business	Education	Entertainment	Health
Proposed	**0.203 ± 0.002**	**0.132 ± 0.023**	**0.148 ± 0.004**	**0.197 ± 0.003**	**0.157 ± 0.023**
GA	0.208 ± 0.004	0.139 ± 0.024	0.159 ± 0.003	0.221 ± 0.009	0.167 ± 0.026
NSGA	0.209 ± 0.004	0.141 ± 0.021	0.159 ± 0.004	0.219 ± 0.009	0.167 ± 0.026
MPSOFS	0.211 ± 0.002	0.168 ± 0.023	0.159 ± 0.004	0.233 ± 0.003	0.166 ± 0.025
Method	Reference	Science	Social	Society	Avg. rank
Proposed	**0.157 ± 0.023**	**0.182 ± 0.004**	**0.127 ± 0.010**	**0.242 ± 0.007**	**1.21**
GA	0.177 ± 0.022	0.215 ± 0.005	0.140 ± 0.011	0.257 ± 0.006	2.86
NSGA	0.171 ± 0.021	0.213 ± 0.006	0.136 ± 0.014	0.261 ± 0.008	2.43
MPSOFS	0.179 ± 0.022	0.215 ± 0.006	0.145 ± 0.011	0.317 ± 0.007	3.50

TABLE 6: Summary of the Friedman statistics F_F $(k = 4, N = 14)$ and critical value in terms of each evaluation measure.

Evaluation measure	F_F	Critical value $(\alpha = 0.05)$
Multilabel accuracy	30.333	
Hamming loss	65.642	2.845
Ranking loss	75.472	
Normalized coverage	16.355	

5. Conclusion

To handle multilabel sensor datasets, we proposed an effective search based on a promising feature subset generation method for multilabel feature selection problem. The main contribution of this work is to propose and validate a new feature subset generation method. Specifically, the proposed method generates candidate feature subsets using important features and chooses promising subsets of features without consuming significant computational cost. Experimental results show that our method converges faster than other conventional methods. In the future, we would like to investigate a new feature subset generation that is more effective because the proposed feature subset generation is strongly dependent on the employed filter method, and it may result redundant feature subsets during the search process. In addition, we would like to apply the proposed method to various sensor datasets and compare the performance with conventional feature selection methods considered from sensory data analysis.

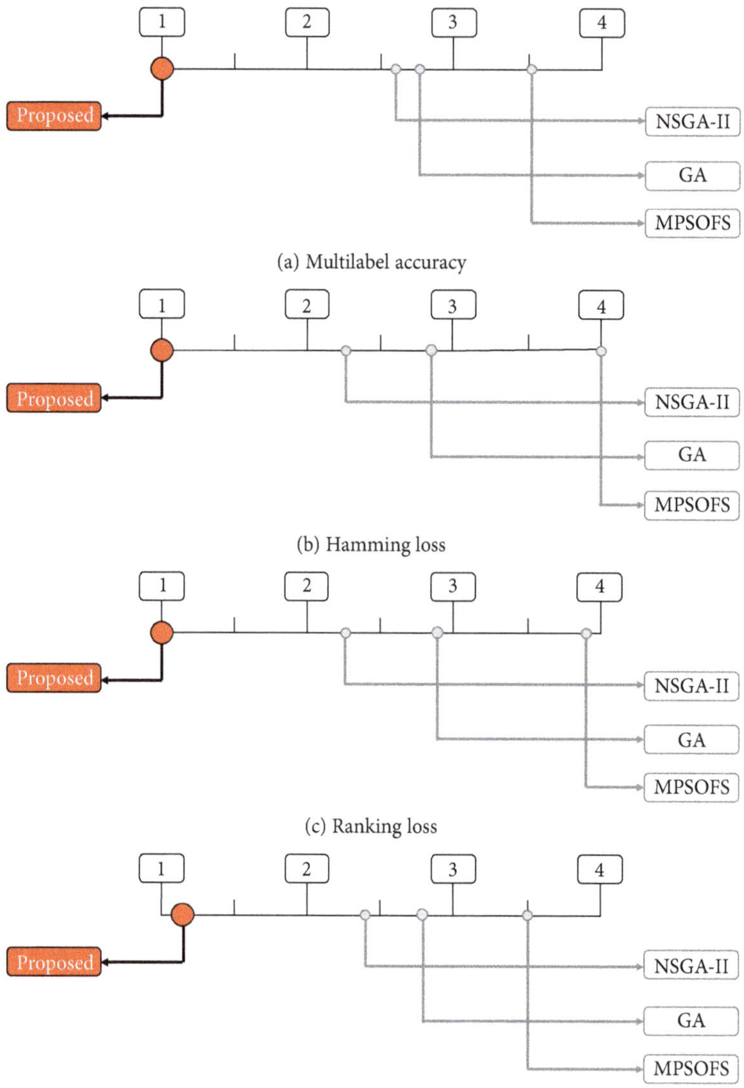

(a) Multilabel accuracy

(b) Hamming loss

(c) Ranking loss

(d) Normalized coverage

FIGURE 2: Bonferroni-Dunn test results of four comparison methods with four evaluation measures (significance level $\alpha = 0.05$).

Conflicts of Interest

The authors declare that they have no conflicts of interest.

Acknowledgments

This research was supported by Basic Science Research Program through the National Research Foundation of Korea (NRF) funded by the Ministry of Science, ICT and Future Planning (NRF-2016R1C1B1014774).

References

[1] S. Qadri, D. M. Khan, S. F. Qadri et al., "Multisource data fusion framework for land use/land cover classification using machine vision," *Journal of Sensors*, vol. 2017, Article ID 3515418, 8 pages, 2017.

[2] A. Alhamoud, V. Muradi, D. Bohnstedt, and R. Steinmetz, "Activity recognition in multi-user environments using techniques of multi-label classification," in *Proceedings of the 6th International Conference on the Internet of Things - IoT'16*, pp. 15–23, Stuttgart, Germany, 2016.

[3] R. Gravina, P. Alinia, H. Ghasemzadeh, and G. Fortino, "Multi-sensor fusion in body sensor networks: state-of-the-art and research challenges," *Information Fusion*, vol. 35, pp. 68–80, 2017.

[4] B. Du, Z. Wang, L. Zhang, L. Zhang, and D. Tao, "Robust and discriminative labeling for multi-label active learning based on maximum correntropy criterion," *IEEE Transactions on Image Processing*, vol. 26, no. 4, pp. 1694–1707, 2017.

[5] H. Ghasemzadeh, N. Amini, R. Saeedi, and M. Sarrafzadeh, "Power-aware computing in wearable sensor networks: an optimal feature selection," *IEEE Transactions on Mobile Computing*, vol. 14, no. 4, pp. 800–812, 2015.

[6] D. Li, Y. Zhou, G. Hu, and C. J. Spanos, "Optimal sensor configuration and feature selection for AHU fault detection and diagnosis," *IEEE Transactions on Industrial Informatics*, vol. 13, no. 3, pp. 1369–1380, 2017.

[7] K. Yan and D. Zhang, "Feature selection and analysis on correlated gas sensor data with recursive feature elimination," *Sensors and Actuators B: Chemical*, vol. 212, pp. 353–363, 2015.

[8] S. Cheng, Z. Cai, J. Li, and H. Gao, "Extracting kernel dataset from big sensory data in wireless sensor networks," *IEEE Transactions on Knowledge and Data Engineering*, vol. 29, no. 4, pp. 813–827, 2017.

[9] R. Kumar, I. Qamar, J. S. Virdi, and N. C. Krishnan, "Multi-label learning for activity recognition," in *2015 International Conference on Intelligent Environments*, pp. 152–155, Prague, Czech Republic, 2015.

[10] J. Read, L. Martino, P. M. Olmos, and D. Luengo, "Scalable multi-output label prediction: from classifier chains to classifier trellises," *Pattern Recognition*, vol. 48, no. 6, pp. 2096–2109, 2015.

[11] M.-L. Zhang and Z.-H. Zhou, "A review on multi-label learning algorithms," *IEEE Transactions on Knowledge and Data Engineering*, vol. 26, no. 8, pp. 1819–1837, 2014.

[12] T. Liu, Y. Chen, D. Li, and M. Wu, "An active feature selection strategy for DWT in artificial taste," *Journal of Sensors*, vol. 2018, Article ID 9709505, 11 pages, 2018.

[13] X. Teng, H. Dong, and X. Zhou, "Adaptive feature selection using v-shaped binary particle swarm optimization," *PLoS One*, vol. 12, no. 3, article e0173907, 2017.

[14] M.-L. Zhang, J. M. Pena, and V. Robles, "Feature selection for multi-label naive Bayes classification," *Information Sciences*, vol. 179, no. 19, pp. 3218–3229, 2009.

[15] J. Lee and D.-W. Kim, "SCLS: multi-label feature selection based on scalable criterion for large label set," *Pattern Recognition*, vol. 66, pp. 342–352, 2017.

[16] Z. Michalewicz and D. B. Fogel, *How to Solve It: Modern Heuristics*, Springer Science & Business Media, 2013.

[17] J. Lee and D.-W. Kim, "Memetic feature selection algorithm for multi-label classification," *Information Sciences*, vol. 293, pp. 80–96, 2015.

[18] J. Lee, W. Seo, and D.-W. Kim, "Effective evolutionary multi-label feature selection under a budget constraint," *Complexity*, vol. 2018, Article ID 3241489, 14 pages, 2018.

[19] A. Zhou, J. Sun, and Q. Zhang, "An estimation of distribution algorithm with cheap and expensive local search methods," *IEEE Transactions on Evolutionary Computation*, vol. 19, no. 6, pp. 807–822, 2015.

[20] J. Lee and D.-W. Kim, "Feature selection for multi-label classification using multivariate mutual information," *Pattern Recognition Letters*, vol. 34, no. 3, pp. 349–357, 2013.

[21] J. Lee and D.-W. Kim, "Mutual information-based multi-label feature selection using interaction information," *Expert Systems with Applications*, vol. 42, no. 4, pp. 2013–2025, 2015.

[22] J. Read, "A pruned problem transformation method for multi-label classification," *Proceedings of New Zealand Computer Science Research Student Conference*, 2008, pp. 143–150, Christchurch, New Zealand, 2008.

[23] N. Spolaor, E. A. Cherman, M. C. Monard, and H. D. Lee, "A comparison of multi-label feature selection methods using the problem transformation approach," *Electronic Notes in Theoretical Computer Science*, vol. 292, pp. 135–151, 2013.

[24] B. Xue, M. Zhang, W. N. Browne, and X. Yao, "A survey on evolutionary computation approaches to feature selection," *IEEE Transactions on Evolutionary Computation*, vol. 20, no. 4, pp. 606–626, 2016.

[25] J. Yin, T. Tao, and J. Xu, "A multi-label feature selection algorithm based on multi-objective optimization," in *2015 International Joint Conference on Neural Networks (IJCNN)*, pp. 1–7, Killarney, Ireland, 2015.

[26] Y. Zhang, D.-w. Gong, X.-y. Sun, and Y.-n. Guo, "A PSO-based multi-objective multi-label feature selection method in classification," *Scientific Reports*, vol. 7, no. 1, p. 376, 2017.

[27] X. Chen, W. Liu, F. Su, and G. Zhou, "Semisupervised multiview feature selection for VHR remote sensing images with label learning and automatic view generation," *IEEE Journal of Selected Topics in Applied Earth Observations and Remote Sensing*, vol. 10, no. 6, pp. 2876–2888, 2017.

[28] P. Gupta and T. Dallas, "Feature selection and activity recognition system using a single triaxial accelerometer," *IEEE Transactions on Biomedical Engineering*, vol. 61, no. 6, pp. 1780–1786, 2014.

[29] A. Jain and D. Zongker, "Feature selection: evaluation, application, and small sample performance," *IEEE Transactions on Pattern Analysis and Machine Intelligence*, vol. 19, no. 2, pp. 153–158, 1997.

[30] K. Yan, L. Ma, Y. Dai, W. Shen, Z. Ji, and D. Xie, "Cost-sensitive and sequential feature selection for chiller fault detection and diagnosis," *International Journal of Refrigeration*, vol. 86, pp. 401–409, 2018.

[31] Y. Luo, Y. Duan, W. Li, P. Pace, and G. Fortino, "A novel mobile and hierarchical data transmission architecture for smart factories," *IEEE Transactions on Industrial Informatics*, vol. 14, no. 8, pp. 3534–3546, 2018.

[32] R. Islam, S. A. Khan, and J.-m. Kim, "Discriminant feature distribution analysis-based hybrid feature selection for online bearing fault diagnosis in induction motors," *Journal of Sensors*, vol. 2016, Article ID 7145715, 16 pages, 2016.

[33] M. Perez, D. M. Rubin, T. Marwala, L. E. Scott, and W. Stevens, "A population-based incremental learning approach to microarray gene expression feature selection," in *2010 IEEE 26-th Convention of Electrical and Electronics Engineers in Israel*, pp. 10–14, Eilat, Israel, 2010.

[34] S. Baluja, "Population-based incremental learning: a method for integrating genetic search based function optimization and competitive learning," Technical Report Carnegie-Mellon University Pittsburgh PA Department of Computer Science, 1994.

[35] Z. Zhu, S. Jia, and Z. Ji, "Towards a memetic feature selection paradigm [application notes]," *IEEE Computational Intelligence Magazine*, vol. 5, no. 2, pp. 41–53, 2010.

[36] M. Pelikan, D. E. Goldberg, and F. G. Lobo, "A survey of optimization by building and using probabilistic models," *Computational Optimization and Applications*, vol. 21, no. 1, pp. 5–20, 2002.

[37] A. Lipowski and D. Lipowska, "Roulette-wheel selection via stochastic acceptance," *Physica A: Statistical Mechanics and its Applications*, vol. 391, no. 6, pp. 2193–2196, 2012.

[38] J. Lee, H. Lim, and D.-W. Kim, "Approximating mutual information for multi-label feature selection," *Electronics Letters*, vol. 48, no. 15, pp. 929-930, 2012.

[39] A. Cano, J. M. Luna, E. L. Gibaja, and S. Ventura, "LAIM discretization for multi-label data," *Information Sciences*, vol. 330, pp. 370–384, 2016.

[40] J. Demsar, "Statistical comparisons of classifier over multiple data sets," *Journal of Machine Learning Research*, vol. 7, pp. 1–30, 2006.

[41] O. J. Dunn, "Multiple comparisons among means," *Journal of the American Statistical Association*, vol. 56, no. 293, pp. 52–64, 1961.

Pencil-on-Paper Sensor for Water Detection

Zhiyuan Zhu,[1] Kequan Xia,[1] Zhiwei Xu,[1] and Hongze Zhang[2]

[1]*Ocean College, Zhejiang University, Zhoushan 316021, China*
[2]*Nanjing Electronic Devices Institute, Nanjing 210016, China*

Correspondence should be addressed to Zhiwei Xu; xuzw@zju.edu.cn and Hongze Zhang; zhanghongze007@pku.edu.cn

Academic Editor: Salvatore Pirozzi

This paper sets out a simple and affordable method for the manufacture of a water sensor involving drawing marks on paper using a flexible pencil. The sensor indicates flexible endurance variations upon exposure to water or a drying surface. It can be employed for the detection of water substances in ethanol. In addition, an experiment suggests the usefulness of the sensor for the sensing of breath. This shows the immense potential applications of the sensor for wearable and also health-inspecting electronic gadgetry.

1. Introduction

Extensive research is being carried out into the development of affordable and adaptable sensing gadgets [1–5]. Owing to their minimalism, affordability, and the rapidity with which they can be manufactured, paper sensors have been garnering more attention [6–9]. For instance, Gandhiraman et al. are working on scalable and inexpensive single-use paper sensors for use in the field of DNA detection on the basis of chemical vapor deposition [9]. In addition, microfluidic construction, jet printing, and photolithography have been studied with an eye to the manufacture of a cheap and straightforward paper sensor. However, an increase in the need to utilize special inks, instruments, and templates with these methods has also been noted.

Fortunately, there is a technique whereby a pencil-on-paper sensor can be made through hand drawing on paper using a graphite pencil. The process makes the manufacture of sensors immensely simple and inexpensive. In addition, pencil-on-paper sensors have been used for a variety of types of analysis, for instance, bacterial [10], chemical compound [11], and gas [12] analysis. Nonetheless, pencil-on-paper sensors can potentially be used in other areas too.

Conversely, the water distribution system requires supplying essential amount of drinkable water at the demand points. Water leaks from the system are likely to give rise to issues and substantial water losses. It is substantially pivotal for water sensing that appears intensively demanding because of the polar character of water; moreover, what currently is being put to use is quite costly, together with being impractically implementable. Dichiara et al. performed a successful integration of nanomaterials in the paper, which are capable of electronic conductance, together with sensing water [13]. Nevertheless, there are necessaries for exploring extensively accessible low-cost commercial substances, together with a straightforward affordable manufacture, in particular, with the use of the pencil-on-paper methodology.

A flexible pencil can leave a graphite/polymer composite trace on paper [11] that can be used to sense the bases of variations in endurance. In this article, a straightforward and affordable approach a water-analytical sensor is hand-drawn onto paper using a flexible pencil. On its exposure to water, the sensor indicates the reversible endurance charges, making it utilizable for the detection of water in ethanol. Moreover, since it can detect water vapor in respiration, the sensor can be used to sense breathing. This suggests its large potential applicability in the area of wearable electronic gadgetry used for health inspection.

2. Experimental, Results, and Discussions

To manufacture our sensor, we drew on office paper with a graphite flexible pencil (see Figure 1(a)). The resultant drawing, ruler-guided in the orthogonal direction, formed a thin layer of graphite/polymer on the office paper. The pencil-trace was 2 cm wide and 4 cm long. Figure 1(b) is an image

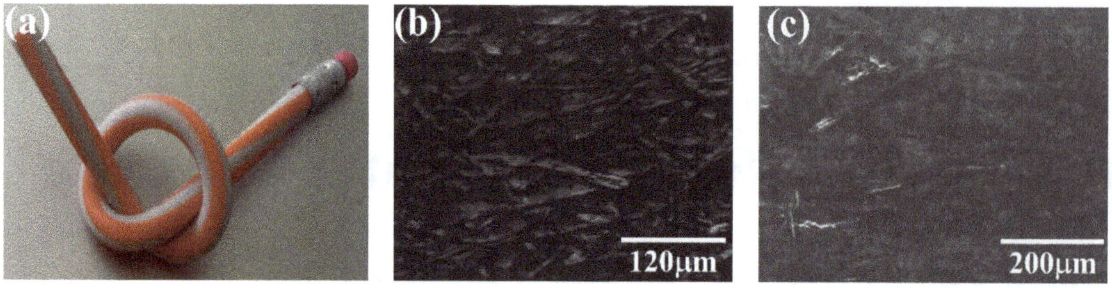

(a) (b) (c)

FIGURE 1: (a) Photograph of graphite flexible pencil, (b) SEM image of the graphite/polymer traces, and (c) SEM image of the graphite/clay trace.

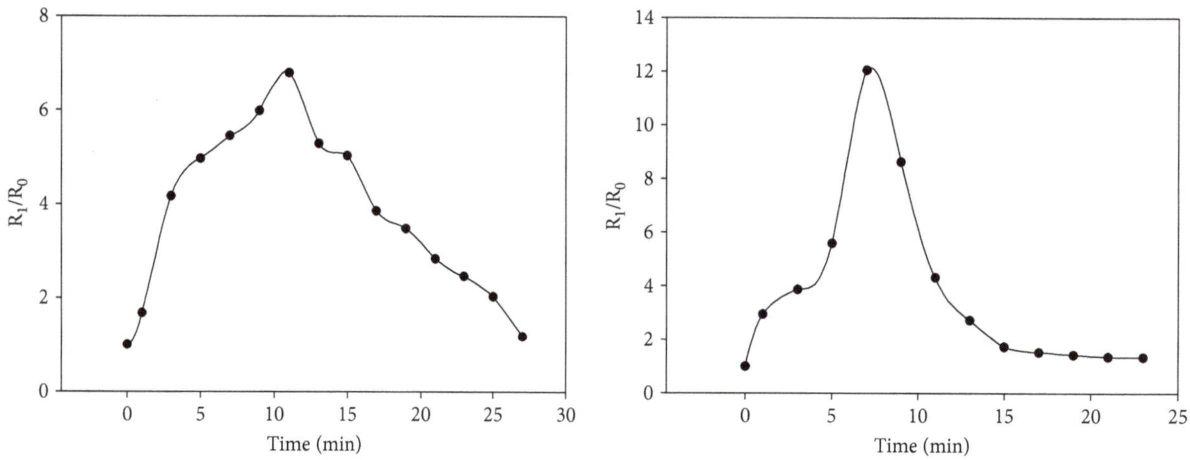

(a) (b)

FIGURE 2: R_1/R_0 versus time for (a) flexible pencil and (b) commercial 2B pencil.

of the graphite/polymer traces on the paper obtained using a scanning electron microscope (SEM).

By comparison, the pencil-on-paper sensor uses a commercial 2B pencil (with pencil leads comprising both graphite and clay) of the same parameters. Figure 1(c) is an image of the graphite/clay traces on the paper obtained using SEM.

The edges of the rectangular copper conductive tape were then used to fix the contact wires annexed to the pencil-trace. Measurement of the electronic characteristics of the strain sensors was performed using the FLUKE 8808A multimeter. Resistance was also measured. R_0 shows this measured resistance. Tap water was dropped onto the central area (2 cm × 2 cm) of the pencil-on-paper sensor. It was then left to dry. R_1 shows the respective resistances subsequent to the application of water. Figures 2(a) and 2(b) show the dependence of R_1/R_0 on the flexible pencil and the commercial 2B pencil, respectively. The sensor indicated the reversible endurance variations produced upon exposure to water or drying. Compared with the clay/graphite pencil sensor, the water sensor method showed larger endurance variations, in addition to a shorter recovery time.

The measured resistance R_1 subsequent to water trigger can be expressed as follows:

$$R_1 = 2 \times R_c + R_g, \tag{1}$$

where R_c is the contact resistance of the wire/copper tape with pencil graphite for one electrode; R_g is the resistance of the graphite resistor.

The total contact resistance comes from the constriction resistance and tunnel resistance. If the contact area is circular with a radius α, the calculation of contact resistance can be concluded from [14, 15].

$$R_c = \frac{\rho_1}{4\alpha} + \frac{\rho_2}{4\alpha} + \frac{\sigma}{\pi\alpha^2}, \tag{2}$$

where ρ_1 and ρ_2 are, respectively, the bulk resistivity of two contact members; σ is the tunnel resistivity.

It is believed that the resistance of the graphite resistor originates from graphite grain resistance and associated boundary resistance [16]. It is assumed that free carriers are trapped in the boundaries between graphite grains, creating a potential barrier which hinders the motion of carriers between graphite grains. Since the resistance of graphite grains is far smaller than that of the boundaries [17], the resistance of the graphite resistor can be approximated as boundary resistance, which is described as follows [18]:

$$R \sim \exp\left(\frac{\varphi}{kT}\right), \tag{3}$$

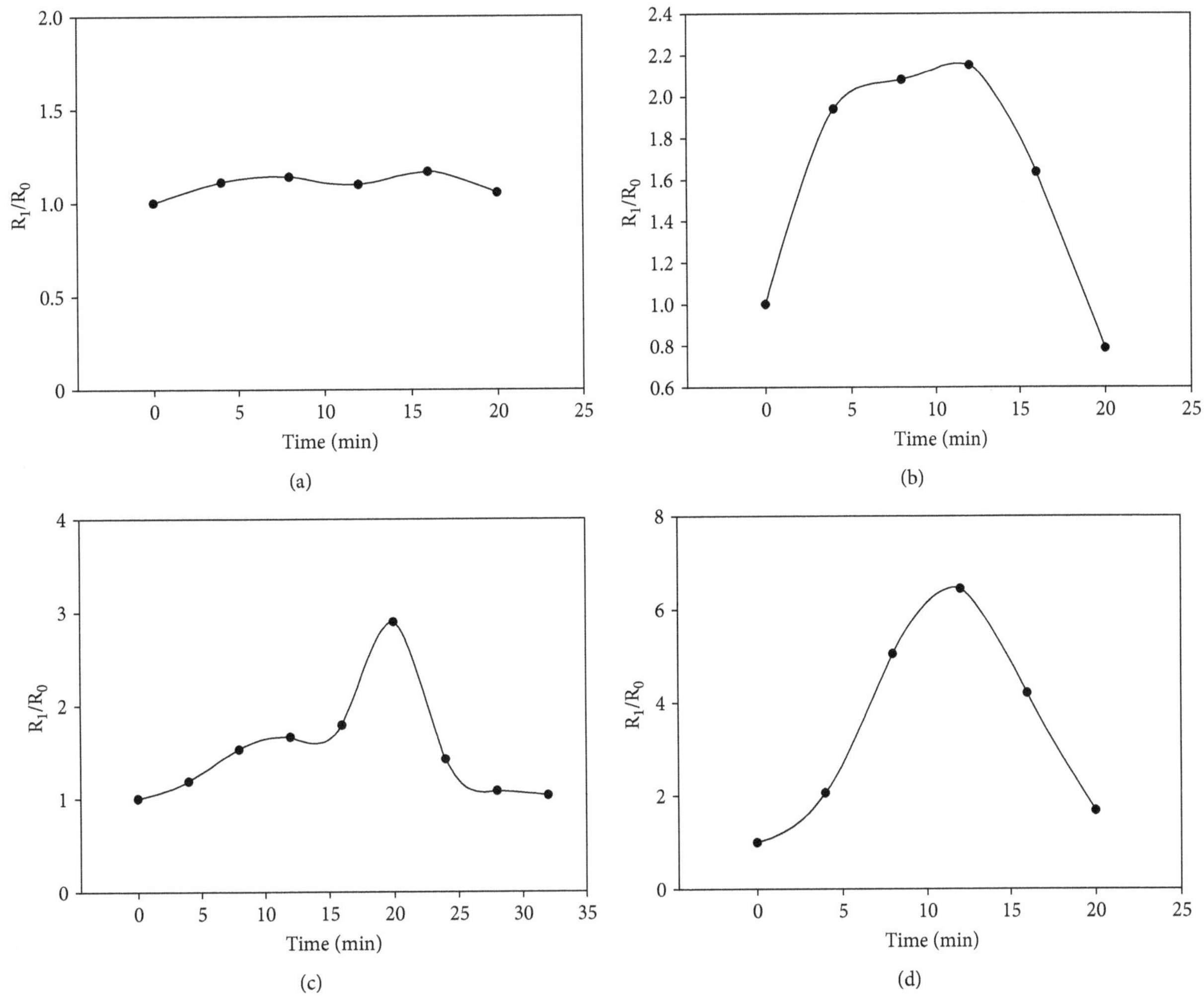

FIGURE 3: Resistance response with water volume fractions of (a) 0%, (b) 5%, (c) 30%, and (d) 60% of ethanol solution.

where k is the Boltzmann constant, φ is the potential barrier which impedes the motion of carriers between graphite grains, and T is the temperature. It is indicated that the application of the chemical substance increased the potential barrier.

It was observed that upon application of the pencil to the paper, graphite elements flaked off and remained attached to the paper fibers. Pencil traces may therefore explain the conductive thin layers produced out of the percolated graphite particle/blinder network on the paper. They may be resistant to both arbitrary formations and layouts. On the other hand, water uptake mainly takes place only in the amorphous areas of the paper cellulose microfibers [13]. The swelling of the paper substrate alters the conductive network of graphite particles, which decreases the potential barrier φ of neighboring graphite particles. In addition, the bulk resistivity of paper is decreased due to water absorption. When water molecules enter or exit the amorphous area of the paper, the cellulose chains either shift separately or move closer together, altering the electron shift through variation in the intertube space that exists between surrounding graphite particles over or under the tunneling space. This gives rise to the exclusive reversibility of the electrical endurance of the compounds.

Therefore, the measured resistance R_1 is decreased according to equation (1)–(3). It can be concluded that, in comparison with the graphite particle/clay networks, the graphite particle/polymer networks undergo larger graphite particle movement. This can be ascribed to the limited bonding between polymer blinder and paper fiber.

Figure 3 shows the endurance response demonstrated at water volume fractions of 0%, 5%, 30%, and 60% of the ethanol solution, respectively. As Figure 3 shows, no substantial resistance changes were observed in the application of the unadulterated ethanol. Nevertheless, on the increase of water content, the resistance exhibited a dramatic increase. The key findings show that the sensor can be utilized in one-time and affordable applications, for example, in the biomedical field.

The sensor sensitivity is defined as the maximum R_1/R_0 per unit change in the water content in ethanol. The data contained in Figures 2 and 3 show sensitivity of approximately 0.0581 per percent of water content in ethanol. The results also show that the sensor responses as expected when water percentage is 5%, indicating that water percentage resolution (the minimal water percentage which can be detected) of our sensor is less than 5%.

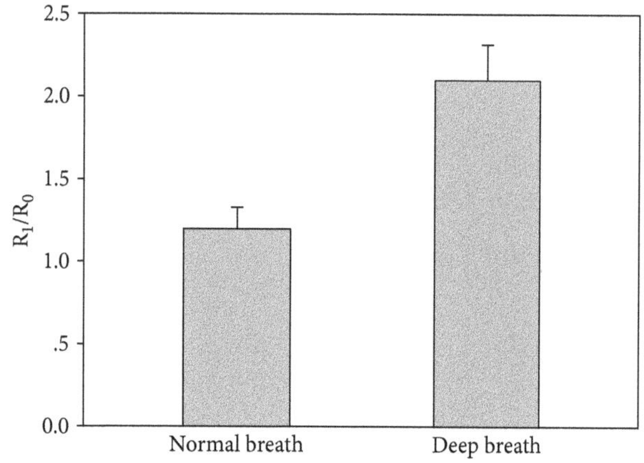

FIGURE 4: Resistance changes subject to normal breaths and deep breaths.

We positioned the paper-based water sensor in the textile mechanism masks for the purpose of demonstrating their application in the breathing sensing. The functional textile mask was joined with the multimeter, tracking the breathing function of the subjects after 2 minutes of breathing. On exhalation, full humidification of the breath was observed. This correspondingly augmented the quantity of water on the sensor. Figure 4 shows the resistance changes subjected to the standard and deep breaths. It can be concluded that there appear to be variations in R_1/R_0 between the standard breath and deep breath. The latter can be applied in human breathing inspection.

3. Conclusion

In this paper, a flexible pencil and paper were used to manufacture a graphite-based sensor for the purpose of water detection. The sensor manifested reversible resistance variations upon exposure to water or drying, making it utilizable in the detection of water in ethanol. Combining paper sensor and textile procedure masks made it possible to detect breathing. This provides a feasible solution to the recording and analysis of breathing patterns.

Conflicts of Interest

The authors declare that they have no conflicts of interest.

Acknowledgments

This investigation received support from the Fundamental Research Funds for the Central Universities and National Natural Science Foundation of China (Grant no. 61674218, Grant no. 61731019). The authors feel obliged towards the Animal Science Experimental Teaching Center of Zhejiang University for SEM characterization. Moreover, we would like to thank the unnamed reviewer for the worthwhile judgment of this paper.

References

[1] G. Wang, W. Huang, E. Manley et al., "Aggregation control in shear-printed conjugated polymer films: implications for enhancing charge transport," *Proceedings of the National Academy of Sciences of the United States of America*, vol. 114, no. 47, pp. E10066–E10073, 2017.

[2] C. Lin, K. Davami, Y. Jiang et al., "Enhancing the stiffness of vertical graphene sheets through ion beam irradiation and fluorination," *Nanotechnology*, vol. 28, no. 29, p. 295701, 2017.

[3] G. Wang, N. Persson, P.-H. Chu et al., "Microfluidic crystal engineering of π-conjugated polymers," *ACS Nano*, vol. 9, no. 8, pp. 8220–8230, 2015.

[4] X. Ge, C. Gu, X. Wang, and J. Tu, "Deep eutectic solvents (DESs)-derived advanced functional materials for energy and environmental applications: challenges, opportunities, and future vision," *Journal of Materials Chemistry A*, vol. 5, no. 18, pp. 8209–8229, 2017.

[5] X. Zhang, M.-D. Han, R.-X. Wang et al., "Frequency-multiplication high-output triboelectric nanogenerator for sustainably powering biomedical microsystems," *Nano Letters*, vol. 13, no. 3, pp. 1168–1172, 2013.

[6] M. Fan, B. Cao, and G. Tian, "Enhanced measurement of paper basis weight using phase shift in terahertz time-domain spectroscopy," *Journal of Sensors*, vol. 2017, Article ID 3520967, 14 pages, 2017.

[7] X. Tian, H. Peng, Y. Li, C. Yang, Z. Zhou, and Y. Wang, "Highly sensitive and selective paper sensor based on carbon quantum dots for visual detection of TNT residues in groundwater," *Sensors and Actuators B: Chemical*, vol. 243, pp. 1002–1009, 2017.

[8] X. S. Zhang, M. Su, J. Brugger, and B. Kim, "Penciling a triboelectric nanogenerator on paper for autonomous power MEMS applications," *Nano Energy*, vol. 33, pp. 393–401, 2017.

[9] R. P. Gandhiraman, D. Nordlund, V. Jayan, M. Meyyappan, and J. E. Koehne, "Scalable low-cost fabrication of disposable paper sensors for DNA detection," *ACS Applied Materials & Interfaces*, vol. 6, no. 24, pp. 22751–22760, 2014.

[10] J. C. Cunningham, P. R. Degregory, and R. M. Crooks, "New functionalities for paper-based sensors lead to simplified user operation, lower limits of detection, and new applications," *Annual Review of Analytical Chemistry*, vol. 9, no. 1, pp. 183–202, 2016.

[11] C. W. Lin, Z. Zhao, K. Jaemyung, and J. Huang, "Pencil drawn strain gauges and chemiresistors on paper," *Scientific Reports*, vol. 4, no. 1, article 3812, 2014.

[12] J. Zhang, L. Huang, Y. Lin et al., "Pencil-trace on printed silver interdigitated electrodes for paper-based NO2 gas sensors," *Applied Physics Letters*, vol. 106, no. 14, article 2866, 2015.

[13] A. B. Dichiara, A. Song, S. M. Goodman, D. He, and J. Bai, "Smart papers comprising carbon nanotubes and cellulose microfibers for multifunctional sensing applications," *Journal of Materials Chemistry A*, vol. 5, no. 38, pp. 20161–20169, 2017.

[14] B. Su, *Electrical, thermomechanical and reliability modeling of electrically conductive adhesives, [Ph. D thesis]*, Georgia Institute of Technology, Atlanta, GA, USA, 2005.

[15] A. Duraj and P. Mach, "Analysis and prediction of electrical contact resistance for anisotropic conductive adhesives," in *2008 31st International Spring Seminar on Electronics Technology*, pp. 358–362, Budapest, Hungary, May 2008.

[16] Y. Gan, W. Chu, and L. Qiao, "STM investigation on interaction between superstructure and grain boundary in graphite," *Surface Science*, vol. 539, no. 1–3, pp. 120–128, 2003.

[17] J. Li, R. Jia, X. Tang, X. Zhao, and S. Li, "Enhanced electric breakdown field of $CaCu3Ti_4O_{12}$ ceramics: tuning of grain boundary by a secondary phase," *Journal of Physics D: Applied Physics*, vol. 46, no. 32, article 325304, 2013.

[18] T. Dinh, H. P. Phan, D. V. Dao, P. Woodfield, A. Qamar, and N. T. Nguyen, "Graphite on paper as material for sensitive thermoresistive sensors," *Journal of Materials Chemistry C*, vol. 3, no. 34, pp. 8776–8779, 2015.

A Novel Feature Selection Scheme and a Diversified-Input SVM-Based Classifier for Sensor Fault Classification

Sana Ullah Jan and Insoo Koo

School of Electrical Engineering, University of Ulsan, Ulsan 44610, Republic of Korea

Correspondence should be addressed to Insoo Koo; iskoo@ulsan.ac.kr

Academic Editor: Giuseppe Maruccio

The efficiency of a binary support vector machine- (SVM-) based classifier depends on the combination and the number of input features extracted from raw signals. Sometimes, a combination of individual good features does not perform well in discriminating a class due to a high level of relevance to a second class also. Moreover, an increase in the dimensions of an input vector also degrades the performance of a classifier in most cases. To get efficient results, it is needed to input a combination of the lowest possible number of discriminating features to a classifier. In this paper, we propose a framework to improve the performance of an SVM-based classifier for sensor fault classification in two ways: firstly, by selecting the best combination of features for a target class from a feature pool and, secondly, by minimizing the dimensionality of input vectors. To obtain the best combination of features, we propose a novel feature selection algorithm that selects m out of M features having the maximum mutual information (or relevance) with a target class and the minimum mutual information with nontarget classes. This technique ensures to select the features sensitive to the target class exclusively. Furthermore, we propose a diversified-input SVM (DI-SVM) model for multiclass classification problems to achieve our second objective which is to reduce the dimensions of the input vector. In this model, the number of SVM-based classifiers is the same as the number of classes in the dataset. However, each classifier is fed with a unique combination of features selected by a feature selection scheme for a target class. The efficiency of the proposed feature selection algorithm is shown by comparing the results obtained from experiments performed with and without feature selection. Furthermore, the experimental results in terms of accuracy, receiver operating characteristics (ROC), and the area under the ROC curve (AUC-ROC) show that the proposed DI-SVM model outperforms the conventional model of SVM, the neural network, and the k-nearest neighbor algorithm for sensor fault detection and classification.

1. Introduction

The sensors in industrial systems are listed as a second major source of faults after the rolling elements (e.g., bearings) on top [1–3]. These faults lead to intolerable consequences including an increase in maintenance costs, compromising the reliability of products and, even more critical, safety [3, 4]. These issues can be avoided significantly by detecting fault appearance instantly. Therefore, the output of a sensor is monitored to promptly identify the anomaly. After detecting, it is mandatory to find out the primary reason of the fault occurrence in order to implement safety measures. For this purpose, faults are categorized into predefined classes obtained from historical data, a process referred to as classification.

Recently, machine learning (ML) techniques, such as neural networks (NN), support vector machines (SVM), and k-nearest neighbors (KNN), are favored for classification problems due to an efficient performance [5–12]. However, they require a feature extraction method to overcome the curse of high dimensionality of input signals. This method characterizes the hundred- or even thousand-dimensional input signal by extracting a few features. Then, these features are used as inputs to the classifiers instead of the raw signal. Although this technique may increase the efficiency in some cases, mostly there is no or very little improvement in the performance of the classifier. A major reason is the use of features with low discriminating power between the samples of different classes. The good features are the ones that characterize the signals from all classes in a dataset such that a signal of one class is easily discriminated from others. To pick such good features from a feature pool, a feature selection (FS) algorithm is used. Therefore, the feature selection step has a dominant role in reliable performance of classifiers in pattern recognition and classification applications. Nevertheless, a combination of individually good features does not necessarily lead to a good classification performance [13]. Therefore, a good FS algorithm tries to find the best combination of features and not a combination of individually good features. Moreover, as mentioned earlier, the complexity of the system is dependent on the dimensions of input vectors in direct relations. Therefore, selecting the fewest possible features when selecting good features can lead to a classifier performance enhancement.

FS algorithms are classified as wrapper-based, embedded-based, and filter-based [14–16]. The wrapper-based methods select features by analyzing the performance of a classifier after each selection. The features that optimize the performance of the classifier are selected. These techniques require high computational resources and a long time to get the best features; even so, optimality is not ensured. Embedded-based feature selection optimizes the classifier and feature selection simultaneously. The problem with these approaches is that the features are selected for the classifier under consideration and may not be able to merge with any other classifier. In contrast, the filter-based feature selection approach selects features irrespective of classifier optimization. In this approach, mutual information (MI) is a widely used measure to select the features most relevant to the target class.

The objective of the current work is to develop an efficient detection and classification algorithm for sensor faults using a supervised ML-based classifier. For this purpose, we first propose a novel MI-based FS scheme to select the exclusive relevant features with high discriminating power from a feature pool. Then, we propose a diversified-input SVM (DI-SVM) model to reduce the dimensions of input vectors to the classifiers ultimately improving the performance.

1.1. Contributions. The major contributions of this work are summarized as follows.

(i) A filter-based FS algorithm is proposed to select a combination of features exclusively discriminating one class from others. For a target class, the measure of MI of the appointed feature on the nontarget class is subtracted from the measure of MI of the same feature on the target class. This technique selects the features having high capability of discriminating target class and, at the same time, having less sensitivity to nontarget classes. The results show the efficiency of the proposed FS scheme over other schemes which utilize the redundancy between the features in addition to relevance to measure the goodness of features.

(ii) Furthermore, we propose a DI-SVM model to reduce the dimensions of input vectors to classifiers. This model inputs a diversified combination of features to different SVMs utilized for multiclass classification. These combinations of features input to any classifier are selected by the feature selection algorithm. For example, the cth classifier is fed with a combination that is selected for the cth class. On the other hand, in a conventional way of using SVMs for multiclass classification, all classifiers are trained with the exact similar number and set of features selected for all classes. The experimental analysis shows that the proposed DI-SVM model further improves the classification performance of the conventional SVM.

(iii) The performance of the proposed methodology is analyzed using two datasets. In the first dataset, the faulty signals are obtained by keeping the index of fault insertion point fixed at 500 during simulation, whereas the second dataset is obtained using faulty signals with fault insertion point varying from 0 to 1000 in each sample. The latter case is used to replicate a practical scenario in industrial systems. The results show that it is more challenging to detect and classify faults in the second case. However, the framework of the proposed FS scheme and DI-SVM achieves satisfactory results in this case also.

(iv) A series of five experiments are performed to compare the performance of the proposed DI-SVM model with those of the conventional SVM, NN, and KNN classifiers. The first two experiments are performed using the conventional SVM, NN, and KNN classifiers without applying a feature selection. In the next experiment, we prove the efficiency of the proposed FS algorithm using measures of accuracy, receiver operating characteristics (ROC), area under the ROC (AUC-ROC), and scatter plots in selected feature spaces. In the last two experiments, we deploy the proposed FS and DI-SVM model along with the abovementioned classifier and compare the performances. The results show that DI-SVM outperforms all the counter three classifiers.

This paper is organized as follows. A review of related works and the proposed feature selection algorithm are presented in Section 2. The theory of the classification model and proposed DI-SVM model is presented in Section 3. The experimental results are illustrated in Section 4. Section 5 presents the discussion about the experimental results, and finally, the study is concluded in Section 6.

2. Mutual Information-Based Feature Selection

2.1. Preliminaries. Before presenting a review of the related works, we present fundamental background knowledge about the MI-based feature selection schemes. For two random variables, x and y, the MI is defined as follows [17]:

$$I(x;y) = \iint p(x,y) \log \frac{p(x,y)}{p(x)p(y)} dxdy, \tag{1}$$

where $p(x)$ and $p(y)$ are probability density functions of continuous random variables x and y, respectively, and $p(x,y)$ is their joint probability density function. For discrete random variables, the integration is replaced with summation as follows:

$$I(x;y) = \sum_{x\in X}\sum_{y\in Y} p(x,y) \log \frac{p(x,y)}{p(x)p(y)}, \tag{2}$$

where $p(x)$ and $p(y)$ are probability mass functions of discrete random variables x and y, respectively, and $p(x,y)$ is their joint probability mass function. The MI can be expressed in terms of entropy as follows:

$$I(x;y) = H(x) - H(x \mid y) = H(y) - H(y|x). \tag{3}$$

$H(x)$ and $H(x \mid y)$ represent the entropy and conditional entropy, defined as follows:

$$H(x) = -\sum_{r\in x} p(r) \log p(r), \tag{4}$$

$$H(x|y) = -\sum_{s\in y}\sum_{r\in x} p(r,s) \log p(r \mid s). \tag{5}$$

2.2. Related Works. In this paper, we focus on the selection criteria of features based on the measure of MI. Different FS algorithms have been proposed in the past using different criteria to measure the goodness of features using MI. Battiti [18] selected features by calculating the MI $I(x_i;c)$ of individual feature x_i with class c. To avoid the selection of redundant features, the MI measure $I(x_i;x_j)$ between two features x_i and x_j is calculated. The MI-based feature selection (MIFS) is given as [18]

$$I(x_i;c) - \beta \sum_{x_j\in S} I(x_i;x_j), \tag{6}$$

where β is the regularization parameter to weight the redundancy $I(x_i;x_j)$ between a candidate feature x_i and the already-selected features $x_j \in S$. Kwak and Choi [19] proved that a large value of β can lead to the selection of suboptimal features. They improved the MIFS scheme to propose MIFS-U by making modifications as follows

$$I(x_i;c) - \beta \sum_{x_j\in S} \frac{I(x_j;c)}{H(x_j)} I(x_j;x_i). \tag{7}$$

In both of these algorithms, the selection of a feature is dependent on the user-defined parameter β weighting the importance of redundancy between features. If β is selected too large, the feature selection algorithm will be dominated by the redundancy factor. The authors of [13] proposed parameter-free criteria of feature selection, named as minimal-redundancy-maximal-relevance (mRMR), given by

$$I(x_i;c) - \frac{1}{|S_{m-1}|} \sum_{x_j\in S_{m-1}} I(x_j;x_i), \tag{8}$$

where m is the number of features to be selected.

Estevez et al. [20] pointed out that the right-hand sides of MIFS and MIFS-U, in (6) and (7), will increase with the increase in the cardinality of the selected feature subset. This will result in the dominating right-hand side, thus forcing the FS to select nonredundant features. This may lead to the selection of irrelevant features before relevant features. Another problem is that there is no technique to optimize β and its value depends highly on the problem under consideration. Although the mRMR partly solves the first problem in MIFS and MIFS-U, the performance is still comparable with those of MIFS and MIFS-U [21].

A normalized MIFS (NMIFS) is proposed to address the above problems and is given as [20]

$$I(x_i;c) - \frac{1}{|S_{m-1}|} \sum_{x_j\in S_{m-1}} \frac{I(x_j;x_i)}{\min\{H(x_j), H(x_i)\}}. \tag{9}$$

An improved version of NMIFS (I-NMIFS) is given in [21] as follows:

$$\frac{I(x_i;c)}{\log_2(|\Omega_c|)} - \frac{1}{|S_{m-1}|} \sum_{x_j\in S_{m-1}} \frac{I(x_j;x_i)}{\min\{H(x_j), H(x_i)\}}. \tag{10}$$

The problem with NMIFS and I-NMIFS is that they both rely on the measure of entropy of both the selected and under-observation features. The entropy of a feature is totally dependent on the number of samples from each class in the dataset. If one class has a high number of samples than the other, this situation will have an effect on the value of entropy of a feature leading to degradation of performance of the classifiers.

Furthermore, all these algorithms rely on the measure of mutual information of a to-be-selected feature with a target class (relevance) and with an already-selected feature (redundancy). In this paper, we focus only on the relevance of a feature with a target class and nontarget class to assess the goodness of a candidate feature. Indeed, a feature having high MI (i.e., relevance) with a target class might not be a good choice due to a high relevance to at least one nontarget class as well. This may lead to degrading rather than improving the performance of the classifier. Therefore, we propose a feature selection scheme that relies only on the measures of relative relevance of a feature with different classes in the dataset. The measure of redundancy is eliminated making the scheme simplified yet achieving satisfactory performance compared to that of the FS schemes utilizing the redundancy factor also in the selection criteria.

2.3. The Proposed Feature Selection Scheme. Given a training dataset composed of N samples and M features, the aim is to select m out of M features for each class c, individually, where $c \in \{1, 2, \dots, C\}$ and C is the number of total classes. The features having maximum relevance to class c and minimum relevance to remaining $c-1$ classes are most suitable for discriminating class c. Relevance is usually described by MI or correlation, of which the MI is a widely adopted measure to indicate the dependence between two random variables. Generally, the mutually independent variables have zero MI between them. The higher the dependency of two variables, the higher the MI.

The proposed feature selection algorithm selects features by considering their measures of MI with both target class and nontarget class. The m features having maximum relevance with a target class are fit for target class, but it may also have high relevance with a nontarget class, making this feature less discriminating. Therefore, the features having maximum relevance with a target class and minimum relevance with nontarget classes are selected. This approach selects the subtlest features among the feature sets for the target class exclusively. The relevance of features exclusive for the target class is obtained by calculating the difference of the MI of the selected feature vector $\mathbf{x}_i$, where $i = 1, 2, \dots, M$, on target class c and the MI of $\mathbf{x}_i$ on nontarget class c'. Mathematically,

$$d = I(x_i\,; c) - I\left(x_i\,; c'\right), \tag{11}$$

where $c' = C/c$ is the set of nontarget classes. The m among M features with maximum values of d is selected for target class c. The MI of x_i on class c is calculated in terms of entropy as follows:

$$I(x_i\,; c) = H(c) - H(c|x_i), \tag{12}$$

where $H(c)$ is the entropy of class c and $H(c \mid x_i)$ is the conditional entropy of c given x_i. Assuming $0.\log 0 = 0$ from continuity, $H(c)$ and $H(c \mid x_i)$ are defined as

$$\begin{aligned} H(c) &= -p(c)\log p(c) - p\left(c'\right)\log p\left(c'\right), \\ H(c|x_i) &= -\sum_{x \in x_i} p(x) \sum_{c=1}^{C} p(c|x)\log p(c|x), \end{aligned} \tag{13}$$

where $p(c)$ and $p(c')$ are the probabilities of the target class and nontarget class, respectively, and can be calculated as $p(c) = n_c/N$ and $p(c') = (N - n_c)/N$, respectively, where n_c is the number of samples corresponding to class c in the training dataset. The conditional probability $p(c \mid x)$ is given as

$$p(c|x) = \frac{p(c)p(x|c)}{\sum_{s=1}^{C} p(s)p(x|s)}, \tag{14}$$

where $s = 1, 2, \dots, C$ correspond to the set of all C classes.

The probability mass function $p(x \mid c)$ can be estimated using the Parzen window method as follows [13]:

$$\hat{p}(x|c) = \frac{1}{N_c} \sum_{i \in I_c} \Phi(x - x_i, h), \tag{15}$$

where I_c represents the set of indices of training samples corresponding to class c and $\Phi(\cdot)$ is the window function. The commonly used Gaussian window function is expressed as

$$\Phi(z, h) = \frac{\exp\left(-\left(z^T \Sigma^{-1} z/2h^2\right)\right)}{(2\pi)^{M/2} h^M |\Sigma|^{1/2}}, \tag{16}$$

where Σ is the matrix covariance of the M-dimensional vectors of random variables, $z = x - x_i$, and the width parameter h is given by $h = \beta/\log N$ for positive constant β. The appropriate selection of $\Phi(\cdot)$ and a large h can converge (16) to a true density [13]. Using (5), (6), and (7), the estimated conditional probability mass function can be obtained as

$$\hat{p}(c \mid x) = \frac{\sum_{i \in I_c} \exp\left(-\left((x - x_i)^T \Sigma^{-1} (x - x_i)/2h^2\right)\right)}{\sum_{k=1}^{C} \sum_{i \in I_k} \exp\left(-\left((x - x_i)^T \Sigma^{-1} (x - x_i)/2h^2\right)\right)}. \tag{17}$$

A pseudocode of steps in the proposed FS algorithm is given in Algorithm 1. A training dataset $\{x_{ij}, c_i\}_{i=1}^{N}$ with N samples and M features, where x_{ij} is the ith element of the jth feature vector, is given input to the selection scheme. A matrix $\mathbf{X}$ of size $C \times m$ is the output of the algorithm where each cth row contains the m features selected for targeting class c. In the first step, a target class c is selected among the sets of all classes $1, 2, \dots, C$. Then, the respective set of nontarget classes is obtained. In the next step, each feature vector is selected simultaneously. For each jth feature vector, the elements from class c and class c' are stored in x_j^c and x_j^t, respectively. The mean values of these vectors are stored in x_j^c and x_j^t. After a series of calculation of the estimated conditional probability, entropy, and conditional entropy, d is calculated for given class c using (2). Then, the m features having maximum values of d in the final vector are selected and stored in the cth row of the matrix $\mathbf{X}$. Similarly, m features are selected for each class and stored in the final matrix $\mathbf{X}$.

3. Classification Models

A generic framework of classification methodology using supervised machine learning-based classifiers is given in Figure 1. This methodology is performed in two phases: a training phase and a testing phase. In the training phase, a classifier is trained using a set of historical (or training) data.

INPUT: $\{x_{ij}, c_i\}_{i=1}^{N}$, Training dataset; $j = \{1, 2, \ldots, M\}$, Number of features; $c \in \{1, 2, \ldots, C\}$, Number of classes; m, Number of features to select by algorithm.
OUTPUT: $X_{C\times m}$, Matrix of C vectors of m features selected for each class c.

for c in C **do**
  $c' \leftarrow C/c$
  for j in M **do**
    $x_j^c \leftarrow$ values of jth feature from class c
    $x_j^t \leftarrow$ values of jth feature from class c'
    $x_j^c \leftarrow$ mean of x_j from class c
    $x_j^t \leftarrow$ mean of x_j from class c'
    Obtain $\hat{p}(c \mid x_j^c)$ and $\hat{p}(c' \mid x_j^t)$ using Eq(8)
    Obtain $H(c)$, $H(c')$ using Eq(4)
    Obtain $H(c \mid x_j)$ and $H(c' \mid x_j)$ using Eq(4)
    Obtain $I(x_j ; c)$ and $I(x_j ; c')$ using Eq(3)
    Obtain $d_j^c \leftarrow I(x_i ; c) - I(x_i ; c')$ using Eq(2)
  end for
  Indices of m maximum values of $d^c \rightarrow X_{c\times m}$
end for

Algorithm 1: Proposed feature selection algorithm.

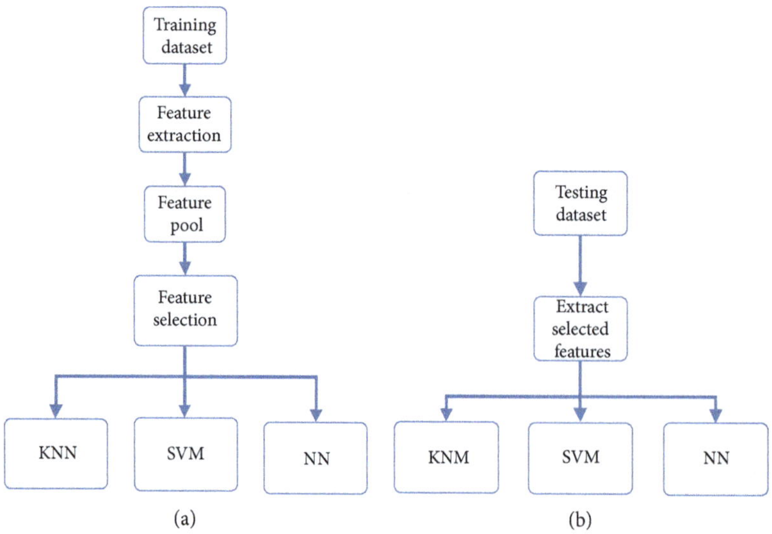

Figure 1: (a) Training phase and (b) testing phase of the generic fault classification framework for machine learning-based classifiers.

The steps include feature extraction from raw signals to obtain a feature pool. Then, an FS algorithm selects the best features in terms of discriminating power between data from different classes. Finally, the sets of the selected features are used to train the classifier, as illustrated in Figure 1(a). The performance of the trained classifier is evaluated using a set of unobserved test signals in the testing phase. The classifier is loaded with the particular features, selected in the training phase and extracted from the test signals as shown in Figure 1(b).

The commonly adopted supervised machine learning-based classifiers include NN, KNN, and SVM-based classifier. A short introduction to each one of these classifiers is given here for completeness. Then, the proposed DI-SVM-based classifier model is presented.

3.1. Neural Network. A neural network (NN), also known as feedforward NN or multilayer perceptron (MLP), is a machine learning tool for classification and regression problems. The structure of NN is inspired from the biological nervous system. It consists of an input layer, an output layer, and at least one hidden layer. An input signal is propagated forward from the input layer to the output layer to obtain the weight vectors of each layer. During this stage, the neurons in hidden and output layers learn about the input patterns from different classes. The neurons are triggered with a differentiable nonlinear activation function. A commonly used sigmoid activation function is given as

$$y_i = \frac{1}{1 + \exp(-v_i)}, \tag{18}$$

where v_i is the weighted sum of all synaptic inputs and the bias to the ith neuron and y_i is the output of the same neuron.

A backpropagation algorithm is used to estimate the difference between the predicted output of the network and the true result. An estimate of the gradient vector containing the gradients of the error surface with respect to the input weights of neurons in hidden and output layers is computed. The gradient vector is then passed backward from the output layer to the input layer to update the weight vectors of each layer. In this paper, a Levenberg-Marquardt (LM) algorithm is used to update the weights of neurons in backpropagation [22].

3.2. k-Nearest Neighbor. A k-nearest neighbor (KNN) is a simple classifier that takes a decision about the test input by simply looking at the k-nearest points in the training set [23]. The classifier counts the number of members from each class in this set of k-nearest neighbors to test the input and classifies the test signal in the class having the highest number of members. The Euclidean distance is commonly used as a distance metric to obtain the set of the nearest neighbors.

3.3. Support Vector Machine. The SVM is a well-known classifier primarily addressing two-class classification problems. A linear hyperplane is drawn as a decision boundary between the data points of the two classes. The optimal decision boundary is the one that maximizes the weight vector $\|\mathbf{w}\|$ [3, 24].

For a given set of training samples, x_i, and discrete class labels, c_i, the SVM tries to obtain the optimal weight vector, $\mathbf{w}$, using the following optimization problem:

$$\begin{aligned} &\min \quad \Phi(\mathbf{w}, \xi) = \frac{1}{2}\|\mathbf{w}\|^2 + \alpha \sum_{i=1}^{N} \xi_i, \\ &\text{such that} \\ &c_i\left(\mathbf{w}^T x_i + b\right) \geq 1 - \xi_i, \quad \text{for } i = 1, 2, \ldots, N, \xi_i \geq 0 \ \forall\, i, \end{aligned} \tag{19}$$

where α is the cost parameter and ξ_i represent the slack variables. A test observation, $\mathbf{x}_j$, is classified by the decision function, given as follows:

$$f(\mathbf{x}_j) = \operatorname{sign}\left(\sum_i \alpha_i K(\mathbf{x}_i, \mathbf{x}_j) + b\right), \tag{20}$$

where $\mathbf{x}_i$ represents the set of support vectors obtained in the training phase and $K(\mathbf{x}_i, \mathbf{x}_j)$ is the kernel function explained in the next section.

For a multiclass (more than two classes) classification problem, an equal number of binary SVMs to the number of classes in a dataset are utilized, where each SVM is used in a one-versus-rest approach [10, 25–27]. The cth classifier is trained with the samples from the cth class as a positive class and the samples from the remaining classes as a negative class.

3.3.1. Kernel Functions. To reduce the complexity of the machine in linearly nonseparable patterns, the data points are mapped on a higher-dimensional feature space where a linear decision surface can be drawn to classify the datasets of different classes. Inner-product kernels are used to perform this job. This is the key technique of SVM to solve the optimization problems of the input feature space in a higher-dimensional feature space. In such a way, the SVM deals with linearly nonseparable problems. For mapping function $\varphi(\cdot)$, the kernel function of two vectors $\mathbf{x}$ and $\mathbf{y}$ is given as $K(\mathbf{x}, \mathbf{y}) = \varphi(x)^T \varphi(y)$. A radial-basis function is a commonly used kernel function, defined as

$$K(\mathbf{x}, \mathbf{y}) = \exp\left(-\frac{\|x - y\|^2}{2\sigma^2}\right). \tag{21}$$

3.4. Proposed Diversified-Input SVM Model. The training and testing phases of the proposed diversified-input SVM (DI-SVM) model for sensor fault detection and classification are illustrated in Figures 2(a) and 2(b), respectively. Given a training dataset composed of N labelled raw signals from a sensor, the first step is to extract M features to obtain the feature pool of size $N \times M$. The term raw signal refers to the signal obtained from a sensor, which is not preprocessed (i.e., any preprocessing technique, such as feature extraction and standardization, is not applied). In other words, the signals from the sensor in the earliest form are the raw signals, whereas the M-dimensional row vector in the feature pool, which is composed of the M features and extracted from a raw signal, is interchangeably referred to as a sample or an observation. After feature extraction to obtain a feature pool, the FS algorithm selects m out of M features for each class, separately. To perform this, the FS algorithm first selects a target class, c, from the given set of classes $1, 2, \ldots, C$, where C is the total number of classes. Then, the goodness of each feature is assessed using a selection criterion to take a decision about selecting the features for the cth class. Subsequently, the next class is selected as the target class, and the feature selection steps are repeated. In this way, m features are selected for each class in the dataset. Finally, the column vectors in the feature pool of training data corresponding to the features selected for each single class are given inputs to C different SVMs, separately, for training. For example, if the two features x and y are selected targeting class c, then the vectors $\mathbf{x}$ and $\mathbf{y}$ containing the values of x and y from the feature pool are input to the SVM to specifically identify class c. Furthermore, the cth classifier is trained with observations of the cth class labelled as a positive class and the remaining $c - 1$ classes labelled as a negative class.

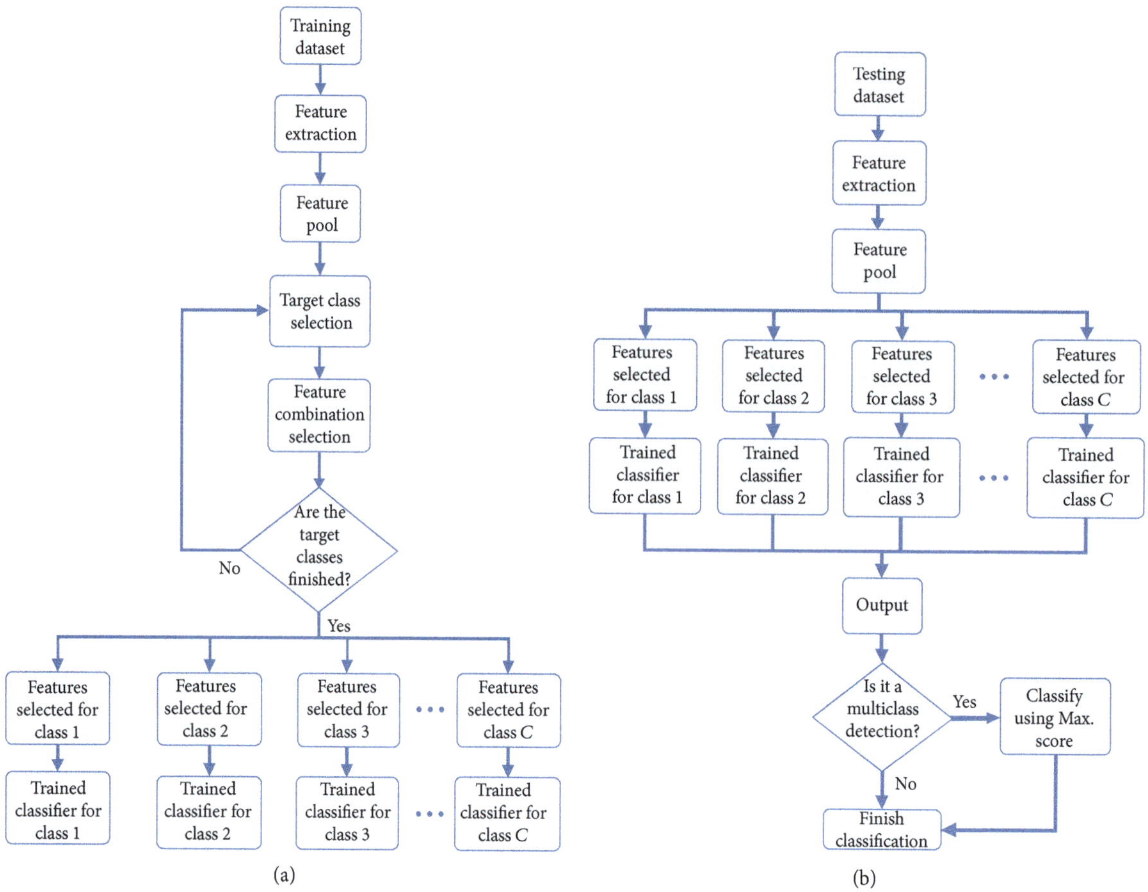

FIGURE 2: Framework of (a) the training phase and (b) testing phase of the proposed diversified-input support vector machine (DI-SVM) model.

In such a way, C different SVMs are trained for C-class classification problems, where the observations input to each SVM are composed of diversified-feature combinations selected by the FS algorithm.

To classify a test raw signal, first, the features selected in the training phase are extracted from the raw signal to obtain the feature pool. The trained C SVMs are used to classify the test observation. Input to the SVMs is given with the same feature combinations as in the training phase. The output class labels and respective scores of all the SVMs are stored in the output vector. It is checked for any observation classified into multiclasses, that is, if any observation is classified positive by more than one SVM in the classifier. In this case, the decision of the SVM with the maximum output score is taken into account, as shown in Figure 2(b).

3.5. *Classifier Parameters*

3.5.1. Cross-Validation. The cross-validation (CV) techniques assess generalization and overcome the overfitting problem of the classifier in the training phase. The k-fold CV technique partitions the training data into k complementary subsets to train and validate SVM k times. In each round, a new subset is used for training, and the remaining samples are used for validation. This cycle is repeated until each sample is used for training at least once. Consequently, each subset is used once for training while being used $k-1$ times for validating the classifier.

3.5.2. Standardization. Standardizing the training data prior to training the SVM may improve the classification performance [28]. This preprocessing technique is applied to prevent large valued features from dominating the small valued features in the dataset. The two commonly used ways of standardizing a dataset include the min–max (also known as normalization) and the z-score formula. In the min–max data standardization, the values of the data vector are scaled from 0 to 1. In contrast, the z-score formula scales the data vector to a standardized vector having zero mean and unit variance. Some other standardization techniques derived from these two basic methods are proposed in the literature [28, 29].

In fact, there is no obvious answer to which technique is the best. However, a basic difference between the two approaches is that the min–max method bounds the standardized data between 0 and 1. In this approach, a single outlier in the data may compress the remaining feature values toward zero. On the other hand, there is no limit on the range of standardized data using the z-score formula. This makes z-score standardization more robust

TABLE 1: Time-domain features.

Name	Definition	Name	Definition
Mean	$\mu = \frac{1}{N}\sum_{i=1}^{N} x_i$	Standard deviation	$Y_{\mathrm{STD}} = \sqrt{\frac{1}{N}\sum_{i=1}^{N}(x_i-\mu)^2}$
Root mean square	$Y_{\mathrm{RMS}} = \sqrt{\frac{1}{N}\sum_{i=1}^{N} x_i^2}$	Variance	$Y_{\mathrm{VAR}} = \frac{1}{N}\sum_{i=1}^{N}(x_i-\mu)^2$
Sign function	$Y_{\mathrm{Sign}} = \mathrm{sign}\,(\varepsilon-\mu)$	Square root of amplitude	$Y_{\mathrm{SRA}} = \left(\frac{1}{N}\sum_{i=1}^{N}\sqrt{\lvert x_i\rvert}\right)^2$
Kurtosis value	$Y_{\mathrm{KV}} = \frac{1}{N}\sum_{i=1}^{N}\left(\frac{x_i-\mu}{Y_{\mathrm{STD}}}\right)^4$	Skewness value	$Y_{\mathrm{SV}} = \frac{1}{N}\sum_{i=1}^{N}\left(\frac{x_i-\mu}{Y_{\mathrm{STD}}}\right)^3$
Crest factor	$Y_{\mathrm{CF}} = \max\frac{\lvert x_i\rvert}{Y_{\mathrm{RMS}}}$	Impulse factor	$Y_{\mathrm{IF}} = \frac{\max\lvert x_i\rvert}{(1/N)\sum_{i=1}^{N}\lvert x_i\rvert}$
Margin factor	$Y_{\mathrm{MF}} = \max\frac{\lvert x_i\rvert}{Y_{\mathrm{SRA}}}$	Form factor	$Y_{\mathrm{FF}} = \frac{Y_{\mathrm{RMS}}}{\mu}$
Kurtosis factor	$Y_{\mathrm{KF}} = \frac{Y_{\mathrm{KV}}}{\left((1/N)\sum_{i=1}^{N}x_i^2\right)^2}$	Peak-to-peak value	$Y_{\mathrm{PPV}} = \max(x_i) - \min(x_i)$

TABLE 2: Frequency-domain features.

Name	Definition	Name	Definition
Center frequency	$Y_{\mathrm{FC}} = \frac{1}{N}\sum_{i=1}^{N} X_i$	Sum FFT	$Y_{\mathrm{SFFT}} = \sum_{i=1}^{r} X_i$
RMS frequency	$Y_{\mathrm{RMSF}} = \sqrt{\frac{1}{N}\sum_{i=1}^{N} X_i^2}$	Root variance frequency	$Y_{\mathrm{RVF}} = \sqrt{\frac{1}{N}\sum_{i=1}^{N}(X_i - Y_{\mathrm{CF}})^2}$

to the outliers in the data as compared to the min–max standardization method.

In z-score standardization, the input feature vectors are scaled to a zero mean by subtracting the statistical mean from each element of the vector and normalized to have a standard deviation of 1 by dividing by the standard deviation of the input vector. Given a vector, $\mathbf{x}_i$, the formula of z-score standardization is given as

$$\mathbf{x}_i^* = \frac{x_i - \mu_{x_i}}{\sigma_{x_i}}, \tag{22}$$

where $\mathbf{x}_i^*$ is the vector of standardized values and μ_{x_i} is the mean of the feature vector $\mathbf{x}_i$ with standard deviation σ_{x_i}.

3.6. Feature Pool. A feature pool was obtained by extracting 14 time-domain and 4 frequency-domain features. The respective names and mathematical definition of these features are illustrated in Tables 1 and 2.

4. Experimental Analysis

4.1. Dataset Acquisition. The types of sensor faults considered in this work include drift, erratic, hardover, spike, and stuck faults [3]. A plot of a raw signal from each of these classes and normal classes is given in Figure 3. The fault is considered a drift fault if the output of the sensor increases linearly. The hardover fault occurred when the output of the sensor is increased from a normal value, represented by a red line in Figure 3. If the erratic fault occurs, the variance of the output signal of the sensor is increased from a routine value. In the case of spike fault occurrence, spikes are observed in the output signal of the sensor. When the output of a sensor sticks to a fixed value, then this fault type is named as stuck fault. These fault types are divided based on the data measurements and can be considered generic in different types of sensors such as the pressure sensor. These faults are referred to by many researchers, with some using different nomenclature. For instance, Yu et al. [4] and Kullaa [30] referred to erratic, stuck, and hardover faults as precision degradation failure, complete failure, and bias failure, respectively.

The training and testing datasets utilized in this work are obtained using a healthy temperature-to-voltage converter TC1047/TC1047A as follows. A set of 100 raw time signals is acquired from a sensor using an Arduino Uno microcontroller board with a serial communication between the Arduino and personal computer. Each raw signal is

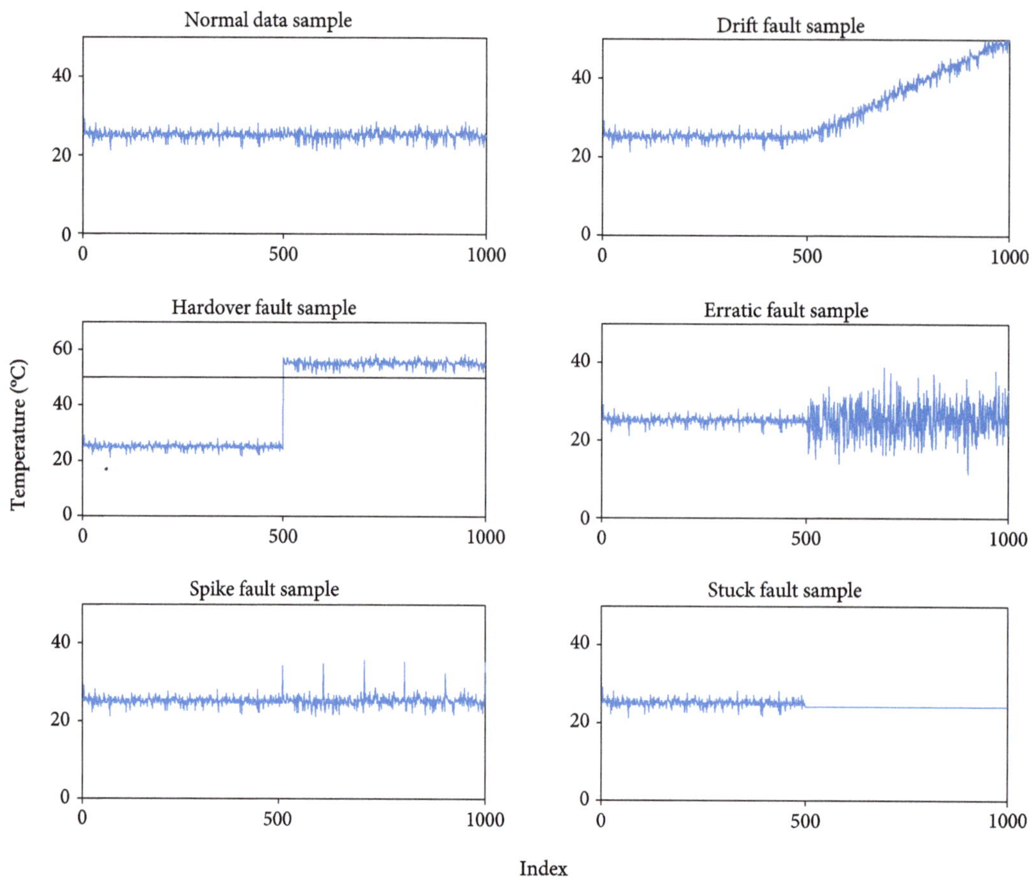

FIGURE 3: Plots of samples of all six classes from dataset 1 (with fixed fault insertion point).

composed of 1000 raw data elements from the sensor. Then, each fault type, that is, drift, hardover, erratic, spike, and stuck faults, is simulated in each stored signal obtained from a healthy sensor. In such a way, a set of 100 signals from each class, including normal and faulty classes, is obtained. The details of fault simulation in a normal data signal are as follows: a drift fault signal s_n^{Drift} is obtained by adding a linearly increasing bias term in a normal signal s_n^{Normal}, where the bias added to the nth element is n times the constant initial bias b_0. A hardover fault signal s_n^{Hardover} is obtained by adding a large constant bias value b to all elements of the normal signal. To obtain an erratic fault signal s_n^{Erratic}, a signal $\dot{s}_n$ of mean 0 and high variance, $\dot{\delta}^2 \gg \delta^{2^{\text{Normal}}}$, where $\delta^{2^{\text{Normal}}}$ is the variance of the normal signal, is added to the raw normal signal. To obtain spike fault signals s_n^{Spike}, a constant bias is added periodically to the uth elements of the normal signal, where $u = v \times \eta$ is the index of elements in the signal with $v = \{1, 2, \ldots,\}$ as a set of natural numbers and $\eta \geq 2$ as a positive integer. Finally, the stuck fault signals s_n^{Stuck} are obtained by keeping a fixed value at all indices of the normal signal. The statistical representations of these faults are given in Table 3, where $s_n^{\text{Normal}}(n = 1, 2, \ldots, 1000)$ is the nth element of the raw signal corresponding to the normal class. In conclusion, the final dataset is composed of 100 raw signals of each of the six classes. This means that 100 raw time signals of each class are included resulting in a total of 6 ∗ 100 raw time signals in the final dataset. As mentioned earlier, a raw signal is comprised of 1000 data elements. Therefore, the final dataset of raw signals has a size of 600 ∗ 1000 data points.

TABLE 3: Statistical representation of fault simulations in the normal signal.

Fault	Statistical Representation
Drift	$s_n^{\text{Drift}} = s_n^{\text{Normal}} + b_n, b_n = nb_0, b_0 = \text{constt}$
Hardover	$s_n^{\text{Hardover}} = s_n^{\text{Normal}} + b, b = \text{constt}$
Erratic	$s_n^{\text{Erratic}} = s_n^{\text{Normal}} + \dot{s}_n, \dot{s}_n \sim N(0, \dot{\delta}^2), \dot{\delta}^2 \gg \delta^{2^{\text{Normal}}}$
Spike	$s_n^{\text{Spike}} = s_n^{\text{Normal}} + b_n,$ $b_n = \begin{cases} b, n = v \times \eta, v = \{1, 2, \ldots,\}, \eta = \text{constt} \\ 0, \text{otherwise} \end{cases}$
Stuck	$s_n^{\text{Stuck}} = \alpha, \alpha = \text{constt}$

In dataset 1, the fault insertion point is kept fixed at index = 500 of each fault sample, illustrated in Figure 3. The

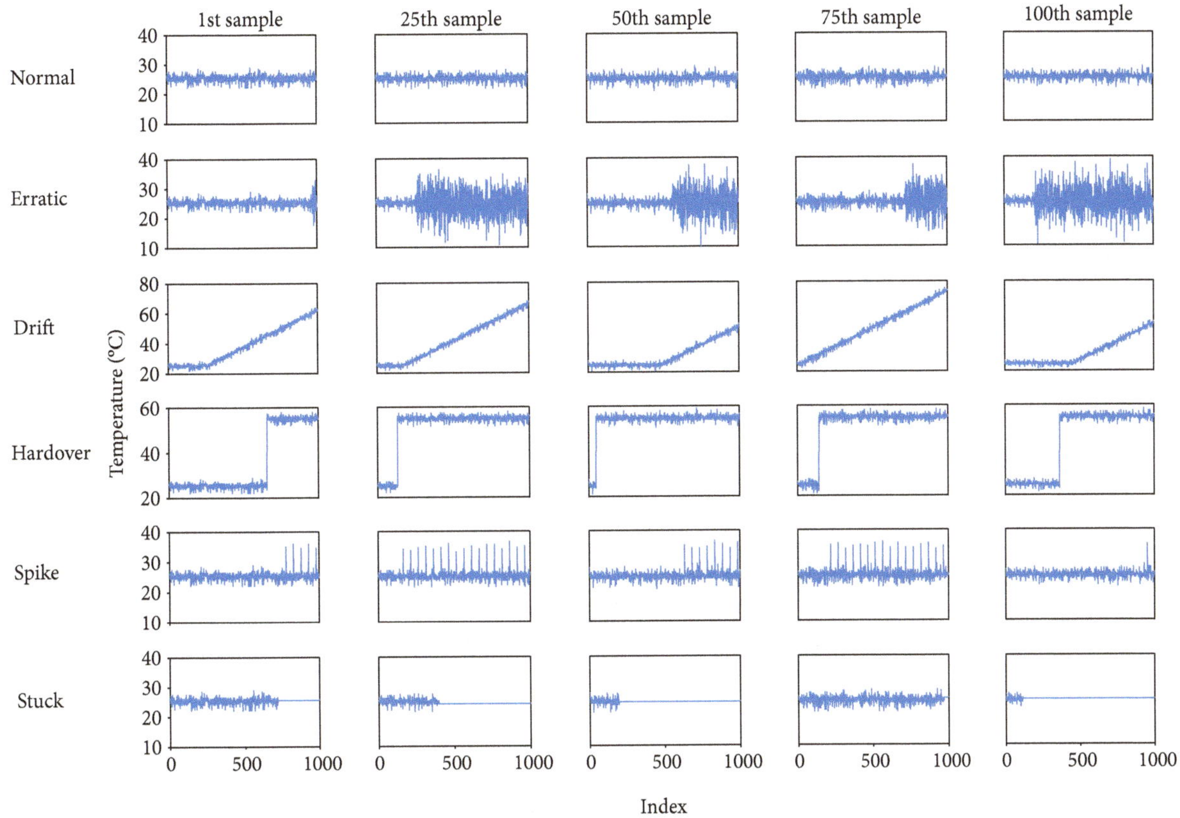

FIGURE 4: Plots of the 1st, 25th, 50th, 75th, and 100th sample of all six classes from dataset 2 (with a variable fault insertion point).

fault insertion point is the index of the element of the signal from where the fault occurrence starts. A similar way of fault simulation is adopted to obtain dataset 2 with the same size as the first dataset; however, the fault insertion point is a variable index, chosen randomly, of the signal. This case is considered to replicate a more practical scenario of the fault occurrence time in sensors. The time-domain plots of the 1st, 25th, 50th, 75th, and 100th sample from each class from the second dataset with a random fault insertion point are given in Figure 4.

4.2. Performance Evaluation Metrics. The performance of the classification models is analyzed and compared using the following metrics.

4.2.1. Accuracy. The accuracy of a classifier is the ratio of the number of true predictions to the total number of observations in the test set. Mathematically,

$$\text{accuracy} = \frac{\text{TP} + \text{TN}}{N}, \tag{23}$$

where TP and TN denote the number of true positive and true negative predictions and N is the total number of observations under the test.

4.2.2. Receiver Operating Characteristics. A receiver operating characteristic (ROC) graph is used to visualize, analyze, and select classifiers based on the performance. A two-dimensional graph represents a tradeoff between the true positive rate plotted on the y-axis and the false positive rate on the x-axis.

4.2.3. Area under the ROC. A comparison of two-dimensional ROC curves of two classifiers becomes difficult if the difference in the performances is too small. It may be needed to represent the performance with a scalar value. Calculating the area under the ROC curve (AUC-ROC) is a common technique used for this purpose [31].

4.3. Classifier Implementation. The SVM-based classifiers in both conventional and proposed models are trained and tested using MATLAB in a one-versus-rest manner. The cost and box constraints of misclassifications in the SVM are set to 1. The solution to the optimization problem is obtained using a sequential minimal optimization (SMO) algorithm.

The NN is composed of an input layer, three hidden layers, and an output layer. Unfortunately, there is no way to compute the optimized number of layers or the number of neurons per layer in the NN. However, the number of nodes in the input layer is equal to the number of input variables. We used 12, 6, and 2 nodes in the hidden layers with the hidden layer of 12 nodes being the nearest to the input layer. The output layer is composed of one node, giving output in the form of an integer number ranging from 1 to 6 corresponding to the normal, erratic, drift, hardover, spike, and stuck faults, respectively. A Levenberg-Marquardt (LM) algorithm is used in the training phase to optimize the weights of the neurons. The mean squared error (MSE) is utilized to update the weights of the network in backpropagation. A sigmoid function is utilized to activate the nodes in the hidden layers.

In the KNN classifier, the nearest 3 neighbors are weighted using the Euclidean distance parameter to classify a test observation. The cost of wrongly identifying an observation of a class in any other class is set to 1.

For all experiments, a 10-fold CV technique is applied in the training phase to improve the generalization of each classifier. Furthermore, a z-score standardization is applied in preprocessing of the test dataset to normalize the input vectors.

4.4. Experimental Results. To analyze the respective efficiency of the proposed feature selection scheme and the proposed DI-SVM model, a series of five experiments are performed. The first and second experiments are performed without applying a feature selection scheme in the fault detection and diagnosis procedure, using datasets 1 and 2, respectively. In these experiments, the performance comparison of the conventional SVM, NN, and KNN classifiers is presented. In experiment 3, the performance of the proposed feature selection scheme is analyzed and compared to the existing schemes. Lastly, the first two experiments are repeated after adding the proposed feature selection algorithm to the system to observe the effect on the performance of classifiers. The performance of the proposed DI-SVM model is also added in comparison with the conventional SVM, NN, and KNN in last two experiments.

4.4.1. Experiment 1: Fault Diagnosis without Feature Selection Using Dataset 1. The first experiment is performed without applying feature selection in the fault detection and classification methodology. The dataset contained 1000-dimensional normal and fault samples having a constant fault insertion point indexed at 500. All the features given in Tables 2 and 3 are extracted and input to the classifiers.

The accuracy results obtained from this experiment using the conventional SVM, NN, and KNN classifiers are given in Table 4. An individual accuracy of each class is obtained by assuming the given class as a positive class and the remaining classes as a set of negative classes. The total accuracy is the percent ratio of the sum of individual accuracies to the sum of the highest possible accuracies of all classes.

As given in the table, an SVM attains higher accuracies than NN and KNN for classifying all classes except for the normal and the stuck faults. The reason is the small difference between the range of features in normal and stuck fault classes. However, the SVM outperforms the NN and the KNN classifiers achieving a total accuracy of 81%. Furthermore, the NN performs worst among the three classifiers achieving a total accuracy of 78%, a difference of almost 1% from 79% achieved by the KNN classifier. However, the NN, as compared to SVM and KNN, distinctly outperforms in classifying the normal class. Moreover, a KNN classifier achieves slightly higher accuracy when classifying the stuck fault class than do NN and SVM.

Table 4: Accuracies (%) of classifiers obtained in experiment 1.

Classifier	Normal	Erratic	Drift	Hardover	Spike	Stuck	Total
SVM	43.05	85.55	92.50	95.83	85.83	83.33	81.01
NN	83.33	63.05	75.55	86.11	81.38	83.33	78.79
KNN	70.83	63.05	82.77	83.61	92.50	83.88	79.44

Table 5: Accuracies (%) of classifiers obtained in experiment 2.

Classifier	Normal	Erratic	Drift	Hardover	Spike	Stuck	Total
SVM	40.83	90.27	84.44	86.94	93.33	85.83	80.27
NN	83.33	60.00	71.11	82.22	77.22	83.33	76.20
KNN	53.33	81.66	65.00	83.05	80.27	83.33	74.44

4.4.2. Experiment 2: Fault Diagnosis without Feature Selection Using Dataset 2. In this experiment, the fault detection and classification methodology is similar to that of experiment 1; however, the dataset contains samples with variable fault insertion between 0 and 1000 index, as shown in Figure 4.

The results in Table 5 show that the samples considered in this case put a higher challenge to classifiers than those considered in the previous case. The total accuracy of all classifiers is degraded as compared to experiment 1. Nevertheless, the SVM still outperforms the NN and KNN classifiers. However, the total accuracy is degraded from 81% to around 80%. Similarly, the total accuracies of the NN and KNN classifier also reduced from 78% and 79% to 76% and 74%, respectively. The NN is successful in achieving an equal accuracy for normal and stuck fault classes, although a reduction is observed in the accuracies of other classes. The performance of the KNN classifier in classifying most of the classes is reduced with a considerable difference. On the other hand, an increased accuracy is reported for the erratic fault class for both SVM and KNN classifiers.

4.4.3. Experiment 3: Feature Selection. The mean values of 14 time-domain and 4 frequency-domain features extracted from the training samples are given in Table 6. The proposed FS scheme is supposed to select the exclusive most relevant features for each class among the given sets of features in the pool.

To obtain an optimal value of m for the FS scheme, the ROCs along with the respective AUC-ROC values are given in Figures 5, 6, 7, and 8. A conventional SVM classifier was used to obtain the discriminating efficiency of the classifier for the normal class in these figures. The results of Figure 5 are obtained using dataset 1. The normal class samples are trained as a positive class whereas the remaining class samples are trained as a single negative class. Figure 6 shows the results obtained using dataset 2 for a similar positive class, that is, normal class, and negative class. In Figures 7 and 8, datasets 1 and 2 are utilized, respectively; however, the set of faulty classes is trained as a positive class and the normal class as a negative class.

As illustrated in the figures, the classifier performs better when dataset 1 is used as compared to the counter case of

TABLE 6: Mean values of features of all samples.

Features	Normal		Erratic		Drift		Hardover		Spike		Stuck	
Fault insertion point	Dataset 1	Dataset 2	Dataset 1	Dataset 2	Dataset 1	Dataset 2	Dataset 1	Dataset 2	Dataset 1	Dataset 2	Dataset 1	Dataset 2
Mean	25.14	25.14	25.14	25.13	26.39	33.67	40.14	39.42	25.20	25.24	25.05	25.14
STD	0.97	0.97	2.98	2.88	1.88	8.02	15.03	11.97	1.24	1.38	0.81	0.76
VAR	0.95	0.95	8.93	9.00	3.57	86.64	226.08	151.99	1.55	1.96	0.71	0.67
RMS	795.07	795.07	795.16	794.95	834.68	1064.80	1269.41	1246.57	796.97	798.28	792.25	795.11
PPV	7.45	7.45	24.91	24.45	10.86	30.23	36.93	36.68	14.85	15.04	6.97	6.63
Sign	1	1	1	1	1	1	−1	−1	1	1	1	1
SRA	25.13	25.13	25.05	25.04	26.36	33.12	38.68	38.43	25.18	25.22	25.04	25.13
KV	4.82	4.82	5.40	6.82	2.77	3.19	1.01	5.99	27.67	25.67	7.37	23.48
SV	−0.3226	−0.3226	−0.0041	−0.0698	0.5293	0.6979	−0.0001	0.0942	2.9003	3.0904	−0.2370	−0.4364
CF	114	1.14	1.48	1.46	1.22	1.47	1.36	1.47	1.43	1.44	1.14	1.12
IF	1.14	1.14	1.49	1.47	1.23	1.52	1.45	1.55	1.44	1.44	1.14	1.12
MF	1.14	1.14	1.49	1.48	1.23	1.54	1.51	1.59	1.44	1.44	1.14	1.12
FF	1.0007	1.0007	1.0070	1.0070	1.0025	1.0286	1.06778	1.0513	1.0012	1.0015	1.0006	1.0004
KF	$2.62e-48$	$2.62e-48$	$2.10e-53$	$2.11e-50$	$1.14e-52$	$2.12e-49$	$8.24e-55$	$8.76e-53$	$9.53e-49$	$7.63e-49$	$1.11e-46$	$5.48e-43$
FC	52.50	52.50	108.91	106.05	59.67	104.61	130.43	135.85	59.35	60.47	45.20	43.87
SFFT	126.00	126.00	126.00	126.00	126.00	126.00	126.00	126.00	126.00	126.00	126.00	126.00
RMSF	795.67	795.67	800.76	800.58	836.81	1099.13	1355.46	1307.12	797.94	799.51	792.70	795.54
RVF	793.93	793.93	793.31	793.20	834.68	1094.06	1349.17	1299.89	795.73	797.21	791.41	794.30

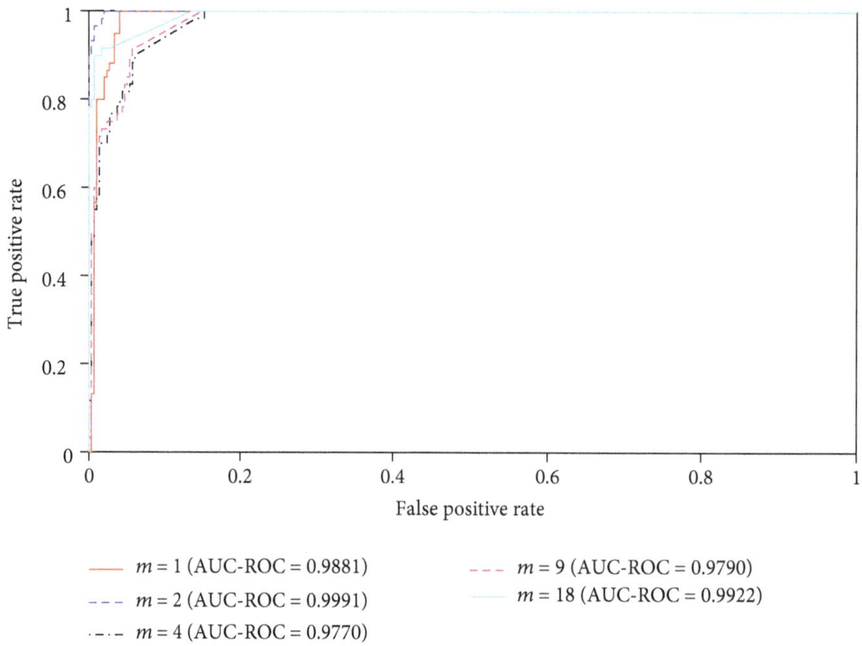

FIGURE 5: ROC and AUC-ROC comparison of the conventional SVM model for various numbers of features (m) selected by the feature selection scheme using dataset 1 when the normal class is selected as a positive class.

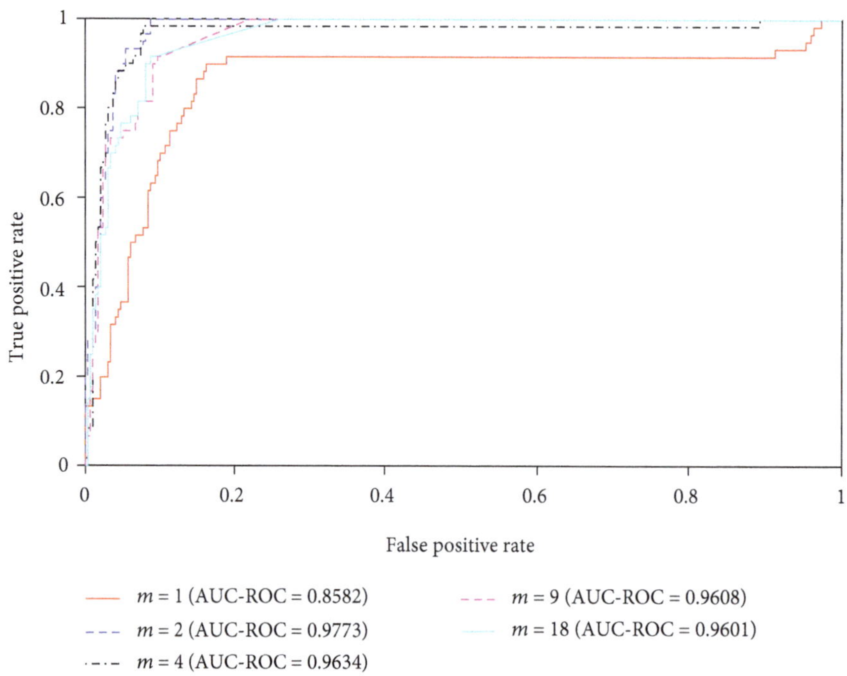

FIGURE 6: ROC and AUC-ROC comparison of the conventional SVM model for various numbers of features (m) selected by the feature selection scheme using dataset 2 when the normal class is selected as a positive class.

using dataset 2. These results show that it is more challenging for a classifier to perform classification when the dataset with a variable fault insertion point is used. Furthermore, it can be observed that $m = 2$ feature selection for the normal class gives the optimal results in all cases as compared to $m = 1$, 4, 9, and 18.

The respective AUC-ROCs for each class are given in Figure 9 using dataset 1 and Figure 10 using dataset 2. The AUC-ROC measures in these figures also illustrate that $m = 2$ is the optimal value for all classes with a given set of features and fault types. The hardover fault type has an almost similar value of AUC-ROC for $m = 1$ and $m = 2$.

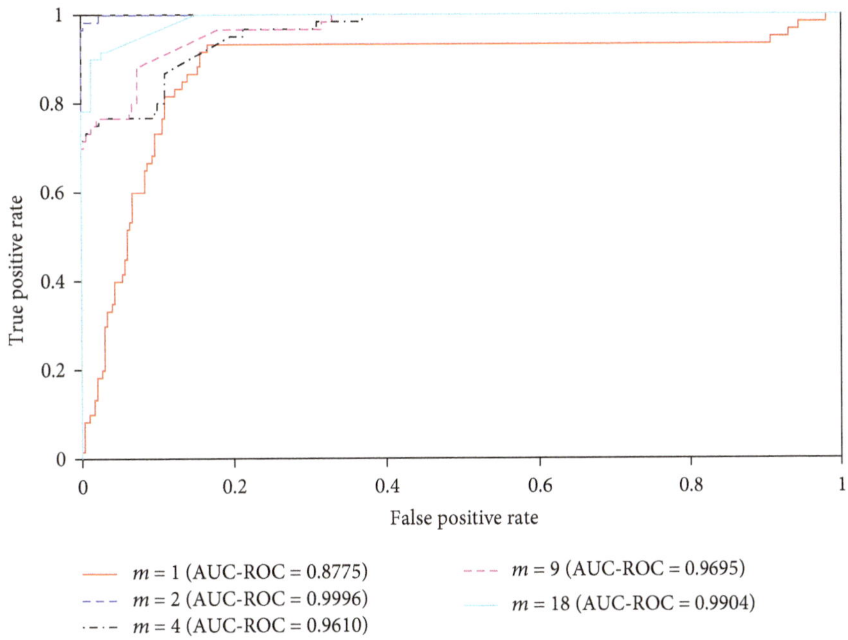

FIGURE 7: ROC and AUC-ROC comparison of the conventional SVM model for various numbers of features (m) selected by the feature selection scheme using dataset 1 when all fault classes are selected as a positive class.

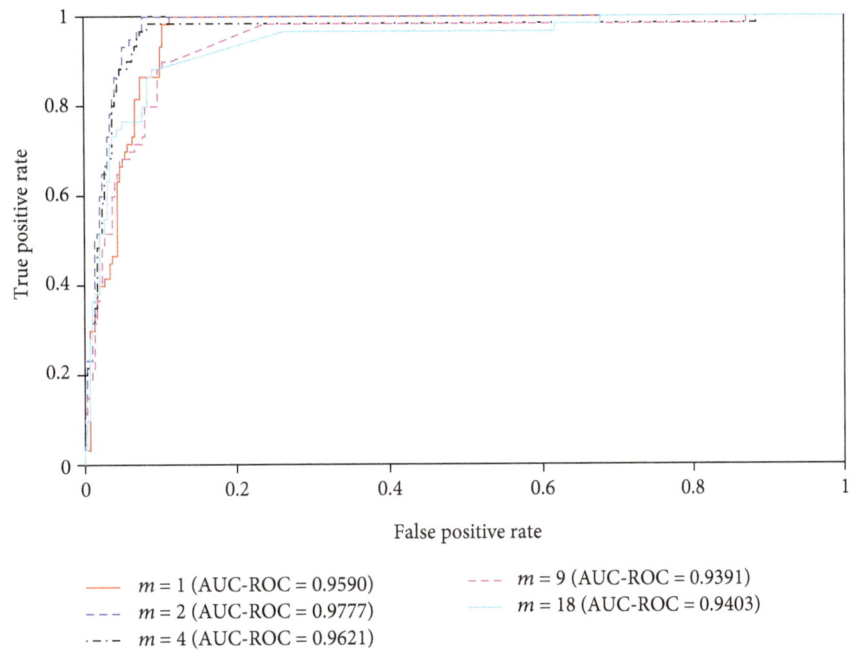

FIGURE 8: ROC and AUC-ROC comparison of the conventional SVM model for various numbers of features (m) selected by the feature selection scheme using dataset 2 when all fault classes are selected as a positive class.

The reason is that only a sign function is a good enough feature to discriminate the hardover fault class. A further increase in the number of feature selection m from 2 shows a degrading AUC-ROC measure for this class as well.

The efficiency of the proposed FS scheme is shown in Figures 11 and 12. The performance is compared with those of the MIFS, MIFS-U, mRMR, NMIFS, I-NMIFS, and exhaustive search algorithm. The exhaustive search algorithm uses all combinations of features in the feature pool to assess the performance of the classifier and then selects the best combination among all. The total accuracy obtained from the conventional SVM classifier is used as evaluation criteria. Datasets 1 and 2 are utilized in Figures 11 and 12, respectively.

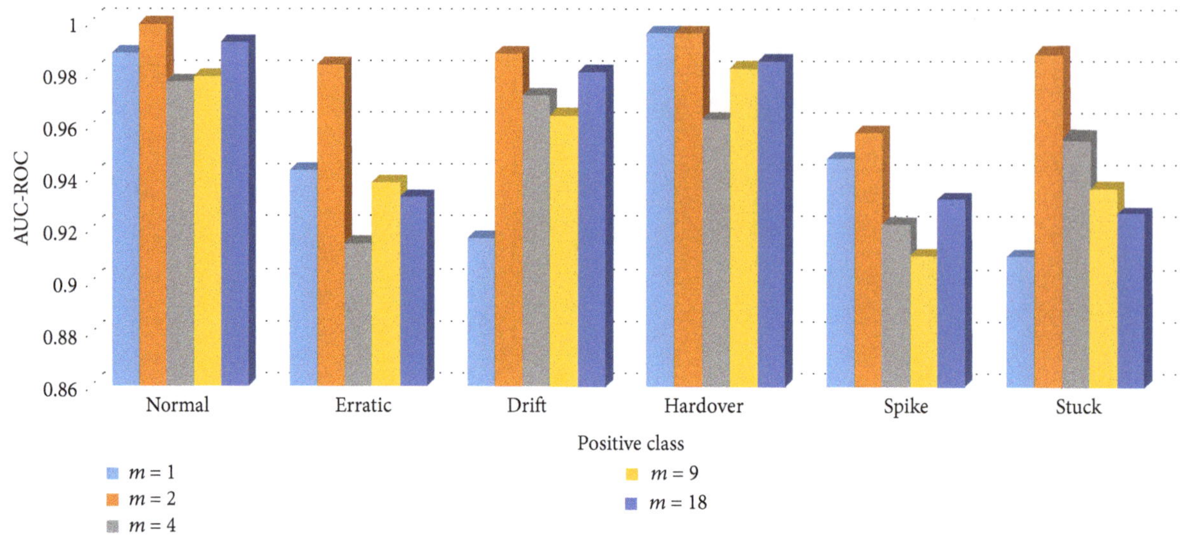

FIGURE 9: AUC-ROCs obtained for different values of m with respective positive classes using dataset 1.

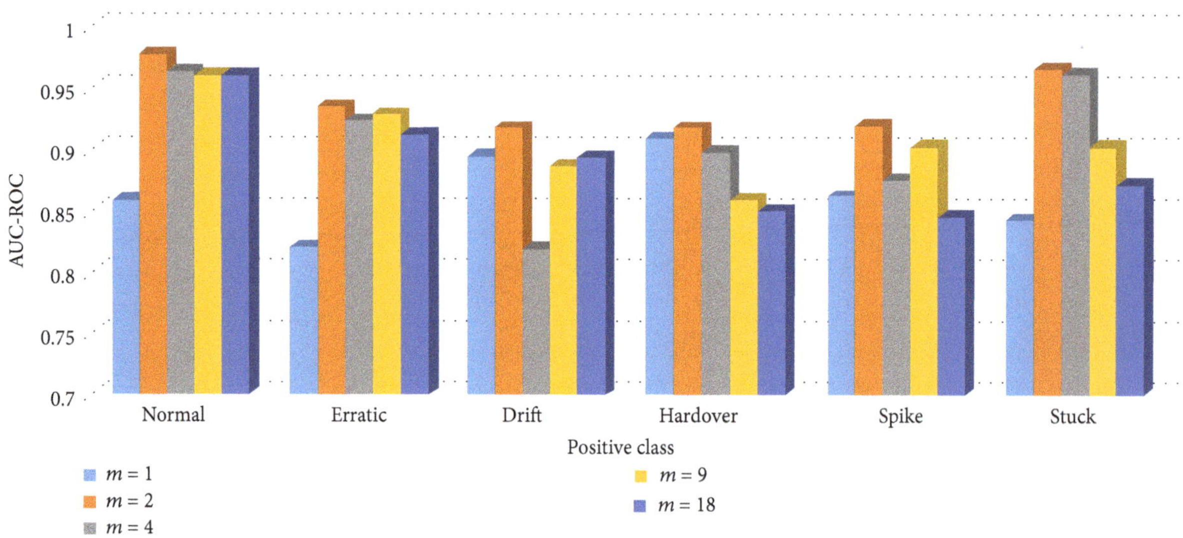

FIGURE 10: AUC-ROCs obtained for different values of m with respective positive classes using dataset 2.

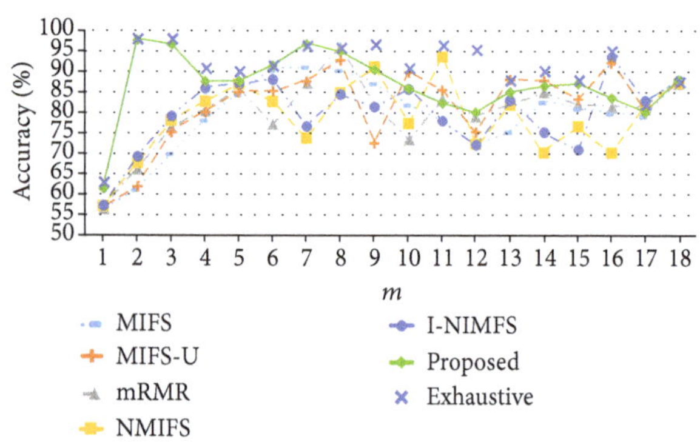

FIGURE 11: Accuracy comparison of feature selection schemes with different values of m and dataset 1.

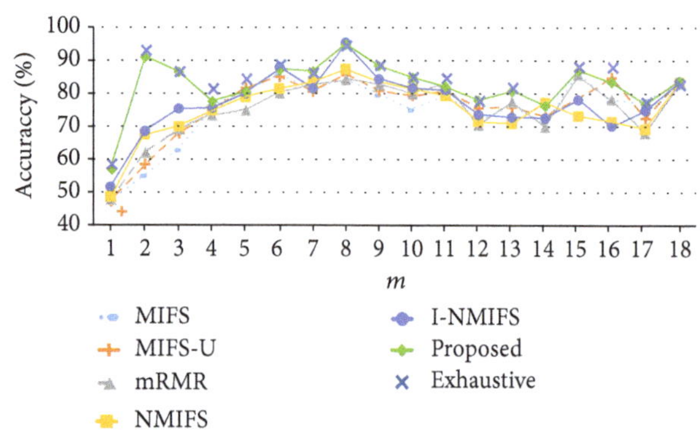

FIGURE 12: Accuracy comparison of feature selection schemes with different values of m and dataset 2.

Table 7: Maximum total accuracies (%) of respective FS with the number of features selected for all classes.

FS scheme	Dataset 1		Dataset 2	
	Number of selected features	Accuracy (%)	Number of selected features	Accuracy (%)
MIFS	12	95.34	13	89.98
MIFS-U	11	95.88	12	90.23
mRMR	10	96.00	11	90.02
NMIFS	10	96.55	10	90.92
I-NMIFS	9	97.01	10	91.00
Proposed	9	97.77	10	91.10
Exhaustive	8	98.02	10	91.98

Table 8: Features selected using the proposed feature selection algorithm for each class.

Class	Normal	Erratic	Drift	Hardover	Spike	Stuck
Selected features	CF	FC	FF	Sign	KV	FC
	FC	STD	Mean	RMS	SFFT	Mean

While selecting the first feature, all the abovementioned schemes have similar criteria of selecting a feature, that is, the measure of mutual information of the candidate feature with the target class. This results in a selection of the same feature by all FS schemes, hence giving comparable performance, as shown in both figures. To select the next features, the criteria of all FS schemes are almost the same, except for the proposed FS scheme. These features rely on both mutual information and redundancy measures, that is, assessing the mutual information of the candidate feature with the target class as well as the redundancy with the already-selected feature. This technique may sometime result in losing a good discriminating feature due to a high redundancy with the already-selected feature. Therefore, these schemes follow a similar trend of increase in the accuracy of the classifier with the increase in the number of the selected features. Furthermore, the selection of the next feature depends highly on the nature of the currently selected feature. It is possible that one of the already-selected features is not a good choice for the target class and it may result in losing a good feature in hand due to a high-redundancy factor. However, the figures show that the proposed FS scheme is able to achieve a good performance by selecting only the 2 most discriminating features per class from the feature pool. Furthermore, including all 18 features in the input to the classifier results in a similar performance irrespective of the FS scheme.

Furthermore, Table 7 shows the optimal accuracy obtained by the classifier with a given total number of features selected by different FS schemes. The results show that the proposed scheme can achieve the results almost comparable to the exhaustive search algorithm with the need of the lowest number of features compared to other techniques. For dataset 1, the proposed scheme can achieve optimal results by selecting only 9 features from the given feature pool. Similarly, 10 features are selected by the proposed FS scheme to obtain optimal results in the case of dataset 2.

The two features selected for each class by the proposed FS scheme are given in Table 8. To show the efficiency of the proposed feature selection algorithm, the scatter plots of all samples are shown in Figure 13 in the selected feature spaces. The red squares represent the data elements of the class for which the respective features are selected. The black circles illustrate the data points of the remaining classes in the same feature space. For instance, the two features selected for the normal class are crest factor (CF) and center frequency (FC). Therefore, the red squares represent the data samples of the normal class in Figure 13(a) in the CF-versus-FC feature space, whereas the black circles represent the data elements of all the remaining classes in this subfigure. A similar trend is followed in the all subfigures.

To observe the feature spaces CF-versus-FC and FC-versus-mean, the scatter plots are redrawn in Figures 14 and 15 with the data elements from normal, stuck, and spike fault classes only. The CF and FC features are selected targeting the normal class. Therefore, most of data elements of the normal class can be distinguished from the nearest classes, that is, spike and stuck fault classes, as shown in Figure 14. Similarly, the FC-versus-mean feature space targets the stuck fault class, and thus the data from the stuck fault class are easily discriminated from the normal and spike fault classes as compared to the CF-versus-FC space, as illustrated in Figure 15. This figure shows the efficiency of the proposed feature space graphically.

A value of $m = 2$ is used to obtain the results of experiments 4 and 5.

4.4.4. Experiment 4: Fault Diagnosis with Feature Selection Using Dataset 1. This experiment is performed applying the proposed FS scheme in the classification methodology and using dataset 1. The performance of the proposed diversified-input (DI-) SVM model is compared with those of the conventional SVM, NN, and KNN classifiers.

The accuracy results of this experiment for all four classifiers are given in Table 9. The proposed DI-SVM achieves a 100% accuracy for erratic, drift, and hardover fault classes. The accuracy of classifying normal, spike, and stuck classes is slightly lower with 98.33%, 98.88%, and 99.44%,

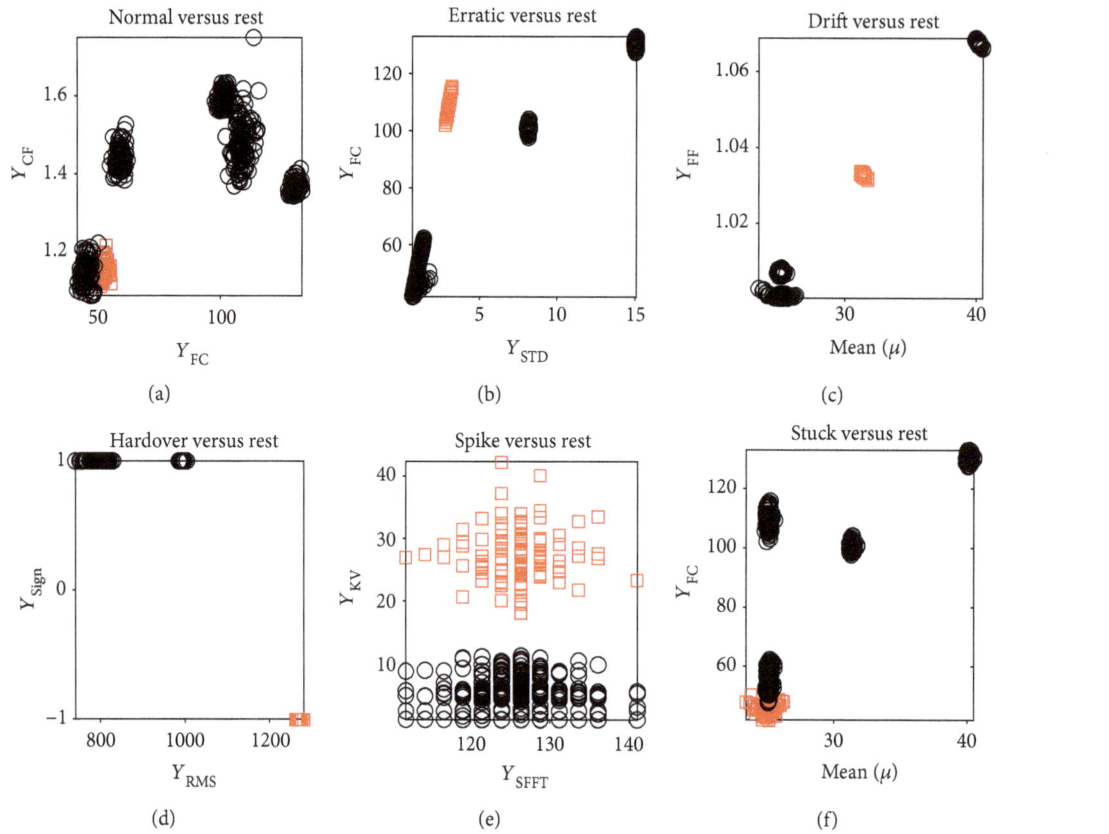

FIGURE 13: Scatter plot data from all classes in the respective feature spaces selected for each class by the feature selection algorithm.

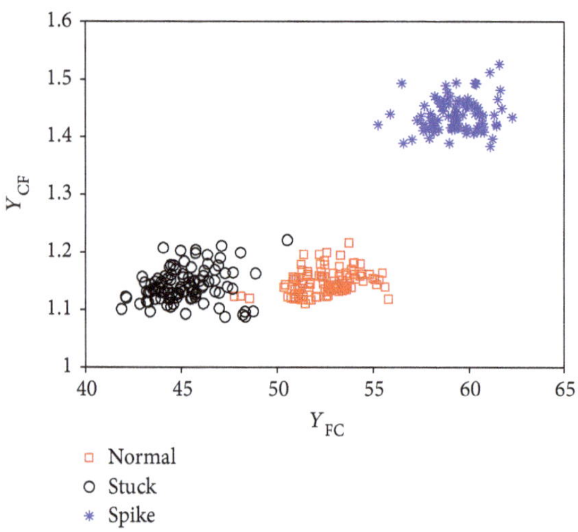

FIGURE 14: Scatter plots of normal, spike, and stuck fault samples in the CF-versus-FC feature space.

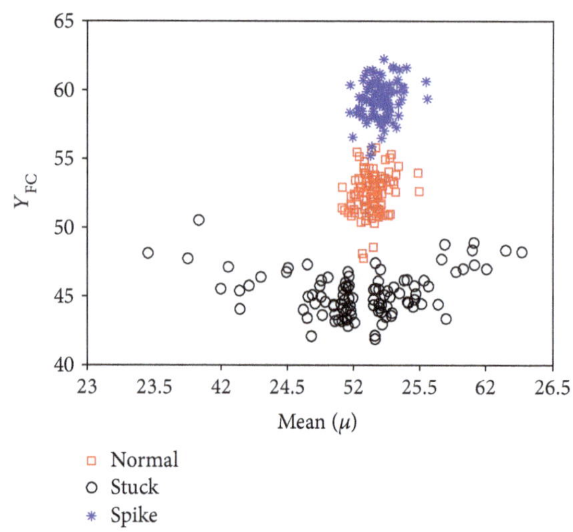

FIGURE 15: Scatter plots of normal, spike, and stuck fault samples in the FC-versus-mean feature space.

resulting in a total accuracy of 99.44% for DI-SVM. The conventional model of SVM is next on the list to perform efficiently achieving a total accuracy of 97.87%. The NN and KNN classifiers perform comparatively worse with a total accuracy of 94% and 85%, respectively.

Table 10 shows the AUC-ROC measures of the four classifiers for the same dataset. The individual AUC-ROC of each class is obtained from the ROC curve attained with the respective class trained as a positive class and the remaining classes trained as a set of negative classes. These results also show that the proposed DI-SVM model outperforms the conventional SVM, NN, and KNN classifiers.

To analyze the effect of the proposed FS scheme and the proposed DI-SVM model, the accuracy of the conventional SVM without applying the FS scheme, the conventional

Table 9: Accuracies (%) of classifiers obtained in experiment 4.

Classifier	Normal	Erratic	Drift	Hardover	Spike	Stuck	Total
DI-SVM	98.33	100.00	100.00	100.00	98.88	99.44	99.44
SVM	93.61	99.16	99.72	99.72	97.22	97.77	97.87
NN	98.61	98.33	98.05	93.33	88.61	90.27	94.53
KNN	83.33	92.77	74.72	83.33	98.05	83.33	85.92

Table 10: AUC-ROC of classifiers obtained in experiment 4.

Classifier	Normal	Erratic	Drift	Hardover	Spike	Stuck
DI-SVM	0.9995	1.0000	1.0000	1.0000	1.0000	0.9997
SVM	0.9994	1.0000	1.0000	1.0000	1.0000	0.9990
NN	0.9697	0.8100	0.6067	0.5966	0.8000	0.9897
KNN	0.9967	1.0000	1.0000	1.0000	1.0000	0.9832

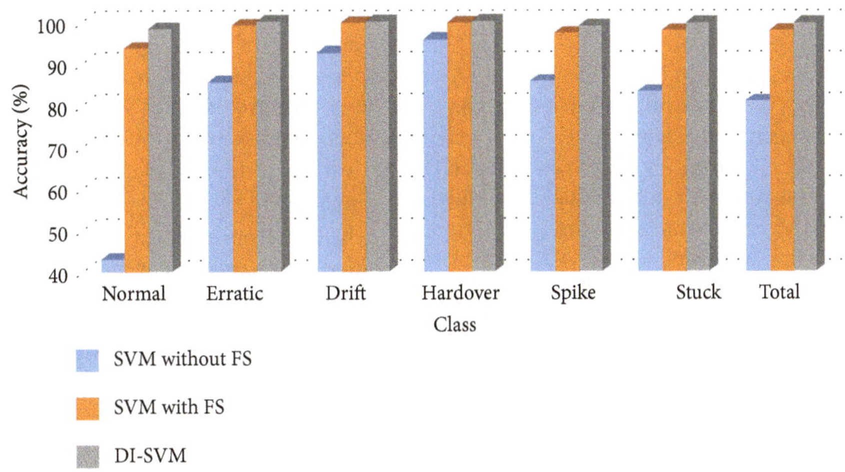

Figure 16: Accuracy comparison of the conventional SVM without FS, SVM with FS, and proposed DI-SVM schemes with the dataset of fixed fault insertion points.

SVM with applying the FS scheme, and the proposed DI-SVM model with FS applied using $m = 2$ is shown in Figure 16. The figure shows that the FS scheme successfully improves the performance of the conventional SVM classifier in classifying all classes. Furthermore, utilizing SVM in the proposed DI-SVM model is able to further improve the classifier's performance.

4.4.5. Experiment 5: Fault Diagnosis with Feature Selection Using Dataset 2. Experiment 5 is a repetition of experiment 4 with a difference in the dataset. In this experiment, dataset 2 is utilized to assess the performance of the proposed FS scheme and the classifiers. As shown in Table 11, the DI-SVM is successful in outperforming the counter three classifiers in classifying the classes in dataset 2 as well. However, the total accuracy is reduced to 93% in this case from 99% in the previous experiment. A similar trend of reduction in accuracy is observed for all classifiers. The intuitive reason is that dataset 2 with a variable fault

Table 11: Accuracies (%) of classifiers obtained in experiment 5.

Classifier	Normal	Erratic	Drift	Hardover	Spike	Stuck	Total
DI-SVM	82.22	99.16	93.32	98.05	99.17	92.21	93.89
SVM	73.89	95.83	92.21	95.82	98.32	89.71	91.10
NN	88.32	88.60	94.71	89.71	91.10	94.72	91.19
KNN	61.38	82.21	76.39	79.17	81.39	83.32	77.30

Table 12: AUC-ROC of classifiers obtained in experiment 5.

Classifier	Normal	Erratic	Drift	Hardover	Spike	Stuck
DI-SVM	0.9767	0.9925	0.9383	1.0000	0.9996	0.8820
SVM	0.9766	0.9829	0.8997	0.8948	0.9663	0.8676
NN	0.9223	0.7933	0.5845	0.6077	0.8103	0.8820
KNN	0.9422	0.9565	0.9913	1.0000	0.9983	0.9269

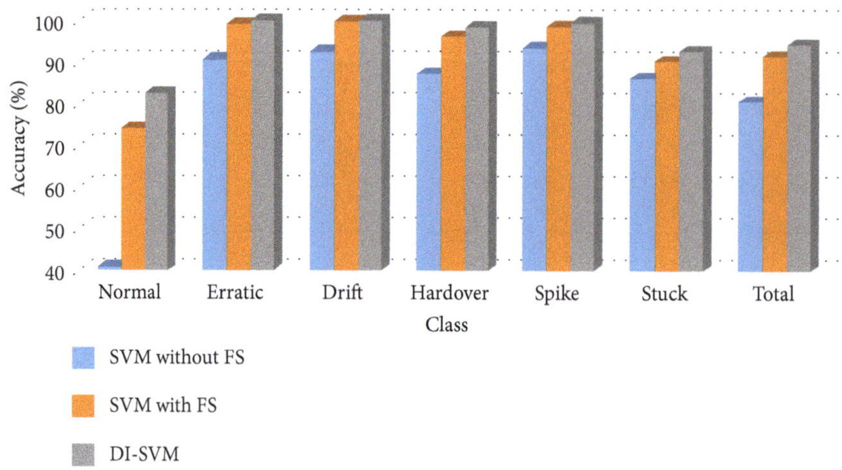

FIGURE 17: Accuracy comparison of the conventional SVM without FS, SVM with FS and proposed DI-SVM schemes with the dataset of variable fault insertion points.

insertion point put a higher challenge in the classification of samples as compared to the classification of samples in dataset 1 with a fixed fault insertion point.

The AUC-ROCs of this experiment are given in Table 12. Once again, the DI-SVM is successful in achieving the optimal AUC-ROC measures for all classes.

For experiment 5, the accuracy comparison of the conventional SVM without applying the FS scheme, the conventional SVM with the FS scheme, and the proposed DI-SVM model with FS applied using $m = 2$ is shown in Figure 17. The figure also strengthens our claim that the proposed DI-SVM model outperforms the conventional SVM model.

5. Discussion

The MI-based FS schemes select features for a given target class based on the measure of MI of the candidate feature with a target class and the mutual redundancy with already-selected features, irrespective of nontarget classes. The proposed FS algorithm considers the MI of a feature on both target and nontarget classes to select the best combination of features discriminating the target class from nontarget classes. As this feature selection scheme is regarded as a filter-based feature selection, the features are selected independently of the classifier performance and the nature of raw signals. Therefore, it is safe to claim that the proposed FS algorithm will efficiently select exclusive most relevant features for a target class among the pooled features in any application. However, it should be noted that the efficiency of the classifier highly depends on the nature of features in the feature pool used to characterize the signal. Using the most characterizing set of features will increase the efficiency of the system and vice versa. From the results of our work, it can be concluded that taking into account the measurement of relevance of the candidate feature with a target class as well as with a nontarget class can help in selecting exclusive discriminating features.

Furthermore, in a conventional model of applying SVMs for multiclass classification, the features selected for each class are concatenated to train all SVMs. Intuitively, reducing the dimension of input observations leads to improved performance of SVMs given that the features are selected properly. Keeping this in mind, we reduced the dimension of the input vector of all SVMs by inputting diverse feature combinations selected for different classes. This means that the feature combination chosen for targeting the cth class is input to the cth SVM, particularly, trained with the data in which the observations of the cth class are labeled as a positive class and vice versa. In this way, the dimension of the input vectors is reduced to simplify the optimization of the classifier. Moreover, similar feature combinations might be selected for more than one class if this combination is the most discriminating feature combination for each class.

Our previous work [3] proved from experimental results that increasing the size of a raw signal, that is, increasing the number of data elements per one raw signal of the sensor output, and increasing the size of the training dataset (increasing the number of training samples) increase the accuracy of the classifier. Therefore, the purpose of selecting a small size for training and testing sets is to observe the lower bound performance of the proposed fault diagnosis methodology.

6. Conclusions

To select the discriminating feature combinations targeting each class, we propose a feature selection scheme that takes into account the measure of relevance of the feature with both target and nontarget classes. The relevance of two variables is characterized by the measure of mutual information between them. The set of m features having the maximum mutual information on the target class and the minimum mutual information on the nontarget classes is chosen for a target class. The results show that this technique can achieve good results as compared to other feature selection schemes that take mutual redundancy measure between the selected features in addition to relevance with a target class.

Furthermore, a DI-SVM model is proposed to improve the performance of the classifier. Each SVM is trained with a single feature combination selected, specifically, for a single target class, unlike the conventional SVM model of delivering all combinations to all SVMs. This approach reduces the complexity of classifier optimization by minimizing the dimensions of the input vectors. The performance of the classifier was analyzed using two different datasets: (1) with a constant fault insertion point and (2) with a variable fault insertion point. A series of 5 experiments were performed to analyze the performance of the proposed fault diagnosis methodology. The evaluation metrics including the accuracy, the ROC, and the AUC-ROC show the efficiency of the proposed FS scheme and proposed DI-SVM model. A further comparison of the performances shows the efficiency of the proposed DI-SVM over the conventional model of SVM-, NN-, and KNN-based classifiers in multiclass sensor fault classification problems.

Conflicts of Interest

The authors declare that they have no conflicts of interest.

Acknowledgments

This work was supported by the Business for Cooperative R&D between Industry, Academy, and Research Institute funded by the Korean Small and Medium Business Technology Administration in 2016 under Grant C0398156.

References

[1] B. P. Duong and J.-M. Kim, "Non-mutually exclusive deep neural network classifier for combined modes of bearing fault diagnosis," *Sensors*, vol. 18, no. 4, p. 1129, 2018.

[2] M. Sohaib, C.-H. Kim, and J.-M. Kim, "A hybrid feature model and deep-learning-based bearing fault diagnosis," *Sensors*, vol. 17, no. 12, p. 2876, 2017.

[3] S. U. Jan, Y.-D. Lee, J. Shin, and I. Koo, "Sensor fault classification based on support vector machine and statistical time-domain features," *IEEE Access*, vol. 5, pp. 8682–8690, 2017.

[4] Y. Yu, W. Li, D. Sheng, and J. Chen, "A novel sensor fault diagnosis method based on modified ensemble empirical mode decomposition and probabilistic neural network," *Measurement*, vol. 68, pp. 328–336, 2015.

[5] S. U. Jan and I. S. Koo, "Sensor faults detection and classification using SVM with diverse features," in *2017 International Conference on Information and Communication Technology Convergence (ICTC)*, pp. 576–578, Jeju, South Korea, October 2017.

[6] P. Huang, Y. Jin, D. Hou et al., "Online classification of contaminants based on multi-classification support vector machine using conventional water quality sensors," *Sensors*, vol. 17, no. 3, p. 581, 2017.

[7] B. Scholkopf, "Support vector machines: a practical consequence of learning theory," *IEEE Intelligent Systems*, vol. 13, 1998.

[8] V. N. Vapnik, "An overview of statistical learning theory," *IEEE Transactions on Neural Networks*, vol. 10, no. 5, pp. 988–999, 1999.

[9] L. Ren, W. Lv, S. Jiang, and Y. Xiao, "Fault diagnosis using a joint model based on sparse representation and SVM," *IEEE Transactions on Instrumentation and Measurement*, vol. 65, no. 10, pp. 2313–2320, 2016.

[10] S. S. Y. Ng, P. W. Tse, and K. L. Tsui, "A one-*versus*-all class binarization strategy for bearing diagnostics of concurrent defects," *Sensors*, vol. 14, no. 1, pp. 1295–1321, 2014.

[11] P. Thanh Noi and M. Kappas, "Comparison of random forest, k-nearest neighbor, and support vector machine classifiers for land cover classification using sentinel-2 imagery," *Sensors*, vol. 18, no. 2, p. 18, 2018.

[12] P. V. Klaine, M. A. Imran, O. Onireti, and R. D. Souza, "A survey of machine learning techniques applied to self-organizing cellular networks," *IEEE Communications Surveys & Tutorials*, vol. 19, no. 4, pp. 2392–2431, 2017.

[13] H. Peng, F. Long, and C. Ding, "Feature selection based on mutual information criteria of max-dependency, max-relevance, and min-redundancy," *IEEE Transactions on Pattern Analysis and Machine Intelligence*, vol. 27, no. 8, pp. 1226–1238, 2005.

[14] G. Ditzler, R. Polikar, and G. Rosen, "A sequential learning approach for scaling up filter-based feature subset selection," *IEEE Transactions on Neural Networks and Learning Systems*, vol. 29, no. 6, pp. 2530–2544, 2018.

[15] M. Kudo and J. Sklansky, "Comparison of algorithms that select features for pattern classifiers," *Pattern Recognition*, vol. 33, no. 1, pp. 25–41, 2000.

[16] Z. M. Hira and D. F. Gillies, "A review of feature selection and feature extraction methods applied on microarray data," *Advances in Bioinformatics*, vol. 2015, Article ID 198363, 13 pages, 2015.

[17] T. M. Cover and J. A. Thomas, *Elements of Information Theory*, John Wiley & Sons, 2012.

[18] R. Battiti, "Using mutual information for selecting features in supervised neural net learning," *IEEE Transactions on Neural Networks*, vol. 5, no. 4, pp. 537–550, 1994.

[19] N. Kwak and C.-H. Choi, "Input feature selection for classification problems," *IEEE Transactions on Neural Networks*, vol. 13, no. 1, pp. 143–159, 2002.

[20] P. A. Estévez, M. Tesmer, C. A. Perez, and J. M. Zurada, "Normalized mutual information feature selection," *IEEE Transactions on Neural Networks*, vol. 20, no. 2, pp. 189–201, 2009.

[21] L. T. Vinh, S. Lee, Y.-T. Park, and B. J. d'Auriol, "A novel feature selection method based on normalized mutual information," *Applied Intelligence*, vol. 37, no. 1, pp. 100–120, 2012.

[22] G. Lera and M. Pinzolas, "Neighborhood based Levenberg-Marquardt algorithm for neural network training," *IEEE Transactions on Neural Networks*, vol. 13, no. 5, pp. 1200–1203, 2002.

[23] J. Yang, Z. Sun, and Y. Chen, "Fault detection using the clustering-kNN rule for gas sensor arrays," *Sensors*, vol. 16, no. 12, p. 2069, 2016.

[24] S. Haykin, *Neural Networks, a Comprehensive Foundation*, Tech. Rep., Macmilan, 1994.

[25] J. Weston and C. Watkins, "Support vector machines for multi-class pattern recognition," *Proceedings of European Symposium on Artificial Neural Networks ESANN'99*, 1999, Brugs, Belgium, 1999.

[26] C. J. C. Burges, "A tutorial on support vector machines for pattern recognition," *Data Mining and Knowledge Discovery*, vol. 2, no. 2, pp. 121–167, 1998.

[27] Z. Wang and X. Xue, "Multi-class support vector machine," in *Support Vector Machines Applications*, pp. 23–48, Springer, 2014.

[28] D.-C. Luor, "A comparative assessment of data standardization on support vector machine for classification problems," *Intelligent Data Analysis*, vol. 19, no. 3, pp. 529–546, 2015.

[29] C.-W. Chu, J. D. Holliday, and P. Willett, "Effect of data standardization on chemical clustering and similarity searching," *Journal of Chemical Information and Modeling*, vol. 49, no. 2, pp. 155–161, 2009.

[30] J. Kullaa, "Detection, identification, and quantification of sensor fault in a sensor network," *Mechanical Systems and Signal Processing*, vol. 40, no. 1, pp. 208–221, 2013.

[31] T. Fawcett, "An introduction to roc analysis," *Pattern Recognition Letters*, vol. 27, no. 8, pp. 861–874, 2006.

9

Metal Detection Sensor Utilizing Magneto-Impedance Magnetometer

Kyoo Nam Choi

Department of Information and Telecommunication Engineering, Incheon National University, Incheon 22012, Republic of Korea

Correspondence should be addressed to Kyoo Nam Choi; knchoi@inu.ac.kr

Academic Editor: Lucio Pancheri

A magnetometer with a longitudinal sensing line using the MI effect was examined experimentally to find an application to replace conventional metal sensors based on an LVDT. A sampling to the second peak signal from the pickup coil on the sensing head with a long, thin soft magnetic ribbon and driven with an ultrashort pulse train with a pulse width of 5~7 ns formed the front end of the metal sensor. Unmatched degaussing and geomagnetic field compensation methods were effective not only to minimize the remnant magnetic field but also to prevent the saturation of the magnetic ribbon and expand the dynamic range. The active noise reduction process reduced the inherent noise to the 1/3 level with a minimum influence on the detection signal. The proposed metal sensor with unmatched degaussing, geomagnetic field compensation, and active noise reduction techniques demonstrated the detection of a 0.2 mm diameter magnetized ferrous ball.

1. Introduction

Metal detection sensors that use a linear variable differential transformer (LVDT) are widely used to detect foreign metal objects in food and in the security industry. This type of metal detector requires a strict balance between two differential inputs in various environmental conditions and complete shielding from external electromagnetic waves. Also, the sensitivity of detecting a foreign metallic object is influenced by the surrounding medium. It is hard for an LVDT-type sensor to detect a metallic object depending on placement angle. In the case of ferrous sphere, the detection resolution of an LVDT-type sensor is ~0.7 mm diameter [1]. The sensitivity of this metal sensor is dependent on the orientation of the metallic object, and it requires multiple sensors to detect ferromagnetic metal independent of object orientation [2]. A magnetometer is an alternative way to detect a magnetized metallic object inside an electric shield medium [3, 4]. A hall sensor is commonly used for magnetic field detection, because it has a simple structure; however, the sensitivity is not sufficient to detect a tiny magnetized metal object [5]. A magneto-resistance (MR) sensor has better resolution than a hall sensor [6], and a giant magneto-resistance (GMR) sensor has better resolution than an MR sensor [7]. Overall, the three sensors above have similar power consumption. Fluxgate [8], magneto-impedance (MI) [9], and giant magneto-impedance (GMI) [10] sensors have higher resolution in this magnetic sensor group. The fluxgate sensor has a lower internal noise level than MI or GMI sensors, but it consumes more power. Thus, a GMI sensor has the advantages of higher sensitivity, lower power consumption, and simple sensor head structure for metal detection applications, despite its relatively higher internal noise level. GMI sensors use a magneto-inductive effect at low frequency ranging from 1 to 10 kHz [11], skin effect at high frequency ranging from 10 kHz to 1 GHz [12], and a ferromagnetic resonance effect at ultrahigh frequency that is >1 GHz [13]. Among these, the GMI sensor that uses a skin effect has the advantage of low power consumption and high resolution. The sampling method [14] or the detection of higher harmonic frequency [15] has been attempted to reduce power consumption or increase sensitivity. A high-resolution micro-GMI sensor has been achieved using a pulse magneto-impedance effect inside a microamorphous wire [16]. This microsensor has shown pinpoint sensitivity at the sensor tip, but multiple sensors are required to cover the longitudinal sensing length. In this paper, a method to enhance the sensitivity of a GMI sensor based on a single longitudinal amorphous ribbon is analyzed, and a method to reduce the noise component is investigated.

2. Design of MI Sensor Head

A magnetometer sensor element consists of a thin longitudinal amorphous ribbon and surrounding wound pickup coil. The impedance of a magnetic ribbon comes from the addition of DC resistance and inductive reactance that is determined by the skin effect. The DC resistance has a negligible effect, and the driving frequency and maximum differential permeability become the major determining factors for sensitivity. The magneto-inductive effect induces an inductive voltage across the pickup coil that is proportional to sqrt $(f\mu(H_{ext}))$, where f is the driving frequency, μ is the maximum differential permeability, and H_{ext} is the external magnetic field. In the case of pulse driving, the driving frequency corresponds to $1/2t_r$, where t_r is the pulse rising time. Also, a demagnetizing field prior to pulse excitation is necessary to prevent the accumulation of a magnetic field inside the core material. The magneto-impedance effect is measured by using the formula $\Delta Z/Z$, where Z is the impedance of the core material and ΔZ is the impedance difference when the external magnetic field is applied. However, in the case of driving with short pulses with a low duty cycle, the change of impedance due to an external magnetic field occurs only within a short duration that is proportional to the applied pulse width, which is in the ns range. Thus, the impedance variation due to the external magnetic field is captured at the instance when the short pulse is applied. And the captured signal response is for the real part of impedance variation due to the magneto-impedance effect. Thus, for sensor application, it is practical to directly measure the magneto-impedance effect as $\Delta V_o(H_{ext})/\min(V_o)$, where $\min(V_o)$ is the minimum output voltage at pickup coil, H_{ext} is the applied external magnetic field, and $\Delta V_o(H_{ext})$ is the output voltage difference when the external magnetic field, H_{ext}, is applied.

2.1. Sensor Head Structure. The sensor was intended to cover a longitudinal sensing length of 90 mm. A soft magnetic material was used as the sensor head core, as shown in Table 1. The pickup coil was wound on a bobbin surrounding the core, and the compensating coil was also wound on a circular bobbin surrounding the pickup coil to cancel the geomagnetic effect. The overall dimension of the core material, pickup, and compensating coils is shown in Tables 1 and 2. The shape of the sensing head is shown in Figure 1.

2.2. Sensitivity Distribution on a Sensor Head. The sensitivity of the sensor was tested along the sensing axis. The peak magnitude of a driving pulse with a width of 4.5 ns was fixed to 3.5 V. Output voltage variation in the first peak signal from the pickup coil responding to geomagnetic and external magnetic fields was measured, as shown in Figure 2. The total field strength of the geomagnetic field at the measurement location was 51 μT or 0.51 G (horizontal intensity of 29 μT, north component of 29 μT, east component of $-4\,\mu$T, and vertical component of 41 μT). The output voltage was derived by sampling and holding a peak value. The tip of the sensor head for positive driving pulse input was rotated along the earth's magnetic poles to evaluate influence by geomagnetic field. A rare earth pressed cube permanent magnet with dimensions of $L2 \times W2 \times H1$ mm and with 500 μT at a distance of 1 cm moved along the longitudinal direction of sensor head. When the tip of the sensor head for positive driving pulse input pointed to the north and the west, one end of the sensor head had the maximum sensitivity and gradually decreased toward the other end, as shown in Figure 2(a). This phenomenon was reversed when the tip of the sensor head for positive driving pulse input was reversed, as shown in Figure 2(b). When the tip of the sensor head for positive driving pulse input pointed to the northeast, both ends of the sensor head had sensitivity. This implies that the sensitivity of the sensor head is sufficient to be influenced by a geomagnetic field. Also, the sensor head has sensitivity not only at both tips but also on the side. The slightly distorted trend in Figure 2 is regarded to be due to a magnetic field from nearby electronic equipment.

2.3. Bipolar Driving. A soft magnet material in a core is magnetized in one direction if it is continuously driven by a unipolar pulse train, as shown in Figure 3(a). In the case of an MI sensor, this accumulates a magnetic field and results in either subtracting or adding to the magnetic field to be measured. One way to solve this erroneous result is by alternatively driving with a matched pulse pair with equal amplitude, but opposite polarity, as shown in Figure 3(b). This scheme is suited for degaussing purpose, but the response by one driving pulse is partially influenced by the other following degaussing pulse unless the two pulses are separated to a sufficient time interval within an acceptable influence level, because a degaussing pulse also induces an MI effect in the negative direction. The disadvantage of this scheme can be alleviated by programming with two different pulses, a driving pulse with a major MI effect and a degaussing pulse with a small MI effect, as shown in Figure 3(c). The physical adjustment is to set a driving pulse duration that induces a maximum MI effect and a degaussing pulse duration with more than 10 times the driving pulse duration. A driving pulse duration from 4.5 to 7 ns was tested to find the optimum pulse width for this experiment.

2.4. Reception Sensitivity. The core of the sensor head was driven by a positive polarity pulse with 4.5 ns width, t_{Drv}, with the degaussing pulse having opposite polarity from the width, t_{Deg}, as Figure 3(c). The peak magnitude of the driving pulse was measured as 3.5 V. The responding signal was measured from the signal pickup coil and had an oscillation waveform with more than 10 decaying peaks. Among them, the first four peaks had the significant responses against external magnetic field variation. A typical reception signal waveform is shown in Figure 4.

The same rare earth pressed magnet, with 500 μT at 1 cm distance, was moved to the center of the sensor head, and the variation of the reception signal waveform was measured, as shown in Figures 5(a) and 5(b). The waveform in channel 1

Table 1: Characteristics of core.

Layer	R	L	Dimension (mm)	Permeability	Material
1	4.1 Ω	4 μH	$L97 \times W1 \times T0.018$	$\mu_r = 5000$	VITROVAC 6205X (FeCo)SiB

Table 2: Characteristics of coil.

Coil type	Bobbin (mm)	Turns	Diameter	R	L
Signal pickup	$L87 \times W2 \times T3$	200	0.15 mm	3.1 Ω	247 μH
Compensating	$L18 \times D9$	100	0.15 mm	3.0 Ω	143 μH

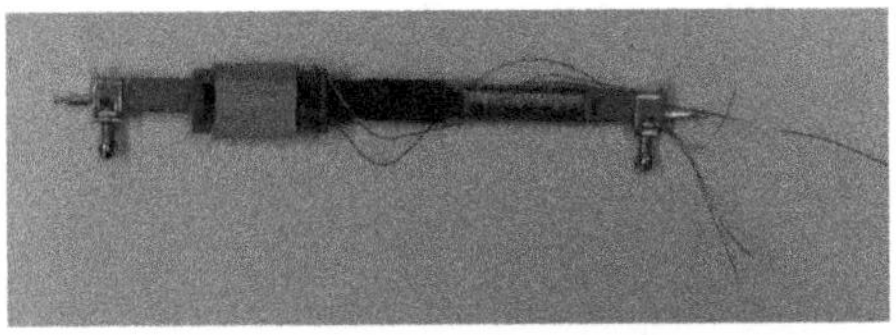

Figure 1: Shape of sensor head.

shows a reference pulse, and the waveform in channel 2 shows a signal waveform at the pickup coil. The reference pulse in Figure 5 indicates the starting point of the signal waveform from the pickup coil. Where a geomagnetic field was present and the tip of the sensor head for positive driving pulse input oriented to the west, the output voltage from the pickup coil was marked as a minimum output voltage, min (V_o), as shown in Figure 5(a). When an external magnetic field was approached, the output voltage from the pickup coil increased to a maximum output voltage, max (V_o), as shown in Figure 5(b). The effect on the real part of magneto-impedance was calculated as 600% for the first peak and 1180% for the second peak waveforms by using the formula in (1). This implies that the second peak has the biggest MI effect for a point magnetic field source in a longitudinal sensor geometry.

$$\frac{\Delta Z}{Z} = \frac{|\max (V_o) - \min (V_o)|}{|\min (V_o)|}. \quad (1)$$

Another experiment to investigate the effect of a uniform magnetic field such as a geomagnetic field was performed. The peak magnitude of driving pulse was fixed to 3.5 V, and the tip of the sensor head for positive driving pulse input was located to point the east and then rotated to point south. The output voltage variation in each peak signal waveform at the pickup coil was measured by varying the pulse width from 5 ns to 7 ns, as shown in Figure 6. The total magnetic field strength of the geomagnetic field at the measurement location was same as 51 μT or 0.51 G. The measurement results showed that the second peak had the biggest output signal variation, and the driving pulse with a 6.5 ns FWHM width had the greatest sensitivity. For the driving pulse of 6.5 ns FWHM, the effect of magneto-impedance was calculated as 2900% for the first peak and 1700% for the second peak waveforms. This indicates that the MI effect is the biggest for the first peak even though the variation of output signal voltage is the biggest for the second peak in a uniform geomagnetic field in a longitudinal sensor geometry.

The peak magnitude of the driving pulse was increased to 5.2 V for the purpose of investigating the effect of driving pulse amplitude, and the responding output signal showed a similar waveform, but with increasing amplitude. The same rare earth pressed magnet was moved to the center of the sensor head, and the output signal waveform was measured with and without an external magnetic field. The tip of the sensor head for positive driving pulse input was oriented to the west, and the output voltage was marked as minimum amplitude, min (V_o). A sample magnet was moved to a 10 mm distance from the center of the sensor head, and the output voltage was marked as maximum amplitude, max (V_o). The maximum output voltage, max (V_o), and a minimum output voltage, min (V_o), were compared, and the effect of magneto-impedance was calculated as 1160% for the first peak and 2100% for the second peak waveforms. This implies that the effect of magneto-impedance is almost proportional to the square of driving pulse voltage.

The output signal voltage at the pickup coil was monitored by continuously varying the driving pulse voltage to the soft magnet. The first peak output signal was found to be matched with the time at a peak of the driving pulse, and each peak value had a tendency to increase proportionally to a square of the driving pulse voltage, as shown in Figure 7. This implies that the output signal voltage is a function of driving pulse power.

The effect of driving pulse voltage on the MI index was measured by varying the driving pulse voltage to the soft magnet. The MI index had a tendency to increase proportionally to a square of the driving pulse voltage, as shown in Figure 8. Thus, the higher the driving voltage is, the better the magneto-impedance index is. However, an external magnetic field far beyond 500 μT at 1 cm distance resulted in the same or less output signal voltage due to saturation of the soft magnet.

Driving with a matched degaussing pulse, as shown in Figure 9(a), was compared with an unmatched degaussing pulse, as shown in Figure 9(b), in a constant geomagnetic field. The driving pulse width was set to 5 ns, and the product of pulse width and pulse magnitude between driving and degaussing pulses was kept equal. The measurement results, as shown in Figure 10, showed that the response of the first peak signal was bigger than the response of the second peak signal in the case of matched degaussing. On the contrary,

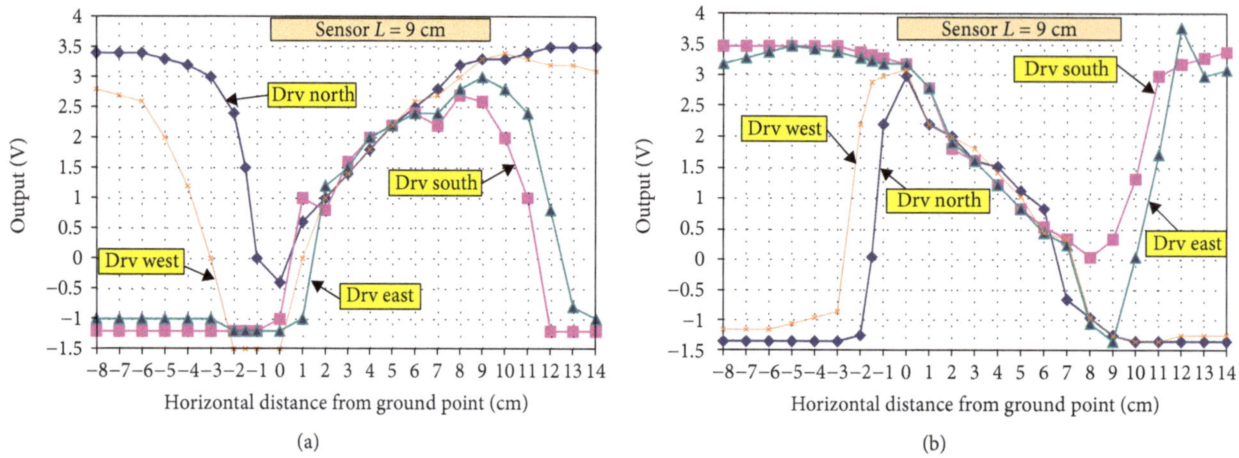

FIGURE 2: Sensitivity of the sensor head along longitudinal direction. (a) Forward driving. (b) Reverse driving.

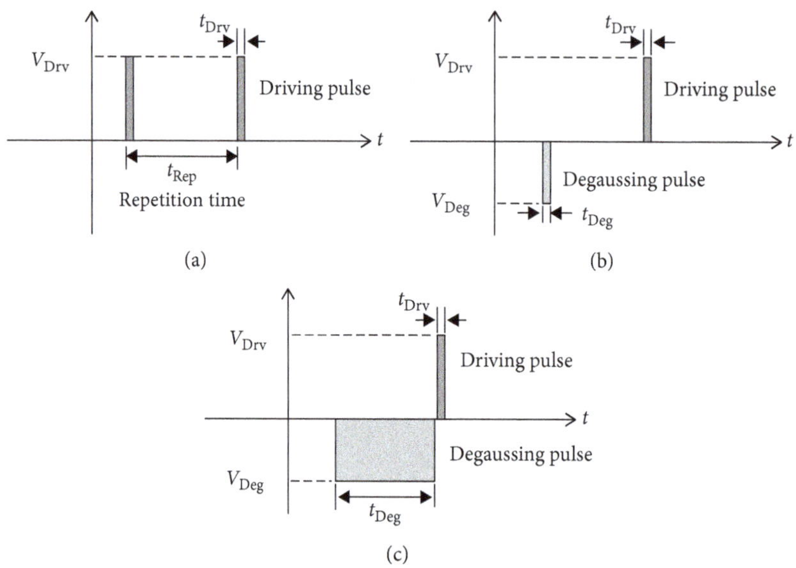

FIGURE 3: Driving pulse waveform of soft magnet with driving amplitude V_{Drv} and degaussing amplitude V_{Deg}. (a) Unipolar driving. (b) Driving with matched degaussing. (c) Driving with unmatched degaussing.

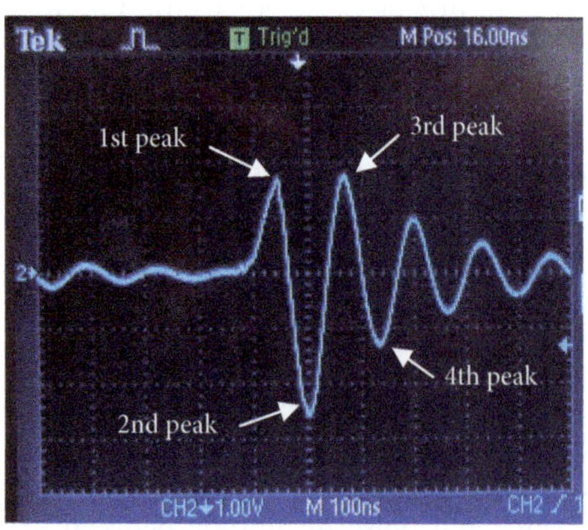

FIGURE 4: Signal waveforms at the pickup coil when driving with a pulse amplitude of 3.5 V.

the second peak signal was bigger than the first peak signal in the case of unmatched degaussing. The dynamic range of the output signal was found to be bigger in the case of unmatched degaussing for both first and second peak signals. This indicates that driving with an unmatched degaussing pulse induces a bigger response against external magnetic field variation.

The effect changing the number of soft magnet layers was investigated using the same sensor head structure. The driving pulse voltage was kept at 3.5 V, and the driving pulse width was set to 5 ns FWHM in a uniform geomagnetic field. An unmatched degaussing pulse was followed by a driving pulse for comparison under the same experiment conditions. The output swing range was reduced to half in the case of two layers for the first peak signal. For the second peak signal, this reduction rate was decreased to one-third in comparison to one layer of soft magnet, as shown in Figure 11(a). In the

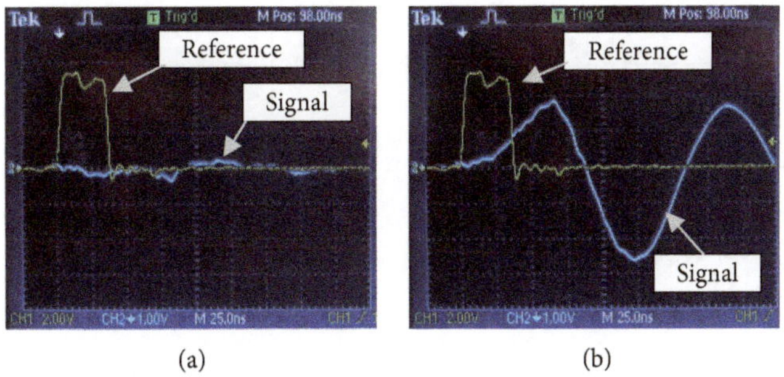

FIGURE 5: Signal waveforms at the pickup coil when driving with 3.5 V pulse amplitude. (a) Without external magnetic field. (b) With external magnetic field.

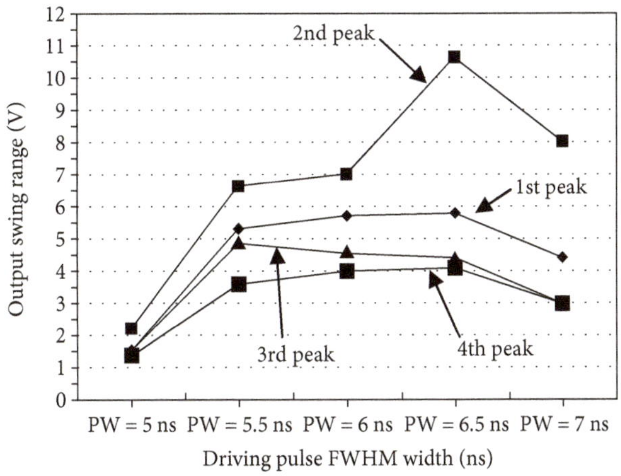

FIGURE 6: Variation of output signal voltage responding to uniform geomagnetic field when driving with 3.5 V pulse amplitude.

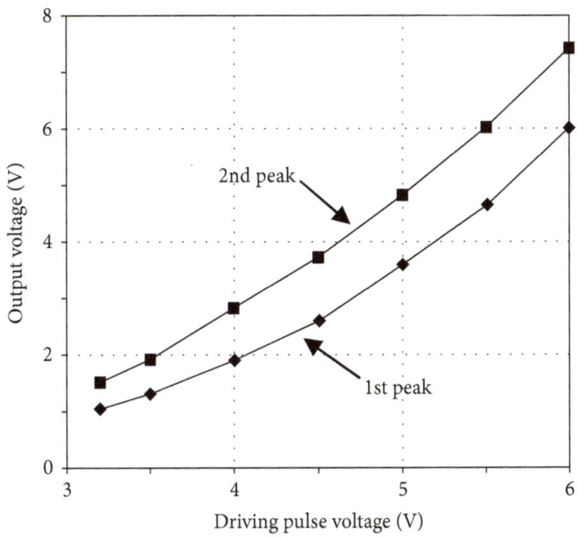

FIGURE 7: Signal output voltage at the pickup coil as a function of driving pulse voltage.

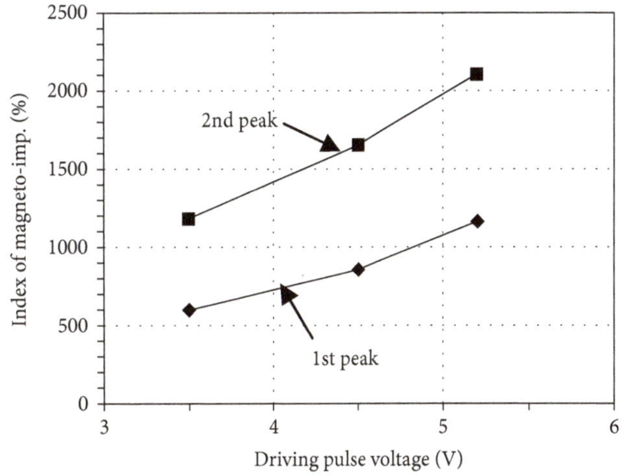

FIGURE 8: Magneto-impedance index of output signal at the pickup coil as a function of driving pulse voltage.

following experiment, the driving pulse voltage was increased to 5.2 V, and a sample magnet was moved to a 10 mm distance from the center of the sensor head. With this experiment setup, the effect on the MI index was investigated by changing the number of soft magnet layers. The MI index was reduced to 30% in the case of two layers for the first peak signal. For the second peak signal, this index reduction was cut to one-tenth in comparison to one layer of soft magnet, as shown in Figure 11(b). This indicates that increasing the number of soft magnet layers results in reduced sensitivity.

3. Metal Detection Sensor

3.1. Metal Sensor Circuit. A metal detection sensor using a magneto-impedance effect was fabricated. The soft magnetic strip of the sensor head was driven through the soft magnet driver by using a pulse train generator, which is composed of a driving pulse and unmatched degaussing pulse. The soft magnet driver is an electronic amplifier that supplies more current to the soft magnet with low impedance by means of impedance matching. The pulse widths were 6.5 ns and 60 ns, respectively. The effect due to geomagnetic field was canceled by adding a null current to the compensating coil of the sensor head. The sampling signal was delayed to sample the desired peak amplitude from the pickup coil by using a variable delay circuit. A driver circuit block diagram of the soft magnetic strip is shown in Figure 12. The reception signal was sampled and held to effectively recover the variation of second peak amplitude from the pickup coil. The product of pulse

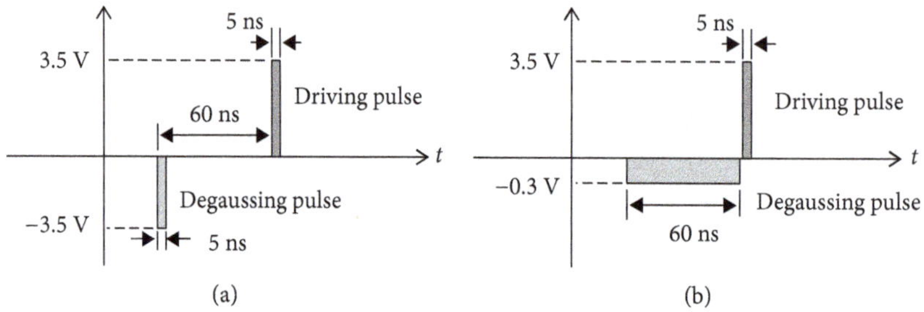

FIGURE 9: Driving pulse waveform of soft magnet with 5 ns width and degaussing pulse waveform. (a) Driving with matched degaussing. (b) Driving with unmatched degaussing.

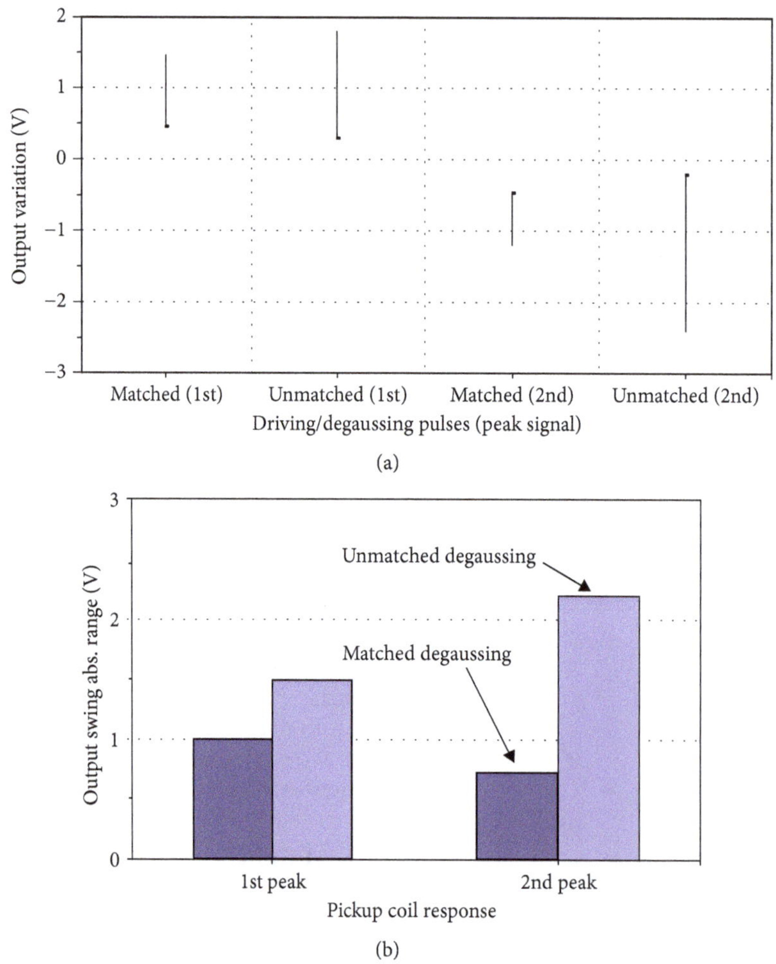

FIGURE 10: Comparison of output signal voltage between matched and unmatched degaussing, responding to geomagnetic field when driving with 3.5 V pulse amplitude. (a) Signal voltage variation. (b) Absolute value of signal voltage variation.

amplitude by pulse width between the driving pulse and the degaussing pulse was kept equal. The feedback signal from the receiver was used to minimize the effect by residual magnetic field by adjusting the amplitude of the degaussing pulse.

A high impedance buffer was used to preserve the waveform from the pickup coil, and a low leakage capacitor was used to hold the sampling signal. The DC level of the sample-and-hold signal was measured and converted to drive a compensation coil for the purpose of geomagnetic field cancelling. The external noise component was attenuated by using an 8th order Bessel passive noise filter with a variable cutoff frequency. Amplification followed, typically 20,000 times, but with a variable range of 40~50,000 times. The residual noise component, which was inherent in the MI sensor, was reduced by using an active noise filter. Finally, the presence of metal was determined by adjusting the threshold level in the detection circuit. A block diagram of the receiver part is shown in Figure 13.

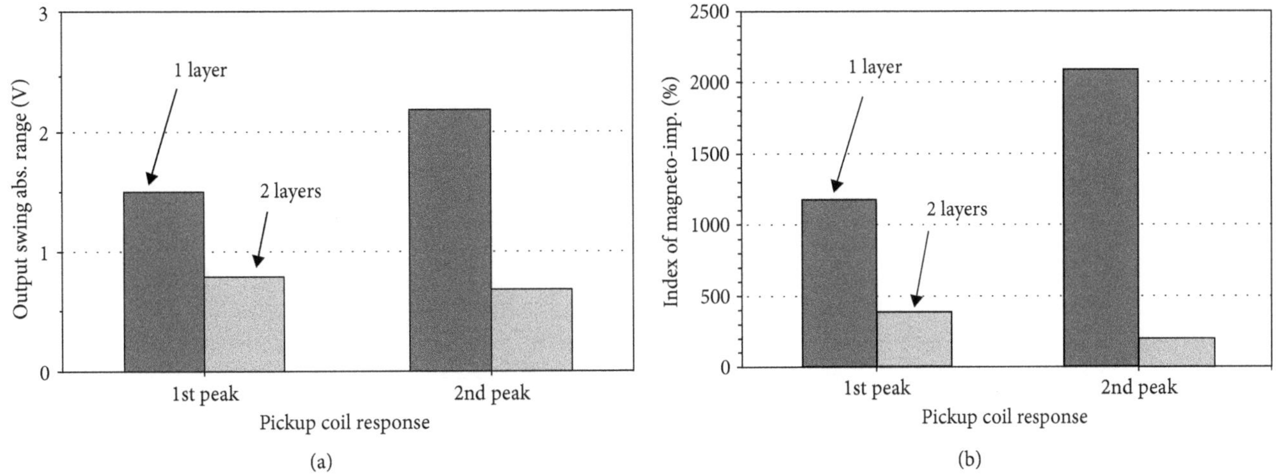

FIGURE 11: Comparison of output signal voltage between one and two soft magnet layers when driving with 5 ns pulse width. (a) Output voltage swing range. (b) MI index.

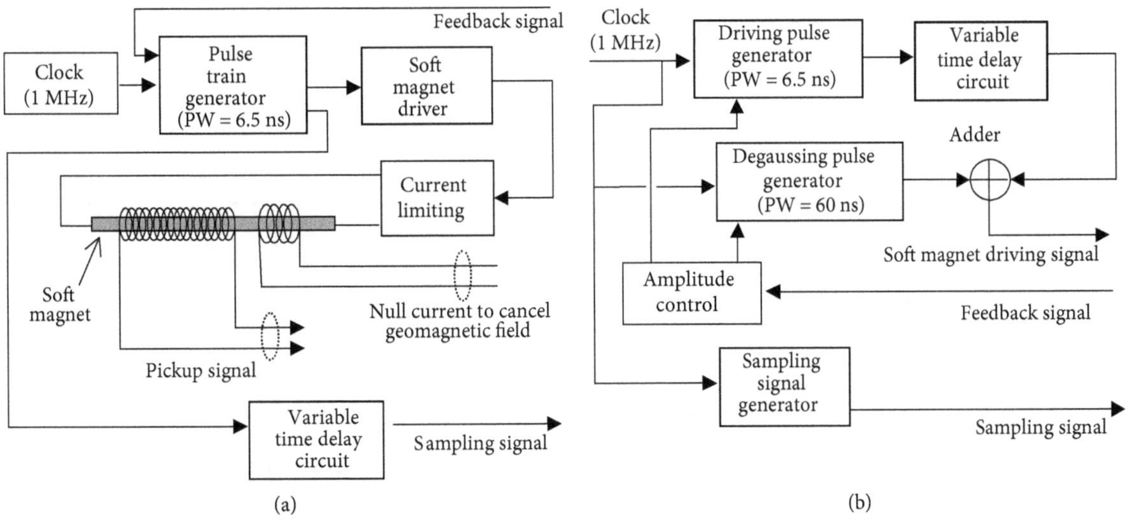

FIGURE 12: Sensor head driver circuit. (a) Block diagram of soft magnet driver. (b) Block diagram of pulse train generator.

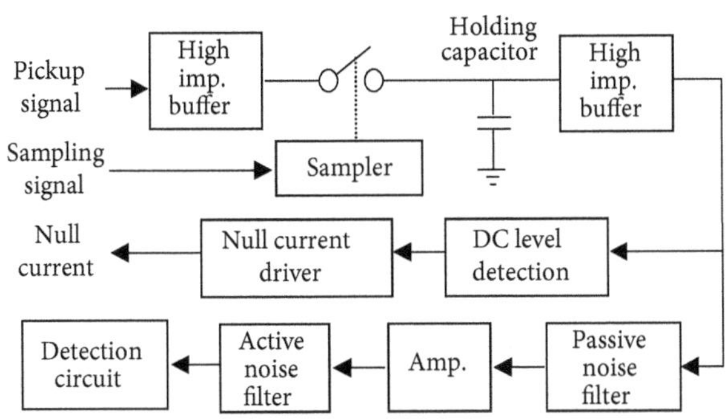

FIGURE 13: Block diagram of receiver circuit.

3.2. Active Noise Reduction. The reception signal from the MI sensor head contains irregular background noise with a wide frequency spectrum, which is hard to reduce by using a conventional passive filter. The characteristic of this noise is not a simple Fourier series frequency component but a mixture of multiple frequency and phase reversal components. Also, if we decrease the threshold point low enough to detect tiny metal objects at micron scale, the magnitude of noise contains a burst component that can be greater than the detection signal and lies in a higher frequency band. The movement of metal objects to be detected is within a lower frequency band of background noise, but it is slower than the period of mixed frequencies. If a metal object triggers an amplitude signal that is bigger than the noise level but wider in time period, the detection signal contains a big wave in a lower frequency band and small noise waves with mixed frequency. The active noise reduction (ANR) method cancels the noise signal with mixed frequency and even bigger burst component, but recovers a detection signal in a lower frequency band and even smaller amplitude than a noise burst. A diagram showing the ANR methodology is shown in Figure 14(a), and a controller in Figure 14(b). Inside the controller, a slow-varying component is removed, and a fast-varying component is recovered by using a differentiator.

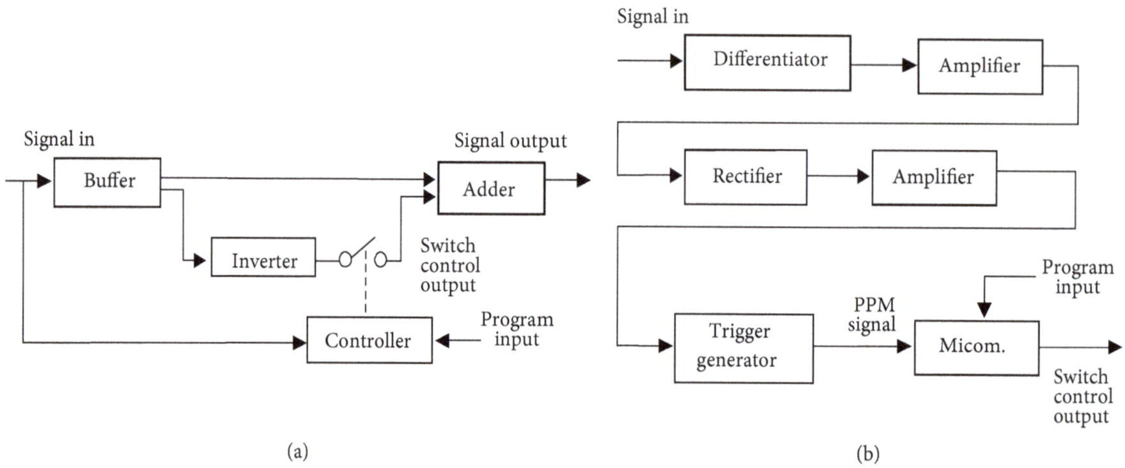

FIGURE 14: Method of active noise reduction. (a) Block diagram of ANR. (b) Controller of ANR.

This method is effective in removing low-frequency components in noise with a mixed frequency, while a detection signal with a longer time interval is also filtered out by using a conventional filter. After amplification and rectification followed by a trigger generator, it is converted to a pulse position modulation (PPM) signal. By using this PPM signal, a microprocessor determines the presence of a metal object by counting the pulse interval. An example showing the effect by active noise reduction is shown in Figure 15.

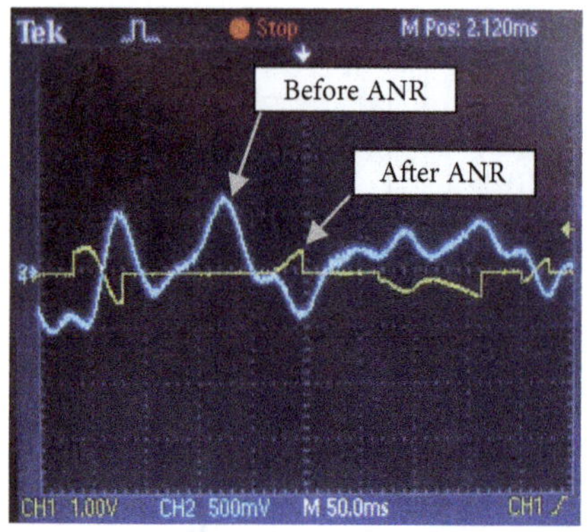

FIGURE 15: Noise waveform. CH1: after ANR; CH2: before ANR.

3.3. Metal Detection. The metal detection sensor with a longitudinal sensor geometry has a varying DC response depending on the longitudinal position at the sensor head and the angle of the sensor head with a geomagnetic field axis. However, an AC small signal response has sensitivity independent of the longitudinal position at the sensor head and the position angle with a geomagnetic field axis, because it senses a difference in signal variation. The DC response was kept to a null point by flowing a current in a compensation coil; thus, it prevented saturation in the soft magnet and enabled a dynamic range expansion. An AC small signal response can be amplified within the margin of the signal-to-noise ratio (SNR), and the ANR method can be applied to increase this SNR. The detection signal responding to a 0.2 mm diameter magnetized ferrous ball is shown in Figure 16. The first peak represents the presence of the magnetized metallic object, and the second small peak is due to unfiltered noise after ANR, which limits the minimum resolution of this magnetometer. The captured analog signal waveform before digital signal processing showed a discernible response to a magnetized metal object.

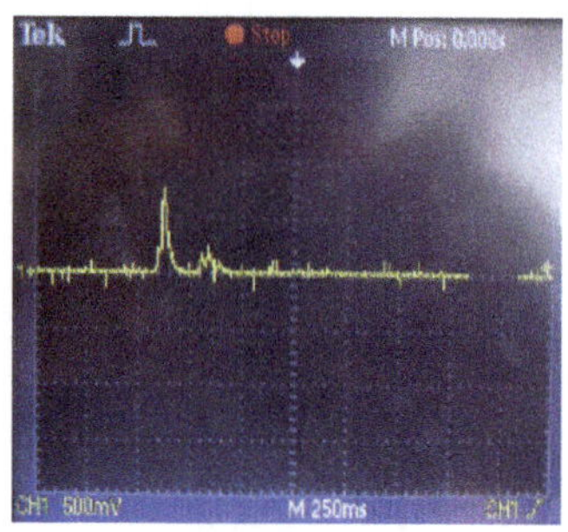

FIGURE 16: Metal detection waveform responding to 0.2 mm diameter ferrous ball.

4. Conclusion

A magnetometer with a longitudinal sensing line using the MI effect was examined experimentally to find an application to replace conventional metal sensors based on an LVDT. The sensing head consisted of a thin, 90 mm ribbon and surrounding wound pickup coil and was driven with an ultrashort pulse train with a pulse width of 5~7 ns. The unmatched degaussing methods were used to minimize the remnant magnetic field, and a compensation coil in the sensor head prevented the saturation of the sensor head and contributed to expand the dynamic range to external magnetic field variation. A noise reduction process using an ANR algorithm was effective in removing noise with mixed frequency components without influencing the detection signal with a longer time interval. A magnetized ferrous ball was used to examine the metal detection capability. The metal sensor in this experiment showed discernible responses to detect a 0.2 mm diameter magnetized metal object.

Conflicts of Interest

The author declares that there is no conflict of interests regarding the publication of this paper.

Acknowledgments

This work was supported by Incheon National University Research Grant in 2017.

References

[1] K. N. Choi, "Characteristics of metal sensor using variable frequency," *The Journal of the Korea Institute of Electronic Communication Sciences*, vol. 9, no. 2, pp. 161–166, 2014.

[2] K. N. Choi, "Two-channel metal detector using two perpendicular antennas," *Journal of Sensors*, vol. 2014, Article ID 412621, 11 pages, 2014.

[3] E. Arribas, I. Escobar, C. P. Suarez, A. Najera, and A. Beléndez, "Measurement of the magnetic field of small magnets with a smartphone: a very economical laboratory practice for introductory physics courses," *European Journal of Physics*, vol. 36, no. 6, article 065002, 2015.

[4] A. Bergen, H. J. van Weers, C. Bruineman et al., "Design and validation of a large-format transition edge sensor array magnetic shielding system for space application," *Review of Scientific Instruments*, vol. 87, no. 10, article 105109, 2016.

[5] M. Zhelamskij, "The active magnetic tracking with scalable coverage: indoor navigation for smartphones," *Journal of Sensors and Sensor Systems*, vol. 5, no. 2, pp. 355–371, 2016.

[6] T.-H. Tran, P. Chao, and P.-C. Chien, "The front-end readout as an encoder IC for magneto-resistive linear scale sensors," *Sensors*, vol. 16, no. 9, article 1416, 2016.

[7] M. Sakthivel, B. George, and M. Sivaprakasam, "A novel GMR-based eddy current sensing probe with extended sensing range," *IEEE Transactions on Magnetics*, vol. 52, no. 4, pp. 1–12, 2016.

[8] M. F. Snoeij, V. Schaffer, S. Udayashankar, and M. V. Ivanov, "Integrated fluxgate magnetometer for use in isolated current sensing," *IEEE Journal of Solid-State Circuits*, vol. 51, no. 7, pp. 1684–1694, 2016.

[9] T. Wang, Y. Zhou, C. Lei et al., "The disturbing effect of the stray magnetic fields on magnetoimpedance sensors," *Sensors*, vol. 16, no. 10, article 1723, 2016.

[10] A. Talaat, M. Churyukanova, J. M. Blanco, M. Ipatov, V. Zhukova, and A. Zhukov, "Simultaneous detection of giant magnetoimpedance and fast domain wall propagation in Co-based glass-coated microwires," *IEEE Magnetics Letters*, vol. 7, pp. 1–4, 2016.

[11] E. M. Kardoulaki, R. R. A. Syms, I. R. Young, and M. Rea, "SNR in MI catheter receivers for MRI," *IEEE Sensors Journal*, vol. 16, no. 6, pp. 1700–1707, 2016.

[12] S. Y. Liu, Q. P. Cao, X. D. Wang et al., "Effects of thickness on structure and magnetic property of Fe-Y-B thin films," *Thin Solid Films*, vol. 616, pp. 608–617, 2016.

[13] K. Lenz, E. Kosubek, K. Baberschke et al., "Magnetic properties of Fe_3Si/GaAs(001) hybrid structures," *Physical Review B*, vol. 72, no. 14, 2005.

[14] S. Tajima, Y. Okuda, T. Watanabe, H. Aoyama, M. Yamamoto, and T. Uchiyama, "High-resolution magneto-impedance sensor with TAD for low noise signal processing," *IEEE Transactions on Magnetics*, vol. 50, no. 11, pp. 1–4, 2014.

[15] J. J. Beato-López, J. I. Pérez-Landazábal, and C. Gómez-Polo, "Magnetic nanoparticle detection method employing nonlinear magnetoimpedance effects," *Journal of Applied Physics*, vol. 121, no. 16, article 163901, 2017.

[16] S. Sandacci, D. Makhnovskiy, L. Panina, K. Mohri, and Y. Honkura, "Off-diagonal impedance in amorphous wires and its application to linear magnetic sensors," *IEEE Transactions on Magnetics*, vol. 40, no. 6, pp. 3505–3511, 2004.

10

Design and Implementation of Acoustic Sensing System for Online Early Fault Detection in Industrial Fans

Cihun-Siyong Alex Gong,[1,2,3] Huang-Chang Lee,[1,4] Yu-Chieh Chuang,[1] Tien-Hua Li,[5] Chih-Hui Simon Su,[5] Lung-Hsien Huang,[5] Chih-Wei Hsu,[6] Yih-Shiou Hwang,[3,7] Jiann-Der Lee,[1,4,8] and Chih-Hsiung Chang[5]

[1]*Department of Electrical Engineering, School of Electrical and Computer Engineering, College of Engineering, Chang Gung University, Taoyuan, Taiwan*
[2]*Portable Energy System Group of Green Technology Research Center, Chang Gung University, Taoyuan, Taiwan*
[3]*Department of Ophthalmology, Chang Gung Memorial Hospital, Linkou, Taoyuan, Taiwan*
[4]*Department of Neurosurgery, Chang Gung Memorial Hospital, Linkou, Taoyuan, Taiwan*
[5]*AI-GTG Group, Taoyuan, Taiwan*
[6]*Chiu Chau Enterprise Co. Ltd., Taoyuan, Taiwan*
[7]*Graduate Institute of Clinical Medical Sciences, College of Medicine, Chang Gung University, Taoyuan, Taiwan*
[8]*Department of Electrical Engineering, Ming Chi University of Technology, New Taipei City, Taiwan*

Correspondence should be addressed to Jiann-Der Lee; jdlee@mail.cgu.edu.tw

Academic Editor: Armando Ricciardi

Industrial fans play a critical role in manufacturing facilities, and a sudden shutdown of critical fans can cause significant disruptions. Ensuring early, effective, and accurate detection of fan malfunctions first requires confirming the characteristics of anomalies resulting from initial damage to rotating machinery. In addition, sensing and detection must rely on the use of sensors and sensing characteristics appropriate to various operational abnormalities. This research proposes an online industrial fan monitoring and fault detection technique based on acoustic signals as a physical sensing index. The proposed system detects and assesses anomalies resulting from preliminary damage to rotating machinery, along with improved sensing resolution bandwidth features for microphone sensors as compared to accelerometer sensors. The resulting Intelligent Prediction Integration System with Internet (IPII) is built to analyze rotation performance and predict malfunctions in industrial fans. The system uses an NI cRIO-9065 embedded controller and a real-time signal sensing module. The kernel algorithm is based on an acoustic signal enhancement filter (ASEF) as well as an adaptive Kalman filter (AKF). The proposed scheme uses acoustic signals with adaptive order-tracking technology to perform algorithm analysis and anomaly detection. Experimental results showed that the acoustic signal and adaptive order analysis method could effectively perform real-time early fault detection and prediction in industrial fans.

1. Introduction

Industrial fans are critical components in industrial facilities and are used to remove exhaust emissions, ventilate, compress air, and drive air-conditioning systems. Such fans typically operate continuously for long durations, and improper assembly or maintenance can result in malfunctions, including vibrations and audible noise. Common mechanical faults are mostly caused by bearing failures, looseness, shaft cracks, poor balance, misalignment, and resonance. Figure 1 shows general maintenance procedures for rotating machinery. Predictive maintenance of such machinery minimizes unexpected failure and potential downtime but relies on accurate performance and measurement techniques

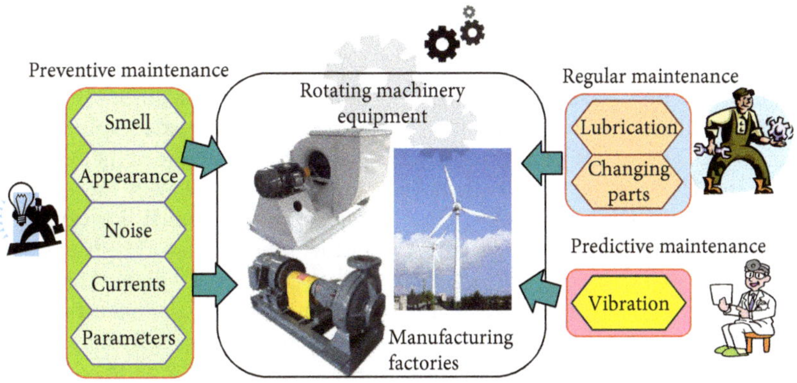

FIGURE 1: The maintenance category of rotating machinery.

through the integration of various sensors to provide real-time data on equipment conditions.

Problems in rotating machinery are typically indicated by the presence of abnormal vibration, noise, or temperature. Accelerometers are typically used to perform vibration signal analysis [1], with portable and online real-time monitoring methods based on the ISO 13373 standards [2].

The traditional signal processing involves transforming the measured vibration time-domain signal by fast Fourier transform (FFT) into frequency domain feature values, a process also known as Frequency Spectrum Analysis [3]. However, the shaft rotational speed of industrial fans changes with factory production conditions, loading demands, and many external environmental factors. Therefore, the measured vibration signals of bearings produce a nonlinear, nonstationary time-domain signal, while the traditional Fourier transform is only appropriate for linear and stationary time-domain signals and cannot accurately reflect time-variant factors to respond to the nonstationary vibration behavior and complex vibration signals of industrial fans. In response to vibration signals indicating early malfunctions, the common spectrum merely presents the concentrated energy at a certain frequency peak value. Spectrum analyses do not show or resolve the vibration energy distributions of the harmonics when other harmonics change with the rotating speed.

Order-tracking analysis is an effective method for the signal analysis of nonstationary vibration signals from rotating machinery. However, the use of order-tracking methods alone while performing dynamic signal analysis results in inadequate signal resolution following Fourier transform spectrum analysis. Due to equipment complexity and large speed variation in rotating machinery, component interrelations raise a nonstationary cross-correlation function problem. Thus, the traditional Fourier transform frequency domain signal analysis can easily cause undesirable smearing while traditional order analysis cannot improve the resolution of each harmonic order. Hence, order tracking integrates the recursive least-square method and the recursive Kalman filter [4, 5] to provide adaptive order tracking. Using the adaptive algorithm overcomes the frequency smearing effect in adjacent and cross orders of traditional order analysis. The adaptive algorithm's high-resolution signal and order curve change with rotation speed characteristics make it very suitable for signal analysis in conditions of high rotating speed or variable speeds. Adaptive order-tracking analysis provides good precision for the real-time vibration signal fault diagnosis of rotating machinery.

Wavelet transform (WT) is another method of analyzing nonstationary signals from machinery rotating at variable speeds. WT provides good time and frequency resolution, overcoming the limitations of time-domain and frequency-domain analysis in Fourier transform. The WT-tracking algorithm also provides effective vibration signal tracking in conditions of variable speed. Therefore, WT is often used to diagnose faults of rotating machinery based on vibration signal analysis [6–9]. It can also be used for spectrum analysis of acoustic signals and to identify signal signatures as a means of diagnosing faults from acoustic signals [10]. However, the major disadvantage of WT is that it applies a fixed wavelet base to decompose signals. Thus, an inappropriate selection of the basis function and a failure due to mismatched signal characteristics can significantly reduce signal analysis performance [11, 12]. In other words, WT is not generally adaptive, and the transform efficiency is highly dependent on the basis function selection.

We have surveyed similar research regarding fault diagnostic methods for various types of rotating equipment, each with their own theoretical foundations, advantages, and solutions. These methods and theories are largely similar and are based largely on vibration sensors [13, 14]. Each approach selects diagnostic subjects and algorithms based on specific research purposes and considerations. This paper starts over from scratch with a reconsideration of sensor type, algorithm simulation, and continuous experimental adjustment in an attempt to improve early anomaly detection in factory-based industrial fans. Our review of the previous three years' studies in related fields found no mention of the use of acoustic sensors and similar research methods (i.e., algorithms) for use in industrial fan fault diagnosis. Using vibration signals to perform fault prediction in rotating machinery can result in weak extraction of early fault features. The weak abnormality features of early faults are generally masked by background noise or other interference, making them difficult to detect. However, early detection is critical to prevent continuous deterioration or sudden failure. In [15, 16], the authors focused on the early identification of fault signals. Industrial fans most commonly suffer from bearing failures which typically proceed through four failure stages [17–19]. In the first stage, slight defects and cracks appear with abnormalities manifesting at ultrasound frequencies (20~60 kHz). The faults will gradually deteriorate until the device fails. However, vibration sensors typically are more sensitive to low frequencies (below 2 kHz roughly) [20], limiting their utility in early fault detection for rotating machinery.

In addition to vibration signals, rotating machinery produces acoustic signals. Certain equipment faults produce specific vibration signals along with corresponding acoustic signals. The acoustic signal from rotating machinery includes considerable meaningful information. Unlike vibration signals, acoustic signals are usually measured by easily applied noncontact methods. Furthermore, acoustic signals have better frequency resolution at medium and high frequencies [20], and acoustic sensors provide advantages for preliminary fault detection in rotating machinery [18].

An industrial fan involves complex acoustic signals in operation. An acoustic sensor possesses superior high-frequency range sensing characteristics to match the signal characteristics of early failure in industrial fans and process fault detection through the order-tracking analysis and adaptive Kalman filter algorithms with good detection results. For the nonstationary signal of the rotation apparatus, the commonly used FFT spectrum analysis cannot be used effectively to distinguish the frequency composition. With regard to the wavelet transformation, its applications are limited to its nonadaptivity. The Kalman filter has been widely applied to various technical fields, featuring its fast signal processing and high performance for big data. The studies in [21–23] are good examples with it, where the Kalman filter has been used to deal with equipment failures. Despite its fast convergence speed, the Kalman filter suffers a little bit from poor stability and a higher filtering threshold in the target. As a result, we have added the tachometer signal of the industrial fan into the proposed system designed to be used in conjunction with the order-tracking technique, where the model-based adaptive Kalman filter inspired from [24–26] has also been utilized to estimate the best status of the dynamic system.

We use high-resolution-order-analysis technology to analyze the characteristics of acoustic and vibration signals in terms of rotation speed to capture information related to the operation of rotating equipment. This information is then combined with the adaptive Kalman filter algorithm to produce an early fault detection method for rotating equipment. The proposed method has particular advantages for fault detection in variable speed conditions. The key is the integration of the industrial fan's tachometer signal and to use the motor shaft angle ($\Delta\theta$) for resampling of nonstationary acoustic signals. We then use the adaptive Kalman filter in a recursive manner to eliminate background noise and various interference signals to estimate the dynamic system's optimal state. This structure is based on variable speeds, and the proposed algorithm provides advantages for noise elimination.

The ASEF and AKF are both excellent kernel algorithms for rotating machinery, and order-tracking techniques can be applied to acoustic signals to differentiate various fault conditions. The most significant thing of this study is to propose an idea using an acoustic sensor to carry out signal detection. By taking the advantage of the superior high-frequency response and combining it with the ASEF and AKF high-resolution algorithm, the built IPII is capable of predicting early failure of the industrial fan. The user can therefore be warned early to use contingency plans. For the periodic dynamic signal of the fan, the IPII uses acoustic signal and fiber-optical signal to obtain the phase. By combining both the order-tracking and AKF algorithms, the IPII can recognize the features resulting in system failure early. For the nonperiodic dynamic signal, it relies mainly on the ASEF algorithm along with the acoustic signal to perform abnormal feature analysis for early diagnosis of the failure. It is an innovative method for predictive maintenance in factories. This study can contribute significantly not only to industrial fan systems but also to any other periodic and nonperiodic rotation systems. The IPII is capable of carrying out real-time analysis and prediction of rotating machinery conditions before failure occurs. Unexpected failure of crucial rotating machinery can be prevented through the development of a fault prediction system better able to identify preliminary damage. Adjusting or modifying the parameters during production processes can prevent production shutdowns and reduce maintenance costs.

2. Signal Feature Extraction Theory

2.1. ASEF

2.1.1. Acoustic Signal Enhancement. Nonstationary signals such as online acoustics and vibrations generated by industrial fans are difficult to calculate, and signal enhancement is a crucial preprocessing step to eliminate noise interference. The subspace signal enhancement method removes noise from the extracted signals, thus improving SNR quality.

A Karhunen-Loeve Transform (KLT) is performed on the noisy zero-mean normalized vector $\widehat{\mathbf{S}} = [\widehat{\mathbf{S}}_1, \widehat{\mathbf{S}}_2, \ldots, \widehat{\mathbf{S}}_N]^T$. The vector $\widehat{\mathbf{S}}$ has a symmetric nonnegative autocorrelation matrix $\mathbf{R} = \varepsilon[\widehat{\mathbf{S}}^T\widehat{\mathbf{S}}]$. Since $\mathbf{R}$ is an $r \times r$ matrix, it determines a complete set of orthogonal eigenvectors associated with the real, nonnegative eigenvectors. These eigenvalues can be ordered as $\lambda_1 \geq \lambda_2 \geq \cdots \geq \lambda_r \geq 0$, and the statistical variance of the data set in the direction of the kth eigenvector β_k corresponds to the eigenvalue λ_k. $\widehat{\mathbf{S}}$ can be expressed using the following orthogonal transformation [27]:

$$\widehat{\mathbf{S}} = \sum_{k=1}^{r} a_k \beta_k, \quad k = 1, \ldots, r, \tag{1}$$

and a_k coefficients, which are called the principal components, can be shown by projecting the data vectors onto each eigenvector as follows:

$$a_k = \widehat{\mathbf{S}}^T \beta_k, \quad k = 1, \ldots, r. \tag{2}$$

Noise reduction can be achieved by reconstructing the initial data using only the r-weighted eigenvectors of the signal-plus-noise subspace, and the linear estimation of the clean vector **S** is used as an estimation criteria as follows [28]:

$$\tilde{\mathbf{S}} = \sum_{k=1}^{r} g_k a_k \beta_k, \quad k = 1, \dots, r, \tag{3}$$

where g_k is a weighting function. An optimal parameter choice for g_k which results in more aggressive noise suppression is given by

$$g_k = \exp\left\{\frac{-v\sigma_n^2}{\lambda_k}\right\}, \quad k = 1, \dots, r, \tag{4}$$

where σ_n^2 is the noise variance and the value of the parameter v is fixed experimentally.

Signal enhancement is achieved by removing the noise subspace and estimating the clean signal from the remaining signal space. Our ASEF features acoustic-sensor-based nonperiodic acoustic signal processing for an industrial fan. It begins with the KLT filtering, followed by the MFCC filtering. The results of the MFCC filtering underwent abnormal feature analysis.

2.1.2. Analysis of Cepstral Acoustic Features. Cepstrum analysis is widely used in automatic speech recognition, acoustic echo cancellation, and acoustic filtering. The cepstrum analysis technique is defined as the inverse Fourier transform of the logarithm of the short-term power spectrum of measured data in the time domain. Cepstrum separates and enhances the periodic signal so the significant correlation function can be identified.

When processing signals [27, 29], we first extracted the sensed signals and then proceeded to signal preprocessing. We then turned the signal into a series of spectrum sequences by short-time FFT. The magnitude spectrum of the signal was then passed through a set of 20 triangular bandpass filters to retrieve the output log-energy of each filter. These triangular filters are spaced on the mel-frequency scale. The relation of mel-frequency and general frequency f is as follows:

$$\mathrm{mel}(f) = 2595 \log_{10}\left(1 + \frac{f}{700}\right), \tag{5}$$

where f corresponds to the linear frequency scale.

To transform the logarithm energy value E_k of each filter to discrete cosine transform (DCT), retrieve the mel-scale frequency cepstral coefficient (MFCC) and convert the frequency domain signal to the time domain. The DCT equation is as follows:

$$\mathrm{MFCC}_n = \sum_{k=1}^{B} E_k \cos\left(\frac{\pi n}{B}(k - 0.5)\right), \quad n = 1, 2, \dots, N, \tag{6}$$

where N is the number of the cepstral coefficients, B is the analysis order, and E_k, $k = 1, 2, \dots, B$ represents the log-energy output of the kth filter.

2.2. AKF

2.2.1. Resampling Theory. This research applies the order analysis of the resampling theorem to analyze and diagnose the acoustic and vibration signal features generated by industrial fans. Particularly in the case of varying rotating shaft speeds, using this theory to extract the acoustic and vibration signals achieves a high resolution from the sampling intervals. A Fourier transform can be used to calculate the order spectrum generated by the periodic vibration/acoustic time sequences from the rotating information. Thus, the acoustic and vibration signal energy corresponds to each phase angle of the rotating information from the fiber-optical switch. Discrete-time signals are resampled according to equal rotating shaft angle intervals. Converting the signals from time domain to the angle domain $(\Delta t \rightarrow \Delta\theta)$ is a function of resampling theory [30].

The proposed experiment collects synchronous acoustic, vibration, and optical signals. We apply interpolation and curve-fitting to the impulse signal of the fiber-optical switch to retrieve the relative angular displacement at each time point. We then resample the acoustic and vibration signals according to equal angular displacement intervals. The resampled signals are called order signals, and the spectrogram after FFT or STFT of the order signals is called the order spectrogram, which can be used to provide a precise analysis of the fundamental frequency and its relative harmonics, thus facilitating analysis of the online acoustic and vibration signal features of rotating machinery.

2.2.2. Adaptive Kalman Filter. An industrial fan's acoustic signal is essentially a frequency-modulated signal that can be expressed as the superposition of sinusoidal signals. The acoustic signal $x(t)$ containing k orders generated by the rotating shaft can be written as

$$x(t) = A_k \cos\left[k\theta(t) + \phi_k\right], \tag{7}$$

where A_k and ϕ_k respectively express the amplitude and phase of the kth order. The variable $\theta(t)$ represents the angular displacement of the rotating spindle that can be computed as

$$\theta(t) = \int_0^t 2\pi f(t)\, d\tau. \tag{8}$$

A proposed adaptive order-tracking analysis based on Kalman filtering is presented. The Kalman-filtering problem entails two equations [4, 5, 24–26]. The first equation is called the process equation and is expressed as

$$x(n+1) = \mathbf{F}(n+1, n)x(n) + \mathbf{v}_1(n), \tag{9}$$

where $\mathbf{F}(n+1, n)$ is a known $M \times M$ state transition matrix related to the state of the system at times $n+1$ and n. The $M \times 1$ vector $\mathbf{v}_1(n)$ represents process noise. The vector $\mathbf{v}_1(n)$ is modeled as a zero-mean white-noise whose correlation matrix is defined as

$$E[\mathbf{v}_1(n)\mathbf{v}_1^H(k)] = \begin{cases} \mathbf{Q}_1(n), & n = k, \\ 0, & n \neq k. \end{cases} \tag{10}$$

The second equation is called the measurement equation and is expressed as

$$\mathbf{y}(n) = \mathbf{C}(n)x(n) + \mathbf{v}_2(n), \tag{11}$$

where $\mathbf{C}(n)$ is a known $N \times M$ measurement matrix. The $N \times 1$ vector $\mathbf{v}_2(n)$ is the measurement noise modeled as a zero-mean white-noise process whose correlation matrix is defined as.

$$E[\mathbf{v}_2(n)\mathbf{v}_2^H(k)] = \begin{cases} \mathbf{Q}_2(n), & n = k, \\ 0, & n\mathbf{Q}_2(n), \end{cases} \tag{12}$$

where the noise vectors $\mathbf{v}_1(n)$ and $\mathbf{v}_2(n)$ are statistically independent so that we have

$$E[\mathbf{v}_1(n)\mathbf{v}_2^H(k)] = 0, \tag{13}$$

for all n and k.

We define the Kalman gain expressed as

$$\mathbf{G}(n) = \mathbf{F}(n+1,n)\mathbf{K}(n,n-1)\mathbf{C}^H(n)\mathbf{R}^{-1}(n), \tag{14}$$

where the $\mathbf{R}(n)$ is called the correlation matrix. The $M \times M$ matrix $\mathbf{K}(n, n-1)$ is called the predicted state-error correlation matrix.

A filtered estimation error vector is shown in

$$\begin{aligned} \mathbf{e}(n) &= \mathbf{y}(n) - \mathbf{C}(n)\hat{\mathbf{x}}(n|\mathbf{y}_n) \\ &= \mathbf{y}(n) - \mathbf{C}(n)\hat{\mathbf{x}}(n|\mathbf{y}_{n-1}) - \mathbf{C}(n)\mathbf{F}(n,n+1)\mathbf{G}(n) \\ &= \boldsymbol{\alpha}(n) - \mathbf{C}(n)\mathbf{F}(n,n+1)\mathbf{G}(n)\boldsymbol{\alpha}(n), \end{aligned} \tag{15}$$

where $\boldsymbol{\alpha}(n)$ is called the innovation vector into the filtered estimation error vector $\mathbf{e}(n)$.

The aforementioned parameter identification problem can be recursively solved using the Kalman-filtering algorithm. The procedure of the recursive Kalman filtering is summarized as follows [23]:

$$\begin{aligned} \mathbf{G}(n) &= \mathbf{F}(n+1,n)\mathbf{K}(n,n-1)\mathbf{C}^H(n) \\ &\quad \cdot \left[\mathbf{C}(n)\mathbf{K}(n,n-1)\mathbf{C}^{\mathrm{H}}(n) + \mathbf{Q}_2(n)\right]^{-1}, \end{aligned} \tag{16}$$

$$\boldsymbol{\alpha}(n) = \mathbf{y}(n) - \mathbf{C}(n)\hat{\mathbf{x}}(n|\mathbf{y}_{n-1}), \tag{17}$$

$$\begin{aligned} \hat{\mathbf{x}}(n+1|\mathbf{y}_n) &= \mathbf{F}(n+1,n)\hat{\mathbf{x}}(n|\mathbf{y}_{n-1}) \\ &\quad + \mathbf{G}(n)\boldsymbol{\alpha}(n), \end{aligned} \tag{18}$$

$$\begin{aligned} \mathbf{K}(n) &= \mathbf{K}(n,n-1) \\ &\quad - \mathbf{F}(n,n+1)\mathbf{G}(n)\mathbf{C}(n)\mathbf{K}(n,n-1), \end{aligned} \tag{19}$$

$$\begin{aligned} \mathbf{K}(n+1,n) &= \mathbf{F}(n+1,n)\mathbf{K}(n)\mathbf{F}^H(n+1,n) \\ &\quad + \mathbf{Q}_1(n). \end{aligned} \tag{20}$$

For the above procedure, all the variables are summarized in Table 1. To initialize the adaptive Kalman-filtering process, the initial conditions are generally $\hat{\mathbf{x}}(1 \mid \mathbf{y}_0) = \mathbf{0}_{(4k+2)\times 1}$ ($4k+2$ is the number of parameters $A_i(n)$) and $\mathbf{K}(1,0) = \mathbf{I}$ ($\mathbf{I}$ is a $(4k+2) \times (4k+2)$ identity matrix). Figure 2 shows the block diagram of the recursive Kalman filter.

Although the adaptive Kalman-filtering algorithm has greater computation and time requirements because of its

Table 1: Summary of the adaptive Kalman-filtering variables.

Variable	Definition	Dimension
$x(n)$	State vector at time n	M-by-1
$y(n)$	Observation vector at time n	N-by-1
$\mathbf{F}(n+1,n)$	State transition matrix from time n to $n+1$	M-by-M
$\mathbf{C}(n)$	Measurement matrix at time n	N-by-M
$\mathbf{Q}_1(n)$	Correlation matrix of process noise vector $\mathbf{v}_1(n)$	M-by-M
$\mathbf{Q}_2(n)$	Correlation matrix of measurement noise vector $\mathbf{v}_2(n)$	N-by-N
$\hat{\mathbf{x}}(n+1 \mid \mathbf{y}_n)$	Predicted estimate of the state vector at time $n+1$, given the observation vectors $\mathbf{y}(1), \mathbf{y}(2), \ldots, \mathbf{y}(n)$	M-by-1
$\hat{\mathbf{x}}(n \mid \mathbf{y}_n)$	Filtered estimate of the state vector at time n, given the observation vectors $\mathbf{y}(1), \mathbf{y}(2), \ldots, \mathbf{y}(n)$	M-by-1
$\mathbf{G}(n)$	Kalman gain at time n	M-by-N
$\boldsymbol{\alpha}(n)$	Innovation vector at time n	N-by-1
$\mathbf{R}(n)$	Correlation matrix of the innovation vector $\boldsymbol{\alpha}(n)$	N-by-N
$\mathbf{K}(n+1,n)$	Correlation matrix of the error in $\hat{\mathbf{x}}(n+1 \mid \mathbf{y}_n)$	M-by-M
$\mathbf{K}(n)$	Correlation matrix of the error in $\hat{\mathbf{x}}(n \mid \mathbf{y}_n)$	M-by-M

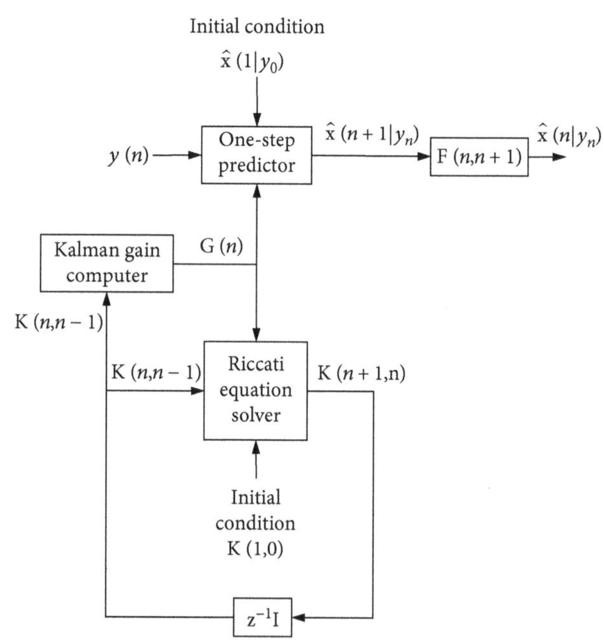

Figure 2: Block diagram of the recursive Kalman filter.

TABLE 2: The specifications of the microphone sensor.

Type	PCB 130E20 (precision condenser)
Frequency response	20 to 10k Hz
Sensitivity	45 mV/Pa

TABLE 3: The specifications of accelerometer sensor.

Type	PCB 603C01 (piezoelectric)
Frequency response	0.5 to 10k Hz
Sensitivity	100 mV/g

recursive calculation structure, we chose it for its advantages of high estimation precision, fast convergence, and robustness. All of these features are crucial to the equipment fault prediction system which requires immediate response to effectively and accurately detect abnormalities.

3. Experimental Structure

3.1. Sensor Specifications and Installation. Common fault signs of industrial fans consist of abnormal vibrations, audible noise, and temperature. To build a complete and effective monitoring system, this research proposes an Intelligent Prediction Integration System with Internet (IPII) to perform online monitoring and fault prediction of industrial fans [31]. The system uses an NI cRIO-9065 embedded controller and a real-time synchronizing module. The specifications of the microphone and accelerometer sensors used are shown in Tables 2 and 3. The sensor installation structure is shown in Figure 3.

3.2. Online Monitoring and Fault Prediction of Industrial Fans with IPII. In addition to the main acoustic signal, the IPII monitoring system also uses vibration and temperature signals to assist diagnosis. The system structure includes three layers based on different functions:

(1) Sensing layer: microphone, accelerometer, thermocouple, and fiber-optical sensors are used to extract various operating signals from industrial fans. This system also uses a wireless sensor network (WSN) to perform wireless sensing transmission.

(2) Smart control layer: it performs computation, analysis, and diagnosis from the extracted data and implements smart controls according to practical requirements, such as emergency measures in response to output alarms or automatically switching to spare machinery.

(3) Monitoring layer: monitors system status through a graphical user interface (GUI) which can be operated at a central control room or on mobile devices via wireless networks for remote monitoring. Smart devices can also be used to remotely shut down potentially abnormal machinery. The system structure is shown in Figure 4.

Signal detection relies mainly on the microphone sensor, assisted by the accelerometer with an adaptive order-tracking algorithm to more precisely distinguish certain featured signals. For example, the acoustic signal feature of the blade pass frequency can be analyzed by signal order analysis to obtain a more precise identification. Meanwhile, it can also obtain abnormal feature order energy signals of bearings, blades, or motors according to accelerometer characteristics. The system includes a thermocouple to detect bearing temperature, along with acoustic and vibration signals to serve as the basis for executing smart controls in response to fault signs.

System hardware consists of an embedded NI RIO-9065 dual-core controller, FPGA, real-time module, signal synchronization A/D module, and WSN module. The kernel algorithm is an adaptive Kalman filter. Acoustic and vibration signals were individually combined with tachometer signals for simultaneous algorithm analysis and abnormal feature extraction. Big data techniques are used to process multiple sensed signals from collection, storage, extraction, and analysis to decision making as a looping execution. The monitoring and control flow chart is shown in Figure 5.

Our ASEF features acoustic-sensor-based nonperiodic acoustic signal processing for an industrial fan. It begins with the KLT filtering, followed by the MFCC filtering. The results of the MFCC filtering underwent abnormal feature analysis. Anomaly detection is just one process of the system, and the goal of anomaly detection is to ensure smooth production line operation. Thus, we combine diagnostics with controls to determine an optimal operation strategy based on current equipment status, while using fault diagnostics to prevent future equipment failure. In response to anomalies, the system uses an integrated smart control to maintain normal production and safety through implementing corresponding contingency measures such as reducing operation speed, activating spare machinery, or adjusting production parameters. Effective early fault detection requires preprocessing of the various signals to extract important features. Signal preprocessing seeks to separate the signal of interest from noise to ensure the effective detection of abnormal features.

The first step is to enhance the sensed signal. Karhunen-Loeve transform (KLT) high-pass filtering eliminates the remaining noise. The Hamming window is then applied with a certain bandwidth. Next, following inverse transform, the signal is subjected to mel-frequency cepstral coefficient (MFCC) filtering to process the signal features. When the signal preprocessing procedures are completed, the signal is then subjected to Kalman filtering for online real-time fault prediction of industrial fans.

This process can also be used for spectrum feature filtering to confirm the sensing characteristics of different sensors. Spectrum feature filtering analyzes the sensing resolution frequency bandwidth feature in a free field, using consistent ambient noise levels, along with environmental and measured conditions for different sensors to assess differences in the practical measured sensing resolution bandwidth features for two or more sensors and can be used as the selected reference for this research. The analysis flow chart is shown in Figure 6.

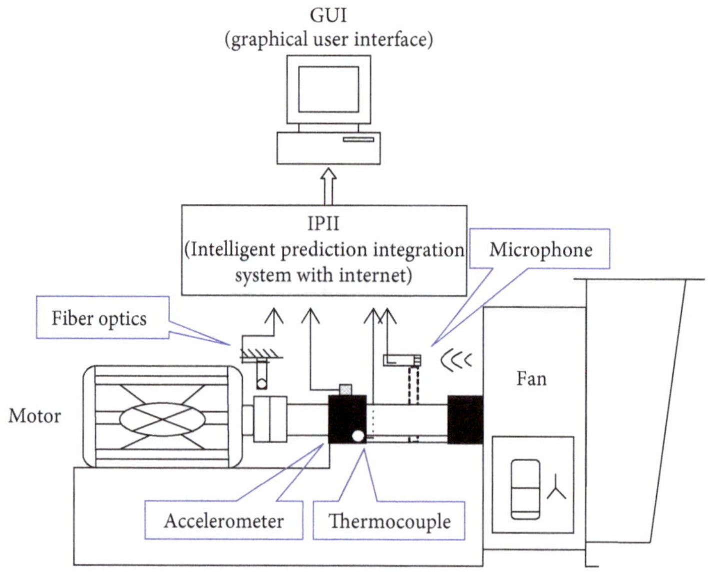

FIGURE 3: Sensor installation structure.

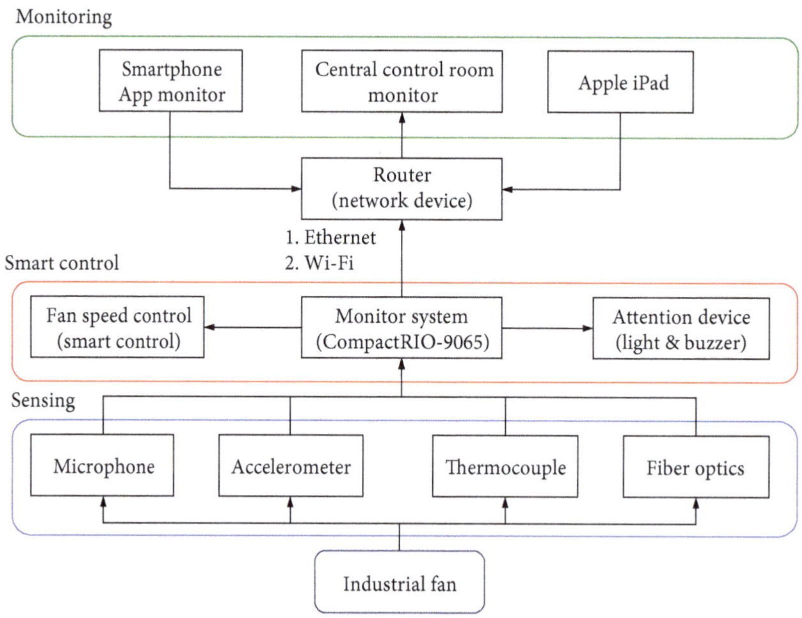

FIGURE 4: IPII system structure diagram.

The IPII incorporates tachometer signal along with the order-tracking technique in it. It is also with the model-based adaptive Kalman filter [24–26]. The filter possesses a recursive least-square algorithm and recursive Kalman filtering in the adaptive signal processing to seek an optimal solution for the problems, where the background noise and any other interference can be effectively filtered out so that we can come out with the best status of the dynamic system. The setup of the IPII is dedicated to signal processing of the collected data from the acoustic and tachometer sensor. The data first underwent short-time Fourier transform (STFT). The results of STFT are resampled and combined with order tracking to perform recursive Kalman filtering, followed by the order feature analysis to obtain the output (Figure 7).

4. Result and Verification

4.1. Setup. The monitoring system uses a high-resolution order-tracking analysis, combining the big data collected from a microphone, accelerometer, and tachometer with the bearing temperature signals to execute the algorithm to implement fault prediction for industrial fans based on acoustic signals. Figure 8 shows the experimental framework based on an industrial fan with an induction motor. The GUI

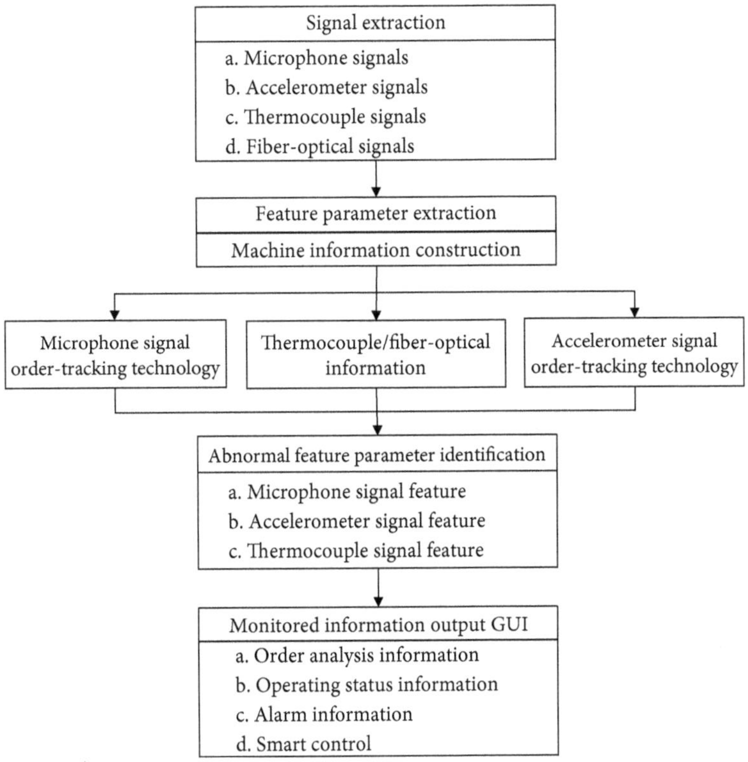

FIGURE 5: Monitoring and control flow chart.

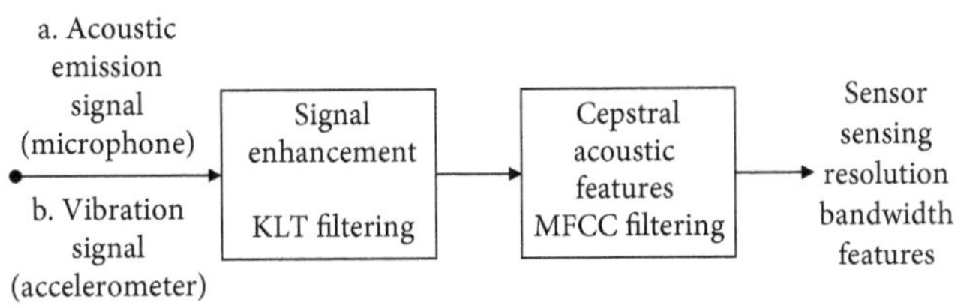

FIGURE 6: Microphone- and accelerometer-sensor-sensing resolution bandwidth feature flow chart.

in Figure 9 can be operated in the central control room or on smart networked devices for remote monitoring.

4.2. Acoustic Signal Preprocessing

4.2.1. Spectrum Feature Filtering of Acoustic Signals. We realize the proposed signal preprocessing and spectrum feature filtering of acoustic signal by means of the filtering algorithm and the experiment structure presented in Section 4.1 on the basis of Section 3. The signal preprocessing procedures and their individual signal waveform variations are described as follows. Figure 10(a) shows the original acoustic time-domain signal from an industrial fan. Figure 10(b) shows the signal following noise reduction using KLT high-pass filtering. We applied the Hamming window with a certain bandwidth and obtained the waveform shown in Figure 10(c). Lastly, the waveform following inverse transform and MFCC signal feature processing is shown in Figure 10(d). The sampling frequency, extraction time, FFT record length, and window of the IPII system were respectively set as 5 kHz, 12 seconds, 2048 points, and the Hamming window.

4.2.2. A Comparison of Acoustic and Vibration Signals in Spectrum Feature Filtering. To choose an appropriate sensor for early fault detection and to confirm the differences in sensing characteristics between acoustic and vibration signals, we compare the different signal preprocessing methods.

The practical resolution bandwidth features of acoustic and vibration sensors after MFCC filtering are shown in Figure 11. The differences in resolution bandwidth features clearly show that the vibration sensor has an excellent resolution within low-frequency fields. In contrast, the acoustic sensor has a greater sensing resolution bandwidth, giving it a higher resolution advantage, especially in medium- and high-frequency fields. That is, the acoustic sensor is better able to distinguish abnormal features for entire frequency fields. Moreover, the typical mechanical properties of rotating machinery suggest that initial component damage would be indicated by abnormalities within medium- and high-

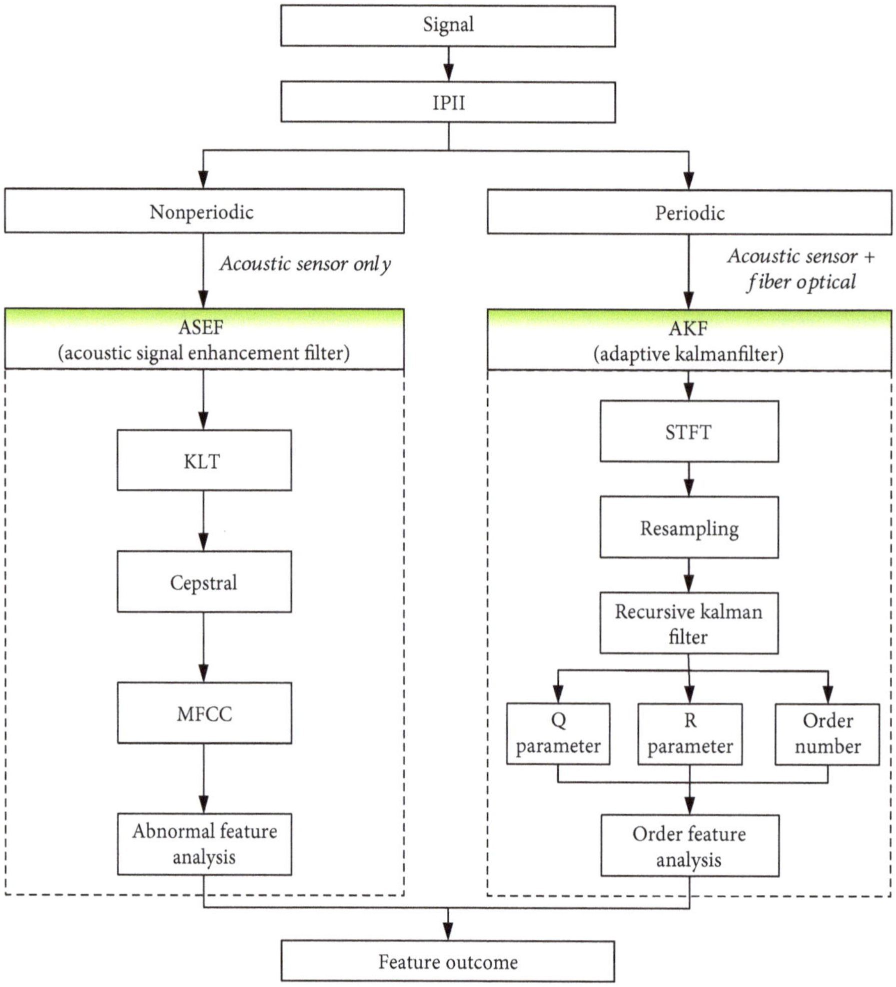

FIGURE 7: Scheme of digital filtering.

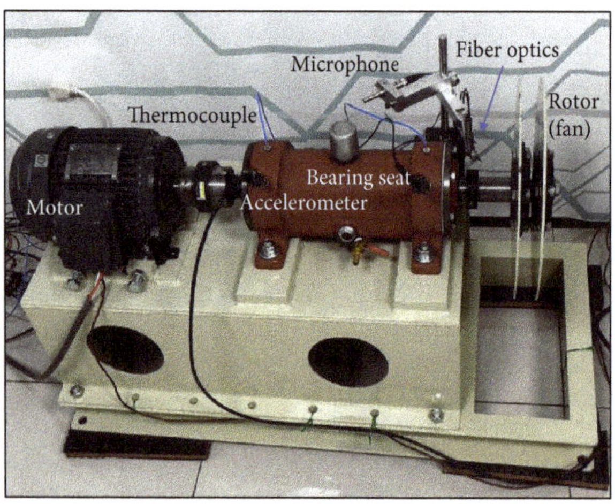

FIGURE 8: The practical experiment structure of an industrial fan.

frequency fields. The sensing resolution bandwidth feature of the acoustic sensor is matched with the abnormality generated by the initial damage. Therefore, the acoustic sensor provides better early fault detection. The results also confirm

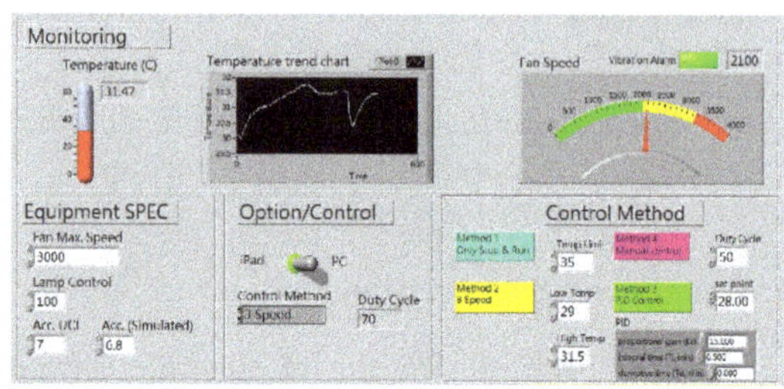

FIGURE 9: Graphical user interface.

the superior effectiveness of acoustic signals in predicting faults online.

4.3. Practical Digital Filter Implementation. Using the adaptive Kalman filter for analysis requires some parameter settings. $\mathbf{Q}_1$ and $\mathbf{Q}_2$ in (16) and (20) are the most critical parameters, in that they reflect actual conditions and thus have a direct impact on algorithm performance. $\mathbf{Q}_1$ is the main correlation convergence factor for the process equation, while $\mathbf{Q}_2$ is the correlation convergence factor for the measurement equation. The initial parameter settings are derived

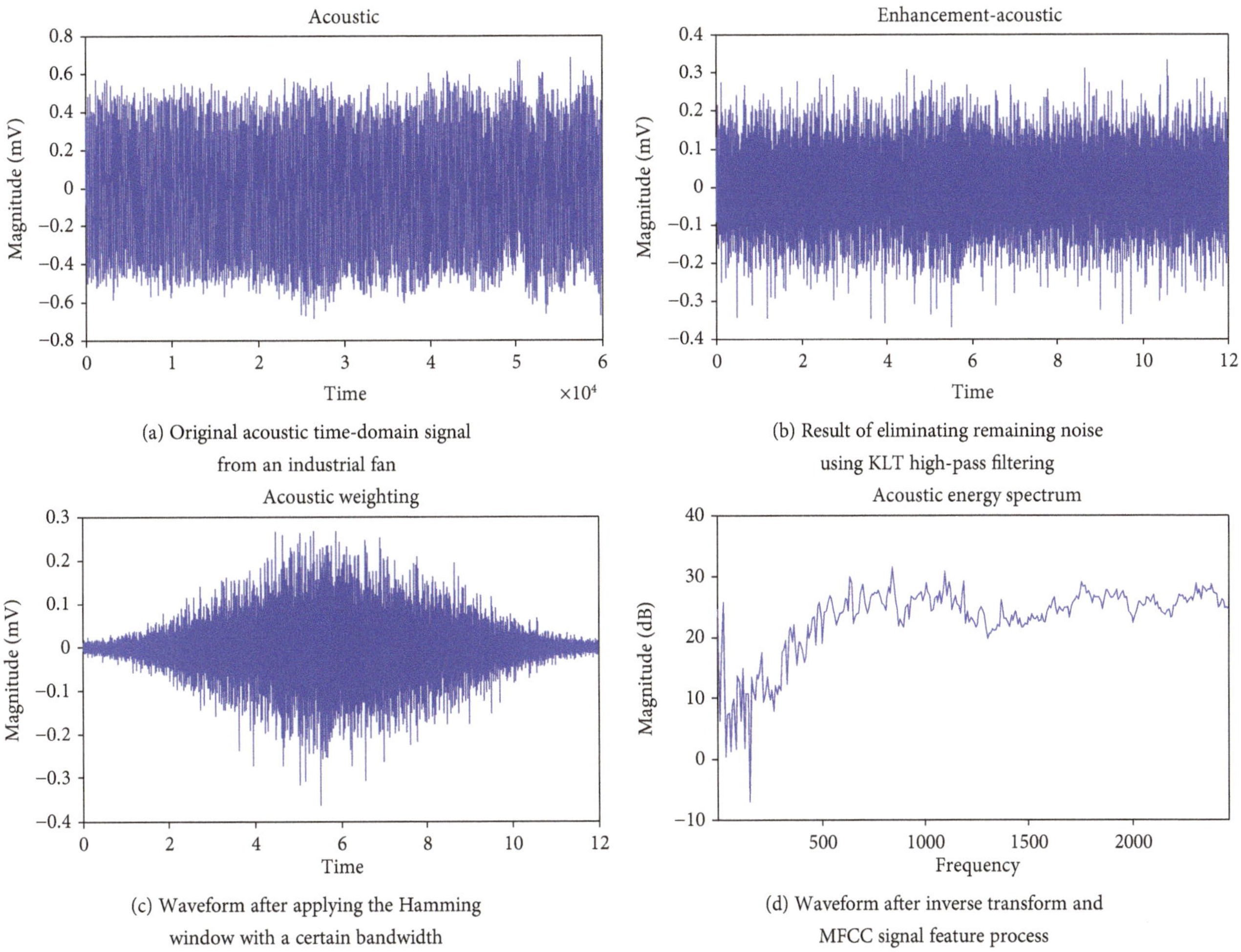

(a) Original acoustic time-domain signal from an industrial fan

(b) Result of eliminating remaining noise using KLT high-pass filtering

(c) Waveform after applying the Hamming window with a certain bandwidth

(d) Waveform after inverse transform and MFCC signal feature process

FIGURE 10: Spectrum feature filtering of acoustic signal.

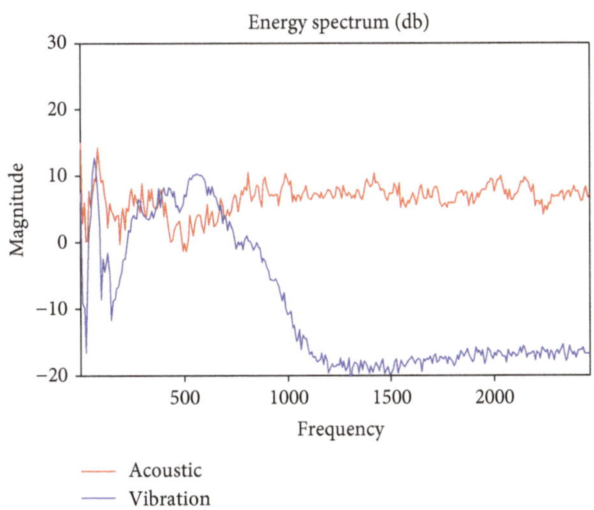

FIGURE 11: The sensing resolution bandwidth features of acoustic and vibration sensors.

from experience and adjusted to actual test conditions. This paper selects different convergence parameters for $\mathbf{Q}_1$ and $\mathbf{Q}_2$. The filter eliminates noise, and the system achieves a high convergence rate; thus, the order energy characteristics of the acoustic signal feature can achieve high-resolution performance. $\mathbf{Q}_1$ is 10^{-9} and 10^{-8}, and the four parameters in $\mathbf{Q}_2$ (10^4, 10^6, 10^8, and 10^{12}) are used in the actual test, with Figure 12 showing filter convergence performance. Due to the use of different $\mathbf{Q}_1$ parameters, the test results for the four $\mathbf{Q}_2$ parameters show that the two $\mathbf{Q}_1$ parameter sets can be used. However, when $\mathbf{Q}_1$ is set at 10^{-9}, it produces better convergence than at 10^{-8}. Thus, 10^{-9} is selected as the main design parameter for filter implementation. Based on industrial fan rotational characteristics and anomaly detection results, 10^8 and 10^{12} are selected as the main filter settings for $\mathbf{Q}_2$, where 10^8 is used for low fan rotation speeds and 10^{12} is used for high rotation speeds. That is, our example system implementation follows the actual convergence conditions using two parameter sets, $\mathbf{Q}_1 = 10^{-9}$ and $\mathbf{Q}_2 = 10^8$ and 10^{12}, to implement the digital filter.

4.4. Malfunction Detection. The proposed system focuses on early fault detection. Using an adaptive order-tracking algorithm, the system obtains the order energy at the time of rotation which can be used to clearly identify equipment abnormalities, making it very suitable for early fault detection.

This paper identifies faults based on the order energy distribution for abnormal states and normal operation. That is,

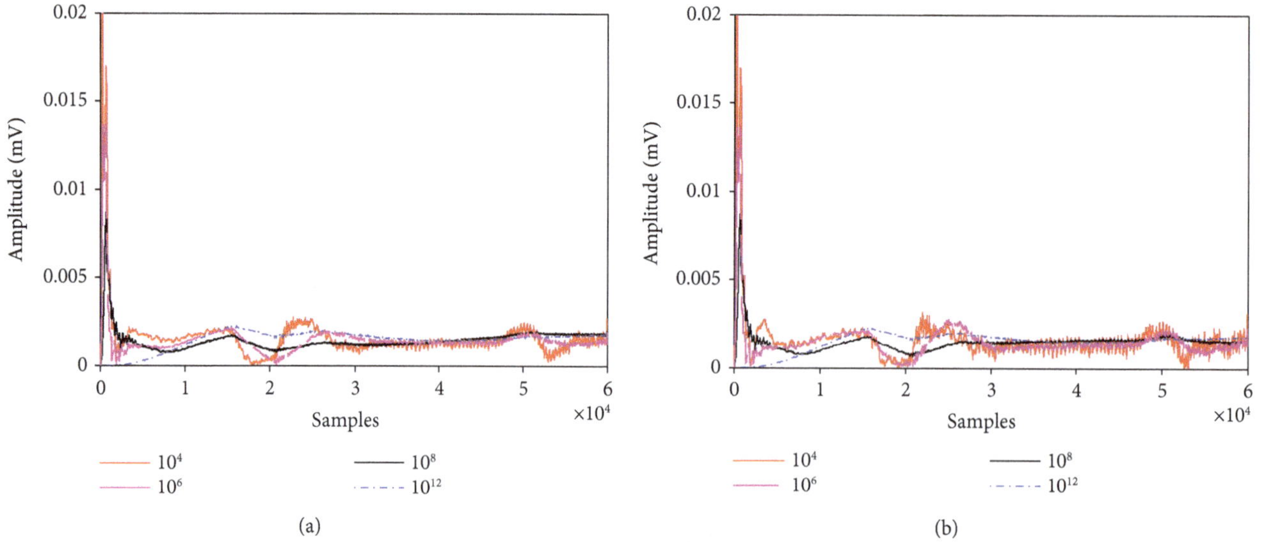

FIGURE 12: Actual test filter convergence performance: (a) $\mathbf{Q}_1 = 10^{-9}$ and $\mathbf{Q}_2 = 10^4, 10^6, 10^8$, and 10^{12}; (b) $\mathbf{Q}_1 = 10^{-8}$ and $\mathbf{Q}_2 = 10^4, 10^6, 10^8$, and 10^{12}.

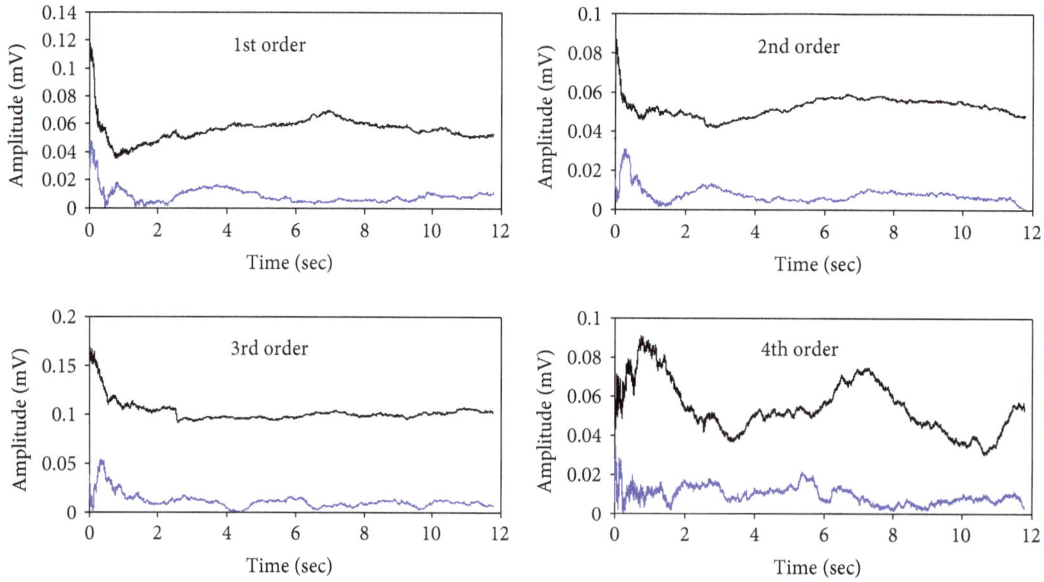

FIGURE 13: Order energy of acoustic signal (the bold line represents a fault condition while the thin line represents a normal condition).

the degree of difference in the order energy ranking is used to determine whether abnormalities exist or not. The threshold values used in the experiments take the normal (mean) value plus 30% as the order energy standard. In practical applications, the sensitivity of this threshold value can be adjusted according to the importance of the particular piece of equipment. In practice, a threshold value can be derived from new machinery after running for 48 hours, and if a fan requires replacement or repair, the system will automatically self-tune the key indicators for the replacement parts.

4.5. Industrial Fan Verification

4.5.1. Under Fixed Rotation Speed Conditions. Accumulated maintenance experience suggests that the motor's inner race is the component most likely to fail first under normal wear, followed by the outer race, ball spin, and fundamental train. Therefore, the experiment simulates an inner race failure on the bearing in an industrial fan after the IPII algorithm computes and observes the signal changes reflecting initial abnormalities. Figure 13 shows the order energy of the acoustic signal. The bold line represents the energy distribution under the broken bearing condition while the thin line represents the energy distribution under normal fan operating conditions. In the 1st- to 4th-order energy charts, the microphone sensor can effectively provide an early diagnosis of the abnormal signal. The system uses synchronizing signal sampling processing; thus, the order energy changes of the vibration signal in Figure 14 show that the accelerometer sensor detected a slight fault sign when the microphone sensor

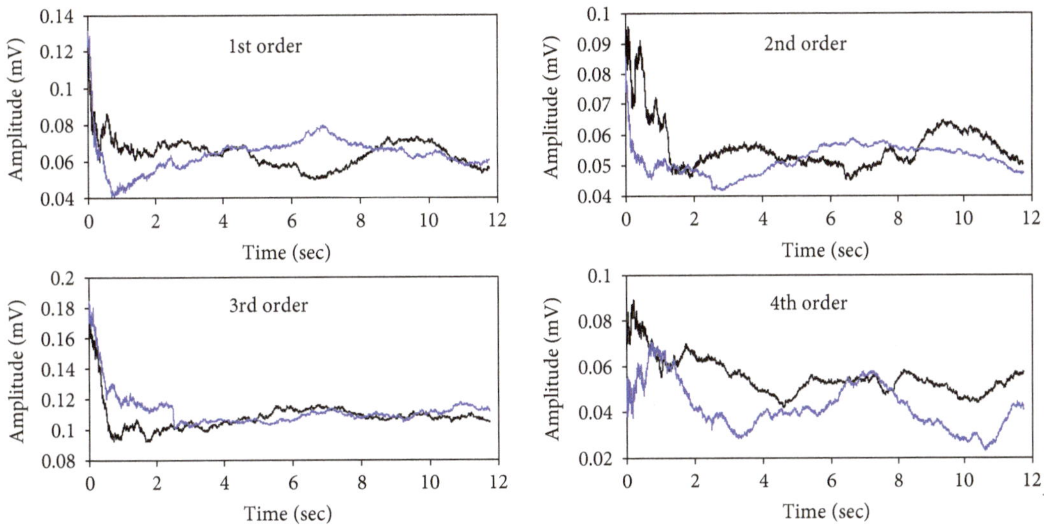

FIGURE 14: Order energy of vibration signal (the bold line represents a fault condition while the thin line represents a normal condition).

TABLE 4: Industrial fan-monitoring parameters.

Parameter	Options
Fan type	Coupling, belt, gear box, direct.
Detection type	Motor housing, bearing seat, fan housing, fan impeller
Sensor type	Microphone, accelerometer, microphone + accelerometer

detects an abnormality. However, there was no significant difference between the fault signal and normal signals. As a result, the microphone outperforms the accelerometer sensor in providing early fault diagnosis of industrial fans.

Different malfunction conditions result in different order energy configurations. Figure 13 shows the 1st- to 4th-order energies for the inner race failure in an industrial fan. In the event of an outer race malfunction, the distribution for each order energy shows different trends, such as loose or unbalanced blades. As shown in the order energy diagrams of Figures 13 and 14, large diagnostic samples are stored in the feature data bank (FDB) for application to actual fault analysis, allowing the system to diagnose various abnormalities. Through this type of condition monitoring, and the subsequent collection and analysis of big data, we can create a reasonable component maintenance and replacement schedule and adjust production parameters to ensure equipment reliability, thus improving product quality and reducing maintenance costs.

4.5.2. Under Transient Rotation Condition. The user interface design accounts for various fan types and installation environments, allowing for parameter adjustments for monitoring of a wide variety of industrial fans, detection types, and sensors. The parameter settings are presented in Table 4. The proposed algorithm can be adjusted to account and compensate for these hardware architecture, sensor, and fault signal characteristics to provide for optimal online test performance. The user interface also provides real-time anomaly trend analysis and monitoring, allowing for real-time anomaly diagnosis.

At low rotation speeds, fan anomaly diagnosis is mainly derived from acoustic order energy. In addition to acoustic order energy, under transient or variable speed conditions, energy spectrum trends produced by MFCC filtering can be used to assist in anomaly detection. In addition, the system simultaneously compares temperature changes to relevant bearings to improve the accuracy of anomaly detection and diagnosis. Figure 15 shows the acoustic order energy and energy spectrum trend for a coupling-type fan with an acoustic sensor in the motor housing at low-speed (50% of the motor's normal speed) and variable-speed (i.e., variable frequency control) conditions. The anomaly signal (bold line) clearly shows that the anomaly can be effectively detected during low-speed operation or under transient or variable speed conditions.

4.6. Smart Control. The proposed system is designed for integration into Industry 4.0-type smart factories and integrates diagnostic, control, and production systems to achieve monitoring, diagnostics, control, and management functions. Figure 16 shows the smart factory architecture based on rotating equipment. The architecture is centered on the proposed Intelligent Prediction Integration System with Internet (IPII), including sensing, intelligent control, MES, and ERP four levels. The sensing level collects operational information, which is then through IPII passed up to the factory's MES and ERP level for production management and resource integration. The system's intelligent control level provides overall control planning to implement individualized smart control programs for various rotating machinery.

Diagnosis is combined with control to determine appropriate operating modes or policies according to the current state of specific equipment. The ultimate goal of these diagnostic techniques is early malfunction prediction, thus preventing unexpected failure. IPII integrates acoustic, vibration, temperature, and rotating speed data. In addition to

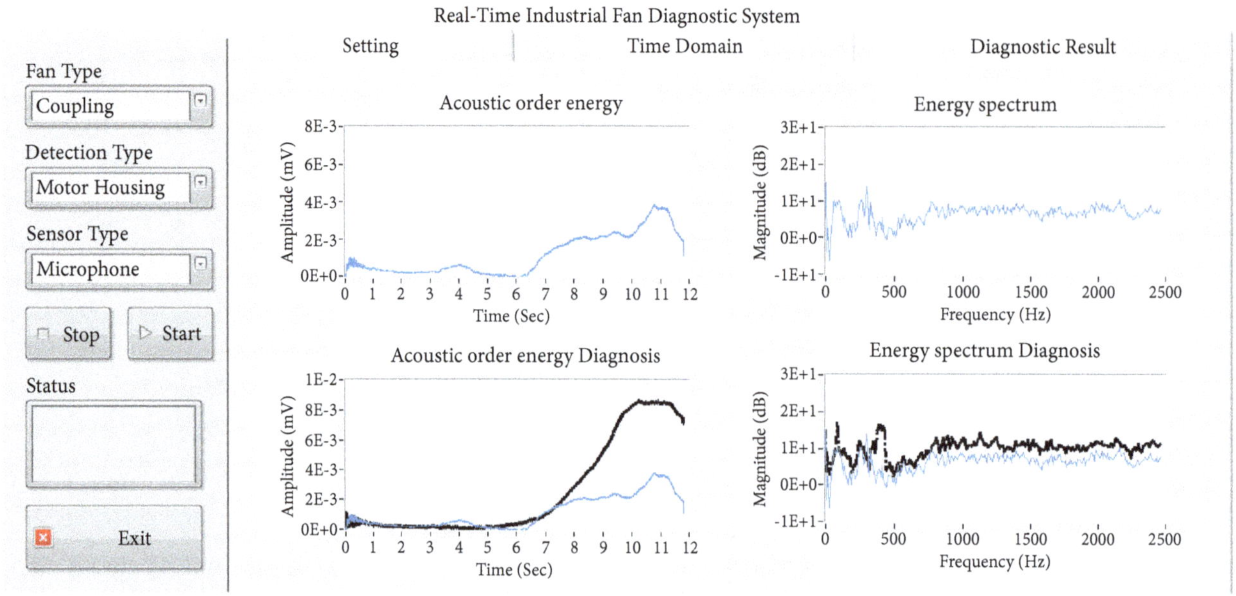

FIGURE 15: Acoustic order energy and energy spectrum trends under variable rotation speed conditions (the bold line represents a fault condition while the thin line represents a normal condition).

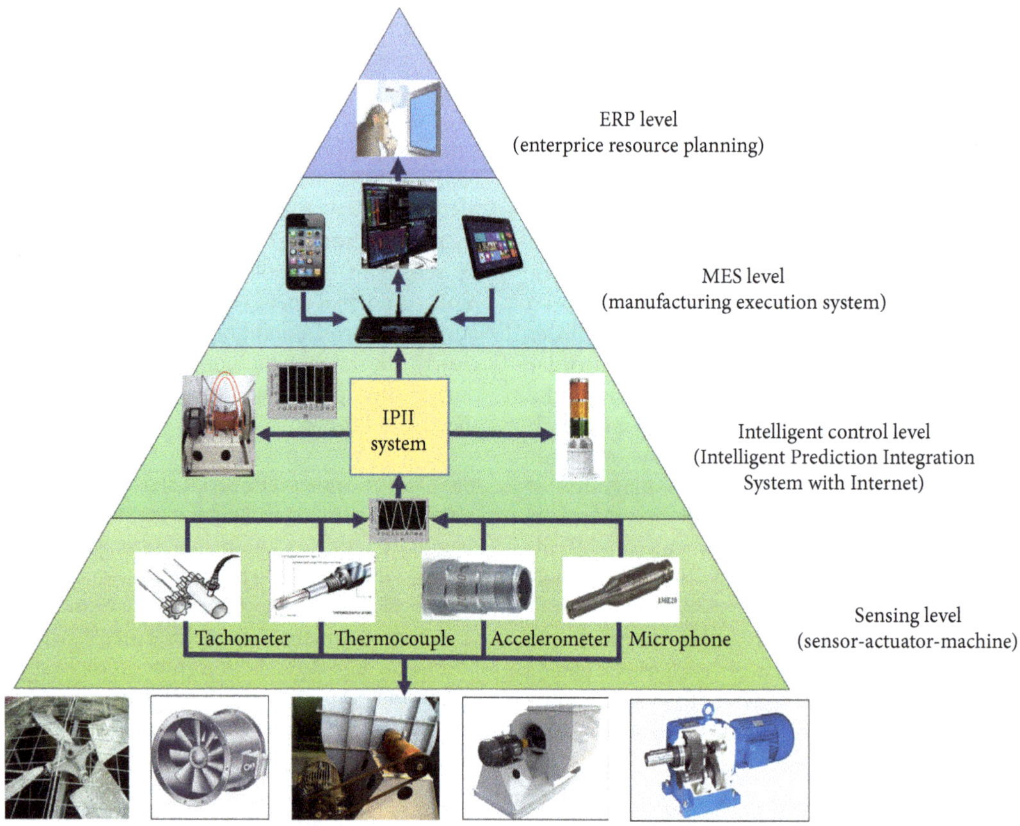

FIGURE 16: Smart factory architecture based on rotating equipment.

fault prediction and diagnosis, it can also be integrated with smart controls. The system can detect abnormalities and integrates smart controls to initiate appropriate responses. If abnormal vibrations, noise, or temperatures are detected, the system can not only immediately alert engineering supervisors but automatically execute appropriate countermeasures. Smart controls are designed to effectively maintain optimal production conditions while preventing unexpected shutdowns by automatically adjusting rotating speed, bringing backup equipment online, or adjusting production parameters to ensure that fan malfunctions do not negatively impact normal production line operations or affect equipment or personnel safety.

5. Conclusion

The difference between the proposed ideal and the state-of-the-art works is a different choice of the sensor. We believe that sensor selection is the key to diagnostic results. In the present studies demonstrated in the literature, most of them directly select vibration sensors for use in a variety of different diagnostic methods with markedly different results. First, this study uses MFCC filtering, commonly used in automatic speech recognition, to analyze the differences between the sensing characteristics of the acoustic and vibration sensors. Secondly, the study seeks to provide a better understanding of signal characteristics for diagnosing equipment failure and the corresponding characteristics of the selected sensor. That is, we first conduct a comparison analysis of the relative advantages of the sensing characteristics of acoustic and vibration sensors in the high-frequency range, to better understand anomalous signals occurring in the high-frequency range for early failure detection of industrial fans. Based on the results, we select the acoustic sensor to replace the vibration sensor. In factory settings, this approach is highly different from traditional vibration-based diagnostic methods. Experimental results verify that the acoustic sensor can provide earlier and more effective failure diagnosis. A review of the literature regarding failure diagnosis for rotating machinery found no mention of previous efforts to use differences in equipment signal anomalies and sensor-sensing characteristics as the basis for sensor selection. Also, we believe that this approach has reference value for sensor selection for early failure detection in other specific types of equipment.

Although the proposed diagnostic method can be applied to most types of rotating machinery, the approach is mainly designed to diagnose potential failure in industrial fans in factory settings. Failure to detect industrial fan failure early can have a significant impact on production; thus, we combine the knowledge of industrial fan experts to elucidate the operational characteristics and fault signal characteristics for industrial fans and analyze the results to select an appropriate sensor and diagnostic method. Experimental results show that the proposed method is very suitable for the early diagnosis of industrial fan failure. A literature review for the previous three years has also been conducted to discover other examples of useful methods for fault diagnosis for industrial fans. The system allows for the fault diagnosis of various types of industrial fans. The original system design considers various types of industrial fans, installation environments, and operating characteristics, and thus, system monitoring accounts for various fan types (e.g., coupling, belt, gear box, and direct), detection types, and sensor types. The proposed algorithm uses these various hardware structures, sensor characteristics, and fault signal characteristics as conditions for adjusting and compensating for the corresponding parameters to optimize actual online test performance.

To summarize, in the technology development of the predictive maintenance for the rotating equipment of factories, IPII possesses three major contributions as compared with the ideas demonstrated in the literature. The first is to replace the "vibration" sensor having only better low-frequency sensitivity to the "acoustic" one. The system equipped with the vibration sensor is subject to poor diagnosis in early failure. On the contrary, the acoustic sensor gives a system better high-frequency response advancing the early diagnosis in system failure. The second is the "property" of the detected dynamic signals. In this regard, we have proposed to adopt the order-tracking technology which can be used to analyze the relationship between the order and rotation speed. The most important thing in this design is the distribution in the order map rather than only the changes in the spectrum. In other words, our way is to fetch each of the order energy calculated from the combination of instant rotation speed and its corresponding sound (or vibration signals). This idea has been proven to be quite effective while being applied to low-rotation-speed equipment, especially for those systems requiring signal analysis of the variable rotation speeds. Last but not least, combining the ASK with the ASEF to deal with, respectively, the periodic and nonperiodic signals is capable of predicting the features causing system failure, where filtering and abnormal feature analysis are carried out. In addition to fault prediction, IPII can also allow for the use of smart controls and remote monitoring via networked devices. The control mode and process parameters of the critical rotating machinery can be instantly adjusted in response to fault signs to prevent production downtime and to reduce maintenance costs. The proposed system can be practically applied in Industry 4.0 smart factories.

Conflicts of Interest

The authors declare no conflict of interest.

Authors' Contributions

Cihun-Siyong Alex Gong generated the research ideas, supervised the entire work, polished the article, and proofread the technical details. Huang-Chang Lee assisted in the design. Yu-Chieh Chuang performed the experiments and wrote the manuscript based on his master's thesis archived domestically in Chinese in Taiwan. Tien-Hua Li, Chih-Hui Su, Lung-Hsien Huang, and Chih-Hsiung Chang technically supported the project. Yih-Shiou Hwang served as technical consultant for signal processing and application. Jiann-Der Lee supervised this work and provided funding support.

Acknowledgments

The authors appreciate the support from the National Science Council (NSC), Taiwan, and Ministry of Science and Technology (MOST), Taiwan, under Grants MOST

106-2221-E-182-005-, MOST 105-2221-E-182-039-, MOST 103-2221-E-182-070-, MOST 103-2815-C-182-012-E, MOST 104-2815-C-182-052-E, MOST 104-2221-E-182-044, and MOST 104-2221-E-182-023-MY2. This work is also supported in part by the Linkou Chang Gung Memorial Hospital (CGMH) under Contracts CMRPD2F0103, CMRPD2G0331, and CMRPD2H0041.

References

[1] S. Patidar and P. K. Soni, "An overview on vibration analysis techniques for the diagnosis of rolling element bearing faults," *International Journal of Engineering Trends and Technology*, vol. 4, no. 5, pp. 1804–1809, 2013.

[2] D. Hansen and A. H. Olsson, "ISO Standard 13373-2: 2005: condition monitoring and diagnostics of machines–vibration condition monitoring–part 2: processing, analysis and presentation of vibration data," *International Standards Organization*, 2009.

[3] S. A. Ansari and R. Baig, "A PC-based vibration analyzer for condition monitoring of process machinery," *IEEE Transactions on Instrumentation and Measurement*, vol. 47, no. 2, pp. 378–383, 1998.

[4] J. D. Wu, M. R. Bai, F. C. Su, and C. W. Huang, "An expert system for the diagnosis of faults in rotating machinery using adaptive order-tracking algorithm," *Expert Systems with Applications*, vol. 36, no. 3, pp. 5424–5431, 2009.

[5] M. Bai, J. Huang, M. Hong, and F. Su, "Fault diagnosis of rotating machinery using an intelligent order tracking system," *Journal of Sound and Vibration*, vol. 280, no. 3-5, pp. 699–718, 2005.

[6] G. G. Yen and K. C. Lin, "Wavelet packet feature extraction for vibration monitoring," *IEEE Transactions on Industrial Electronics*, vol. 47, no. 3, pp. 650–667, 2000.

[7] H. Bendjama, S. Bouhouche, and M. S. Boucherit, "Application of wavelet transform for fault diagnosis in rotating machinery," *International Journal of Machine Learning and Computing*, pp. 82–87, 2012.

[8] Q. Miao, C. Tang, W. Liang, and M. Pecht, "Health assessment of cooling fan bearings using wavelet-based filtering," *Sensors*, vol. 13, no. 1, pp. 274–291, 2013.

[9] K. Li, P. Chen, and H. Wang, "Intelligent diagnosis method for rotating machinery using wavelet transform and ant colony optimization," *IEEE Sensors Journal*, vol. 12, no. 7, pp. 2474–2484, 2012.

[10] P. Sharma and N. Saraswat, "Diagnosis of motor faults using sound signature analysis," *International Journal of Innovative Research in Electrical, Electronics, Instrumentation and Control Engineering*, vol. 3, no. 5, 2015.

[11] R. Yan and R. X. Gao, "Base wavelet selection for bearing vibration signal analysis," *International Journal of Wavelets, Multiresolution and Information Processing*, vol. 7, no. 04, pp. 411–426, 2009.

[12] B. Liu, "Selection of wavelet packet basis for rotating machinery fault diagnosis," *Journal of Sound and Vibration*, vol. 284, no. 3-5, pp. 567–582, 2005.

[13] Q. Zhang, A. Qin, L. Shu, G. Sun, and L. Shao, "Vibration sensor based intelligent fault diagnosis system for large machine unit in petrochemical industries," *International Journal of Distributed Sensor Networks*, vol. 11, no. 8, Article ID 239405, 2015.

[14] G. Tang, W. Hou, H. Wang, G. Luo, and J. Ma, "Compressive sensing of roller bearing faults via harmonic detection from under-sampled vibration signals," *Sensors*, vol. 15, no. 10, pp. 25648–25662, 2015.

[15] W. Guo, P. W. Tse, and A. Djordjevich, "Faulty bearing signal recovery from large noise using a hybrid method based on spectral kurtosis and ensemble empirical mode decomposition," *Measurement*, vol. 45, no. 5, pp. 1308–1322, 2012.

[16] R. G. Sorte and P. Ram, "Vibration based blind identification of bearing failures in rotating machinery," *International Journal of Science and Research (IJSR)*, 2015.

[17] L. C. K. Reuben and D. Mba, "Bearing time-to-failure estimation using spectral analysis features," *Structural Health Monitoring: An International Journal*, vol. 13, no. 2, pp. 219–230, 2013.

[18] M. Elforjani and D. Mba, "Assessment of natural crack initiation and its propagation in slow speed bearings," *Nondestructive Testing and Evaluation*, vol. 24, no. 3, pp. 261–275, 2009.

[19] Bearing Specialists Association, "Vibration analysis," 2014, August 2015 http://www.bsahome.org/tools/pdfs/Vibration_Analysis_web.pdf.

[20] A. Ukil, M. Zlatanski, and M. Hochlehnert, "Monitoring of HV generator circuit breaker contact ablation based on acoustic emission," *IEEE Transactions on Instrumentation and Measurement*, vol. 62, no. 10, pp. 2683–2693, 2013.

[21] E. Farjah, S. R. Khayam Hoseini, T. Ghanbari, and H. Givi, "Extended Kalman filter-based method for inter-turn fault detection of the switched reluctance motors," *IET Electric Power Applications*, vol. 10, no. 8, pp. 714–722, 2016.

[22] S. Dey, P. Pisu, and B. Ayalew, "A comparative study of three fault diagnosis schemes for wind turbines," *IEEE Transactions on Control Systems Technology*, vol. 23, no. 5, pp. 1853–1868, 2015.

[23] F. Lu, Y. Huang, J. Huang, and X. Qiu, "A hybrid Kalman filtering approach based on federated framework for gas turbine engine health monitoring," *IEEE Access*, vol. 6, pp. 9841–9853, 2018.

[24] S. Haykin, *Adaptive Filter Theory*, Englewood Cliffs, NJ: Prentice-Hall, 4th edition, 2001.

[25] M. R. Bai, J. Jeng, and C. Chen, "Adaptive order tracking technique using recursive least-square algorithm," *Journal of Vibration and Acoustics*, vol. 124, no. 4, p. 502, 2002.

[26] M. R. Bai, J. Jeng, and C. Chen, "Adaptive order tracking technique using recursive Kalman filtering," *Journal of Vibration and Acoustics*, vol. 124, no. 4, p. 502, 2002.

[27] S. A. Selouani, *Speech Processing and Soft Computing*, Springer Science & Business Media, 2011.

[28] Y. Ephraim and H. L. Van Trees, "A signal subspace approach for speech enhancement," *IEEE Transactions on Speech and Audio Processing*, vol. 3, no. 4, pp. 251–266, 1995.

[29] S. Davis and P. Mermelstein, "Comparison of parametric representations for monosyllabic word recognition in continuously spoken sentences," *IEEE Transactions on Acoustics, Speech, and Signal Processing*, vol. 28, no. 4, pp. 357–366, 1980.

[30] P. D. Tafti, S. Shirani, and X. Wu, "On interpolation and resampling of discrete data," *IEEE Signal Processing Letters*, vol. 13, no. 12, pp. 733–736, 2006.

[31] Z. G. Liang, Q. Yang, Y. D. Wan, F. He, X. C. Wang, and M. Wang, "The study of IOT architecture for steel strip production process," in *Advanced Materials Research*, vol. 572, pp. 364–370, Trans Tech Publications, 2012.

11

Polymeric Foams as the Matrix of Voltammetric Sensors for the Detection of Catechol, Hydroquinone, and Their Mixtures

C. Fernandez-Blanco,[1] M. Mugica,[2] M. A. Rodriguez-Perez,[2] C. Garcia-Cabezón,[1] and M. L. Rodriguez-Mendez[1]

[1]*GroupUVaSens, Department of Inorganic Chemistry, Engineers School, Universidad de Valladolid, Valladolid, Spain*
[2]*Cellular Materials Laboratory (CellMat), Condensed Matter Physics Department, University of Valladolid, Campus Miguel Delibes, Paseo de Belén n7, Valladolid, Spain*

Correspondence should be addressed to M. L. Rodriguez-Mendez; mluz@eii.uva.es

Academic Editor: Andrea Cusano

Porous electrodes based on polymethylmethacrylate and graphite foams (PMMA_G_F) have been developed and characterized. Such devices have been successfully used as voltammetric sensors to analyze catechol, hydroquinone, and their mixtures. The presence of pores induces important changes in the oxidation/reduction mechanism of catechol and hydroquinone with respect to the sensing properties observed in nonfoamed PMMA_graphite electrodes (PMMA_G). The electropolymerization processes of catechol or hydroquinone at the electrode surface observed using PMMA_G do not occur at the surface of the foamed PMM_G_F. In addition, the limits of detection observed in foamed electrodes are one order of magnitude lower than the observed in the nonfoamed electrodes. Moreover, foamed electrodes can be used to detect simultaneously both isomers and a remarkable increase in the electrocatalytic properties shown by the foamed samples, produces a decrease in the oxidation potential peak of catechol in presence of hydroquinone, from +0.7 V to +0.3 V. Peak currents increased linearly with concentration of catechol in presence of hydroquinone over the range of $0.37{\cdot}10^{-3}$ M to $1.69{\cdot}10^{-3}$ M with a limit of detection (LOD) of 0.27 mM. These effects demonstrate the advantages obtained by increasing the active surface by means of porous structures.

1. Introduction

The assessment of antioxidants in foods is of great importance due to the health benefits associated to their intake, to their influence in the organoleptic characteristics of foods and beverages, and to their role in their preservation. Catechol (1,2-dihydroxybenze, CT) and hydroquinone (1,4-dihydroxybenze, HQ) are two dihydroxybenzene isomers that occur naturally in many plant foods (onions, coffee, tea leaves, etc.) and beverages such as wines [1].

Simultaneous detection of dihydroxybenzene isomers as CT and HQ is difficult because they have similar structure and characteristics. The main methods for the determination of CT and HQ mixtures are high-performance liquid chromatography (HPLC) [2–5], gas chromatography coupled with mass spectrometry [5, 6], spectrophotometry [7–9], fluorescence [10, 11], chemiluminescence [12–14], or capillary electrochromatography [15, 16]. Among all these techniques, electrochemical methods represent an advantage due to their low price, high sensitivity, short experimental times, and portability [17, 18].

The simultaneous assessment of CT and HQ at conventional carbon electrodes is difficult because their similar chemical structures lead to resembling responses which produce an overlap of their redox peaks [19–21]. In order to overcome this problem, electrodes have been chemically modified with many materials such as activated glassy carbon [22], modified electrodes with graphene [18, 23–26], carbon nanoparticles [27], carbon nanofibers [28, 29], carbon nanotubes [30], or gold nanoparticles [29, 31].

In addition, electrodes can be modified using different techniques, leading to electrodes with distinct characteristics which depend (among other factors) in the number of active sites at the electrode surface. Thus, the sensitivity and limits of detection can be improved by increasing the number of active sites at the electrode surface. Different strategies can be used to increase the number of active sites such as the use of layered hydroxide films [24], mesoporous materials [25], or nanostructured sensing layers [32, 33]. A different approach could be to use foamed materials as the sensing element. Recently, foams from metals like nickel [34–36] or copper [37–39] have been developed to obtain efficient electrochemical sensors. These electrodes show a high specific active surface, but they are not ideal materials to fabricate electrodes because they have redox activity. Foamed carbon materials are also good candidates, but the electrodes are fragile.

Our group proposes here a new class of foamed electrodes based on a polymeric matrix. The reason is that polymeric foams are easy to obtain and the pore size and structure can be easily controlled. In addition, their conductivity can be improved by doping the foam with a carbonaceous material. In particular, in this work, a composite foam formed by a polymeric matrix of polymethylmethacrylate (PMMA) and graphite as the conductive additive is developed here using the gas dissolution foaming method in which the saturation temperature, foaming temperature, and saturation pressure are controlled parameters during the foaming process [40]. The obtained foams have been employed as electrochemical sensors for the detection of CT and HQ which are two phenols of high interest for food industry. The dynamic properties of the electrodes and the limits of detection have been evaluated. The possibility of distinguishing these similar isomers has been analyzed.

2. Materials and Methods

2.1. Chemicals. Polymethylmethacrylate (PMMA) V 825T was kindly supplied by ALTUGLAS® International (Colombes, France) in the form of pellets. The material used presents a density (ρ) of 1.19 g·cm^{-3} (measured at 23°C and 50% HR) and a glass transition temperature (Tg) of 114.5°C measured by differential scanning calorimetry (DSC). Medical grade CO_2 (99.9% purity) was used as blowing agent.

Catechol (CT), hydroquinone (HQ), and KCl were purchased from Sigma-Aldrich, and graphite (G) (>99%) was purchase from Fluka. Chloroform stabilized with ethanol was purchased from Scharlab. All chemicals were used as received. Aqueous solutions were prepared using high-quality water (resistivity of 18.2 MΩ·cm^{-1}) from a Milli-Q system (Millipore, Billerica, MA, USA).

2.2. Sensor Preparation. 1.4 g of PMMA was dissolved in 32 mL of hot chloroform (50°C). Then, the solution was placed in a Petri dish to evaporate the solvent until a PMMA film was obtained.

Working in a similar way, films of PMMA/graphite (2/1) were prepared. Previously, the mixture was homogenized using a sonicator probe (Vibra Cell sonics) to get a good dispersion of graphite in the PMMA solution. This PMMA/graphite ratio was chosen according to preliminary experiments that showed that using lower amounts of graphite did not produce enough conductivity to be used as sensitive material.

Foamed samples were produced in a high pressure vessel (model PARR 4681) provided by Parr Instrument Company (Moline, IL, USA). The pressure system comprises an accurate pressure pump controller (model SFT-10) provided by Supercritical Fluid Technologies Inc. With this set-up, a set of foaming experiments has been performed through the so-called gas dissolution foaming process. This process consists of two basic steps. The first one is the saturation of the polymeric sample with gas under high pressure. In this step, the gas reaches an equilibrium concentration when given enough time for diffusion of gas into the polymer, giving rise to a gas-polymer system. In the second step, bubbles are nucleated in the gas-polymer system by means of a thermodynamic instability which, in this specific case, can be obtained by a sudden drop in pressure [41]. The foaming experiments have been performed by a single-step foaming process [40] to prevent fast diffusion of the gas during the extraction of the polymer films from the autoclave after saturation. In this process, the gas saturation temperature is higher than the $T_{g,ef}$ of the polymer (which is in a rubbery state), allowing the gas expansion during the pressure release. Hence, the samples are already foamed when extracted from the pressure vessel. In these tests, the saturation pressure has been set at 30 MPa. Saturation temperature and saturation time were fixed at 50°C and 24 h. After saturation, the pressure was released at a high depressurization rate (90 MPa/s) by means of an electrovalve.

Moreover, gas diffusion can provoke heterogeneous cellular structures in the samples even on the single-step foaming method. Consequently, films have been confined in a mold consisting of two aluminum frames (Figure 1), restricting CO_2 surface diffusion and increasing the gas availability for cell nucleation and growth [42]. The mold was closed by means a torque wrench so as the screws reach the same tighten level, achieving the desired constrainment effect.

2.3. Structure Characterization by SEM. A scanning electron microscope (Quanta 200 FEG, Hillsboro, OR, USA) was used to obtain images of both the surface and the inner structure of the foams. Images of the surface were obtained by covering the surface with gold using a sputter coater (model SDC 005, Balzers Union, Balzers, Liechtenstein). Images of the inner structure were obtained by fracturing the samples: samples were cooled down with liquid nitrogen and then fractured and coated with gold using a sputtering.

2.4. Electrochemical Measurements. Electrochemical experiments were carried out using a PARSTAT 2273 potentiostat/galvanostat (AMETEK scientific instrument) in a three-cell configuration. A platinum foil with a size of 2 cm × 1 cm and a Ag/AgCl/KCl 3M electrode was used as counter electrode and reference electrode, respectively. PMMA samples were used as the working electrode including a PMMA film (PMMA), a PMMA foam (PMMA_F), a PMMA film doped with graphite (2/1), which is denoted as

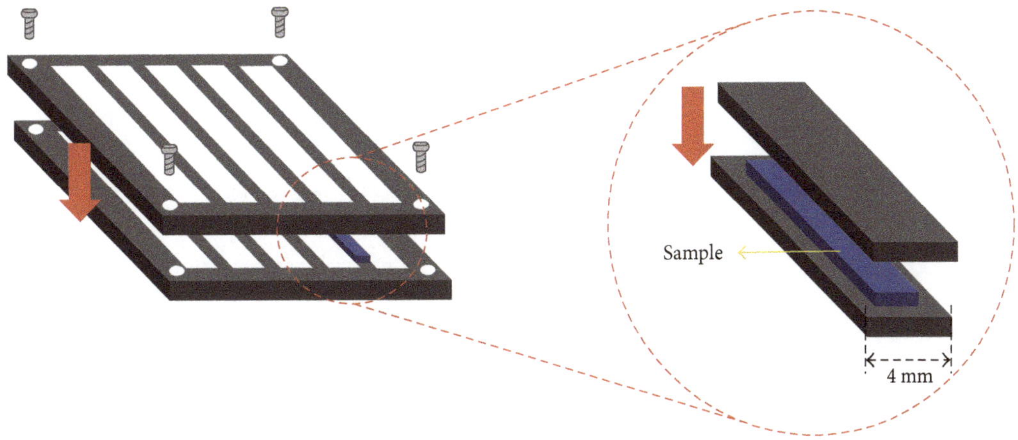

FIGURE 1: Aluminum frame used to constrain the films during the foaming tests.

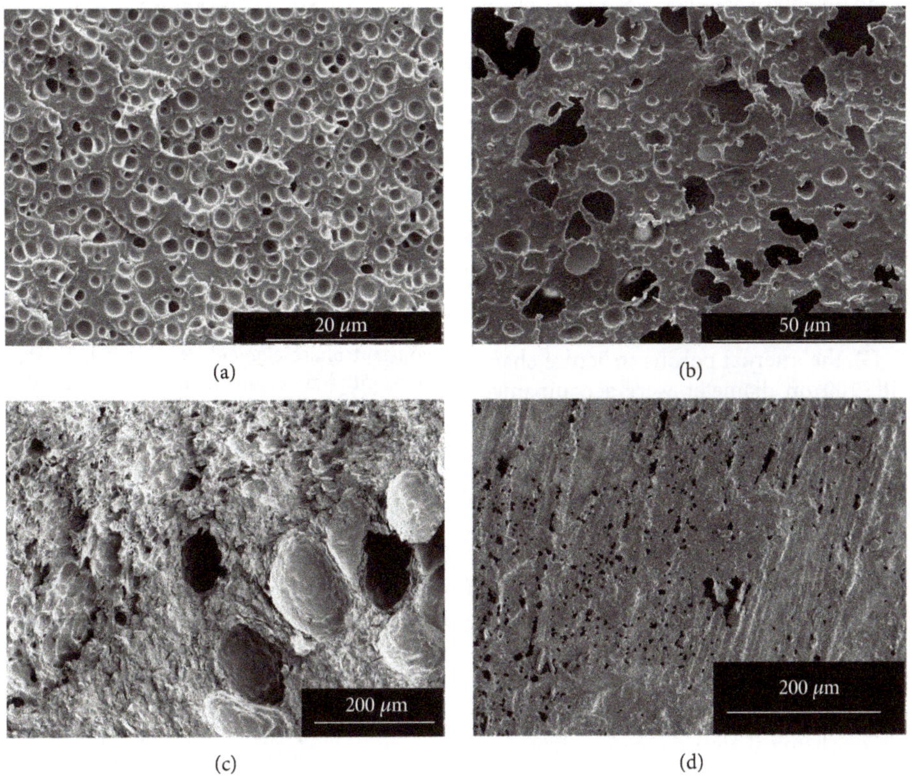

FIGURE 2: SEM images of (a) PMMA_F inner structure, (b) PMMA_F surface structure, (c) PMMA_F_G inner structure, and (d) PMMA_F_G surface structure.

PMMA_G, and a PMMA foam doped with graphite (2/1) which is denoted as PMMA_F_G.

Electrochemical impedance spectroscopy (EIS) was carried out in a 0.1 M KCl solution with a frequency range from 31600 Hz to 0.1 Hz with signal amplitude of 10 mV at a working potential of 0.0 V.

Square wave voltammetry was used for studying the response for our sensors towards phenols. The reason to use this technique is that it reduces the faradaic current, producing lower limits of detection than cyclic voltammetry [43]. Calibration curves were built using square wave voltammetry (SWV) as electrochemical technique from −0.6 V to +1.0 V at a scan rate of 50 mV/s with an amplitude pulse of 50 mV, a duration of pulse of 0.02 s, and a step height of 2 mV. Dynamic properties were evaluated carrying out a cyclic voltammetry in a 0.1 M of KCl solution that contains 10^{-3} M of CT from −0.5 V to +1.5 V for increasing scan rates (from 0.1 V/s to 0.3 mV/s).

3. Results and Discussion

3.1. Structural Characterization of the Electrodes. In a first step, both the internal and the surface structure of the foamed samples were characterized using Scanning Electron Microscope (SEM). This information was used to evaluate the

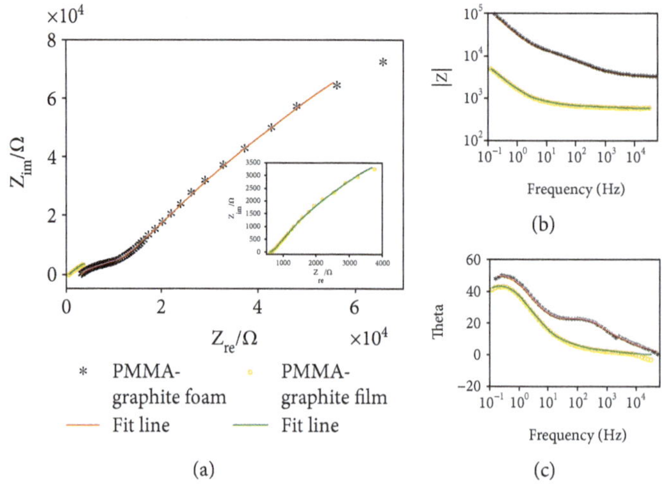

FIGURE 3: (a) Nyquist and (b) impedance Bode plot and (c) phase Bode plot of PMMA_G film and PMMA_F_G in 0.1 M KCl. Frequency swept from 31,600 Hz to 0.1 Hz at 0 V in a 0.1 M KCl solution.

influence of the porosity of the samples in the sensing properties of PMMA/graphite composites.

Figure 2(a) shows the internal structure of the PMMA foam (PMMA_F) where a homogeneous microporous structure with an average pore diameter of 2 μm was observed. The image of the foam surface (Figure 2(b)) showed larger pores with irregular shapes and sizes with an average diameter of 7 μm. When PMMA foam was doped with graphite (PMMA_F_G), the internal porous structure changed. Large pores with 100 μm diameter were accompanied by many small pores with an average diameter of 12 μm (Figure 2(c)). It is also noticeable that the pores were more irregular in shape than those observed in the undoped PMMA_F. This effect can be attributed to the presence of the graphite flakes that modify the foaming process. The surface structure changed drastically in the presence of graphite, and a profusion of small pores of about 4 μm diameter (Figure 2(d)) was observed.

3.2. Electrochemical Impedance Spectroscopy (EIS) Characterization. Electrochemical impedance spectroscopy (EIS) was used to study the changes in the electrochemical properties of PMMA induced by the insertion of graphite and by the foaming.

As expected, the Nyquist plots of nonfoamed PMMA and foamed PMMA_F immersed in 0.1 M KCl showed vertical straight lines, confirming that undoped samples behaved as insulators.

When graphite was added to the structure of a PMMA film (PMMA_G) or a PMMA foam (PMMA_F_G), samples became conductive and the Nyquist diagram (Figure 3(a)) revealed two unfinished capacitive loops. The semicircle in the high frequency region showed a smaller radius than the large arc observed in the low frequency region. As observed in Figure 3(a), the radii increased in foamed samples. Given that radii of the semicircles are related to the efficiency of the charge transfer, it can be concluded that the electron transfer resistance was higher in the foamed sample. In good accordance, the Bode plots showed that the impedance modulus of the foamed PMMA_F_G was more than one order of magnitude larger than in the nonfoamed PMMA_G (Figures 3(b) and 3(c)).

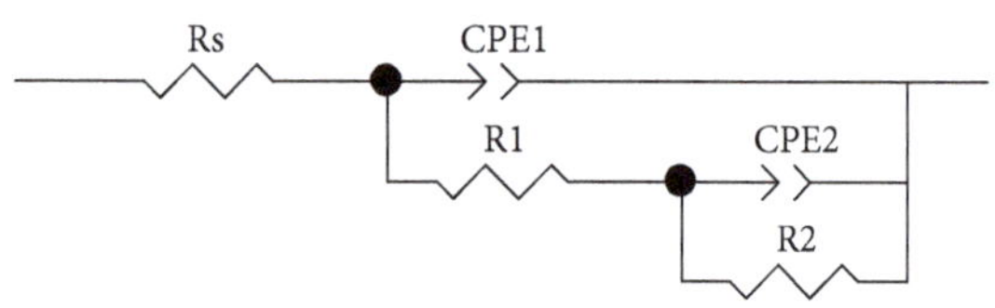

FIGURE 4: Equivalent electrical circuit used to simulate the response of the samples where Rs is the solution resistance, CPE1 is a first constant phase element, R1 is a first resistance, CPE2 is a second constant phase element, and R2 is a second resistance.

In order to calculate the charge transfer resistance, the impedance values were fitted to the electrical circuit shown in Figure 4. In the proposed model, two constant phase elements CPE (representing a nonideal capacitive response of the interface material solution) and two resistances were used. The depression in the Nyquist semicircle and the phase angle of capacitor different from −90° are an indication of this non-pure capacitive behavior. The impedance of a CPE element was defined as:

$$Z_{\mathrm{CPE}} = \frac{1}{C(j\omega)^n}, \qquad (1)$$

where C is a parameter numerically equal to the admittance at $\omega = 1$ rad·s^{-1}, j is the imaginary unit, ω is the frequency, and n is the phase shift. When $n = 1$ the CPE describes an ideal capacitor, when $n = 0$ is an ideal resistor and, when $n \approx 0.5$ is a finite-Warburg element accounting for diffusion of mobile charge. In practice, n is in the range of zero to one, and C can be approximately converted into a capacitance.

The results are presented in Table 1. Chi-squared (χ^2) values were used to evaluate the quality of fitting. The

TABLE 1: Analytical data obtained from the impedance spectroscopy measurements.

Rs ($\Omega\cdot cm^2$)	C_{CPE1} ($s^n\ \Omega^{-1}\cdot cm^{-2}$)	n_{CPE1}	R1 ($\Omega\cdot cm^2$)	C_{CPE1} ($s^n\ \Omega^{-1}\cdot cm^{-2}$)	n_{CPE2}	R2 ($\Omega\cdot cm^2$)	χ^2 (10^{-4})
PMMA_G							
610	$6.7\cdot10^{-5}$	0.67	205.3	$1.5\cdot10^{-4}$	0.70	15270	3.7
PMMA_F_G							
3151	$3.4\cdot10^{-6}$	0.58	13974	$6.8\cdot10^{-6}$	0.75	418250	4.6

obtained values were in the range of 10^{-4}, indicating the suitability of the circuit.

Table 1 shows that *R*1 values were always significantly lower than *R*2. The addition of *R*1 and *R*2 (considered as the electron transfer resistance) clearly increased in the foamed sample and the *CPE-C* values were lower for the foamed sample. The decrease in the capacitance and the increase in the resistance was induced by porosity.

Moreover, the impedance behavior indicated that the electron transfer resistance was higher in the foamed sample due to the air trapped into the foam micropores, increasing thus the resistivity [44]. In spite of this, the conductivity was high enough to use the foams as highly porous electrodes where the increase in the porosity and the subsequent higher superficial area can be an advantage.

3.3. Sensing Behavior. Detection of Antioxidants. Once the structure and the conductivity were characterized by SEM and EIS, graphite-doped samples were used as working electrodes in an electrochemical cell for the detection of CT and HQ, two of the most interesting antioxidants found in foods and beverages.

3.3.1. Catechol and Hydroquinone Determination. Figures 5(a) and 5(b) show the square wave voltammetry (SWV) recorded for increasing catechol concentrations using nonfoamed and foamed sensors. As observed, the foaming process induced important differences in the electrochemical responses. In the case of PMMA_G, low catechol concentrations produced a single redox peak at about +0.35 V (peak 1); as the concentration of catechol increased, a second peak appeared at higher potentials (+0.75 V) (peak 2), which is usually attributed to the dimerization or polymerization of phenols at the electrode surface [45, 46]. In contrast, the response of the foamed PMMA_F_G to catechol was characterized by a single peak centered at around +0.35 V, and polymerization was not observed. The absence of polymerization is an advantage because polymerization is an important cause of electrode fouling.

SWVs recorded at increasing hydroquinone concentrations are shown in Figures 5(c) and 5(d) for PMMA_G and PMMA_F_G sensors, respectively. In the case of the nonfoamed electrode, two peaks were observed even at low concentrations: the first peak appeared at +0.15 V (peak 1) and the second one appeared at +0.55 V (peak 2). When the concentration increased, a new peak at about +0.75 V was observed as a shoulder. In the voltammograms registered using the foamed PMMA_F_G, the first peak was centered about +0.12 V (peak 1). This time, the polymerization peak was observed at high concentrations (peak 2) and shifted to higher potentials as the concentration increased, reaching a final value of +0.9 V.

According to these results, the nonfoamed and the foamed electrodes have different electrochemical behaviors. Thus, the oxidation mechanism for catechol and hydroquinone on both surfaces is different. In addition, foamed electrodes prevent polymerization processes that can cause fouling of the electrodes.

Limits of detection (LOD) were calculated from the calibration curves constructed by representing the current intensity of the redox peaks versus catechol or hydroquinone concentrations. It has to be noticed that in some cases, two linear ranges were observed. Table 2 shows the linear ranges, the analytical linear calibration curves, the analytical parameters obtained from the curves including the regression coefficients (R^2), the residual standard errors (s_{yx}), and the LODs calculated form the curves using R-project as the program for statistical computing.

According to data listed in Table 2, important advantages can be observed in the foamed samples: linear ranges are larger in the foamed sensors, the residual standard errors are smaller, and LODs are lower. For instance, using PMMA_G, limits of detection for CT of 0.47 mM and 0.29 mM were calculated for the first and second oxidation peak, respectively, with a probability of false positive (α) and false negative (β) of 0.05. In contrast, LODs calculated using PMMA_F_G were 0.06 mM and 0.17 mM. This means that LODs are clearly diminished, mainly in the case of the first linear range, where a decrease of one order of magnitude was observed.

Similar results were obtained for HQ determination. In this case, the decrease observed in the LODs calculated using peak 1 was more marked than in peak 2, and a LOD of 0.12 mM was attained for the foamed sensor.

3.3.2. Simultaneous Determination of Hydroquinone and Catechol. Hydroquinone and catechol are two isomers and therefore, their oxidation peaks occur at very close potentials. This makes difficult the simultaneous assessment of both compounds. According to the results shown in Figure 5, the oxidation peak of CT is around +0.35 V and the oxidation peak for HQ is around +0.15 V. In addition, the degree of polymerization in foamed sensors is smaller. These results point to the possible advantage of PMMA foamed sensors to discriminate between both isomers, allowing the simultaneous assessment of both components.

In order to evaluate the capability of our foamed sensors to discriminate between both isomers, voltammograms were registered in solutions containing a fixed concentration of hydroquinone (2 mM) and increasing concentrations of

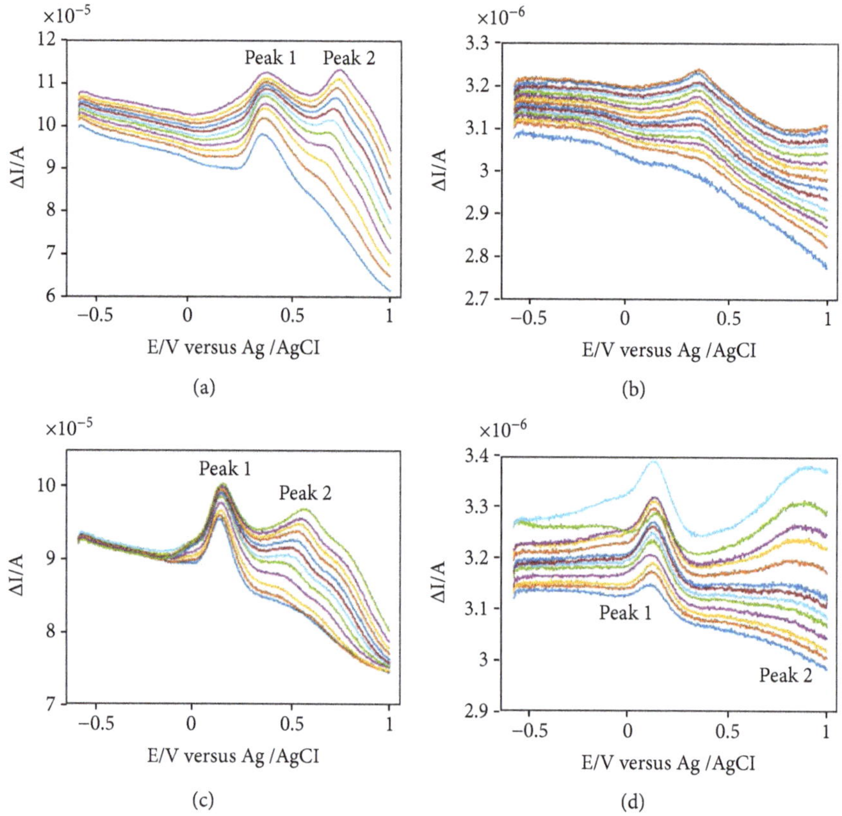

FIGURE 5: SWV of (a) PMMA_G immersed in CT concentrations from 0 to 2.91 mM; (b) PMMA_F_G immersed in CT concentrations from 0 to 1.67 mM; (c) PMMA_G immersed in HQ concentrations from 0 to 3.15 mM; and (d) PMMA_F_G immersed in HQ concentrations from 0 to 2.68 mM. 0.1 M KCl was used as supporting electrolyte.

TABLE 2: Analytical data obtained from the calibration curves.

		Linear range (mM)	Calibration curves	R^2	s_{yx}	LOD (mM)
PMMA_G						
CT	Peak 1	0.99–2.68	$y = 1.00 \cdot 10^{-4} + 3.92 \cdot 10^{-6}x$	0.9812	$3.20 \cdot 10^{-7}$	0.47
	Peak 2	0.50–1.72	$y = 6.77 \cdot 10^{-5} + 1.99 \cdot 10^{-5}x$	0.9890	$1.10 \cdot 10^{-6}$	0.29
	Peak 2	1.96–2.91	$y = 8.58 \cdot 10^{-5} + 9.39 \cdot 10^{-6}x$	0.9906	$3.44 \cdot 10^{-7}$	0.53
HQ	Peak 1	1.23–3.15	$y = 9.77 \cdot 10^{-5} + 8.73 \cdot 10^{-7}x$	0.9786	$8.85 \cdot 10^{-8}$	0.62
	Peak 2	0.50–2.44	$y = 7.86 \cdot 10^{-5} + 6.39 \cdot 10^{-6}x$	0.9982	$1.85 \cdot 10^{-7}$	0.13
PMMA_F_G						
CT	Peak 1	0.15–0.45	$y = 2.94 \cdot 10^{-6} + 3.44 \cdot 10^{-7}x$	0.9931	$3.09 \cdot 10^{-9}$	0.06
	Peak 1	0.45–1.67	$y = 3.04 \cdot 10^{-6} + 1.19 \cdot 10^{-7}x$	0.9921	$4.30 \cdot 10^{-9}$	0.17
HQ	Peak 1	0.37–1.11	$y = 3.09 \cdot 10^{-6} + 1.59 \cdot 10^{-7}x$	0.9941	$3.26 \cdot 10^{-9}$	0.12
	Peak 2	0.37–2.68	$y = 2.95 \cdot 10^{-6} + 1.64 \cdot 10^{-7}x$	0.9991	$3.55 \cdot 10^{-9}$	0.09

catechol, from 0.26 to 1.92 mM. The response of the PMMA_G sensor immersed in a solution that contains 2 mM of hydroquinone and increasing concentrations of catechol (Figure 6(a)) showed a peak at about +0.70 V that was growing progressively with the catechol concentration. This peak could be used to build a robust calibration curve in function of the catechol concentration with a good regression coefficient and low residual standard error (Table 3 collects the regression parameters). The calibration curve showed a linear range, from 0.61 to 1.92 mM. A LOD of 0.34 mM was obtained from the curve with a probability of false positive (α) and false negative (β) of 0.05. Similarly, the response for the PMMA_F_G towards hydroquinone 2 mM and increasing amounts of catechol also showed a peak at +0.25 V that increased linearly with the catechol concentration. In this case, the linear range was larger and ranged from 0.37 to 1.69 mM. Carrying out a robust regression, it was possible to obtain a calibration curve with a good regression

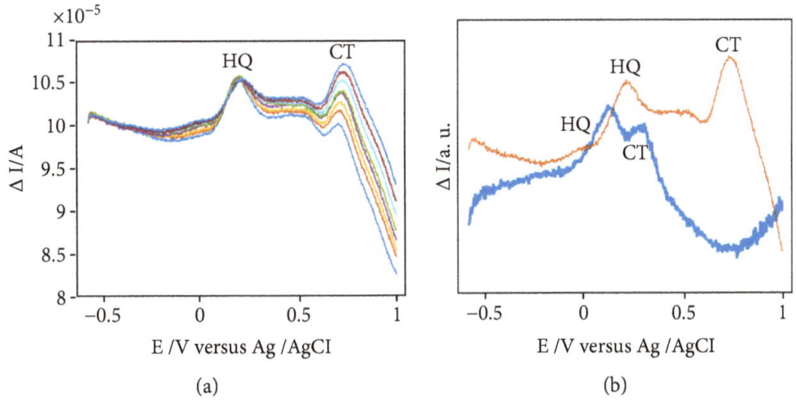

FIGURE 6: (a) SWV recorded using PMMA_G for increasing concentrations of catechol from 0.61 to 1.92 mM in the presence of 2 mM hydroquinone in 0.1 M KCl as supporting electrolyte, and (b) comparison for SWV recorded for PMMA_G (red line) and PMMA_F_G (blue line) for 1.92 mM of catechol in the presence of 2 mM hydroquinone in 0.1 M KCl as supporting electrolyte.

TABLE 3: Analytical data obtained from the calibration curves of mixtures CT/HQ.

	Linear range (mM)	Regression curve	R^2	s_{yx}	LOD (mM)
PMMA_G					
CT	0.61–1.92	$y = 9.73{\cdot}10^{-5} + 5.38{\cdot}10^{-6}x$	0.9806	$3.32{\cdot}10^{-7}$	0.34
PMMA_F_G					
CT	0.37–1.69	$y = 3.28{\cdot}10^{-6} + 7.82{\cdot}10^{-8}x$	0.9804	$4.67{\cdot}10^{-9}$	0.27

coefficient and low residual standard error (Table 3 summarizes the regression parameters). Furthermore, this calibration curve could be used to calculate a limit of detection of 0.27 mM, with a probability of false positive (α) and false negative (β) of 0.05.

According to these results, both the PMMA_G and the PMMA_G_F sensors can be used to assess catechol in presence of hydroquinone with very close limits of detection. However, foamed sensors show two important advantages. First, the foamed sensor has a larger linear range. But the main advantage is the high electrocatalytic activity of the foamed porous electrode that caused an important shift of the catechol peak from +0.70 V to +0.25 V that can reduce considerably the potential that needs to be applied in the working conditions of amperometry. This electrocatalytic effect has been illustrated in Figure 6(b), where the SWV recorded for the same concentration of catechol and hydroquinone (1.92 and 2 mM, resp.) is displayed.

3.4. Dynamic Properties. Once the different electrocatalytic properties of PMMA electrodes were established, the dynamic properties were analyzed by registering cyclic voltammograms at increasing scan rates. The dynamic properties for our PMMA_G and PMMA_F_G sensors were characterized in a catechol solution of 10^{-3} M in 0.1 M KCl as supporting electrolyte from −0.50 V to +1.50 V at scan rates from 0.1 to 0.3 V/s.

Representing the current intensity of the reduction peak (at around +0.40 V) versus the square root of the scan rate, a linear relationship was obtained. Data could be adjusted to the Randles-Sevcik equation (2), indicating a quasi-reversible process controlled by diffusion [43] (Table 4).

TABLE 4: Dynamic character. Analytical data obtained from the dynamic properties using a solution that contains catechol 10^{-3} M and KCl 0.1 M as supporting electrolyte.

	Regression curve	R^2	s_{yx}
PMMA_G			
CT	$y = 4.19{\cdot}10^{-5}\text{-}4.31{\cdot}10^{-4}x$	0.9994	$8.20{\cdot}10^{-7}$
PMMA_F_G			
CT	$y = -2.30{\cdot}10^{-6}\text{-}1.29{\cdot}10^{-5}x$	0.9951	$7.17{\cdot}10^{-8}$

$$i_p = 0.4463 \cdot n \cdot F \cdot A \cdot C \cdot \left(\frac{nFvD}{RT}\right)^{1/2} \qquad (2)$$

Where i_p is the peak current, n is the electron number transferred in the redox process, F is the Faraday constant, R is the gas constant, T is the temperature, A is the area of the working electrode, C is the concentration for the electroactive species, D is the diffusion coefficient for the electroactive species, and v is the scan rate.

As observed in Table 4, the slopes of the curves were one order of magnitude lower in the foamed sensor, indicating that kinetics was clearly modified by the presence of pores in the structure and a foamed surface.

4. Conclusions

Catechol (CT), hydroquinone (HQ), and their mixtures have been successfully analyzed at nonenzymatic electrodes based on PMMA doped with graphite. The porous structure of foamed electrodes induced important advantages including reduction of fouling effects, excellent electrocatalytic activity, and decrease in the limits of detection of CT and HQ that were one order of magnitude lower in foamed electrodes. In the case of mixtures, the electrocatalytic effect played an important role and LODs obtained were in the range of 10^{-4} M. These results confirmed the promising capabilities of foamed polymers as sensitive materials for electrochemical sensors.

Conflicts of Interest

The authors declare no conflict of interest.

Acknowledgments

This paper is dedicated to Professor Jose Antonio de Saja, who passed away on November 20, 2017. This work was supported by MINECO-FEDER (AGL2015-67482-R) and the Junta de Castilla y Leon-FEDER (VA-011U16). C. Fernandez-Blanco would also like to thank Junta de Castilla y León and FEDER for her postdoctoral grant (VA-011U16).

References

[1] G. Bartosz, *Food Oxidants and Antioxidants: Chemical, Biological, and Functional Properties*, Z. E. Sikorski, Ed., CRC Press, Boca Raton, FL, USA, 2013.

[2] W. Gao and C. Legido-Quigley, "Fast and sensitive high performance liquid chromatography analysis of cosmetic creams for hydroquinone, phenol and six preservatives," *Journal of Chromatography A*, vol. 1218, no. 28, pp. 4307–4311, 2011.

[3] G. Marrubini, E. Calleri, T. Coccini, A. F. Castoldi, and L. Manzo, "Direct analysis of phenol, catechol and hydroquinone in human urine by coupled-column HPLC with fluorimetric detection," *Chromatographia*, vol. 62, no. 1-2, pp. 25–31, 2005.

[4] H. Cui, C. He, and G. Zhao, "Determination of polyphenols by high-performance liquid chromatography with inhibited chemiluminescence detection," *Journal of Chromatography A*, vol. 855, no. 1, pp. 171–179, 1999.

[5] D. H. Bremner, A. E. Burgess, D. Houllemare, and K.-C. Namkung, "Phenol degradation using hydroxyl radicals generated from zero-valent iron and hydrogen peroxide," *Applied Catalysis B: Environmental*, vol. 63, no. 1-2, pp. 15–19, 2006.

[6] S. C. Moldoveanu and M. Kiser, "Gas chromatography/mass spectrometry versus liquid chromatography/fluorescence detection in the analysis of phenols in mainstream cigarette smoke," *Journal of Chromatography A*, vol. 1141, no. 1, pp. 90–97, 2007.

[7] A. Nenov, S. Mukamel, M. Garavelli, and I. Rivalta, "Two-dimensional electronic spectroscopy of benzene, phenol, and their dimer: an efficient first-principles simulation protocol," *Journal of Chemical Theory and Computation*, vol. 11, no. 8, pp. 3755–3771, 2015.

[8] L. Zhao, B. Lv, H. Yuan, Z. Zhou, and D. Xiao, "A sensitive chemiluminescence method for determination of hydroquinone and catechol," *Sensors*, vol. 7, no. 4, pp. 578–588, 2007.

[9] P. Nagaraja, R. A. Vasantha, and K. R. Sunitha, "A sensitive and selective spectrophotometric estimation of catechol derivatives in pharmaceutical preparations," *Talanta*, vol. 55, no. 6, pp. 1039–1046, 2001.

[10] M. F. Pistonesi, M. S. Di Nezio, M. E. Centurión, M. E. Palomeque, A. G. Lista, and B. S. Fernández Band, "Determination of phenol, resorcinol and hydroquinone in air samples by synchronous fluorescence using partial least-squares (PLS)," *Talanta*, vol. 69, no. 5, pp. 1265–1268, 2006.

[11] M. Bosco, M. P. Callao, and M. S. Larrechi, "Resolution of phenol, and its di-hydroxyderivative mixtures by excitation–emission fluorescence using MCR-ALS: application to the quantitative monitoring of phenol photodegradation," *Talanta*, vol. 72, no. 2, pp. 800–807, 2007.

[12] H. Cui, Q. Zhang, A. Myint, X. Ge, and L. Liu, "Chemiluminescence of cerium(IV)–rhodamine 6G–phenolic compound system," *Journal of Photochemistry and Photobiology A: Chemistry*, vol. 181, no. 2-3, pp. 238–245, 2006.

[13] J. McCall, C. Alexander, and M. M. Richter, "Quenching of electrogenerated chemiluminescence by phenols, hydroquinones, catechols, and benzoquinones," *Analytical Chemistry*, vol. 71, no. 13, pp. 2523–2527, 1999.

[14] J. McCall and M. M. Richter, "Phenol substituent effects on electrogenerated chemiluminescence quenching," *Analyst*, vol. 125, no. 3, pp. 545–548, 2000.

[15] N. Guan, Z. Zeng, Y. Wang, E. Fu, and J. Cheng, "Open tubular capillary electrochromatography in fused-silica capillaries chemically bonded with macrocyclic dioxopolyamine," *Analytica Chimica Acta*, vol. 418, no. 2, pp. 145–151, 2000.

[16] H. Lü, J. Wang, X. Wang, X. Wu, X. Lin, and Z. Xie, "Single-step preparation and characterization of polymeric monolith for pressurized capillary electrochromatography of typical homologs," *Journal of Separation Science*, vol. 30, no. 17, pp. 2993–2999, 2007.

[17] S. Sakthinathan, S. Palanisamy, S.-M. Chen, P.-S. Wu, L. Yao, and B.-S. Lou, "Electrochemical detection of phenol in industrial pollutant absorbed molecular sieves by electrochemically activated screen printed carbon electrode," *International Journal of Electrochemical Science*, vol. 10, pp. 3319–3328, 2015.

[18] K. Chen, Z. L. Zhang, Y. M. Liang, and W. Liu, "A graphene-based electrochemical sensor for rapid determination of phenols in water," *Sensors*, vol. 13, no. 5, pp. 6204–6216, 2013.

[19] M. F. Barroso, N. de-los-Santos-Álvarez, C. Delerue-Matos, and M. B. P. P. Oliveira, "Towards a reliable technology for antioxidant capacity and oxidative damage evaluation: electrochemical (bio)sensors," *Biosensors and Bioelectronics*, vol. 30, no. 1, pp. 1–12, 2011.

[20] C. Apetrei, I. M. Apetrei, J. A. SajaDe, and M. L. Rodriguez-Mendez, "Carbon paste electrodes made from different carbonaceous materials: application in the study of antioxidants," *Sensors*, vol. 11, no. 2, pp. 1328–1344, 2011.

[21] F. Matemadombo, C. Apetrei, T. Nyokong, M. L. Rodríguez-Méndez, and J. A. de Saja, "Comparison of carbon screen-printed and disk electrodes in the detection of antioxidants using CoPc derivatives," *Sensors and Actuators B: Chemical*, vol. 166-167, pp. 457–466, 2012.

[22] A. J. Saleh Ahammad, S. Sarker, M. A. Rahman, and J.-J. Lee, "Simultaneous determination of hydroquinone and catechol at an activated glassy carbon electrode," *Electroanalysis*, vol. 22, no. 6, pp. 694–700, 2010.

[23] W. Zhang, J. Zheng, Z. Lin et al., "Highly sensitive simultaneous electrochemical determination of hydroquinone, catechol and resorcinol based on carbon dot/reduced graphene oxide composite modified electrodes," *Analytical Methods*, vol. 7, no. 15, pp. 6089–6094, 2015.

[24] M. Li, F. Ni, Y. Wang et al., "Sensitive and facile determination of catechol and hydroquinone simultaneously under coexistence of resorcinol with a Zn/Al layered double hydroxide film modified glassy carbon electrode," *Electroanalysis*, vol. 21, no. 13, pp. 1521–1526, 2009.

[25] T. Gan, J. Sun, K. Huang, L. Song, and Y. Li, "A graphene oxide–mesoporous MnO_2 nanocomposite modified glassy carbon electrode as a novel and efficient voltammetric sensor for simultaneous determination of hydroquinone and catechol," *Sensors and Actuators B: Chemical*, vol. 177, pp. 412–418, 2013.

[26] H. Du, J. Ye, J. Zhang, X. Huang, and C. Yu, "A voltammetric sensor based on graphene-modified electrode for simultaneous determination of catechol and hydroquinone," *Journal of Electroanalytical Chemistry*, vol. 650, no. 2, pp. 209–213, 2011.

[27] M. Amiri, S. Ghaffari, A. Bezaatpour, and F. Marken, "Carbon nanoparticle–chitosan composite electrode with anion, cation, and neutral binding sites: dihydroxybenzene selectivity," *Sensors and Actuators B: Chemical*, vol. 162, no. 1, pp. 194–200, 2012.

[28] Q. Guo, J. Huang, P. Chen, Y. Liu, H. Hou, and T. You, "Simultaneous determination of catechol and hydroquinone using electrospun carbon nanofibers modified electrode," *Sensors and Actuators B: Chemical*, vol. 163, no. 1, pp. 179–185, 2012.

[29] Z. Huo, Y. Zhou, Q. Liu, X. He, Y. Liang, and M. Xu, "Sensitive simultaneous determination of catechol and hydroquinone using a gold electrode modified with carbon nanofibers and gold nanoparticles," *Microchimica Acta*, vol. 173, no. 1-2, pp. 119–125, 2011.

[30] L. Alshahrani, X. Li, H. Luo et al., "The simultaneous electrochemical detection of catechol and hydroquinone with [Cu(Sal-β-Ala)(3,5-DMPz)$_2$]/SWCNTs/GCE," *Sensors*, vol. 14, no. 12, pp. 22274–22284, 2014.

[31] C. Medina-Plaza, C. García-Cabezón, C. García-Hernández et al., "Analysis of organic acids and phenols of interest in the wine industry using Langmuir–Blodgett films based on functionalized nanoparticles," *Analytica Chimica Acta*, vol. 853, pp. 572–578, 2015.

[32] E. G. R. Fernandes, L. C. Brazaca, M. L. Rodríguez-Mendez, J. A. Sajade, and V. Zucolotto, "Immobilization of lutetium bisphthalocyanine in nanostructured biomimetic sensors using the LbL technique for phenol detection," *Biosensors and Bioelectronics*, vol. 26, no. 12, pp. 4715–4719, 2011.

[33] C. Medina-Plaza, L. N. Furini, C. J. L. Constantino, J. A. de Saja, and M. L. Rodriguez-Mendez, "Synergistic electrocatalytic effect of nanostructured mixed films formed by functionalised gold nanoparticles and bisphthalocyanines," *Analytica Chimica Acta*, vol. 851, pp. 95–102, 2014.

[34] W. Lu, X. Qin, A. M. Asiri, A. O. Al-Youbi, and X. Sun, "Ni foam: a novel three-dimensional porous sensing platform for sensitive and selective nonenzymatic glucose detection," *Analyst*, vol. 138, no. 2, pp. 417–420, 2013.

[35] Q. Xiao, X. Wang, and S. Huang, "Facile synthesis of Ni(OH)$_2$ nanowires on nickel foam via one step low-temperature hydrothermal route for non-enzymatic glucose sensor," *Materials Letters*, vol. 198, pp. 19–22, 2017.

[36] X. Wang, X. Guo, J. Chen et al., "Au nanoparticles decorated graphene/nickel foam nanocomposite for sensitive detection of hydrogen peroxide," *Journal of Materials Science & Technology*, vol. 33, no. 3, pp. 246–250, 2017.

[37] H. Feng, Z. Huang, X. Lou, J. Li, and G. Hui, "Study of a sucrose sensor by functional cu foam material and its applications in commercial beverages," *Food Analytical Methods*, vol. 10, no. 2, pp. 407–418, 2017.

[38] C. Zhao, X. Wu, X. Zhang, P. Li, and X. Qian, "Facile synthesis of layered CuS/RGO/CuS nanocomposite on Cu foam for ultrasensitive nonenzymatic detection of glucose," *Journal of Electroanalytical Chemistry*, vol. 785, pp. 172–179, 2017.

[39] X. Liu, W. Yang, L. Chen, and J. Jia, "Three-dimensional copper foam supported CuO nanowire arrays: an efficient nonenzymatic glucose sensor," *Electrochimica Acta*, vol. 235, pp. 519–526, 2017.

[40] J. Pinto, J. A. Reglero-Ruiz, M. Dumon, and M. A. Rodriguez-Perez, "Temperature influence and CO_2 transport in foaming processes of poly(methyl methacrylate)–block copolymer nanocellular and microcellular foams," *The Journal of Supercritical Fluids*, vol. 94, pp. 198–205, 2014.

[41] D. Miller and V. Kumar, "Microcellular and nanocellular solid-state polyetherimide (PEI) foams using sub-critical carbon dioxide II. Tensile and impact properties," *Polymer*, vol. 52, no. 13, pp. 2910–2919, 2011.

[42] S. Siripurapu, J. A. Coughlan, R. J. Spontak, and S. A. Khan, "Surface-constrained foaming of polymer thin films with supercritical carbon dioxide," *Macromolecules*, vol. 37, no. 26, pp. 9872–9879, 2004.

[43] A. J. Bard and L. R. Faulkner, *Electrochemical Methods: Fundamentals and Applications*, Wiley, New York, NY, USA, 2001.

[44] B. Q. Li and X. Lu, "The effect of pore structure on the electrical conductivity of Ti," *Transport in Porous Media*, vol. 87, no. 1, pp. 179–189, 2011.

[45] M. D. Ryan, A. Yueh, and W.-Y. Chen, "The electrochemical oxidation of substituted catechols," *Journal of the Electrochemical Society*, vol. 127, no. 7, pp. 1489–1495, 1980.

[46] F. Pospíšil, M. Hrubcová, and M. Cvikková, "Anodic oxidation of coumarie and caffeic acids and their effects on nitrate uptake and nitrate reducta se in *Nicotia na tabacum* cell suspension," *Biologia Plantarum*, vol. 29, no. 3, pp. 175–182, 1987.

12

A Sensor Network with Embedded Data Processing and Data-to-Cloud Capabilities for Vibration-Based Real-Time SHM

Nicola Testoni,[1] **Cristiano Aguzzi**,[2] **Valentina Arditi**,[1] **Federica Zonzini**,[1] **Luca De Marchi**,[1] **Alessandro Marzani**,[3] **and Tullio Salmon Cinotti**[2]

[1]*Department of Electrical, Electronic and Information Engineering, University of Bologna, Bologna, Italy*
[2]*Department of Computer Science and Engineering, University of Bologna, Bologna, Italy*
[3]*Department of Civil, Chemical, Environmental and Materials Engineering, University of Bologna, Bologna, Italy*

Correspondence should be addressed to Nicola Testoni; nicola.testoni@unibo.it

Academic Editor: Christos Riziotis

This work describes a network of low power/low-cost microelectromechanical- (MEMS-) based three-axial acceleration sensors with local data processing and data-to-cloud capabilities. In particular, the developed sensor nodes are capable to acquire acceleration time series and extract their frequency spectrum peaks, which are autonomously sent through an ad hoc developed gateway device to an online database using a dedicated transfer protocol. The developed network minimizes the power consumption to monitor remotely and in real time the acceleration spectra peaks at each sensor node. An experimental setup in which a network of 5 sensor nodes is used to monitor a simply supported steel beam in free vibration conditions is considered to test the performance of the implemented circuitry. The total weight and energy consumption of the entire network are, respectively, less than 50 g and 300 mW in continuous monitoring conditions. Results show a very good agreement between the measured natural vibration frequencies of the beam and the theoretical values estimated according to the classical closed formula. As such, the proposed monitoring network can be considered ideal for the SHM of civil structures like long-span bridges.

1. Introduction

Real-time structural health monitoring (SHM) of civil and industrial buildings requires a network of smart sensors which are capable to simultaneously acquire signals coming from a plurality of transducers [1]. Meaningful information should then be conveyed to the final user either in a raw or in a processed form by means of easy-to-deploy communication infrastructures and stored in scalable and highly available database without compromising the performance.

Among the variety of monitoring approaches, techniques based on the changes of structural natural frequencies and modes shapes have been largely investigated and adopted in the civil engineering field spanning from historical buildings [2–5] to more modern long-span bridges [6–9]. The continuous analysis of vibration features, in fact, can provide extremely useful information about possible changes in the structural properties and in turn on the overall state of the structure.

As such, advanced sensing networks capable to be permanently installed on the structure to be monitored are currently under investigation. In particular, research developments strive to reduce (i) the weight and cost of the sensing elements, (ii) the sensor network power consumption, and (iii) the cost for the deployment of centralized data acquisition systems, as well as (iv) the amount of cables [10–13] whereas, in the other end, aimed at (v) handling and sharing large amount of collected data.

In this scenario, the use of microelectromechanical (MEMS) accelerometers, thanks to their low cost and low power features, allows to cope with drivers (i) and (ii) [14–17]. However, wireless sensors can neither store/process all raw data locally nor reliably forward the data: wireless technologies are in fact typically incompatible with high-rate and high-resolution data collection and long-term monitoring tasks [18, 19].

Recent advancements in information technologies made possible to integrate SHM with Internet to track sensible data anytime and anywhere: the concepts of Internet of Things (IoT) [20–23] and its most recent evolution, the Social Web of Intelligent Things (SWIT) [24], can be fruitfully applied to SHM to provide almost real-time comprehensive information about monitored structures to all the involved figures, thus allowing to tackle drivers (iii) to (v). At the same time, due to the amount and criticality of the collected data, problems in data collection quickly arise. These problems can be addressed by adopting database management systems designed to handle large amounts of data capable to provide high availability with no single point of failure. These systems are often decentralized, in order to be fault tolerant at a geographic level, scalable, and support replication to ensure data preservation under harsh conditions [25, 26].

In this work, a network of small footprint, low power, and lightweight MEMS-based sensor nodes for modal shape and natural frequency analysis is presented. Network components are an upgraded and repurposed version of a previous work [27] aimed at guided wave detection. The network was designed to be, at the same time, low cost, scalable, and easily reconfigurable to be suitable both for short-term and long-term monitoring tasks. Weight and size reduction in the presented network are crucial to reduce cabling complexity, ease deployment, and demonstrate that a functional SHM system comprehensive of data collection and signal processing can be permanently embedded in structures without negatively impinging on their weight and size.

To achieve this goal, the proposed network replaces the traditional point-to-point connection between sensors and acquisition unit with the sensor area network (SAN) bus based on data-over-power (DoP) communication. To optimize this shared communication resource, techniques similar to that employed in a wireless sensor network can be used, yet sensor node power consumption is much less of a concern, allowing for continuous, real-time sampling of the structure health status. As a result, a sensor near electronics, which allows to preprocess monitoring information, losslessly encode it and transmit it to the cloud via a purposely designed companion gateway device was developed.

This sensor node was designed to be connected to other similar devices to create a network of up to 64 nodes: the total number of nodes on the network can be expanded arbitrarily by means of repeater nodes not discussed in this work. Communication to a notebook PC or to an embedded platform like Raspberry is performed in a half-duplex fashion at an effective speed of 200 ksps by means of a low-voltage, high-speed, half-duplex RS485 transceiver.

The paper is organized as follows. In Section 2 the sensor node, the gateway, and the implemented database interface are discussed; these materials are used as building block for the SHM sensor network. Section 3 discusses sensor calibration and data acquisition, and processing and collection. Finally, in Section 4, after discussing the application of the proposed sensor network to a case of study, the results of the modal shape and natural frequency analysis algorithms applied to the acquired dataset are compared with the theoretical predictions. Conclusions are then drawn in Section 5.

2. Materials

2.1. Sensor Node. The sensor node is built upon a four building block architecture as shown in Figure 1. All the devices in the sensor node are capable to withstand temperatures in the range of −40°C to 85°C.

At its heart is an ST Microelectronics STM32F3 32 bit, 3.3 V low power microcontroller unit (MCU): it belongs to a family of low-voltage 32 bit mixed-signal MCUs with DSP and FPU instructions which feature an integrated 8 MHz oscillator, a 40KiB SRAM, and a 256KiB FLASH, thus reducing the number of external components, hence the area occupied. The two integrated serial peripheral interface (SPI) and universal synchronous/asynchronous receiver/transmitter (USART) contain all the clock generators, shift registers, and data buffers necessary to perform input/output serial data transfers independently of device program execution. The maximum current consumption in active mode at a system clock frequency of 64 MHz is less than 40 mA but it can be decreased to less than 1.37 μA in standby mode.

In order to expand its limited memory capacity, a microchip 23LC1024 128Kibit Serial SRAM device is used as a temporary storage for the processed data samples. The MCU can access it through a dedicated SPI interface. It features an unlimited number of read/write cycles and zero write time, allowing for data rate up to 20 Mbps in sequential access mode. Still, it has a very low power consumption; when in standby, only 2 μA is absorbed from the 3.3 V power supply, while typical operative current is less than 2 mA.

To interface the MCU to the SAN network, an ST Microelectronics ST3485EB 3.3 V low power transceiver (XCVR) for RS-485 and RS-422 communications is used. All transmitter outputs and receiver inputs are protected to ±15 kV IEC 610004-2 air discharge. The MCU can access it through a dedicated USART interface. The driver is short-circuit current limited and is protected against excessive power dissipation by thermal shutdown circuitry that places the driver outputs into a high impedance state. It features a guaranteed 12 Mbps data rate at a very low power consumption; when in current shutdown mode, only 1 μA is absorbed from the 3.3 V power supply, while typical operative current is less than 1.5 mA. A mesh of passive components is used to interface the XCVR to the SAN network: two 10 μF capacitors are used to capacitively couple the XCVR input/outputs to the bus, while a resistive partition network allows for fail-safe biasing at less than 27 μA.

The entire sensor node is powered through a Texas Instrument LM3480 100 mA low-dropout (LDO) regulator. A device with a fixed 3.3 V output voltage was selected to reduce the number of required external components. This IC features operation from an input voltage as high as 30 V and an ensured maximum dropout of 1.2 V at the full 100 mA load, which is compliant with the 5.0 V power supply of the entire SAN bus. Its quiescent current is less than 1.75 mA and is inversely proportional to the load current. A

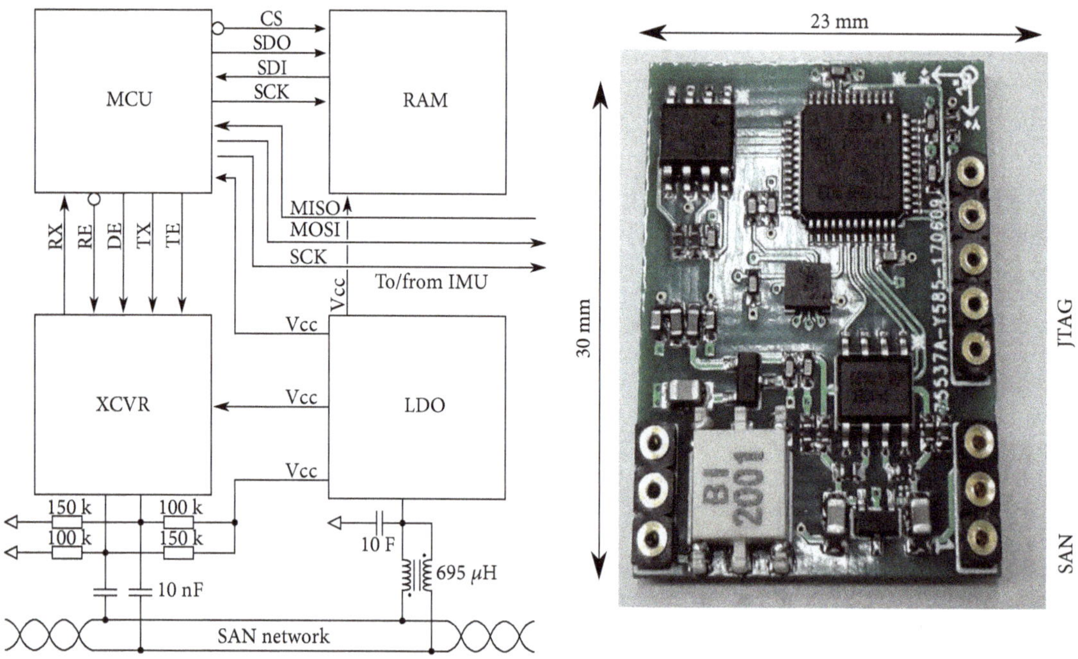

(a) (b)

Figure 1: Schematic diagram of the sensor node (a) and the relative prototype (b).

695 μH differential mode choke is used along with a 10 μF capacitor to extract power from the SAN bus and feed it to the LDO.

Finally, the sensor node is customized to collect accelerations and angular velocities by means of an ST microelectronics LSM6DSL iNEMO inertial measurement unit (IMU). This device is a system-in-package featuring a 3D digital accelerometer and a 3D digital gyroscope performing at 0.65 mA in high-performance mode and enabling always-on low power measurements. The MCU can access it through a dedicated SPI interface. It has a full-scale maximum acceleration range of ± 16 g and a maximum angular rate range of ± 2000 dps. When in current shutdown mode, only 3 μA is absorbed from the 3.3 V power supply.

The chosen triaxial MEMS accelerometer, with a sensitivity of 825 mV/g in a small package (2.5 mm × 3.0 mm × 0.8 mm) compares well versus commercial uniaxial high sensitivity piezoelectric accelerometers with similar sensitivity (1000 mV/g) in a much larger package (48 mm × 48 mm × 37 mm), as well as triaxial high sensitivity piezoelectric accelerometers with lower sensitivity (100 mV/g) in a compact package (14 mm × 14 mm × 14 mm); both these solutions do not offer signal processing capabilities. As such, the proposed solution demonstrates the possibility to integrate within a small package both triaxial sensing and signal processing capabilities.

Structural vibrations are recorded by the IMU and stored in the MCU internal RAM. Program instructions and filter coefficients are stored in the MCU embedded flash memory for the purpose of data processing. After processing, data can be stored in the external SPI serial SRAM, waiting for remote retrieval and collection.

The mean value of the total current consumed by a woken-up sensor node amounts to 44.8 mA when the SAN bus is powered at 5.0 V, whereas a sleeping node consumes less than 2 mA. Each sensor node is roughly 30 mm × 23 mm wide and weighs less than 5 g, making it attractive for fields where size, power, and weight reduction are crucial.

2.2. Gateway Device. A Raspberry Pi is employed as a smart data relay. This device is a low cost, flexible, single-board computer with a small form factor which is popular in the IoT field and starting to get attention from the SHM community [20, 21, 28]. The Pi comes in various models (in date publication order): A, B, A+, B+, 2, Zero, 3, and Zero W. Among them, the current flagship board of Raspberry Pi Foundation, the Raspberry Pi 3(https://www.raspberrypi.org/products/raspberry-pi-3model-b/) was chosen. The board is equipped with a quad-core 1.2GHz Broadcom 64 CPU, 1 GB of RAM and BCM43438 wireless LAN interface. It mounts a Debian-based Linux distribution called Raspbian. Its form factor (85 mm × 56 mm) is not as small as the nodes in SAN but it can still be suitable in SHM without any disadvantages, considering that most likely there will be only one gateway in a monitored structure. About the power consumption, the device also drains considerably higher currents than the sensor nodes. In idle state, as referred in [29], it consumes 260 mA, whereas at runtime operation, current consumptions raise to 480 mA. This difference compensates for the computational and communicational capabilities gain.

The main task of the Pi is to send the sensed data to an online database and serve as an interface to allow remote communication with every SAN node. The adoption of the Pi 3 has cut the cost of development thanks to its easy programming and connectivity interfaces.

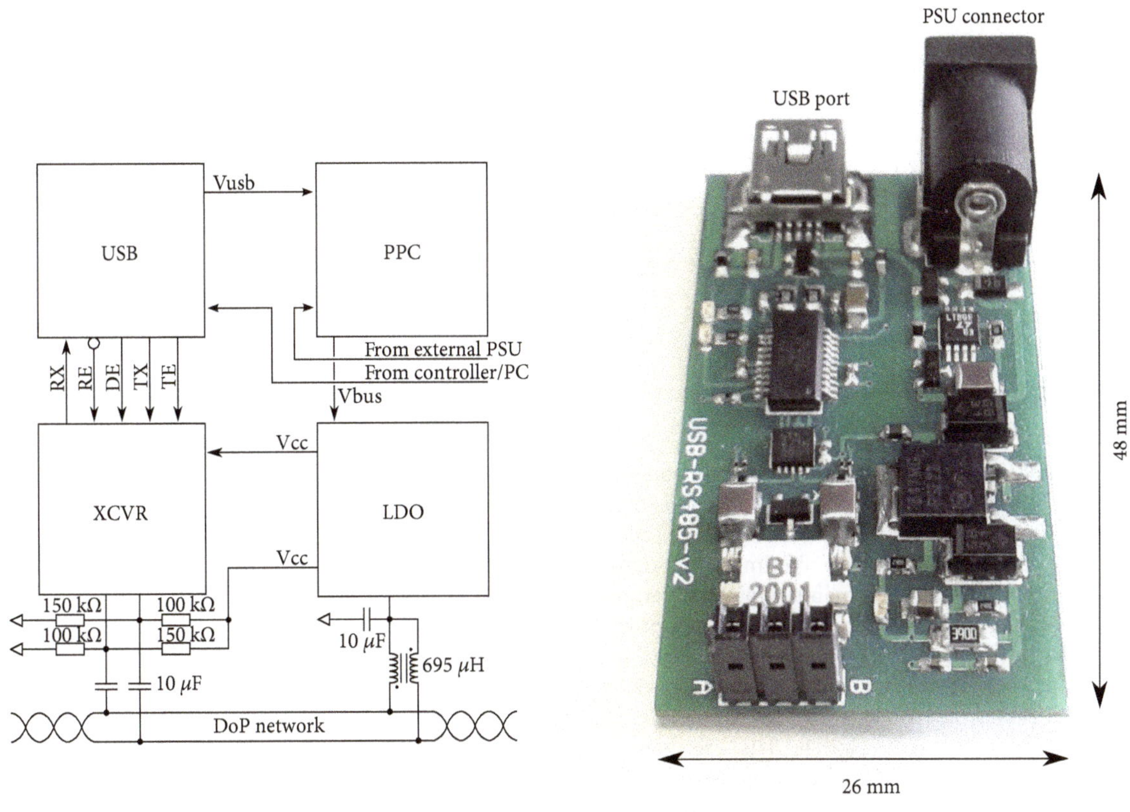

FIGURE 2: Schematic diagram of the SAN interface (a) and the relative prototype (b).

The device is connected to the SAN through ad hoc serial interface and communicates with remotes nodes using its Wi-Fi interface. The physical SAN network interface is built upon a four building block architecture as shown in Figure 2. All the devices in the network interface are capable to withstand temperatures in the range of −40°C to 85°C.

An FTDI FT231X USB to full handshake UART integrated circuit (IC) is used to provide USB connectivity to the Raspberry. The entire USB 2.0 full speed protocol is handled on this chip, allowing for data transfer rates from 300 baud to 3 Mbaud at TTL levels. It is operated by a single supply line taken directly from the USB bus and internally regulated to by integrated low-dropout (LDO) regulators. When active, it typically consumes 8 mA, but it can also enter a suspend state in which power consumption drops to 125 μA.

A Linear Technology LTC4414 low loss PowerPath™ Controller (PPC) is used to control an external P-channel MOSFET in order to create a near ideal diode function for power switchover. This permits the interface to perform a highly efficient management of the available power sources, namely, the 5.0 V, 500 mA USB power supply line, and an external 3.5 V to 36 V, 5A auxiliary power supply unit. When conducting, the measured voltage drop across the MOSFET is typically 20 mV with a typical $R_{DS(ON)}$ of 31 mΩ. The load is automatically disconnected from the USB power line when the auxiliary source is connected. The wide supply operating range of the PPC also supports operations from one to eight Li-ion cells in series to manage battery-operated scenarios. The low quiescent current (30 μA typical) is independent of the load current.

The PPC primarily feeds a Texas Instrument LM317 medium current 1.2 V to 37 V adjustable LDO voltage regulator. When powered through the 5.0 V USB source, quiescent current consumption never exceeds 1.5 mA; conversely, currents up to 4.0 mA can be measured when a higher voltage is supplied through the auxiliary power line. Similarly to the sensor node, the LDO feeds an ST Microelectronic ST3485EB RS485/RS422 transceiver (XVCR), used to interface the UART to the SAN network. Through the LDO section, the SAN network bus is also powered by means of two mutually coupled 695 μH inductors in an antiparallel configuration. This filters out common mode interferences from the SAN network bus, while allowing the flow of power supply current [30].

The mean value of the total current consumed by the interface circuit mounts to 12 mA when the SAN bus is powered at 5.0 V, whereas a sleeping interface consumes less than 2 mA. Each interface node is roughly 48 mm × 26 mm wide and weigh less than 10 g, making it attractive for fields where size, powe,r and weight reduction are crucial.

Concerning power consumption, one of the gateways is about that of 11 sensor nodes: a system comprising 16 nodes and one gateway device powered at 5 V consumes slightly less than 6 VA. This compares very favorably with a commercial system capable to collect data from the same amount of node,

Table 1

SensorId	Value	Timestamp
A40CFF12..	5	1514454710
A40CFF12..	4.2	1514454709
A40CFF12..	4.1	1514454708
A40CFF12..	4.1	1514454707
A40CFF12..	3	1514454706

Partition key: ((sensorId), value, and *timestamp).*

which requires more than 300 VA for data acquisition without offering data processing capabilities.

2.3. Online Database. Data collected in loco by sensors in a monitored structure should be deployed online to implement a remote monitoring system. Furthermore, logs about events occurred during the structure lifetime can provide useful information for planning maintenance operations. Consequently, in our software architecture, we employed a database to store this data. The database technology is a Cassandra (http://cassandra.apache.org/) instance installed in a remote data center.

Cassandra is an open source NoSQL database management system maintained by the Apache foundation. It offers robust support for clusters and it is designed to handle large amount of data. Furthermore, it features high availability due to its distributed nature with no single point of failure [31]. Studies have also shown its scalability and high throughput capabilities [25]. It employs a tailored data model to achieve its performance. The most notable difference from relational databases [32] is how tables are structured. Specifically, how rows are indexed in the tables and how tables are stored throughout the cluster which allows to fastly insert, delete, update, and read data.

The *primary key* is subdivided in two sections. One defines the *partition key* that identifies data locality, which means rows with the same *partition key* are stored inside the same cluster. The other, the so-called *clustering columns*, specifies how the rows are sorted, which allows to fast insert, delete, update, and read data. The combination of *partition key* and *clustering columns* should be unique to obtain a correct data model.

For example, consider to record acceleration values along the x-axis of an accelerometer and to query the database for its top five values as shown in Table 1.

The *sensorId* column is the *PARTITION KEY* ensuring that all the data from the sensor A40CFF12.. are stored in the same node. Then the rows are ordered by *value* (descending order) and finally by *timestamp* (also in descending order) that guarantee uniqueness. A query issued to the database could be

SELECT ∗ FROM acceleration by sensor

WHERE id = 'A40CFF12..' (1)

LIMIT 5

Accordingly to this query, the database engine will search for the node where the id is stored, then thanks to the already ordered rows retrieve the first five.

The features of Cassandra are perfectly suited for the implementation of structural monitoring systems. In particular, its high throughput allows to manage high rates of events generated in critical scenarios (i.e., earthquakes) and high sampling frequencies. High availability and robustness guarantee data collection even in presence of failures. Finally, the capability to handle big amount of data is crucial for large structures which can be sensorized with hundreds of devices.

For the purpose of this work, we deployed a Cassandra 2.2 instance in our testing environment on an Arch Linux 4.14.6-1 single cluster with default settings.

3. Methods

3.1. Sensor Calibration. Accelerometers used in this work, the ultra-low power high-performance three-axis iNEMO by STMicroelectronics, are guaranteed by the producer to be factory calibrated. It means that trimming values are stored inside the device in a nonvolatile memory, downloaded into registers, and used to correct any measure during the activity. However, in order to obtain data as much precise and reliable as possible, we adopted an additional calibration procedure suitable to be performed by the end-user on the field [33–35].

The calibration procedure starts with the acquisition of six values corresponding to the acceleration vector completely projected on every coordinate axis, in both positive and negative directions (e.g., vector (0,0,1) for positive direction of the x-axis). Each acquisition lasts a few seconds, then the values are averaged and normalized by the MCU and a single vector is associated with the corresponding stationary position. Comparing the raw measurements (gathered in matrix $\mathbf{W}$) to ideal values (theoretical stationary positions, in matrix $\mathbf{Y}$), the calibration parameters are calculated as the twelve coefficients that form the so-called calibration matrix ($\mathbf{X}$), so that

$$\mathbf{Y} = \mathbf{W}\mathbf{X}. \tag{2}$$

Finding $\mathbf{X}$ means solving an overdetermined linear system and this can be done through the evaluation of pseudoinverse matrix

$$\mathbf{X} = \left[\mathbf{W}^{\top}\mathbf{W}\right]^{-1}\mathbf{W}^{\top}\mathbf{Y}, \tag{3}$$

when all the columns of $\mathbf{W}$ are independent, as in this case. The obtained calibration parameters are stored on the sensor node flash memory and used to correct every new raw acquisition, until a new calibration procedure or a reset command is sent.

3.2. Data Acquisition. The whole system is able to measure projections of acceleration vectors on the three space directions x, y, and z with different features. Accelerometers were programmed to work, according to specific needs of the users, in a selectable full-scale of ±2 g, ±4 g, ±8 g, or ± 16 g, from a minimum sampling frequency of 1 Hz up to 5.3 kHz and in three different operation modes (high, normal, and low power) that correspond to various data resolutions in bits (12, 10, and 8, resp.). Moreover, to further decrease power consumption, when the device does not collect data, the sensor node is automatically set to a default condition of power-down mode that lasts until a specific command is sent.

The accelerometer embeds a 10 bit, 32-level FIFO buffer for each of the three output channels. This resource allows a consistent power saving because MCU does not need to poll for new data frequently and can be put to sleep mode for a much longer period of time. Due to the capacity of 10 bits for each slot in the stack, when the FIFO is operating, the only data resolutions possible are normal mode and low power mode.

More specifically, data can be downloaded from the accelerometer output registers according to four different working modes of their FIFO buffers: (i) bypass mode, (ii) FIFO mode, (iii) stream mode, and (iv) stream-to-FIFO mode.

In the *bypass mode*, the FIFO is nonoperative: the output registers used by the accelerometer are just six and they always contain the current data.

In *FIFO mode*, data from x, y, and z channels are stored into the FIFO, which is continuously fed until it is full. When all the levels are occupied, the buffer stops collecting new data. The device can generate an interrupt signal when FIFO reaches a given filling (watermark) level, in order to avoid data loss.

The *stream mode* differs from the previous one because when the buffer is full new data overwrite the older ones, even if they have not been read yet. Watermark level handling is applicable in this case too.

Finally, in the *stream-to-FIFO mode*, the transition from stream-to-FIFO mode is provided when a trigger event occurs.

To improve the efficiency of the whole system and reduce the CPU time, all the reading stored data operations were created to be handled by interrupt routines. In this work, the stream mode was chosen because it gives to the user more flexibility concerning the sampling frequency while guaranteeing the availability of fresh acceleration data.

3.3. Parameter Estimation. Modal shape and natural frequency estimation were conducted in this work in two modalities: on-sensor and off-sensor.

3.3.1. On-Sensor Estimation. Modal estimation was developed on-sensor by implementing procedures to detect relevant peaks in the acquired signal spectrum.

More precisely, we exploited functions from CMSIS DSP library to perform fast Fourier transform for floating point, 32-bit data. The magnitude of the complex Fourier transform coefficients is passed as input to an algorithm of peak detection [36]. Such algorithm compares the magnitude of the ith coefficient ($X[i]$) with a given threshold, which can be set independently for each axis and to the previous and following ones, $X[i-1]$ and $X[i+]$, respectively: if $X[i]$ is greater than threshold and its adjacent samples, the peak is detected.

In order to avoid false natural frequencies detection, during the experimental trials, many different values for the threshold levels in the peak detection algorithm were tested. According to the capability of the system to act differently on each axis, it was also possible to optimize the sensor node sensitivity for each channel separately. Moreover, to identify the peak position with more accuracy, a correction factor to the identified quantized frequency i was applied by performing a quadratic interpolation:

$$i - \frac{|X[i+1]| - |X[i-1]|}{2(|X[i+1]| - 2|X[i]| + |X[i-1]|)}. \tag{4}$$

Such correction is applied to all the identified peaks as part from the zero frequency acceleration which is related to gravity.

3.3.2. Off-Sensor Estimation. More sophisticated algorithms are computed off-sensor to extract natural frequencies and modal shapes. It is worth noting that the extraction of the modal parameters in the pristine structure is of extreme importance because it creates a set of benchmark values that must be as accurate as possible to detect abnormalities during the real-time estimation stage. Several procedures were considered to compute the probability density function (PSD) of the sensor acquisitions either based on parametric or nonparametric approaches. Among the algorithms of the first group, we considered autoregressive (AR) [37] and AR + noise models [38] which are suited for operational modal analysis in presence of strong acquisition noise. As concerning the nonparametric approaches, we have considered the frequency domain decomposition (FDD) [39], periodogram, and Welch estimation [38] methods.

It is worth noting that, besides data and power communication, the data-over-power bus also natively allows for data acquisition time base synchronization and consequently the output-only estimation of modal shapes. According to the authors in [40], the synchronization of data recording has a direct impact on the covariance index of the original mode shape to its time-shifted version estimated on-sensor. In the proposed sensor network, the maximum delay between each sensor is 50 μs, which corresponds to a maximum degradation of 1% in the correlation factor of the fourth mode.

Algorithms in the time and frequency domain were developed for the purpose of output-only estimation of modal shapes, followed by the application of the second order blind identification (SOBI) method, a strategy which reveals the independent components hidden within a set of measured signal mixtures [41]. Alternatively, the frequency domain decomposition (FDD) method can be used to identify modal parameters of a dynamic system by applying the singular value decomposition (SVD) technique to the output spectral density matrix, whereas the time domain decomposition (TDD) computes the SVD of the energy correlation matrix referred to bandpass-filtered data around the expected vibrational modes [39].

3.4. Data Collection. As anticipated in Section 2.2, the Raspberry Pi 3 device is in charge of data gathering. A custom made Go language software [42] able to query every node in the SAN network and send the data to the remote Cassandra service is installed on board. The process is running in the background as a service in the Raspbian OS and behaves as described in Algorithm 1.

```
procedure startSensor(address: int)
    command ← new AlwaysRunCommand
    send(address,command)
    waitForAck(address, TIMEOUT)
Init loop:
for all sensor address do
    startSensor(address)
Main loop:
for all sensor address do
    command ← new GetAccCommand
    send(address,command)
    data ← read(TIMEOUT)
    ifvalid(data) then
        send(address,ACK)
        acc ← decode(data)
        timestamp ← decode(data)
        sendToDB(address,acc,timestamp)
    startSensor(address)
gotoMain loop.
```

ALGORITHM 1: Data collection algorithm.

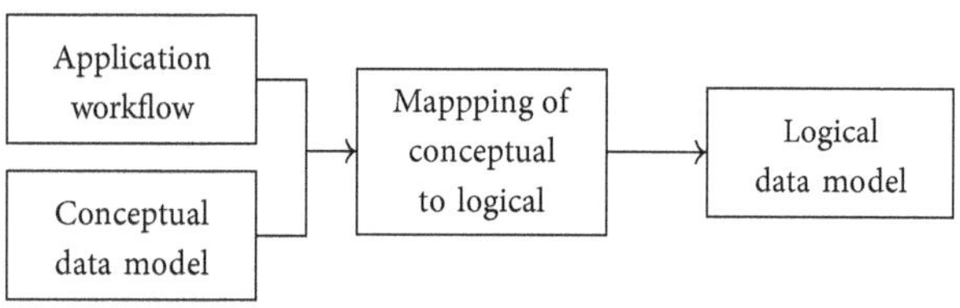

FIGURE 3: Query-driven design process phases.

First, all sensors are programmed to start the sensing phase in the *Init loop*. The *startSensor* procedure sends to a specific address the *AlwaysRunCommand* that defines the triggers in every spatial axis of acceleration sensing space. Each sensor can perform both individual and collective triggering, hardware, and software; in particular, each sensor is capable to receive and transmit trigger information from and to the other sensors in the network with a maximum delay of 50 μs, which is less than the minimum sampling period of 189 μs at 5.3 KHz. If the acceleration sensed in one of the sensors exceeds a given threshold, data is stored in the FIFO buffer and the event is communicated to all the other sensors. This event triggers the *Main loop* where data is retrieved from the buffers in the sensors. After data reading, the sensor must be reprogrammed to sense the acceleration continuously. Finally, the primitives *send* and *read* manage the serial connection with the SAN while *sendToDB* function communicates with the Cassandra cluster over the Wi-Fi interface.

3.4.1. Data Model Design. Using Cassandra as database implies the use of a new data model design created to ensure sound and efficient design [43]. This kind of development process is called query-driven design, and it is based on the definition of tables from the application workflow. As shown in [43], the process is made up of several phases, depicted in Figure 3.

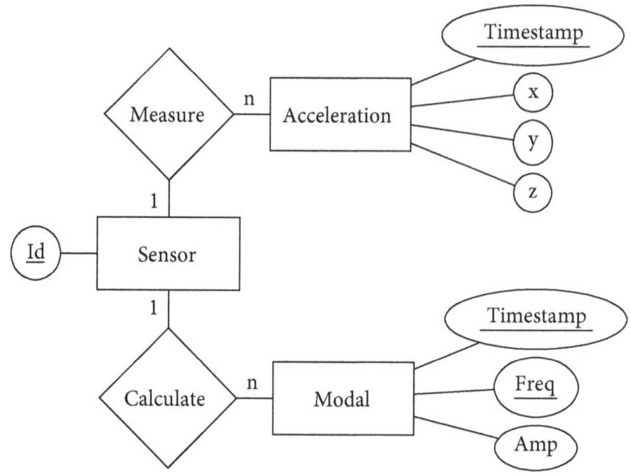

FIGURE 4: Logical data model of the monitoring application; primary and clustering keys are represented by underlined text.

TABLE 2: Tables for the queries in the monitoring application: entries marked with K representing the *primary key*, whereas C stands for the *clustering key*.

Q1: AccBySensor	
SensorId	K
Timestamp	C
Value	
Q2: AccBySensorTH	
SensorId	K
Value	C
Timestamp	C
Q3: ModBySensor	
SensorId	K
Freq	C
Timestamp	C
Amp	
Q4: ModAccBySensor	
SensorId	K
Timestamp	C
Freq	C
Value	
Amp	

In the *application workflow*, we start by defining the queries needed for adequate application behavior.

(Q1) Retrieve acceleration data produced by a sensor in a certain time interval (latest-first).

(Q2) Retrieve acceleration data produced by a sensor above a certain threshold (latest-first).

(Q3) Retrieve modal parameters estimated by a sensor in a certain time interval (highest frequency first).

(Q4) Retrieve combined acceleration and estimated modal parameters (highest frequency first).

These are basic queries that allow application users to get information about the monitored structure. Q1 ensures data enumeration (i.e., list all data acquired by SENSOR 1),

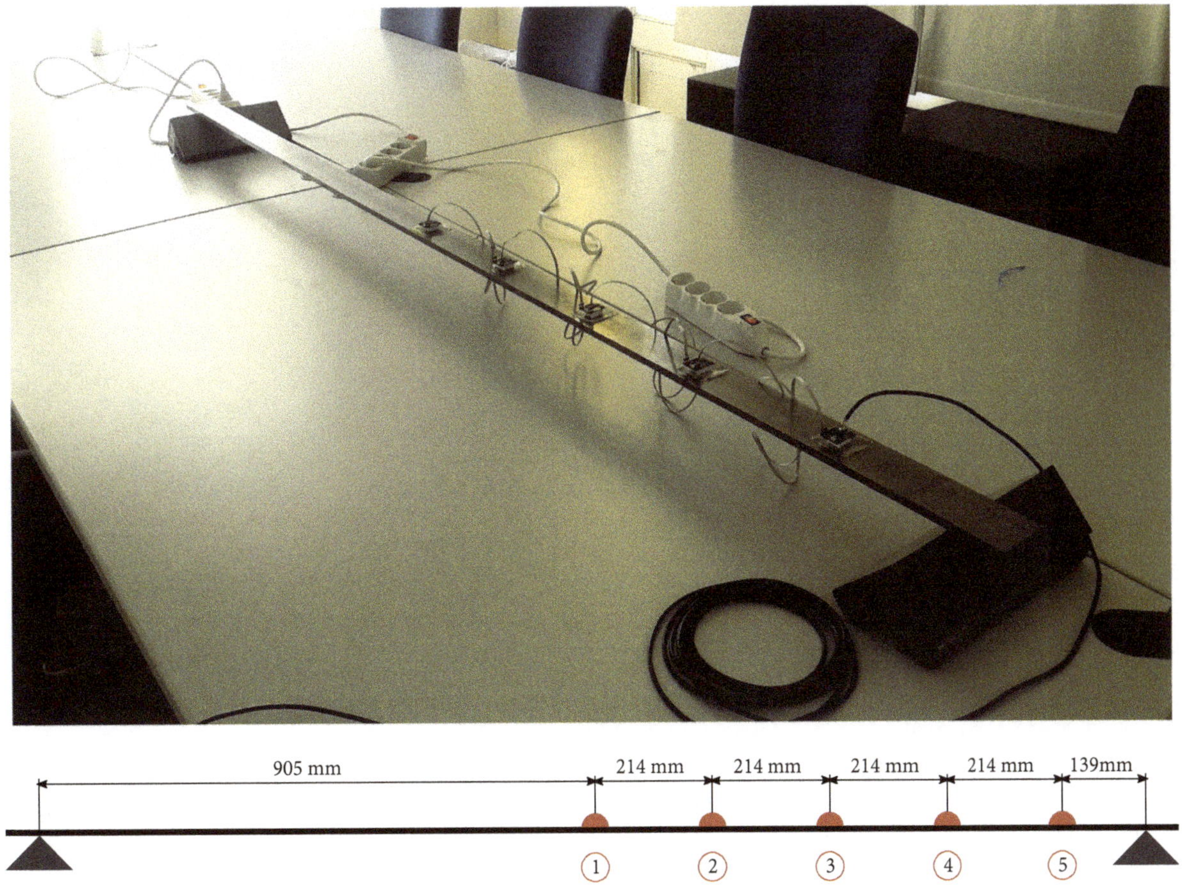

FIGURE 5: Experimental setup for modal shape and natural frequency verification and comparison.

Q2 gives a basic level of filtering (i.e., select only relevant accelerations), Q3 concerns modal parameter data enumeration, and finally Q4 correlates an acceleration value with modal parameters at that timestamp.

Combining the defined queries with the logical data model for the monitoring application shown in Figure 4 and following the rules defined in [43], the tables reported in Table 2 are crafted, which guarantee the correctness of data model stored in the Cassandra database. Notice that if a new kind of sensor data is to be gathered, the system can be scaled by adding a new pair of tables satisfying Q1 and Q2.

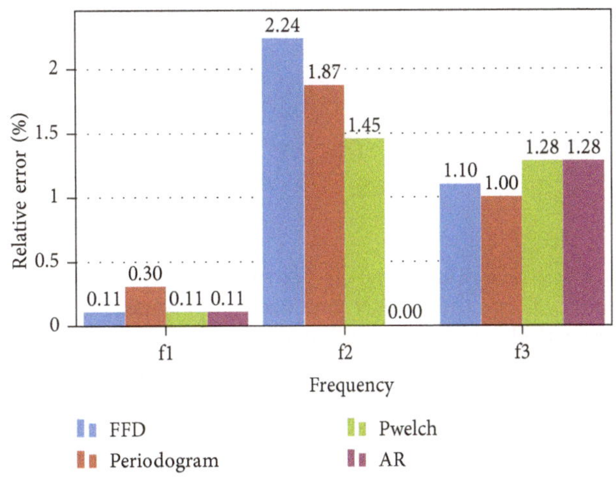

FIGURE 6: Error distribution in vibrational modes extractions comparing different techniques of PSD estimation.

4. Result and Discussion

A network comprising five sensor nodes and one interface connected in a daisy-chain fashion was tested in a setup which consists of a simply supported $L = 1900$ mm steel beam with cross-section base $b = 60$ mm and height $h = 10$ mm in free vibrations conditions, as shown in Figure 5. Sensors node are separated by 214 mm with respect to the first sensor, which is placed close to the beam center.

The beam was excited in correspondence of its center by mean of an impact hammer, thus allowing it to oscillate in condition of free vibrations. The amplitude of the acquired signals ranged from 0.1 g to 2.3 g, respectively, referred to the minimum and maximum value of the recorded induced acceleration along the z-axis. The chosen impact position is very close to the antinode of the first vibrational mode: as a result, this mode is much more energetic than other recorded modes, showing a difference of nearly 40 dB which can intensely affects the optimal performance of modal shapes reconstruction techniques.

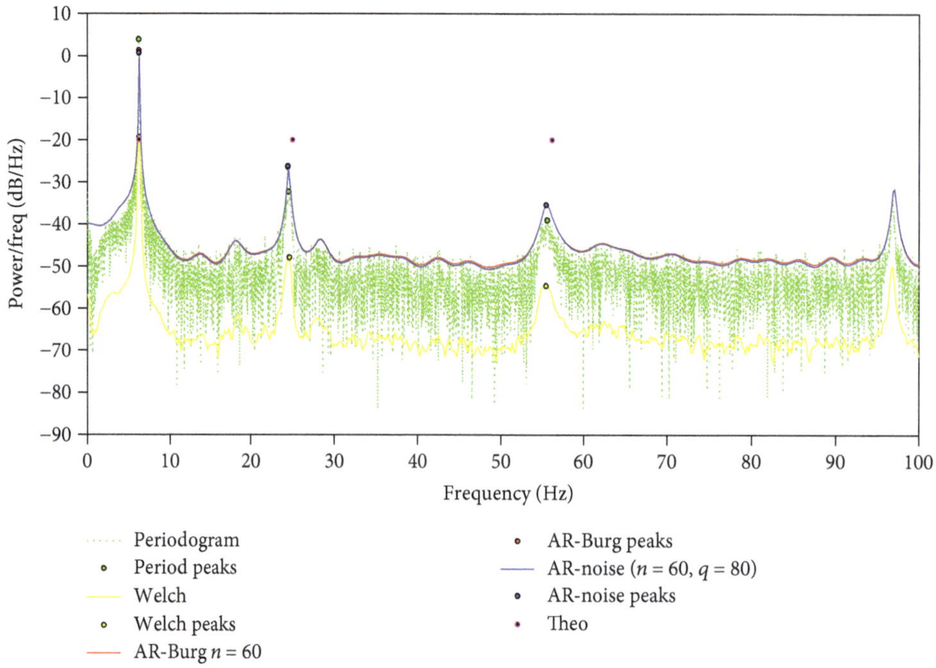

FIGURE 7: Comparison between different PSD estimators.

Signal processing techniques have been applied to the raw acquired signals to extract natural frequencies and modal shapes. The results of the modal shape and natural frequency estimation algorithms applied to the acquired dataset were compared with theoretical predictions obtained from the physical model of the beam to assess the performance of the implemented circuitry. In particular, the extracted first three natural frequencies are compared with the nominal values estimated as 6.25 Hz, 24.99 Hz, and 56.24 Hz through the theoretical formula:

$$f_n = \frac{1}{2\pi}\frac{(n\cdot\pi)^2}{L^2}\sqrt{\frac{E\cdot I}{\rho\cdot A}}, \tag{5}$$

where $\rho = 7880\,\text{kg/m}^3$ is the density of steel, $A = b\times h$ and $I = bh^3/12$ are the cross-section area and moment of inertia, respectively, and n is the frequency index. Here, for accurately measured beam dimensions and weight, a value of Young's modulus $E = 195\text{GPa}$ was assumed to minimize the error on f_1.

As it can be seen in Figure 6, the outcome of the algorithms computed is highly consistent with respect to the physical model, even in presence of high differences in the energy distribution. The relative error in natural frequencies estimation is always below 2.5% even in the worst case, related to the middle frequency. At low frequency, the performance of the spectral estimation methods is almost the same, whereas autoregressive techniques are more accurate for the estimation of the second vibrational component. At higher frequencies (above 30 Hz), there are no relevant differences among parametric and nonparametric algorithms.

Considering the fact that AR models are suited for systems affected by noise, the obtained results suggest that it is necessary to select the spectral estimation methods depending on the SNR. As shown in Figure 7, the PSD computed with the techniques mentioned demonstrates that, at low frequencies, there is an evident vertical alignment between computed and theoretical values, whereas the output is less precise when frequency increases. It is also evident that even the fourth natural frequency can be estimated from each of the tested algorithms, as testified by the peaks around 97.0 Hz.

Coherently with the literature [44], SOBI technique does not perform equally well as FDD or TDD since it is suited for dealing with signals sources in which mode strengths are each other similar and excitation is stationary. For real scenarios, characterized by low SNR values, noise can deteriorate the performance of the SOBI approach, especially while considering that no prior preprocessing is required apart from whitening raw signals.

The impact of sampling frequency on performance is also not negligible: in fact, by reducing the sample time, the number of samples to be used for modal shapes reconstruction is correspondingly augmented. This condition is of fundamental importance when some of the expected natural frequencies of the structure under test are close to Nyquists frequency: in such a scenario a strong correlation between estimation errors and sampling conditions is expected [45].

Results depicted in Figure 8 display the first three mode shapes extracted with three different techniques superimposed to the theoretical mode shapes. Again, the experimental behavior of the structure very well fits the physical model at low frequency, whereas deviation is more remarkable for modes related to higher natural frequencies.

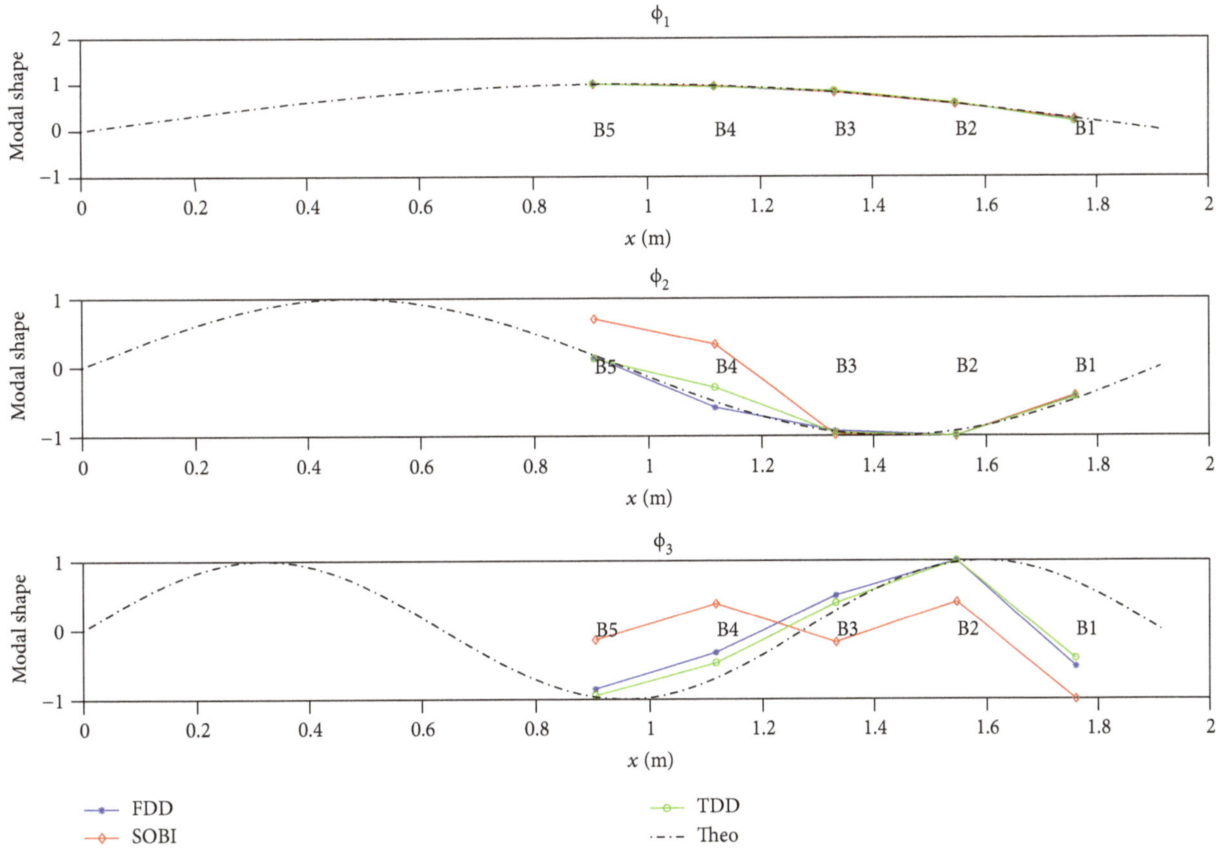

FIGURE 8: Modal shapes extracted with different techniques: FDD, TDD, and SOBI.

Such behavior is due to the fact that the lowest frequency component is the most energetic. It is also worth highlighting that, although the SOBI method is totally unsupervised, its performance is almost equivalent to FDD and TDD, making it suitable for autonomous damage detection systems.

5. Conclusions

This work reports on the implementation of a lightweight SHM sensor network comprising miniaturized sensor nodes and a gateway based on a single board PC. The sensors and the gateway are capable to exchange data and power on the same bus, thus minimizing the burden due to cabling. Also, by exploiting a wired connection, power figures and data bandwidth are fully compatible with the requirements of real-time data acquisition and processing. Two different processing modalities were made available, one exploiting the data processing capabilities of the sensor nodes, the other the ones of the gateway itself. Data processing results can be uploaded to a highly available database system for data collection and further interpretation without compromising the overall system performance. A modal shape and natural frequency estimation experiment has been conducted in order to assess the accuracy of the proposed SHM network at reconstructing the natural oscillation frequencies and shapes of a simply supported steel beam. The discussed results reveal the potentialities of the implemented network for SHM applications thanks to its versatility and high scalability, becoming a suitable candidate for a relatively cheap and low consumption system capable to provide real-time information.

Conflicts of Interest

The authors declare that there is no conflict of interests with any company or organization regarding the material discussed in this paper.

References

[1] J. P. Lynch and K. J. Loh, "A summary review of wireless sensors and sensor networks for structural health monitoring," *The Shock and Vibration Digest*, vol. 38, no. 2, pp. 91–128, 2006.

[2] V. Pazzi, A. Lotti, P. Chiara, L. Lombardi, M. Nocentini, and N. Casagli, "Monitoring of the vibration induced on the arno masonry embankment wall by the conservation works after the May 25, 2016 riverbank landslide," *Geoenvironmental Disasters*, vol. 4, no. 1, p. 6, 2017.

[3] J. Hester, S. Prabhu, S. Atamturktur, and J. Sorber, "Remote and wireless long-term vibration monitoring of historic monuments," *Procedia Engineering*, vol. 199, pp. 3302–3307, 2017.

[3] J. Hester, S. Prabhu, S. Atamturktur, and J. Sorber, "Remote and wireless long-term vibration monitoring of historic monuments," *Procedia Engineering*, vol. 199, pp. 3302–3307, 2017.

[4] G. Boscato, A. Dal Cin, S. Russo, and F. Sciarretta, "Shm of historic damaged churches," *Advanced Materials Research*, vol. 838-841, pp. 2071–2078, 2013.

[5] A. Girolami, D. Brunelli, and L. Benini, "Low-cost and distributed health monitoring system for critical buildings," in *2017 IEEE Workshop on Environmental, Energy, and Structural Monitoring Systems (EESMS)*, pp. 1–6, Milan, Italy, July 2017.

[6] J. Shinae and B. F. Spencer Jr, *Structural Health Monitoring for Bridge Structures Using Smart Sensors, Report No. nsel-035*, Technical report, Department of Civil and Environmental Engineering University of Illinois at Urbana-Champaign, 2015.

[7] Z. Chen, X. Zhou, X. Wang, L. Dong, and Y. Qian, "Deployment of a smart structural health monitoring system for long-span arch bridges: a review and a case study," *Sensors*, vol. 17, no. 9, p. 2151, 2017.

[8] K. Chebrolu, B. Raman, N. Mishra, P. K. Valiveti, and R. Kumar, "Brimon: a sensor network system for railway bridge monitoring," in *Proceeding of the 6th international conference on Mobile systems, applications, and services - MobiSys '08*, pp. 2–14, Breckenridge, CO, USA, June 2008.

[9] F. Moreu, R. E. Kim, and B. F. Spencer Jr, "Railroad bridge monitoring using wireless smart sensors," *Structural Control and Health Monitoring*, vol. 24, no. 2, article e1863, 2017.

[10] X. Liu, S. Tang, and X. Xu, "Smart wireless sensor nodes for structural health monitoring," *Intelligent Sensor Networks*, pp. 77–91, 2012.

[11] F. Federici, R. Alesii, A. Colarieti et al., "Design of wireless sensor nodes for structural health monitoring applications," *Procedia Engineering*, vol. 87, pp. 1298–1301, 2014.

[12] S. Cho, C.-B. Yun, J. P. Lynch, A. T. Zimmerman, B. F. Spencer Jr, and T. Nagayama, "Smart wireless sensor technology for structural health monitoring of civil structures," *Steel Structures*, vol. 8, no. 4, pp. 267–275, 2008.

[13] D.-W. Lim, S. C. Mantell, and P. J. Seiler, "Wireless monitoring algorithm for wind turbine blades using piezo-electric energy harvesters," *Wind Energy*, vol. 20, no. 3, pp. 551–565, 2017.

[14] S. N. Pakzad, G. L. Fenves, S. Kim, and D. E. Culler, "Design and implementation of scalable wireless sensor network for structural monitoring," *Journal of Infrastructure Systems*, vol. 14, no. 1, pp. 89–101, 2008.

[15] T. Torfs, T. Sterken, S. Brebels et al., "Low power wireless sensor network for building monitoring," *IEEE Sensors Journal*, vol. 13, no. 3, pp. 909–915, 2013.

[16] G.-D. Zhou and T.-H. Yi, "Recent developments on wireless sensor networks technology for bridge health monitoring," *Mathematical Problems in Engineering*, vol. 2013, 33 pages, 2013.

[17] M. Z. A. Bhuiyan, G. Wang, J. Cao, and J. Wu, "Deploying wireless sensor networks with fault-tolerance for structural health monitoring," *IEEE Transactions on Computers*, vol. 64, no. 2, pp. 382–395, 2015.

[18] Z. A. B. Md, G. Wang, and K.-K. R. Choo, "Secured data collection for a cloud-enabled structural health monitoring system," in *2016 IEEE 18th International Conference on High Performance Computing and Communications; IEEE 14th International Conference on Smart City; IEEE 2nd International Conference on Data Science and Systems (HPCC/SmartCity/DSS)*, pp. 1226–1231, Sydney, NSW, Australia, December 2016.

[19] M. Z. Alam Bhuiyan, J. Wu, G. Wang, and J. Cao, "Sensing and decision making in cyber-physical systems: the case of structural event monitoring," *IEEE Transactions on Industrial Informatics*, vol. 12, no. 6, pp. 2103–2114, 2016.

[20] A. Abdelgawad and K. Yelamarthi, "Internet of things (iot) platform for structure health monitoring," *Wireless Communications and Mobile Computing*, vol. 2017, 10 pages, 2017.

[21] A. Abdelgawad and K. Yelamarthi, "Structural health monitoring: Internet of things application," in *2016 IEEE 59th International Midwest Symposium on Circuits and Systems (MWSCAS)*, pp. 1–4, October 2016.

[22] Y. Nakamura, Y. Arakawa, and K. Yasumoto, "Senstick: a rapid prototyping platform for sensorizing things," in *2016 Ninth International Conference on Mobile Computing and Ubiquitous Networking (ICMU)*, pp. 1–6, Kaiserslautern, Germany, October 2016.

[23] P. Desai, A. Sheth, and P. Anantharam, "Semantic gateway as a service architecture for iot interoperability," in *2015 IEEE International Conference on Mobile Services*, pp. 313–319, New York, NY, USA, June-July 2015.

[24] L. Console, I. Lombardi, C. Picardi, and R. Simeoni, "Toward a social web of intelligent things," *AI Communications*, vol. 24, no. 3, pp. 265–279, 2011.

[25] T. Rabl, S. Gómez-Villamor, M. Sadoghi, V. Muntés-Mulero, H.-A. Jacobsen, and S. Mankovskii, "Solving big data challenges for enterprise application performance management," *Proceedings of the VLDB Endowment*, vol. 5, no. 12, pp. 1724–1735, 2012.

[26] B. R. Chang, H.-F. Tsai, C.-Y. Chen, and C.-L. Guo, "Empirical analysis of high efficient remote cloud data center backup using hbase and cassandra," *Scientific Programming*, vol. 2015, Article ID 294614, 10 pages, 2015.

[27] N. Testoni, L. De Marchi, A. Ferraro, and A. Marzani, "A coin size, 40mw, 20 grams sensor node for guided waves detection," *Sensors and Smart Structures Technologies for Civil, Mechanical, and Aerospace Systems 2015*, 2015.

[28] W. Flores-Fuentes, M. Rivas-Lpez, D. Hernndez-Balbuena et al., "Online shm optical scanning data exchange," in *2016 IEEE 25th International Symposium on Industrial Electronics (ISIE)*, pp. 940–945, Santa Clara, CA, USA, June 2016.

[29] R. P. Dramble, "Power consumption," December, 2017 https://www.pidramble.com/wiki/benchmarks/power-consumption.

[30] T.-k. Chin and D. Tran, "Combine power feed and data link via cable for remote peripherals," February, 2015 http://www.eetimes.com/document.asp?doc_id=1279219, 2011.

[31] A. Lakshman and P. Malik, "Cassandra: a decentralized structured storage system," *ACM SIGOPS Operating Systems Review*, vol. 44, no. 2, pp. 35–40, 2010.

[32] P. Beynon-Davies, "Relational data model," in *Database Systems*, pp. 23–41, Springer, 1996.

[33] STMicroelectronics, *Parameters and Calibration of a Low-g 3-Axis Accelerometer*, 2014.

[34] S.-h. P. Won and F. Golnaraghi, "A triaxial accelerometer calibration method using a mathematical model," *IEEE Transactions on Instrumentation and Measurement*, vol. 59, no. 8, pp. 2144–2153, 2010.

[35] U. Qureshi and F. Golnaraghi, "An algorithm for the in-field calibration of a mems imu," *IEEE Sensors Journal*, vol. 17, no. 22, pp. 7479–7486, 2017.

[36] M. Abe and I. I. I. Julius O Smith, "Design criteria for simple sinusoidal parameter estimation based on quadratic interpolation of fft magnitude peaks," in *Audio Engineering Society Convention 117*, Audio Engineering Society, 2004.

[37] B. Peeters and G. De Roeck, "Stochastic system identification for operational modal analysis: a review," *Journal of Dynamic Systems, Measurement, and Control*, vol. 123, no. 4, pp. 659–667, 2001.

[38] P. Stoica and R. L. Moses, *Spectral Analysis of Signals*, vol. 452, Pearson Prentice Hall, Upper Saddle River, NJ, 2005.

[39] T.-H. Le, Y. Tamura, A. Yoshida, and N. Dong, "Frequency domain versus time domain modal identifications for ambient excited," *Structure*, vol. 6, 2010.

[40] V. Krishnamurthy, K. Fowler, and E. Sazonov, "The effect of time synchronization of wireless sensors on the modal analysis of structures," *Smart Materials and Structures*, vol. 17, no. 5, article 055018, 2008.

[41] F. Poncelet, G. Kerschen, and J.-C. Golinval, *Experimental Modal Analysis Using Blind Source Separation Techniques*, International Conference on Noise and Vibration Engineering, Leuven, 2006.

[42] A. A. A. Donovan and B. W. Kernighan, *The Go Programming Language*, Addison-Wesley Professional, 2015.

[43] A. Chebotko, A. Kashlev, and S. Lu, "A big data modeling methodology for apache cassandra," in *2015 IEEE International Congress on Big Data*, pp. 238–245, New York, NY, USA, June-July 2015.

[44] B. Swaminathan, B. Sharma, and S. Chauhan, "Utilization of blind source separation techniques for modal analysis," in *Structural Dynamics*, T. Proulx, Ed., vol. 3 of Conference Proceedings of the Society for Experimental Mechanics Series, pp. 189–206, Springer, New York, NY, 2011.

[45] S. Park and N. Stubbs, "Reconstruction of mode shapes using Shannon's sampling theorem and its application to the nondestructive damage localization algorithm," in *Smart Structures and Materials 1995: Smart Systems for Bridges, Structures, and Highways*, vol. 2446, pp. 280–293, San Diego, CA, United States, April 1995.

13

Monitoring and Control Systems in Agriculture using Intelligent Sensor Techniques

Imran Ali Lakhiar,[1] **Gao Jianmin**,[1] **Tabinda Naz Syed**,[1] **Farman Ali Chandio**,[1] **Noman Ali Buttar**,[1] **and Waqar Ahmed Qureshi**[2]

[1]*Key Laboratory of Modern Agricultural Equipment and Technology, Ministry of Education, Institute of Agricultural Engineering, Jiangsu University, Zhenjiang, 212013 Jiangsu, China*
[2]*Research Centre of Fluid Machinery Engineering and Technology, Jiangsu University, Zhenjiang, 212013 Jiangsu, China*

Correspondence should be addressed to Gao Jianmin; gaojianminujs@163.com

Guest Editor: Marco Grossi

In recent years, intelligent sensor techniques have achieved significant attention in agriculture. It is applied in agriculture to plan the several activities and missions properly by utilising limited resources with minor human interference. Currently, plant cultivation using new agriculture methods is very popular among the growers. However, the aeroponics is one of the methods of modern agriculture, which is commonly practiced around the world. In the system, plant cultivates under complete control conditions in the growth chamber by providing a small mist of the nutrient solution in replacement of the soil. The nutrient mist is ejected through atomization nozzles on a periodical basis. During the plant cultivation, several steps including temperature, humidity, light intensity, water nutrient solution level, pH and EC value, CO_2 concentration, atomization time, and atomization interval time require proper attention for flourishing plant growth. Therefore, the object of this review study was to provide significant knowledge about early fault detection and diagnosis in aeroponics using intelligent techniques (wireless sensors). So, the farmer could monitor several paraments without using laboratory instruments, and the farmer could control the entire system remotely. Moreover, the technique also provides a wide range of information which could be essential for plant researchers and provides a greater understanding of how the key parameters of aeroponics correlate with plant growth in the system. It offers full control of the system, not by constant manual attention from the operator but to a large extent by wireless sensors. Furthermore, the adoption of the intelligent techniques in the aeroponic system could reduce the concept of the usefulness of the system due to complicated manually monitoring and controlling process.

1. Introduction

Agriculture has an ancient history nearly dates back to thousands of years. Moreover, its advancement has been pushed by implementing the several new systems, practices, technologies, and approaches with the time. It employs over one-third of the global workforce [1]. The agriculture is the backbone of an economy for many countries and executes a significant contribution to the development of the economy for underdeveloped countries. Besides, it steers the process of economic prosperity in developed countries. Several research studies concluded that overall world agriculture uses approximately seventy percent per year available fresh water to irrigate only seventeen percent of the land. Another side, the total available irrigated land is gradually decreasing due to the rapidly increasing of food requirements and effects of global warming [2, 3]. In other words, agriculture is dealing with new main significant challenges. Foote [4] said FAO reported that world food production must be increased by seventy percent to provide sufficient food production for the fast-growing population and urbanisation. The expected world population growth for the half of the present century is daunting. However, depending on the estimate, it could be expected to rise above the nine billion people by midcentury. As many studies reported that the population is increasing very fast, the global population was one billion in 1800,

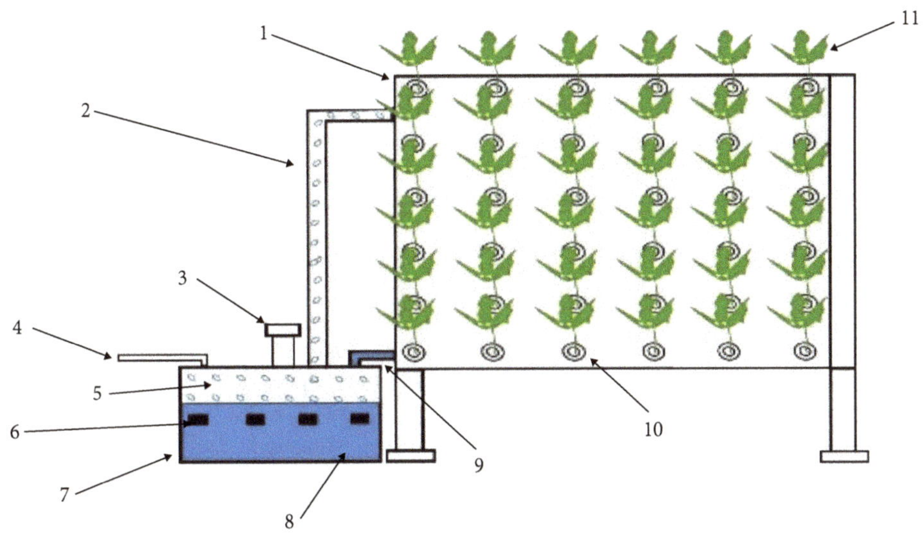

Figure 1: Basic diagram of the aeroponic plant cultivation system by Lakhiar et al. [15]. 1. Growth chamber. 2. Nutrient fog transmission pump. 3. Misting fan. 4. Power supply line. 5. Nutrient fog. 6. Ultrasonic atomizers. 7. Nutrient reservoir. 8. Nutrient solution. 9. Nutrient recycle line. 10. Plant holder. 11. Plant.

and it increased to seven billion people in 2012. However, studies report feared that at the end of the current century, it could be expected to reach eleven billion people and there could be many, many more mouths to feed soon. Thus, the rapid increase of the population, alongside the decrease in agriculture land, intensification of global climate changes, and exacerbation of water resources, declines labour force and energy crunches are posing tremendous challenges and hurdles to the agriculture sector [5, 6]. Furthermore, the developing and developed countries will deal with substantial water crises and issues due to rapid urbanisation and industrialisation. The available fresh water for irrigated agriculture land is supposed to decrease in future [7, 8]. Besides, the unpredictable climate changes include extreme weather conditions, intense storms, heat waves, and floods will have a substantial adverse impact on world agriculture sector. We need more production from agricultural systems to meet the growing food demands. Otherwise, we will suffer from food insecurity problems which will be the biggest threat. Moreover, Qiu and coworkers [9] revealed that the progress of the agriculture production is not only significant for to producing food to feed the population, but it is also essential for the industrial sector. Similarly, the agriculture is the main source to produce the raw material for many industrial sectors. Therefore, it must be understood that industrial and agricultural developments are not alternatives. However, both sectors are complementary to each other on the path to achieving the food security issues.

As the evolution of humankind from hunters and gatherers to agrarian societies, the efforts have mainly focused on improving the plant yield and productivity by either genetic changes, cultural or husbandry, management practices, or by developing and introducing plant protection measures. Accordingly, in the last and present century, peoples have started exploring the possibilities by adopting different modern techniques in agriculture. The adoption of the precision farming methods in agriculture is one of the excellent examples. The purpose is to try and mechanise them in agriculture to prevent the crop losses due to sudden climatic changes, soil-borne diseases, pest attaches, and so on. However, many research studies have been suggested and reported that problems and challenges of agriculture could overcome by adopting the precision farming methods. At present, several countries are increasing their farming productivities by implementing the precision farming methods.

Baudoin et al. [10] reported that the artificial plant growing method (e.g., greenhouse and factory farms) is one of the fundamental types of precision agriculture. Nowadays, the method is receiving importance and gaining the intention of the growers. The method can provide sufficient food supply throughout the year. In the system, the plant grows around the year by artificially adjusting and controlling the surrounding environmental conditions such as temperature, CO_2 (carbon dioxide), humidity, light intensity, airflow, and nutrients supply within the confined facilities [11, 12]. Besides, the system minimises environmental impact and maximises the crop yield with significant results as compared with traditional (open-field) cultivation system [13]. Savvas and team [14] informed that at present soilless plant cultivation is one of the most disruptive inventions ever presented in the field of artificial plant growing system. The soilless system refers to plant cultivation techniques without the use of soil by providing artificial solid material or water nutrient solution as a growing medium instead of soil. However, the water culture is related to the process of hydroponic and aeroponic plant cultivation (Figure 1). In both methods, the roots of the plant are continuously or intervalley nurtured with or within water nutrient solution by providing a specific control environment in artificial supporting structure [16, 17]. Both methods provide many benefits to the grower such as full control of nutrient concentration and supply and prevention of many soil-borne diseases and infections to plant, thus resulting in increased plant yield with significant returns, high quality, and more efficient use of available natural resources [18, 19].

Several studies reported aeroponic and hydroponic systems as a modern and innovative plant cultivation techniques under the soilless system. By adopting these techniques, the growing food crises could be resolved [20, 21]. Moreover, the hydroponic system to grow initially leafy green vegetables was the first to emerge, which started taking commercial exploitation routes in industrialised countries in west and east but eventually was found to have particular defects and problems that forced people to discover and experiment with newer variations and techniques like the aeroponic system. According to the NASA report, the aeroponic system could reduce water, nutrient, and pesticide usage by 98, 60, and 100 percent, respectively, and increase the plant yield by 45 to 75 percent [22].

The primary motivation of this review article is to provide an idea about the use of intelligent sensor techniques in the aeroponic system. It could provide an opportunity for full automation, scalability, anytime-anyplace access monitoring, and fault diagnostics in the aeroponic system. Moreover, it would be helpful for the local farmer and grower to provide timely information about rising problems and influencing factors for successful plant growth in the aeroponic system. The farmers could start to understand their crops at a micro scale and able to communicate with plant through accessible technology. To the best of our knowledge, this is the first work to provide a brief review of the use of intelligent sensor techniques in the aeroponic system. However, the rest of the paper is organised as follows: Section 2 describes the current work in the aeroponic system with intelligent sensor techniques. In Sections 3, 4, and 5, we present the brief description about the aeroponic system, application, and working protocol of wireless sensor network in the aeroponic system. Sections 6, 7, 8, and 9 describe the advantages, future application, application of artificial intelligence in agriculture, and conclusion.

2. Related Work

Aeroponics is the new plant growing technique of modern agriculture. Until now, it is not entirely implicated among the farmers. Mostly, it is practiced by the researchers for performing the experimental studies. Their study reports concluded that it could be well accepted in agriculture as a modern-day plant cultivation activity where the modern farmer does not need soil to grow the plant. However, the aeroponic system has some substantial vulnerability like a failure of water supply pumps, nutrient distribution line and preparation, and atomization nozzle clogging, which require special knowledge and attention to avoid damage, rapid plant death, and failure of the system [23]. Furthermore, the integration of the intelligent agriculture techniques could be the best solution to avoid or deal with the above-mentioned issues without any technical expertise. Xiong and Qiao [24] reported that the integration of the intelligent agriculture systems could be an effective approach for solving complex problems of agriculture domains. Zhai et al. [25] reported that presently, several research studies had been conducted on the use of intelligent techniques in agriculture especially in the last two decades. Besides, several new techniques and application have been introduced and patented to improve the traditional agriculture practices. However, experts mainly focused and monitor the climatic condition, soil properties, water quality, plant development, livestock management, and fertilizer application, pesticide application, and illumination control through various intelligent techniques [26–32]. Meanwhile, it could be concluded that traditional agricultural logistics is improved and upgraded by introducing and implementing the several modern technologies and techniques in agriculture domains [33]. Basnet and Bang [34] reported that the collecting information through sensors and communication technology played a vital role in improving agricultural production. It has shifted agriculture from input-intensive to knowledge-intensive, and agriculture becomes more networked and decision-making. Both small- and large-scale farmer can benefit from introducing this technique into the agriculture value chain, having their productivity increased, quality improved, services extended, and costs reduced. It provides insights into various issues in the agriculture like weather prediction, crop and livestock disease, irrigation management, and supply and demand of agriculture inputs and outputs and helps in solving those problems. Rehman and Shaikh [35] concluded that at present, several information technologies including satellite navigation, grid and ubiquitous computing, and sensor network are exercised in agriculture. However, the application of the sensor network is supporting agriculture practices and activities in a very positive direction [36, 37]. Zhang and coworkers [38] used a sensor network to monitor air temperature, humidity, ambient light, and soil moisture and temperature. Also, the aeroponic system is the new application of the soilless agriculture. Besides, several studies had been successfully designed the aeroponic system by using various information technologies approaches such as Tik and coworkers [39] designed and implemented a wireless sensor network to monitor the aeroponic system. They used temperature, light intensity, pH, and EC monitoring sensors. Moreover, the study reported that the wireless sensor network offers a wide range of information which could be required for the horticulturist to provide a greater understanding of how these environmental and nutrient parameters are correlated with plant growth. The real-time information obtained from sensor nodes can be utilised to optimise strategies to control the temperatures and the other properties of the nutrient solution. A study by Pala et al. [40] proposed an approach to monitor automation and early fault detection tools in the aeroponic system through intelligent techniques. In the protocol, they designed a highly scalable aeroponic system and coded as aero-pot prototype. They developed software based on a genetic algorithm to optimise power consumption of the aeroponic system. Their study concluded that using this software user can define various properties and virtually configure the aeroponic system. The developed software can allow the user to add and remove the lights and pumps and define consumption of added devices with minimum grower effort. Laksono et al. [41] designed a wireless sensor and actuator network for the controlling, monitoring, and conditioning of an aeroponic growth chamber. The designed wireless protocol was based on ZigBee technique. They also designed a data transmission system to transfer the data from the database server to administrator through text message. The proposed system was based on the sensors, actuators, communication system, and database server. The experiment results showed that the proposed wireless protocol based on ZigBee techniques was a useful tool of the wireless sensor network to monitor the aeroponic system. Jonas et al. [42] developed an automatic monitoring system to control the environmental and nutrient supply of the aeroponic system. The designed wireless protocol was based on Arduino development board. Their study concluded that the proposed system can control the nutrient atomization frequency based on the root chamber moisture content. However, the system can automatically transfer all the gathered information to a web server and also share on Twitter. Sani et al. [43] recommended a web-based control and monitoring system for the aeroponic system. Their system was composed of microcontrollers (using Arduino IDE program), actuators (two relays include atomization spray and fan on/off on specific time), the sensor (temperature and pH sensor), LDR (light intensity sensor), and communication modules (GSM/GPRS/3G modem). The present study concluded that our proposed design was able to monitor and measure the temperature, pH, light intensity, atomization time and interval time, and fan activation time and interval time in the aeroponic system. The proposed method was able to directly send the real-time information from the sensor to the server via the Internet using GSM. Anitha and Periasamy [44] designed wireless sensor technique to monitor the aeroponic system. The technique was based on the ZigBee prototype. The proposed network architecture was based on temperature, pressure, humidity, water level, and pH monitoring sensors. The sensors transmitted the gathered data to the GSM (Global System for Mobile) node or coordinator node, whereas the gateway device was used to transfer the data to the personal computer. However, a server was connected to the database where the maximum and minimum threshold values of pH, water level, and temperature were fixed. Furthermore, if the monitored value reaches above or below the threshold values stored in the database, thus, the system was able to start the alarm sound to aware the farmer. Another study in 2016 by Kernahan and Cupertino [45] invented a system to monitor and control the aeroponic system using wireless techniques. They concluded that a reliable aeroponic system provides a wireless connection between its subsystem for the exchange of data and commands. The various subsystems manage one or more plants growing atriums include nutrient atomization on hanged roots, maintenance, control of nutrient solution level, the addition of various nutrients, and control of the light quantity and cycle. A study by Montoya et al. [46] designed a wireless sensor system to monitor the aeroponic system. The system protocol was based on the Arduino development board. They used analog and digital sensors for monitoring temperature, nutrient atomization, EC, and pH fluctuations and level of nutrient solution in the nutrient reservoir. In order to acquire data and automation system, the two Arduinos were managed in a master-slave configuration and connected to each other through wireless by Wi-Fi. All the recorded data was autosaved in microSD memory and sent to a web page. Their study concluded that the proposed protocol could be used for automation and could monitor the aeroponic system. Kerns and Lee [47] firstly designed and introduced an aeroponic system using IoT (Internet of things) to automate the system. The proposed system is comprised of a mobile application, service platform, and IoT device with sensors (pH balance, temperature, and humidity). They used Raspberry Pi Zero device and designed a system to monitor and measure the selected parameters. The gathered data was autosaved into the database server by sending an SQL query. Their study concluded that the proposed system could help farmers to control and monitor the aeroponic system remotely. Furthermore, Karu [48] also designed and implemented a high-precision system for small-scale aeroponic plant cultivation. The system is allowed to precisely control the nutrient solutions, pH, and EC levels and gives data about humidity, temperature, pH, and EC concentration and amount of the nutrient solution in the reservoir. A recent study by Martin and Rafael [49] also proposed and suggested systems, methods, and devices for the aeroponic system. Mithunesh et al. [50] proposed an intelligent control system for an aeroponic system. The system protocol was based on an open-source development board called Raspberry Pi. Their study concluded that the developed system provides the simple management and high availability established by using both the local and global systems. Idris and Sani [51] designed monitoring and control system for the aeroponic system. They concluded that the developed system is able to monitor the aeroponic system working parameter such as temperature and humidity by sending the data in real time from sensor to the display system. Janarthanan et al. [52] concluded that the problems of the aeroponic system could be solved by the use of wireless sensor and actuator system. It allows the user to monitor and interact with the system through mobile app and a web interface. However, Liu and Zhang [53, 54] designed an aeroponic system for automatic control of water-fertilizer and temperature. They concluded that system supplies an experimental platform with features of simple structure and convenient control.

3. The Aeroponic System

The aeroponic system is one of the techniques of the soilless culture, where the plant grows in the air with the assistance of artificial support instead of soil or substrate culture. It is an air-water plant growing technique where lower portions such as the roots of the plant are hanged inside the growth chamber under complete darkness in controlled conditions. However, the upper portions of the plant such as leaves, fruits, and crown portion are extending outside the growth chamber. Usually, the artificial supporting structure (plastic or thermofoam) is provided to support and divide the plant into two parts (roots and leaves). In the system, plant roots are openly exposed in the air and directly irrigated with a small droplet size of the water nutrient at interval basis. The nutrient solution is supplied through different atomization nozzles with or without high air pressure. Moreover, several studies considered aeroponics as a modern-day agricultural activity which is practiced in an enclosed growth chamber under entire controlled conditions, as it could eliminate the external environmental factors as compared with traditional agriculture activity. Hence, it is no longer dependent on large-scale land use, and it could be set up in any place, a building that has lifted global climate without considering the current climate such as rainy season and winter [23, 55–59]. Buer et al. [60] reported that atomization nozzle uses the tiny amount of the water nutrient solution and provides an excellent growth environment for the plant. Zobel and Lychalk [61] said it is a modern-day agricultural research tool which provides several agricultural research opportunities for a researcher with significant results by providing artificial growth conditions. However, Table 1 shows the essential monitoring and control parameters in the aeroponic system. Hessel et al. [62] and Clawson et al. [63] studies discovered that aeroponics contributes to the advances and developments in many areas of plant root studies. It provides an excellent chance for plant researchers to deeply study the behavior of plant root under different conditions and without any complications. Until now, many researchers had conducted plant root research and experimental studies root response to drought [64], effects of different oxygen concentrations on plant root development [65, 66], root microorganism [67–69], arbuscular mycorrhizal fungi production [70], and legume-rhizobia interaction [71]. Furthermore, studies also practiced the technique by growing vegetables, fruits, herbs, and medicinal root-based plant [72–74] such as tomato, potato, soybean, maize, lettuce, *Anthurium andreanum*, and *Acacia mangium* [15, 59, 75–79].

3.1. Present Status of the Aeroponic System. The aeroponic system is one of a holistic production management method in agriculture which promotes and improves agroecosystem, health, and biodiversity. The system has a paramount reputation in the horticulture department, because of its implications on the economic and technical aspects in the agriculture. Among all agriculture systems, only the aeroponic system is receiving the full attention of farmers, policymakers, entrepreneurs, and agricultural researchers. The grower could reduce the requirements of chemical inputs including fertilizers, herbicides, pesticides, and other agrochemicals. The grower could obtain higher cultivated plant yield and quality as compared with other growing methods. However, the aeroponic system is labour-intensive. It offers many opportunities for the farmers to increase rural employment. The farmer can grow a plant in their homes by providing artificial growth environment. Anitha and Periasamy [44] reported that nowadays many families are practicing the aeroponic system on their terrace. Besides, several countries of the world are using aeroponics for making an expansion in nourishment creation managing the monetary issues and making the nation naturally amicable to have their particular food supply. While some years ago, the use of the aeroponic system was limited almost around the world [84]. At present, the system is acquiring more attention from the farmers and several countries are being effectively adopting as an economical and environmentally friendly vegetable and fruit growing system. However, it is practiced in following countries: Abu Dhabi, Australia, Bhutan, Bolivia, Brazil, Bangladesh, Burkina Faso, China, Canada, Colombia, Ecuador, Egypt, Ethiopia, France, Germany, Ghana, Greece, Indonesia, Italy, India, Iran, Japan, Israel, Kenya, Korea, Malaysia, Mongolia, Malawi, New Zealand, Nigeria, Peru, Philippines, Poland, Russia, Rwanda, Saudi Arabia, South Africa, Spain, Singapore, South Korea, Slovakia, Sri Lanka, Taiwan, Thailand, Uzbekistan, and Vietnam. Besides, attempts are made to represent the system in other countries of the world [23].

3.2. Key Problems and Difficulties of the Aeroponic System. Aeroponic cultivation is performed in an outdoor and indoor installation and or in a greenhouse under controlled conditions. It may be carried out within a facility that includes the provision of light for plant growth, the centralised delivery of nutrient solution, and electrical power. The growing plants are set in a growth chamber and periodically soaked with nutrient solution small mist ejecting through atomization nozzle (Figure 2). In addition, the aeroponic system gives the chance to control the entire growth chamber environment precisely. The aeroponic system is the modern technique of the agriculture which is still under development. Until now, limited studies have been performed, and conducted studies concluded that the system has some problems and issues. Studies suggested that aeroponics is performed without soil or any solid media; thus, the main observed problems are water and nutrient buffer, any failure of the water pumps, nutrient solution distribution and preparation, atomization nozzle clogging, and so on, which lead to rapid death of the grown plant [40]. Kernahan and Cupertino [45] reported that the aeroponic system provides better control of the plant growth and nutritional availability and prevents the plant from various diseases and root rot. However, during plant growth from sowing to harvest time, the methods adopted in the aeroponic system require a little hand-operated contribution, interference regarding physical presence, and expertise in domain knowledge of plants, environment control, and operations to maintain and control the growth of the plant.

Moreover, there is a requirement to sustain and keep retain the nutrient solution parameters which include nutrient temperature, pH, and EC concentration in a narrow range of preferred values for optimal growth. If these parameters drift outside the desired range, it will create several problems for plant growth. In addition, some supplemental parameters can adjust to optimise the plant growth further. The additional parameters are atomization time, atomization interval time, air temperature, relative humidity, light intensity, and carbon dioxide (CO_2) concentration which make the system complicated and time-consuming with high human energy and with the higher level of expert training and skill for operating the system. However, the grower has the responsibility to control and monitor the fluctuations of the above parameters in the desired range to achieve the suitable growth conditions for the specific plants. A failure to accurately control and monitor the parameters could

TABLE 1: Basic monitoring and control parameters in the aeroponic system [23, 80–83].

No.	Parameters	Common value	Instruments
1	Nutrient atomization	Mist/spray/aerosol/droplet size at high pressure from 10 to 100, low pressure from 5 to 50, and ultrasonic foggers from 5 to 25 microns, respectively	Atomization nozzle (high and low pressure, atomization foggers)
2	Growing medium	Plant holder	Any artificial root supporting structure
3	Desirable pH of the nutrient solution	The pH value depends on the cultivar (onion 6.0–7.0, cucumber 5.8–6.0, carrot 5.8–6.4, spinach 5.5–6.6, lettuce 5.5–6.5, tomato 5.5–6.5, and potato 5.0–6.0)	pH measuring device
4	Desirable EC of the nutrient solution	The EC value depends on the cultivar (onion 1.4–1.8, cucumber 1.7–2.2, carrot 1.6–2.0, spinach 1.8–2.3, lettuce 0.8–1.2, tomato 2.0–5.0, and potato 2.0–2.5 $ds{\cdot}m^{-1}$)	EC measuring device
5	Humidity	Provide 100% available moisture	Humidity measuring device
6	Temperature	Optimum 15°C–25°C and should not increase to 30°C and less than 4°C	Temperature measuring device
7	The light inside the box	The light inside the growth box must be dark enough	Cover the growth chamber with locally available material
8	Atomization time	Depends on the cultivar growth stage	Manually operating the system with timer
9	Atomization interval time	Depends on the cultivar growth stage	Manually operating the system with timer

High-pressure atomization nozzle Low-pressure atomization nozzle Ultrasonic foggers

FIGURE 2: Aeroponically cultivated lettuce by Lakhiar et al. [23].

significantly affect the growth of the plant and cause financial loss. If any component failure occurs while the operator is not present on site, it may be detected too late to prevent harm, because systems generally include some automated means for periodically providing nutrient mist to the plant roots, refilling a nutrient reservoir, and managing light cycles and intensity. Therefore, the aeroponic cultivation considered hitherto to be somewhat unsuitable for the local grower and due to the above reasons and it is not common to find an installation. However, the main reason for the low acceptability of the aeroponic system is not a cost, but the main drawback is the amount of attention required of the grower with a high level of expertise and judgment. For the above-discussed reasons, more sophisticated and advanced monitoring techniques have implemented in the aeroponic system for early fault detection, real-time monitoring, and control and automation of the system. Hence, it would be advantageous to use artificial intelligent tools (Figure 3) in the aeroponic system to detect fault and diagnosis problems on time. Thus, it could help to avoid rapid damages to grown plants and help to fully automate the aeroponic system.

4. The Aeroponic System and Sensor Network

In recent years, early fault detection and diagnosis using an intelligent agricultural monitoring system is considered as the best tool to monitor plant without any complicated operations and laboratory analysis which required domain expertise and extensive time. The development of these convenient features has attracted much attention in the agriculture. The system is based on a wireless sensor network which comprises of a data server, a wireless convergence node, a plurality of wireless routers, and a plurality of wireless sensor nodes. However, the wireless sensor nodes are used as the

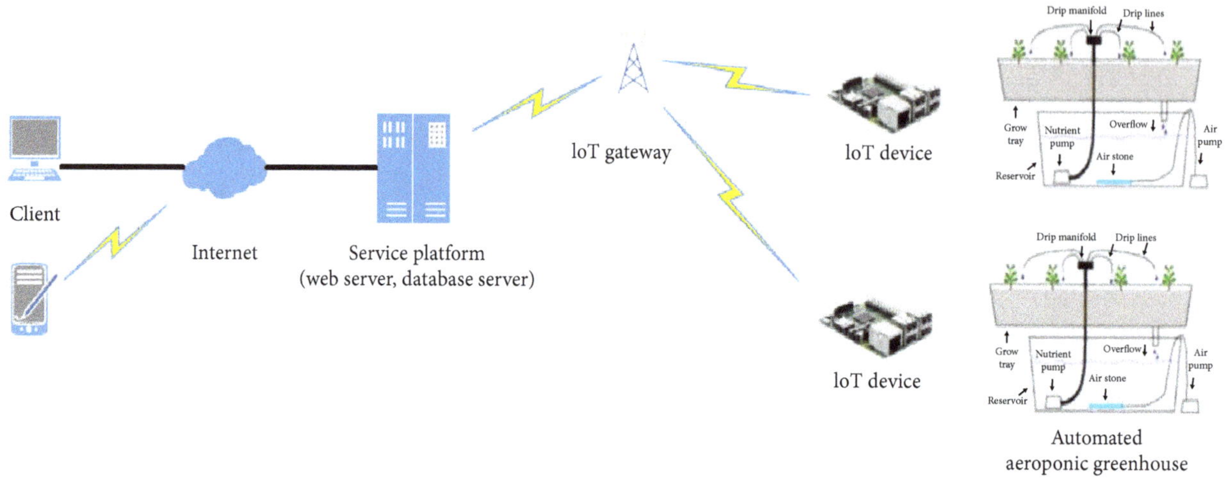

FIGURE 3: The aeroponic system using IoT technology by Kerns and Lee [47].

signal input of the intelligent agricultural monitoring system and are used to collect each selected parameter of farming operations to be monitored. Park et al. [85] stated that wireless sensor network-based systems could be a significant method to fully automate the agriculture system, because the sensors provide real-time significant information and believed to eliminate the considerable costs of just wiring. Another study by Kim [86] said that in agriculture, sensor network technique helps to improve existing systems installed in the greenhouse efficiently and smoothly by forwarding real-time collected information to the operator through the radio signals. The system optimises the transmission protocols more accurate and quick and maximises the application of energy to save the energy and reduce the consumption. Pala and team [40] suggested that the utilisation of artificial intelligence techniques in the aeroponic systems could lead not only to find early fault detection but also to fully automate the system without any or small interventions of human operators. The aeroponic system could gain more popularity among local farmers by deploying this technique in a system for monitoring and controlling purpose. However, it will conserve resources and minimise impacts on the environment. The farmers could start to understand their crops at a micro scale and able to communicate with plant through accessible technology. Therefore, in this article, we explored how wireless sensing technologies wove into the aeroponic system. Thus, the primary motivation of this review article was to provide an idea about different intelligent agriculture monitoring tools used for early fault detection and diagnosis for plant cultivation in the aeroponic system (Figure 4). Additionally, it would be helpful for the local farmer and grower to provide timely information about rising problems and influencing factors for successful plant growth in the aeroponic system. The adoption of the intelligent agriculture monitoring tools could reduce the concept of unsuitable for the amateur.

4.1. Number of Sensor Nodes and Input Parameters. At present, the utilisation of different sensor techniques is almost possible in every field of life due to the sharp progressions in the currently available technologies. Moreover, the sensor is a device that has capabilities to measure physical attributes and convert them into signals for the observer [87]. A WSN (wireless sensor network) traditionally consists of a few to dozens and in some cases thousands of the sensor nodes which are connected to one or more sensors [88]. Generally, it includes a BS (base station), which acts as a gateway between the WSN and the end users. Each sensor node is consisting of five main components, which are a microcontroller unit, a transceiver unit, a memory unit, a power unit, and a sensor unit [89]. Each one of these components is a determinant in designing a WSN for deployment. Furthermore, the microcontroller unit is in charge of the different tasks, data processing, and the control of the other components in the node [88]. Through the transceiver unit, a sensor mode performs its communication with other nodes and other parts of the WSN. It is the most dominant communication unit. The memory unit is another important part of the WSN system, which is used to store the observed data. The memory unit could be RAM, ROM, and their other memory types flash or even external storage devices such as USB. Lastly, the last one unit is the power unit. It is one of the critical components of the system which is for node energy supply. However, the power unit could be any source; it can store in batteries (most common) rechargeable or not on in capacitors. In addition, for extra power supply and recharging the power unit, the available natural resource could be used. The natural sources induce solar power energy in forms of photovoltaic panels and cells, wind power energy with turbines, kinetic energy from water, and so on. Last but not the least is the sensor unit, which includes several sensors for parameter measurements such as temperature, humidity, carbon dioxide, methane, and carbon monoxide [90]. However, in the aeroponic system, the total required number of sensors and actuators depends on the size and requirement of the operator.

4.2. Sensor Types and Monitoring Parameters. In this review study, we reviewed the previous work done on the aeroponic system using wireless sensor network technique. We found that the primary objective of a wireless sensor network

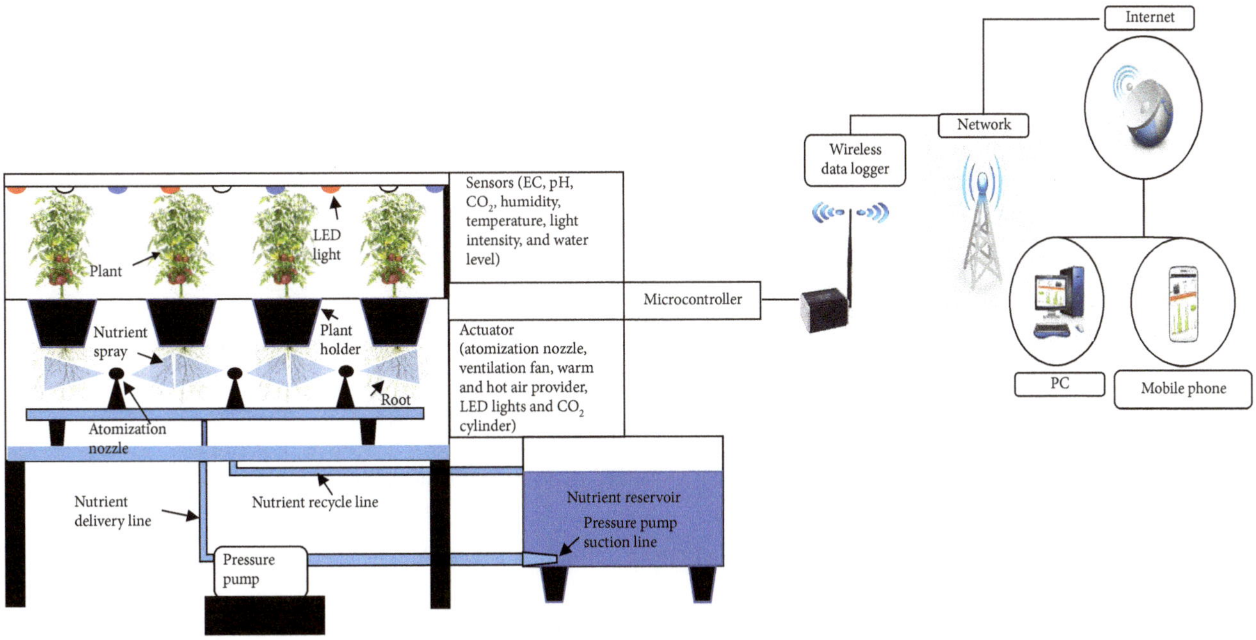

Figure 4: Aeroponic cultivation control system.

system for the aeroponic system is to control the growth chamber climatic condition as per the crop data sheet. However, the basic principle of the aeroponic system is to grow the plant by suspending in the closed, semiclosed, or dark environment in the air with artificial provided support. In the system, the plant stems, leaves, and any fruit grow in a vegetative zone above the suspension medium, and roots dangle below the suspension medium in an area commonly referred as a root zone [46]. Generally, closed cell foam is compressed around the lower stem and inserted into an opening in the aeroponic growth chamber, which decreases labour and expense. However, the trellising is used to suspend the weight of cultivated plant [44]. Ideally, the environment is kept free from pests and diseases so that the plants grow healthier and more quickly than other plants grown on techniques. Furthermore, the key to the success and high yields of the air gardening is a scientific grade monitoring of the conditions and accurate control of the growing environment. Each plant yields and needs a different environmental condition for growth. However, the plant growth is mainly influenced by the surrounding environmental and climatic variables and the amount of water and the fertilizers supplied by irrigation. There is a requirement to monitor and control liquid nutrient parameters in a narrow range of preferred values for optimal growth. The parameters include nutrient temperature, pH, and EC concentration. If the parameters drift outside the desired range, the plants can harm. Besides, there are some additional parameters which can be adjusted to further optimise growth, such as air temperature, relative humidity, light intensity, and carbon dioxide (CO_2) concentration. Idris and Sani [51] reported that the one solution to solve the problems of monitoring and controlling the growing conditions in the space environment is by applying some sensors. The sensor can detect and monitor a number of parameters such as temperature, humidity, light intensity, O_2 and CO_2 levels, direction, and wind speed. Aside from the sensors, there is also a requirement for the actuators to distribute nutrients and waters to plant roots or lower stems (Figure 5). The sensor collects the information of the various environmental conditions and forwards the signals to the actuator to take place and produce the outcome for the collected information to know the status of that parameter. The actuator can control the environment changes. The sensors store information that analyzes the environment and identifies the location, object, people, and their situations. The sensor provides multiple contributions in various domains that depend on a variety of attribute and variant in time [87, 91, 92].

4.2.1. Temperature Sensor. In the aeroponic system, the temperature is one of the critical factors significantly determining plant growth and development. A reduction in temperature below the optimal conditions often results in suboptimal plant growth. A different cultivar requires a different temperature level for the photosynthesis process and growth, which can advance the plant growth stage. It will eventually bring us substantial economic benefits. In the aeroponic system, the optimum growth chamber temperature should not be less and more than 4 and 30°C, respectively, for successful plant growth. The temperature fluctuations of aeroponic growth chamber can significantly affect the root growth, respiration, transpiration, flowering, and dormant period [93]. Therefore, the temperature sensors can be used to monitor the temperature fluctuations of the aeroponic system. At present, temperature sensors are used in many applications like environmental controls, food processing units, medical devices, and chemical handling. The temperature sensor is a device mainly composed of thermocouple or resistance temperature detector. The temperature sensor measures the real-time temperature reading through

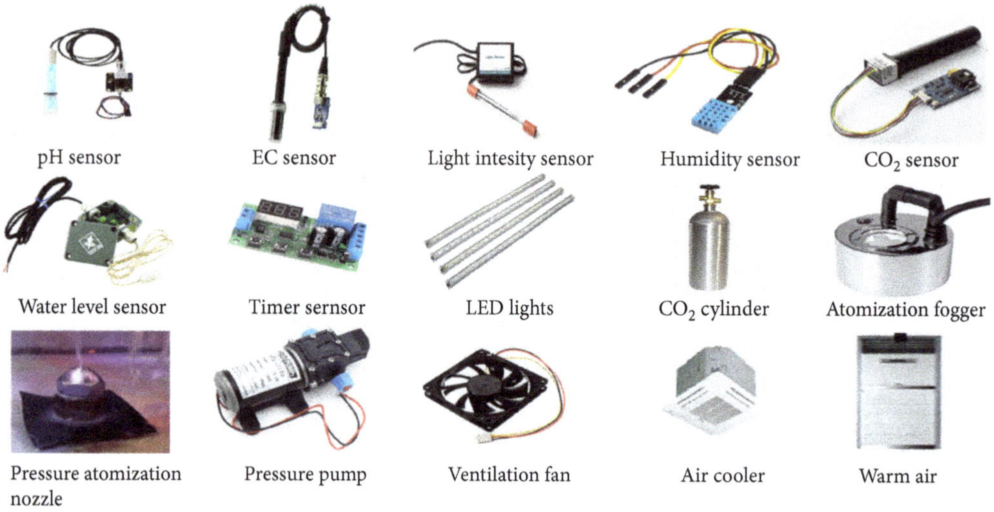

FIGURE 5: Sensors and actuators used in an aeroponic system.

an electrical signal. The sensors collect the data about temperature from a particular source and convert the data into an understandable form for a device or an observer. The temperature sensor accurately measures temperatures slower changing from critical applications such as facilities or rooms and sends them to the user's webpage.

4.2.2. Humidity Sensor. Aeroponics is the technique of cultivating plant by providing the water nutrient small spray in the air. Thus, the humidity is another important parameter of aeroponic growth chamber environments, and its control is recognised to be very important for significant plant growth. In the aeroponic system, the plant gets all available moisture in the growth chamber. Moreover, if the growth chamber has too high or less moisture content, both conditions will create many problems for the plant. Accordingly, an accurate and precise means of testing moisture content in the growth chamber will help farmers to monitor their crops and provide a suitable growth environment for the plant. Wang et al. [94] reported that a humidity sensor is a device that detects and measures water vapour present in the air within a room or enclosure. At present, humidity sensors are widely used in medicine, agriculture, and environmental monitoring. However, the most commonly used units for humidity measurement are relative humidity [95]. The development of humidity sensors has shown remarkable progress because of using various types of sensing materials in recent years. The sensing materials used in humidity sensors can classify into ceramics, polymers, and composites [96]. The humidity sensor could be placed in the growth chamber to maintain the moisture level. If the moisture level becomes less than the plant requirement, the sensors will forward the signals to atomization nozzles to perform their work.

4.2.3. Light Intensity Sensor. As we know, all vegetable plants and flowers require large amounts of sunlight, and each plant group reacts differently and has the different physiology to deal with light intensity. Some plant performs well in low light intensity and some in high light intensity. However, the aeroponic system implements in indoor conditions, so it is necessary for the farmer to provide sufficient light quantity of at least 8 to 10 hours for a day to grow the healthy plant. The artificial lighting is a better option to present enough intensity to produce a healthy plant [97, 98]. In the conventional aeroponic system, the control of the light quantity present in the growth chamber is mostly done by farmer through observing the plant condition. However, it is a time-consuming and challenging task for the farmer to provide the required light concentration accurately. It could be a better option to use intelligent agriculture techniques to monitor the light intensity in the aeroponic system. The intelligent agriculture techniques mean using the sensor system to control the light intensity. The light sensor is an electronic device which is used to detect the presence or nonpresence of light and darkness. There are several types of light sensors including photoresistors, photodiodes, and phototransistors. These light sensors distinguish the substance of light in a growth chamber and increase or decrease the brightness of light to a more comfortable level. Light sensors can be used to automatically control the lights such as on/off. By adopting the sensor network in aeroponics, the farmer could be able to monitor light intensity without any human interference. Because the sensors will perform all work such as if the light intensity in the growth chamber will be less than the required light quantity for plant growth, the sensor will automatically forward the signal to the LED light to turn on until the light quantity reaches to the desired level.

4.2.4. CO_2 Sensor. The appropriate oxygen concentration in the root environment is crucial to keep the root metabolism in nutrition solution. The available oxygen concentration for the root environment is a hugely significant factor since low concentrations affect the root respiration, nutrient absorption, and, consequently, the plant growth [66]. Thus, the CO_2 sensor could be used to monitor the carbon dioxide fluctuations in the aeroponic growth chamber. A carbon dioxide sensor is an instrument which is used for the measurement of carbon dioxide gas concentration. Bihlmayr

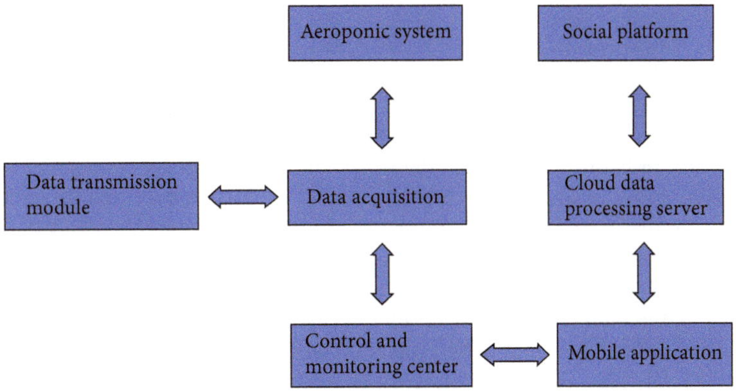

Figure 6: Schematic protocol of wireless network in the aeroponic system.

[99] reported that CO_2 sensors are used to measure indoor air quality in a building to perform demand-based ventilation. However, the CO_2 sensor data measuring range is in between 500 and 5000 parts per million. There are two main types of the CO_2 sensors which include nondispersive infrared carbon dioxide sensors (NICDS) and chemical carbon dioxide sensors (CCDS), whereas the NICDS detected CO_2 in a gaseous environment by its characteristic absorption and composed of an infrared detector, an interference filter, a light tube, and an infrared source. However, the CCDS of sensitive layers are based on polymer or heteropolysiloxane with low-energy consumption [100].

4.2.5. Water Level Sensor. The aeroponics is the method of the plant cultivation by providing a small mist of the nutrient solution in the growth chamber. Thus, there is no any use of soil; just water is required to cultivate the plant throughout the germination to harvest time. Therefore, the water nutrient solution reservoir is one of the major components of the aeroponic system which should be monitored throughout the growth period. In the conventional aeroponic system, the farmer checks the water nutrient level in the nutrient solution reservoir, and if he founds water level less than the desired level, he maintained accordingly. However, by adopting the precision agriculture techniques, the farmer will be able to monitor and control water nutrient level through the intelligent methods such as wireless sensors. The water nutrient level sensors detect the liquid level in the reservoirs and facilitate operator in collecting water nutrient level data in real time. The sensors will alert the operator about any potential property damage that results from any leaks and also allowing to know when a container is nearing empty.

4.2.6. EC and pH Sensor. In the aeroponic system, the plant productivity is closely related to nutrient uptake and the EC and pH regulation of the nutrient solution. The EC and pH concentration of the nutrient solution affects the availability of the nutrients to plants [101]. The pH and EC concentrations are controlled to prevent barrier growth. Their measurement is essential because the solubility of minerals in acidic, alkaline, and ion concentration of all the species in solutions is different and the solution concentration changes with solubility [102, 103]. The unmonitored EC and pH concentration of the nutrient solution will quickly lead to a situation where plants cannot absorb the essential nutrients, if not corrected this will eventually lead to harmful plant growth and poor productivity. Thus, the EC and pH concentration of the nutrient solution is a critical parameter to be measured and controlled throughout the plant growth. Moreover, in the conventional aeroponic system, the EC and pH value of the nutrient solution is mostly monitored manually by performing laboratory analysis or using advanced equipment which is a time-consuming process. For instance, when the EC of the nutrient solution decreased or increased, the control of nutrient solution concentration is mostly achieved by adding more high concentration nutrient solution or the fresh water, respectively, to the nutrient solution to maintain the EC level to the prescribed target range. Similarly, for pH, an acid solution and an alkali solution are used to control the pH fluctuation of the nutrient solution within a specified target range [101]. However, these conventional methods are time-consuming and challenging task for the farmer to maintain the EC and pH value at the desired range accurately. In addition, the EC and pH sensor could be used to deal with the above challenges.

5. Sensor Working Protocol in the Aeroponic System

Today, the world demands automatic tools to do most of the work for them without bothering its user for doing some task. So, the concept is all about a very high level of automation system which will be independent of its users to a very great extent, reduce human efforts, and save all kinds of resource utilisation, as monitoring and controlling will be done by computers leaving very few easily manageable tasks for humans, and it will interest more people to join this field [104–106]. Moreover, the monitoring and control system for the aeroponic system mainly consists of following sections which include the aeroponic system, data acquisition, controlling the equipment, data transmission module, cloud data processing server, social communication platform, and mobile application. A typical architecture of sensor nodes for controlling and monitoring the aeroponic system is shown in Figure 6. Furthermore, in architecture, the data acquisition section refers to some sensor nodes used in the system to establish a data acquisition module. The data acquisition module is placed in the aeroponic system or near the growth chamber to collect the real-time information from selected parameters (temperature, light intensity, humidity, nutrient solution level, atomization quantity, and photos of the growing plants) and transmit the gathered data to the control and management centre. However, the control and management section refers to the central processing unit (CPU) of the system. The CPU of the system consists of some primary functions such as Arduino and WRTnod protocols, whose work is to store, manage gathered data from collection nodes, process, and then accurately and automatically send to the web server in real time [104–109]. Thus, the system can help the farmer and grower to monitor and control the smart aeroponic system remotely using the mobile app. In other words, the plant will be able to talk with the farmer through a mobile app that whether the selected parameters are working well or not.

6. Advantages of Sensor Techniques in the Aeroponic System

The continuously increasing food demands require rapid improvement and development in the food production system. However, to enhance the quality and productivity of the cultivated crop, peoples are moving towards the modern plant cultivation technologies in agriculture. Thus, the aeroponics is one of the rising plant growing technologies in agriculture as a modern-day cultivation technique, where the plant is cultivated in an air environment, and no any soil support is provided. In the aeroponic system, a number of the parameters are required to control for successful plant growth because there is no any growing medium provided to the plant. For example, if the plant has some sudden stress and the farmer is not present at the site that means the plant will die. Therefore, the proper management of the crop is essential. In the conventional aeroponic system, grower uses his knowledge, skills, and judgment to adjust and maintain the parameters such as EC and pH meter, minimum and maximum temperature, light intensity, and humidity level through several instruments and checks the readings which are labour-intensive and time-consuming task. To deal with the above problems, the aeroponic system can be developed with a wireless sensor and actuator network for monitoring the key parameters at lower labour cost, time, and without any technical knowledge. The wireless sensor and actuator network offer several advantages including faster response to confrontational climatic conditions and better quality control of the crop that produces at a lower labour cost. This advancement in the aeroponic system through wireless sensor network for monitoring growth chamber environment is beneficial. However, the monitoring system also offers a range of information which could be required by plant scientists or grower to provide a greater understanding of how these environmental and nutrient parameters correlate with plant growth. It is now recognised that plant grower can perfectly and easily acquire the skills needed to operate an aeroponic system. It provides the full control of the system, not by constant manual attention from the operator but to a large extent by wireless sensors.

7. Future Application

Artificial intelligence agriculture techniques are considered as a high potential, improving technique for decision-making in agriculture. Nowadays, it is quickly getting peoples' intention, more and more visible in our society and dynamically turning our social awareness and lifestyle. The techniques provide several opportunities to monitor the plant growth and development from pre- to postharvest. Aeroponics is the new plant cultivation technology of agriculture which is still under development. However, we reviewed the literature and found that only limited study had been conducted on the implementation of the intelligent agricultural techniques in the aeroponic system. Moreover, until now, most of the studies had been designed the aeroponic system using a wireless sensor network using ZigBee and Arduino system with Bluetooth, global system of mobile and Wi-Fi, and communication modules. During a literature survey, we noted that no any single study had implemented the idea of the cloud computing and big data techniques in aeroponics to collect real-time information via the Internet. The techniques provide many advantages to the user such as reduction of the initial cost, allocation of the resources on demand, and maintenance and upgrade performed in the back-end, easy, and rapid development. The techniques present a chance to the operator to stay connected with the system using mobile accessories like a smartphone, tablet, and PCs at any location via the Internet which is not restricted by conditions, locations, and time. Furthermore, the system would be designed using additional artificial intelligence techniques such as image processing, automatic seedling transplanters, and harvesting and packing robots. The purpose of image processing and analysis is to measure and identify the physiology, growth, development, nutrient deficiencies, diseases, and other phenotypic properties of the plants through automated and nondestructive analysis.

8. Application of Artificial Intelligence Techniques in Agriculture

Agriculture is the primitive and ancient application that humans started first after born on earth. It has an extensive history between numerous industries and very intimately linked to the human development on earth. Moreover, in the past, the agriculture sector was labour-intensive, but the next-generation farmers, researchers, and associated organisations proposed and applied new farming methods, technologies, skills, and knowledge in agriculture as a modern era to reduce labour-intensive task. Presently, the technology is considered as a key tool to overcome many challenges and eases the way how people live. In the past, the many problems of agriculture, especially in irrigation water management, crop yield production, environment predication, and decision-making, were decided by many factors. In addition, the fertilization often decided by the mathematical equations, formulas, or the experiences of the experts. The cultivation is represented by descriptive and causal knowledge, and diagnosis of pests and diseases is represented by uncertain knowledge. Thus, this knowledge and experience are illogically incomplete and imprecise, and the traditional procedures can not handle them. However, artificial intelligence has its superiority. It could be an effective approach for solving complex problems to the levels of experts using imitate experts [110]. The term artificial intelligence (AI) was developed in 1956, as "the science and engineering of making intelligent machines" [111]. It is a abroad discipline, which was developed for the interaction of several types of fields such as computer science, information theory, cybernetics, linguistics, neurophysiology, and psychology [112]. The main purpose of the creation of the intelligence techniques is to find the solutions for complex problems and to work, react, and respond like humans. It performed work better than a well-qualified person and brought positive economic and environmental results [112, 113]. Artificial intelligence (AI) tools have helped to predict the behavior of nonlinear systems and to control variables to improve the operating conditions of a system's environment [114–117]. A recently published report highlighted that artificial intelligence is emerging as part of the solutions towards improved agricultural productivity. The global artificial intelligence in agriculture is expected to grow at a significant level. It is employed to improve the efficiency of daily tasks in agriculture such as the adoption of robots and drones, crop health monitoring protocols, automated irrigation system, and driverless tractors [118]. At present, several research studies have been performed by implementing the artificial intelligent techniques in agriculture. Popa [119] revealed that some of the developed applications for agriculture are expert systems and software, sensors for collecting and transmitting data, and robotic and automation which are adapted from different industries into agriculture, whereas the expert systems and software is the planning process such as strategic or operational; it has benefited substantially, due to the expansion of personal computer and Internet use. The systems are generated through the structured knowledge base and reasoning mechanisms acquired from a human expert but with an enhanced computational power and speed [120]. These systems can demarcate management zones taking in consideration with relevant factors and able to recommend suitable crop rotations, optimal plant density, water requirements, appropriate fertilizer use, diagnosing pests and diseases for crops, and suggesting preventive or curative measures [121]. Huang et al. [122] discussed soft computing and applications in agriculture. The study reported that soft computing is the combination of the computing technologies, such as an artificial neural networks (ANNs), fuzzy logic (FL), and genetic algorithms (GAs). These techniques are opposed to the hard computing method which states to a huge set of stochastic and statistical methods. The hard computing provides inaccurate solutions and results of very complex problems through modelling and analysis with a tolerance of imprecision, uncertainty, partial truth, and an approximation. However, the soft computing techniques are used to achieve tractability and robustness. It provides a low-cost solution with a tolerance of imprecision, uncertainty, partial truth, and approximation [123–127]. Sui and Thomasson [128] developed a BP-trained feedforward ANNs to predict nitrogen status in cotton plants based on a data from a ground-based sensing system. Tumbo and team [129] used an on-the-go system for sensing chlorophyll status in corn using BP-trained feedforward ANNs and fiber optic spectrometry to acquire spectral response pattern data in corn fields. Tang et al. [130] developed a texture-based weed classification method consisting of a low-level Gabor wavelet-based feature extraction algorithm and a high-level ANN-based pattern recognition algorithm. El-Faki and group [131] established and tested ANN-based weed detection algorithms capable of detecting the leading weed species competing with wheat and soybean crops. A study by Krishnaswamy and Krishnan [132] predicted the nozzle wear rates for four fan nozzles using regression and ANN methods. Pearson and Wicklow [133] developed a neural network to identify fungal species that infect single kernels using principal components of the reflectance spectra as input features. Smith et al. [134] developed year-round air temperature prediction models for prediction horizons from 1 to 12 h using feedforward-style ANNs. Zadeh [123] introduced the concept of fuzzy sets as a mean for describing complex systems without the requirements for precision. Fuzzy logic may also be useful for descriptive systems, those that fall somewhere between hard systems and soft systems, such as biology and agriculture [135]. Studies reported that in agriculture, fuzzy logic is used for multicriteria analysis of the image, image classification, vegetation mapping, assessment of soil suitability, and planning forest harvesting [136–143]. Al-Faraj and coworkers [144] established a rule-based FL crop water stress index (CWSI) using growth chamber data and tested this method on tall fescue canopies grown in a greenhouse. Thomson et al. [145] and Thomson and Ross [146] developed a coupled sensor- and model-based irrigation scheduling method. Yang et al. [147, 148] informed on the development of an image capture/processing system to detect weeds and a fuzzy logic decision-making system to determine where and how much herbicide to apply in an agricultural field. Gil and team [149] applied multiple linear regression and FL inference models to evaluate the effects of micrometeorological conditions on pesticide application for two spray qualities (fine and very fine). Qiu et al. [150] established a fuzzy irrigation decision-making system using virtual instrumentation platform of sensors, test instruments, data logger, and LabVIEW. Generally, published studies use on/off controllers where the inherent complexity of irrigation process made it difficult to achieve optimal results [151]. Ali et al. [152] developed temperature and humidity controller inside the greenhouse using fuzzy logic. However, several studies have been conducted to develop many control strategies to optimise the greenhouse environment using artificial intelligence techniques such as neural network, fuzzy logic controller, adaptive predictive control, PID, and nonlinear adaptive PID control [153–159]. Zhu et al. [160] used the remote wireless system for water quality online monitoring in intensive aquaculture using artificial neural networks. The results demonstrate that online monitoring for water quality information could be accurately acquired and predicted by using the remote wireless system. Mahajan et al. [161] reported that agriculture is noteworthy that computer vision applications have grown due to reduced equipment costs, increased computational power, and increasing interest in nondestructive food assessment methods. The use of these techniques presents advantages when compared with traditional methods based on manual work; however, there are still some challenges to be overcome. Moreover, the principle of artificial intelligence in agriculture is one where a machine can perceive its environment, and through a certain capacity of flexible rationality, can act to address a specified goal related to that environment.

9. Conclusion

The objective of our study was to present the information about the use of automated monitoring and controlling technique in the aeroponic system. The aeroponic system is the new plant cultivation method of the modern agriculture. Its existence can allow producing food whole year without any interval. The system could create an excellent set which encourages the sustainable city life for those peoples who want to live in urban area. Moreover, during plant growth from sowing to harvest time, the methods adopted in the aeroponic system require a little hand-operated contribution, interference regarding physical presence, and expertise in domain knowledge of plants, environment control, and operations to maintain and control the growth of the plant. Therefore, the system is considered hitherto to be somewhat unsuitable for the grower, and due to the above reasons, it is not common to find an installation. We reviewed the literature and found that implementation of advanced monitoring technology tools in aeroponics could provide an opportunity for the farmer to monitor and control several paraments without using laboratory instruments, and the farmer can control the entire system remotely. Thus, it could reduce the concept of the usefulness of the system due to the complicated manual monitoring and controlling process. The technology offers incredible opportunities for the aeroponic system to increase the capability, reliability, and availability among the farmers and growers. We believe that our review article will contribute to the adoption of the advanced monitoring technology in the aeroponic system. However, the technique provides a range of information which could be required by plant scientists to provide a greater understanding of how these environmental and nutrient parameters correlate with plant growth.

Conflicts of Interest

The authors declare that they have no conflicts of interest.

Acknowledgments

We acknowledged that this work was financially supported by the Jiangsu Agriculture Science and Technology Innovation Fund (CX (18) 3048), the National Natural Science Foundation of China Program (no. 51275214), and the project funded by the Priority Academic Program Development of Jiangsu Higher Education Institutions (no. 37(2014)).

References

[1] J. James and M. P. Maheshwar, "Plant growth monitoring system, with dynamic user-interface," in *2016 IEEE Region 10 Humanitarian Technology Conference (R10-HTC)*, pp. 1–5, Agra, India, December 2016.

[2] D. Pimentel, B. Berger, D. Filiberto et al., "Water resources: agricultural and environmental issues," *Bioscience*, vol. 54, no. 10, pp. 909–918, 2004.

[3] M. Taher Kahil, J. Albiac, A. Dinar et al., "Improving the performance of water policies: evidence from drought in Spain," *Water*, vol. 8, no. 2, p. 34, 2016.

[4] W. Foote, *To Feed the World in 2050, We Need to View Small-Scale Farming as a Business*, Skoll World Forum, Oxford, UK, 2015.

[5] V. Doknić, *Internet of Things Greenhouse Monitoring and Automation System. Internet of Things: Smart Devices, Processes, Services*, 2014, http://193.40.244.77/idu0330/wpcontent/uploads/2015/09/140605_Internet-of-Things_Vesna-Doknic.pdf.

[6] D. K. Großkinsky, J. Svensgaard, S. Christensen, and T. Roitsch, "Plant phenomics and the need for physiological phenotyping across scales to narrow the genotype-to-phenotype knowledge gap," *Journal of Experimental Botany*, vol. 66, no. 18, pp. 5429–5440, 2015.

[7] E. Playán and L. Mateos, "Modernization and optimization of irrigation systems to increase water productivity," *Agricultural Water Management*, vol. 80, no. 1-3, pp. 100–116, 2006.

[8] L. Levidow, D. Zaccaria, R. Maia, E. Vivas, M. Todorovic, and A. Scardigno, "Improving water-efficient irrigation: prospects and difficulties of innovative practices," *Agricultural Water Management*, vol. 146, pp. 84–94, 2014.

[9] R. Qiu, S. Wei, M. Zhang et al., "Sensors for measuring plant phenotyping: a review," *International Journal of Agricultural and Biological Engineering*, vol. 11, no. 2, pp. 1–17, 2018.

[10] W. Baudoin, R. Nono-Womdim, N. Lutaladio et al., *Good Agricultural Practices for Greenhouse Vegetable Crops: Principles for Mediterranean Climate Areas (No. 217)*, Food and Agriculture Organization of The United Nations, Rome, 2013.

[11] M. Lee and H. Yoe, "Analysis of environmental stress factors using an artificial growth system and plant fitness optimization," *BioMed Research International*, vol. 2015, Article ID 292543, 6 pages, 2015.

[12] S. M. Moon, S. Y. Kwon, and J. H. Lim, "Minimization of temperature ranges between the top and bottom of an air flow controlling device through hybrid control in a plant factory," *The Scientific World Journal*, vol. 2014, Article ID 801590, 7 pages, 2014.

[13] C. Stanghellini, "Horticultural production in greenhouses: efficient use of water," *Acta Horticulturae*, vol. 1034, pp. 25–32, 2014.

[14] D. Savvas, G. Gianquinto, Y. Tuzel, and N. Gruda, "Soilless culture," in *Good Agricultural Practices for Greenhouse Vegetable Crops: Principles for Mediterranean Climate Areas (No. 217)*, W. Baudoin, R. Nono-Womdim, N. Lutaladio, A. Hodder, N. Castilla, C. Leonardi, S. Pascale, and M. Qaryouti,

Eds., pp. 303–354, Food and Agriculture Organization of The United Nations, Rome, 2013.

[15] I. A. Lakhiar, X. Liu, G. Wang, and J. Gao, "Experimental study of ultrasonic atomizer effects on values of EC and pH of nutrient solution," *International Journal of Agricultural and Biological Engineering*, vol. 11, no. 5, pp. 59–64, 2018.

[16] J. P. Beibel, *Hydroponics -The Science of Growing Crops without Soil*, Department of Agriculture. Tallahassee. Bulletin, 1960.

[17] J. L. Reyes, R. Montoya, C. Ledesma, and R. Ramírez, "Development of an aeroponic system for vegetable production," *Acta Horticulturae*, vol. 947, pp. 153–156, 2012.

[18] M. Raviv and L. Lieth, "Significance of soilless culture in agriculture," in *Soilless Culture: Theory and Practice*, M. Raviv and J. H. Lieth, Eds., pp. 117–156, Elsevier, Amsterdam, 2007.

[19] V. Valenzano, A. Parente, F. Serio, and P. Santamaria, "Effect of growing system and cultivar on yield and water-use efficiency of greenhouse-grown tomato," *The Journal of Horticultural Science and Biotechnology*, vol. 83, no. 1, pp. 71–75, 2008.

[20] L. Maharana and D. N. Koul, "The emergence of hydroponics," *Yojana*, vol. 55, pp. 39-40, 2011.

[21] International Center of Applied Aeroponics (ICAA), *Press Release: Golden Potato*, Hanoi, Vietnam, 2014, http://www.icaeroponics.com/press-release.html.

[22] NASA Spinoff, *Innovative Partnership Program*, Publications and Graphics Department NASA Center for Aerospace Information (CASI), 2006.

[23] I. A. Lakhiar, J. Gao, T. N. Syed, F. A. Chandio, and N. A. Buttar, "Modern plant cultivation technologies in agriculture under controlled environment: a review on aeroponics," *Journal of Plant Interactions*, vol. 13, no. 1, pp. 338–352, 2018.

[24] F. Xiong and K. Qiao, "Intelligent systems and its application in agriculture," *IFAC Proceedings Volumes*, vol. 32, no. 2, pp. 5597–5602, 1999.

[25] Z. Zhai, J.-F. Martínez Ortega, N. Lucas Martínez, and J. Rodríguez-Molina, "A mission planning approach for precision farming systems based on multi-objective optimization," *Sensors*, vol. 18, no. 6, p. 1795, 2018.

[26] S. O. Petersen, C. C. Hoffmann, C. M. Schafer et al., "Annual emissions of CH_4 and N_2O, and ecosystem respiration, from eight organic soils in Western Denmark managed by agriculture," *Biogeosciences*, vol. 9, no. 1, pp. 403–422, 2012.

[27] G. Mohammed, F. Trolard, M. Gillon, A. L. Cognard-Plancq, A. Chanzy, and G. Bourrie, "Combination of a crop model and a geochemical model as a new approach to evaluate the sustainability of an intensive agriculture system," *Science of The Total Environment*, vol. 595, pp. 119–131, 2017.

[28] J. Gago, C. Douthe, R. E. Coopman et al., "UAVs challenge to assess water stress for sustainable agriculture," *Agricultural Water Management*, vol. 153, pp. 9–19, 2015.

[29] D. H. Park and J. W. Park, "Wireless sensor network-based greenhouse environment monitoring and automatic control system for dew condensation prevention," *Sensors*, vol. 11, no. 4, pp. 3640–3651, 2011.

[30] A. Barriuso, G. Villarrubia González, J. de Paz, Á. Lozano, and J. Bajo, "Combination of multi-agent systems and wireless sensor networks for the monitoring of cattle," *Sensors*, vol. 18, no. 2, p. 108, 2018.

[31] M. Perez-Ruiz, P. Gonzalez-de-Santos, A. Ribeiro et al., "Highlights and preliminary results for autonomous crop protection," *Computers and Electronics in Agriculture*, vol. 110, pp. 150–161, 2015.

[32] A. Maher, E. Kamel, F. Enrico, I. Atif, and M. Abdelkader, "An intelligent system for the climate control and energy savings in agricultural greenhouses," *Energy Efficiency*, vol. 9, no. 6, pp. 1241–1255, 2016.

[33] A. R. De la Concepcion, R. Stefanelli, and D. Trinchero, "A wireless sensor network platform optimized for assisted sustainable agriculture," in *IEEE Global Humanitarian Technology Conference (GHTC 2014)*, pp. 159–165, San Jose, CA, USA, October 2014.

[34] B. Basnet and J. Bang, "The state-of-the-art of knowledge-intensive agriculture: a review on applied sensing systems and data analytics," *Journal of Sensors*, vol. 2018, Article ID 3528296, 13 pages, 2018.

[35] Aqeel-ur-Rehman and Z. A. Shaikh, "Smart agriculture," in *Application of Modern High-Performance Networks*, pp. 120–129, Bentham Science Publishers Ltd, 2009.

[36] N. Wang, N. Zhang, and M. Wang, "Wireless sensors in agriculture and food industry—recent development and future perspective," *Computers and Electronics in Agriculture*, vol. 50, no. 1, pp. 1–14, 2006.

[37] L. Ruiz-Garcia, L. Lunadei, P. Barreiro, and I. Robla, "A review of wireless sensor technologies and applications in agriculture and food industry: state of the art and current trends," *Sensors*, vol. 9, no. 6, pp. 4728–4750, 2009.

[38] W. Zhang, G. Kantor, and S. Singh, "Integrated wireless sensor/actuator networks in an agricultural application," in *Proceedings of the 2nd international conference on Embedded networked sensor systems - SenSys '04*, p. 317, Baltimore, MD, USA, November 2004.

[39] L. B. Tik, C. T. Khuan, and S. Palaniappan, "Monitoring of an aeroponic greenhouse with a sensor network," *IJCSNS International Journal of Computer Science and Network Security*, vol. 9, no. 3, pp. 240–246, 2009.

[40] M. Pala, L. Mizenko, M. Mach, and T. Reed, "Aeroponic greenhouse as an autonomous system using intelligent space for agriculture robotics," in *Robot Intelligence Technology and Applications 2. Advances in Intelligent Systems and Computing, vol 274*, J. H. Kim, E. Matson, H. Myung, P. Xu, and F. Karray, Eds., pp. 83–93, Springer, Cham, 2014.

[41] P. Laksono, I. Idris, M. I. Sani, and D. N. Putra, "Lab prototype of wireless monitoring and control for seed potatoes growing chamber," *Proceedings of the Asia-Pacific Advanced Network*, vol. 37, pp. 20–29, 2014.

[42] P. Jonas, A. Maskara, A. Salguero, and A. Truong, "Garduino: a cyber-physical aeroponics system," 2015, http://arxiv.org/abs/1011.1669v3, http://ecal.berkeley.edu/files/ce186/projects/truonganders{_}4941954{_}63829747{_}Garduino-1.pdf.

[43] M. I. Sani, S. Siregar, A. P. Kurniawan, R. Jauhari, and C. N. Mandalahi, "Web-based monitoring and control system for aeroponics growing chamber," in *2016 International Conference on Control, Electronics, Renewable Energy and Communications (ICCEREC)*, pp. 162–168, Bandung, Indonesia, September 2016.

[44] P. Anitha and P. S. Periasamy, "Energy efficient green house monitoring in the aeroponics system using Zigbee networks," *Asian Journal of Research in Social Sciences and Humanities*, vol. 6, no. 6, pp. 2243–2250, 2016.

[45] K. Kernahan and C. A. Cupertino, "Aeroponics growth system wireless control system and method of using," US Patent 20160021836A1, 2016.

[46] A. P. Montoya, F. A. Obando, J. G. Morales, and G. Vargas, "Automatic aeroponic irrigation system based on Arduino's platform," *Journal of Physics: Conference Series*, vol. 850, article 012003, 2017.

[47] S. C. Kerns and J. L. Lee, "Automated aeroponics system using IoT for smart farming," in *8th International Scientific Forum, ISF 2017*, UNCP, USA, September 2017.

[48] T. Karu, "High precision farming system based on aeroponics," Master's Thesis. Tallinn University of Technology, Faculty of Information Technology, 2017.

[49] J. A. Martin and S. Rafael, "Systems, methods and devices for aeroponics plant growth," US Patent US20180007845A1, 2018.

[50] P. Mithunesh, K. Gupta, S. Ghule, and P. S. Hule, "Aeroponic based controlled environment based farming system," *IOSR Journal of Computer Engineering (IOSR-JCE)*, vol. 17, no. 6, pp. 55–58, 2015.

[51] I. Idris and M. I. Sani, "Monitoring and control of aeroponic growing system for potato production," in *2012 IEEE Conference on Control, Systems & Industrial Informatics*, Bandung, Indonesia, September 2012.

[52] K. Janarthanan, K. Theviyanthan, F. M. Najath, and I. A. Ahamed, "Cyberponics - a fully automated greenhouse systemBachelors Thesis. Project ID: CTP/2017/01.

[53] J. Liu and Y. W. Zhang, "An automatic aeroponics growth system for bamboo seedling and root observation," *Applied Mechanics and Materials*, vol. 307, pp. 97–102, 2013.

[54] J. Liu and Y. W. Zhang, "An automatic aeroponics growth system based on ultrasonic atomization," *Applied Mechanics and Materials*, vol. 288, pp. 161–166, 2013.

[55] J. Osvald, N. Petrovic, and J. Demsar, "Sugar and organic acid content of tomato fruits (Lycopersicon lycopersicum mill.) grown on aeroponics at different plant density," *Acta Alimentaria*, vol. 30, no. 1, pp. 53–61, 2001.

[56] I. Nir, "Growing plants in aeroponics growth system," *Acta Horticulturae*, vol. 126, pp. 435–448, 1982.

[57] F. Zsoldos, A. Vashegyi, and L. Erdei, "Lack of active K+ uptake in aeroponically grown wheat seedlings," *Physiologia Plantarum*, vol. 71, no. 3, pp. 359–364, 1987.

[58] P. Barak, J. D. Smith, A. R. Krueger, and L. A. Peterson, "Measurement of short-term nutrient uptake rates in cranberry by aeroponics," *Plant Cell and Environment*, vol. 19, no. 2, pp. 237–242, 1996.

[59] M. W. Mbiyu, J. Muthoni, J. Kabira et al., "Use of aeroponics technique for potato (Solanum tuberosum) minitubers production in Kenya," *Journal of Horticulture and Forestry*, vol. 4, no. 11, pp. 172–177, 2012.

[60] C. S. Buer, M. J. Correll, T. C. Smith et al., "Development of a nontoxic acoustic window nutrient-mist bioreactor and relevant growth data," *In Vitro Cellular & Developmental Biology - Plant*, vol. 32, no. 4, pp. 299–304, 1996.

[61] R. W. Zobel and R. F. Lychalk, "Aeroponic growth system with nutrient fog stabilization," US Patent US5937575A, 1999.

[62] M. I. Hessel, G. E. Richert, and J. G. E. Nevill, "Airflow-contained aeroponic nutrient delivery for a microgravity plant growth unit," *Biotronics*, vol. 21, pp. 33–38, 1993.

[63] J. M. Clawson, A. Hoehn, L. S. Stodieck, and P. Todd, *NASA-Review of Aeroponics, Aeroponics for Spaceflight Plant Growth*, Society of Automotive Engineers, Inc, 2000, http://aeroponicsdiy.com/nasa-review-of-aeroponics/.

[64] K. T. Hubick, D. R. Drakeford, and D. M. Reid, "A comparison of two techniques for growing minimally water-stressed plants," *Canadian Journal of Botany*, vol. 60, no. 3, pp. 219–223, 1982.

[65] D. V. Shtrausberg and E. G. Rakitina, "On the aeration and gas regime of roots in aeroponics and water culture," *Agrokhitniia*, vol. 4, pp. 101–110, 1970.

[66] H. Soffer and D. W. Burger, "Effects of dissolved oxygen concentration in aero-hydroponics on the formation and growth of adventitious roots," *Journal of the American Society for Horticultural Science*, vol. 113, pp. 218–221, 1988.

[67] L. L. Hung and D. M. Sylvia, "Production of vesicular-arbuscular mycorrhizal fungus inoculum in aeroponic culture," *Applied and Environmental Microbiology*, vol. 54, pp. 353–357, 1988.

[68] D. M. Sylvia and A. G. Jarstfer, "Sheared—root inocula of vesicular— arbuscular mycorrhizal fungi," *Applied and Environmental Microbiology*, vol. 58, no. 1, pp. 229–232, 1992.

[69] R. E. Wagner and H. T. Wilkinson, "An aeroponics system for investigating disease development on soybean taproots infected with *Phytophthora sojae*," *Plant Disease*, vol. 76, no. 6, pp. 610–614, 1992.

[70] D. M. Sylvia and D. H. Hubbel, "Growth and sporulation of vesicular-arbuscular mycorrhizal fungi in aeroponic and membrane systems," *Symbiosis*, vol. 1, pp. 259–267, 1986.

[71] R. W. Zobel, P. del Tredici, and J. G. Torrey, "Method for growing plants aeroponically," *Plant Physiology*, vol. 57, no. 3, pp. 344–346, 1976.

[72] C. B. Christie and M. A. Nichols, "Aeroponics - a production system and research tool. South pacific soilless culture conference," *Acta Horticulturae*, vol. 648, pp. 185–190, 2004.

[73] A. L. Hayden, "Aeroponic and hydroponic systems for medicinal herb, rhizome and root crops," *HortScience*, vol. 41, no. 3, pp. 536–538, 2006.

[74] S. Chandra, S. Khan, B. Avula et al., "Assessment of total phenolic and flavonoid content, antioxidant properties, and yield of aeroponically and conventionally grown leafy vegetables and fruit crops: a comparative study," *Evidence-Based Complementary and Alternative Medicine*, vol. 2014, Article ID 253875, 9 pages, 2014.

[75] L. J. du Toit, H. W. Kirby, and W. L. Pedersen, "Evaluation of an aeroponics system to screen maize genotypes for resistance to fusarium graminearum seedling blight," *Plant Disease*, vol. 81, no. 2, pp. 175–179, 1997.

[76] D. S. Mueller, S. Li, G. L. Hartman, and W. L. Pedersen, "Use of aeroponic chambers and grafting to study partial resistance to Fusarium solani f. sp. glycines in soybean," *Plant Disease*, vol. 86, no. 11, pp. 1223–1226, 2002.

[77] N. Kacjan-Maršić, and J. Osvald, "Nitrate content in lettuce (*Lactuca sativa* L.) grown on aeroponics with different quantities of nitrogen in the nutrient solution," *Acta Agronomica Hungarica*, vol. 50, no. 4, pp. 389–397, 2002.

[78] G. Fascella and G. V. Zizzo, "Preliminary results of aeroponic cultivation of *Anthurium andreanum* for cut flower production. VIII international symposium on protected cultivation in mild winter climates: advances in soil and soilless

cultivation under protected environment," *Acta Horticulturae*, no. 747, pp. 233–240, 2007.

[79] F. Martin-Laurent, F. Y. Tham, S. K. Lee, J. He, and H. G. Diem, "Field assessment of aeroponically grown and nodulated Acacia mangium," *Australian Journal of Botany*, vol. 48, no. 1, pp. 109–114, 2000.

[80] T. Buckseth, A. K. Sharma, K. K. Pandey, B. P. Singh, and R. Muthuraj, "Methods of pre-basic seed potato production with special reference to aeroponics—a review," *Scientia Horticulturae*, vol. 204, pp. 79–87, 2016.

[81] A. H. Calori, T. L. Factor, J. C. Feltran, E. Y. Watanabe, C. C. Moraes, and L. F. V. Purquerio, "Electrical conductivity of the nutrient solution and plant density in aeroponic production of seed potato under tropical conditions (winter/spring)," *Bragantia*, vol. 76, no. 1, pp. 23–32, 2017.

[82] S. Abdullateef, M. H. Bohme, and I. Pinker, "Potato minituber production at different plant densities using an aeroponic system," *Acta Horticulturae*, vol. 927, pp. 429–436, 2012.

[83] I. Farran and A. M. Mingo-Castel, "Potato minituber production using aeroponics: effect of plant density and harvesting intervals," *American Journal of Potato Research*, vol. 83, no. 1, pp. 47–53, 2006.

[84] H. S. Kim, E. M. Lee, M. A. Lee et al., "Production of high-quality potato plantlets by autotrophic culture for aeroponic systems," *Journal of the Korean Society for Horticultural Science*, vol. 123, pp. 330–333, 1999.

[85] D. H. Park, B. J. Kang, K. R. Cho et al., "A study on greenhouse automatic control system based on wireless sensor network," *Wireless Personal Communications*, vol. 56, no. 1, pp. 117–130, 2011.

[86] Y.-S. Kim, "Expert development for automatic control of greenhouse environment," *Journal of Korean Flower Research Society*, vol. 12, no. 4, pp. 341–345, 2004.

[87] Aqeel-ur-Rehman, A. Z. Abbasi, N. Islam, and Z. A. Shaikh, "A review of wireless sensors and networks' applications in agriculture," *Computer Standards & Interfaces*, vol. 36, no. 2, pp. 263–270, 2014.

[88] A. Arora, R. Ramnath, E. Ertin et al., "ExScal: elements of an extreme scale wireless sensor network," in *11th IEEE International Conference on Embedded and Real-Time Computing Systems and Applications (RTCSA'05)*, pp. 102–108, Hong Kong, China, August 2005.

[89] R. Aquino-Santos, A. Gonzalez-Potes, A. Edwards-Block, and R. A. Virgen-Ortiz, "Developing a new wireless sensor network platform and its application in precision agriculture," *Sensors*, vol. 11, no. 1, pp. 1192–1211, 2011.

[90] S. Yoo, J. Kim, T. Kim, S. Ahn, J. Sung, and D. Kim, "A^2S: automated agriculture systems based on WSN," in *2007 IEEE International Symposium on Consumer Electronics*, pp. 1–5, Irving, TX, USA, June 2007.

[91] G. Abowd, A. K. Dey, P. Brown, N. Davies, M. Smith, and P. Steggles, *Towards a Better Understanding of Context and Context-Awareness, The Workshop on The What, Who, Where, When, and How of Context-Awareness as Part of the 2000 Conference on Human Factors in Computing Systems (CHI 2000)*, Springer, The Netherlands, 1999.

[92] B. N. Schilit and M. M. Theimer, "Disseminating active map information to mobile hosts," *IEEE Network*, vol. 8, no. 5, pp. 22–32, 1994.

[93] V. Otazú, *Manual on Quality Seed Potato Production Using Aeroponics*, vol. 44, International Potato Center (CIP), Lima, Peru, 2010, http://cippotato.org.research/publication/manaul-on-quality-seed-potato-production-using-aeroponics.

[94] C. Wang, A. Zhang, and H. R. Karimi, "Development of La^{3+} doped CeO_2 thick film humidity sensors," *Abstract and Applied Analysis*, vol. 2014, Article ID 297632, 6 pages, 2014.

[95] Z. Chen and C. Lu, "Humidity sensors: a review of materials and mechanisms," *Sensor Letters*, vol. 3, no. 4, pp. 274–295, 2005.

[96] S. D. Zor and H. Cankurtaran, "Impedimetric humidity sensor based on nanohybrid composite of conducting poly (diphenylamine sulfonic acid)," *Journal of Sensors*, vol. 2016, Article ID 5479092, 9 pages, 2016.

[97] I. Ioslovich and P. O. Gutman, "Optimal control of crop spacing in a plant factory," *Automatica*, vol. 36, no. 11, pp. 1665–1668, 2000.

[98] K. Kato, R. Yoshida, A. Kikuzaki et al., "Molecular breeding of tomato lines for mass production of miraculin in a plant factory," *Journal of Agricultural and Food Chemistry*, vol. 58, no. 17, pp. 9505–9510, 2010.

[99] W. Bihlmayr, *Application Note 313. Subject to Modifications*, pp. 2–14, 2011, http://www.enocean.com.

[100] K. Kalwinder, *Carbon Dioxide Sensor. AZO Sensor. Article ID=234*, 2013, http://www.azosensors.com/article.aspx?ArticleID=234.

[101] T. Asao, *Hydroponics - A Standard Methodology for Plant Biological Researches*, Intech, Rijeka, Croatia, 1st edition, 2012.

[102] D. Borgognone, G. Colla, Y. Rouphael, M. Cardarelli, E. Rea, and D. Schwarz, "Effect of nitrogen form and nutrient solution pH on growth and mineral composition of self-grafted and grafted tomatoes," *Scientia Horticulturae*, vol. 149, pp. 61–69, 2013.

[103] S. P. Friedman, "Soil properties influencing apparent electrical conductivity: a review," *Computers and Electronics in Agriculture*, vol. 46, no. 1-3, pp. 45–70, 2005.

[104] L. Yang, V. Sarath Babu, J. Zou, X. C. Cai, T. Wu, and L. Lin, "The development of an intelligent monitoring system for agricultural inputs basing on DBN-SOFTMAX," *Journal of Sensors*, vol. 2018, Article ID 6025381, 11 pages, 2018.

[105] T. L. Dinh, W. Hu, P. Sikka, P. Corke, L. Overs, and S. Brosnan, "Design and deployment of a remote robust sensor network: experiences from an outdoor water quality monitoring network," in *32nd IEEE Conference on Local Computer Networks*, pp. 799–806, Dublin, Ireland, October 2007.

[106] P. Marino, F. P. Fontan, M. A. Dominiguez, and S. Otero, "Environmental monitoring based on emerging wireless technologies," in *Fourth International Conference on Networking and Services (icns 2008)*, pp. 30–34, Gosier, Guadeloupe, March 2008.

[107] I. F. Akyildiz, W. Su, Y. Sankarasubramaniam, and E. Cayirci, "Wireless sensor networks: a survey," *Computer Networks*, vol. 38, no. 4, pp. 393–422, 2002.

[108] M. A. Hebel, "Meeting wide-area agricultural data acquisition and control challenges through ZigBee wireless network technology," in *4th World Congress Conference on Computers in Agriculture and Natural Resources*, Orlando, FL, USA, July 2006.

[109] P. K. Haneveld, *Evading Murphy: A Sensor Network Deployment in Precision Agriculture*, Delft, Netherlands,

2007http://www.st.ewi.tudelft.nl/koen/papers/LOFAR-agro-take2.pdf.

[110] Y.-D. Lee, "Implementation of greenhouse environment monitoring system based on wireless sensor networks," *Journal of the Korea Institute of Information and Communication Engineering*, vol. 17, no. 11, pp. 2686–2692, 2013.

[111] S. L. Andresen, "John McCarthy: father of AI," *IEEE Intelligent Systems Magazine*, vol. 17, no. 5, pp. 84-85, 2002.

[112] P. Lu, S. Chen, and Y. Zheng, "Artificial intelligence in civil engineering," *Mathematical Problems in Engineering*, vol. 2012, Article ID 145974, 22 pages, 2012.

[113] K. de Koning, "Automatic milking–common practice on dairy farms," in *Proceedings of the First North American Conference on Precision Dairy Management and the Second North American Conference on Robotic Milking*, pp. 52–67, Toronto, Canada, 2010.

[114] S. H. Chen, A. J. Jakeman, and J. P. Norton, "Artificial intelligence techniques: an introduction to their use for modelling environmental systems," *Mathematics and Computers in Simulation*, vol. 78, no. 2-3, pp. 379–400, 2008.

[115] W. Batayneh, O. Al-Araidah, and K. Bataineh, "Fuzzy logic approach to provide safe and comfortable indoor environment," *International Journal of Engineering, Science and Technology*, vol. 2, no. 7, pp. 65–72, 2010.

[116] T. K. Das and Y. Das, "Design of a room temperature and humidity controller using fuzzy logic," *American Journal of Engineering Research*, vol. 2, no. 11, pp. 86–97, 2013.

[117] P. G. Lee, "Process control and artificial intelligence software for aquaculture," *Aquacultural Engineering*, vol. 23, no. 1-3, pp. 13–36, 2000.

[118] "Energias market research. Global artificial intelligence (AI) in agriculture market," March 2018, https://www.energiasmarketresearch.com/global-artificial-intelligence-ai-agriculture-market/.

[119] C. Popa, "Adaption of artificial intelligence in agriculture," *Bulletin UASVM Agriculture*, vol. 68, no. 1, pp. 284–293, 2011.

[120] A. Rafea, *Expert System Applications: Agriculture*, Central Laboratory for Expert Systems, Giza Egypt, 2009.

[121] K. S. Wai, A. L. B. A. Rahman, M. F. Zaiyadi, and A. A. Aziz, *Expert System in Real World Applications*, pp. 1–4, 2005, http://citeseerx.ist.psu.edu/viewdoc/download?doi=10.1.1.117.9964&rep=rep1&type=pdf.

[122] Y. Huang, Y. Lan, S. J. Thomson, A. Fang, W. C. Hoffmann, and R. E. Lacey, "Development of soft computing and applications in agricultural and biological engineering," *Computers and Electronics in Agriculture*, vol. 71, no. 2, pp. 107–127, 2010.

[123] L. A. Zadeh, "Fuzzy sets," *Information and Control*, vol. 8, no. 3, pp. 338–353, 1965.

[124] L. A. Zadeh, "Outline of a new approach to the analysis of complex systems and decision processes," *IEEE Transactions on Systems, Man, and Cybernetics*, vol. SMC-3, no. 1, pp. 28–44, 1973.

[125] L. A. Zadeh, "Possibility theory and soft data analysis," in *Mathematical Frontiers of the Social and Policy Sciences*, L. Cobb and R. M. Thrall, Eds., pp. 69–129, Westview Press, Boulder, CO, USA, 1981.

[126] D. E. Rumelhart and J. L. McClelland, *Parallel Distributed Processing: Explorations in the Microstructures of Cognition, vol. I*, MIT Press, Cambridge, MA, USA, 1986.

[127] D. E. Goldberg, *Genetic Algorithms in Search, Optimization, and Machine Learning*, Addison Wesley, Reading, MA, USA, 1989.

[128] R. Sui and J. A. Thomasson, "Ground-based sensing system for cotton nitrogen status determination," *Transactions of the ASABE*, vol. 49, no. 6, pp. 1983–1991, 2006.

[129] S. D. Tumbo, D. G. Wagner, and P. H. Heinemann, "On-the-go sensing of chlorophyll status in corn," *Transactions of the ASAE*, vol. 45, no. 4, pp. 1207–1215, 2002.

[130] L. Tang, L. Tian, and B. L. Steward, "Classification of broadleaf and grass weeds using Gabor wavelets and an artificial neural network," *Transactions of the ASAE*, vol. 46, no. 4, pp. 1247–1254, 2003.

[131] M. S. El-Faki, N. Zhang, and D. E. Peterson, "Weed detection using color machine vision," *Transactions of the ASAE*, vol. 43, no. 6, pp. 1969–1978, 2000.

[132] M. Krishnaswamy and P. Krishnan, "Nozzle wear rate prediction using regression and neural network," *Biosystems Engineering*, vol. 82, no. 1, pp. 49–56, 2002.

[133] T. C. Pearson and D. T. Wicklow, "Detection of corn kernels infected by fungi," *Transactions of the ASABE*, vol. 49, no. 4, pp. 1235–1245, 2006.

[134] B. A. Smith, G. Hoogenboom, and R. W. McClendon, "Artificial neural networks for automated year-round temperature prediction," *Computers and Electronics in Agriculture*, vol. 68, no. 1, pp. 52–61, 2009.

[135] B. Center and B. P. Verma, "Fuzzy logic for biological and agricultural systems," *Artificial Intelligence Review*, vol. 12, pp. 213–225, 1998.

[136] G. C. Vieira, A. R. de Mendonca, G. F. da Silva, S. S. Zanetti, M. M. da Silva, and A. R. dos Santos, "Prognoses of diameter and height of trees of eucalyptus using artificial intelligence," *Science of the Total Environment*, vol. 619-620, pp. 1473–1481, 2018.

[137] N. T. R. Ahamed, K. G. Rao, and J. S. R. Murthy, "GIS-based fuzzy membership model for crop-land suitability analysis," *Agricultural Systems*, vol. 63, no. 2, pp. 75–95, 2000.

[138] M. Boyland, J. Nelson, F. L. Bunnell, and R. G. D'Eon, "An application of fuzzy set theory for seral-class constraints in forest planning models," *Forest Ecology and Management*, vol. 223, no. 1-3, pp. 395–402, 2006.

[139] P. F. Fisher, "Remote sensing of land cover classes as type 2 fuzzy sets," *Remote Sensing of Environment*, vol. 114, no. 2, pp. 309–321, 2010.

[140] H. Jiang and J. R. Eastiman, "Application of fuzzy measures in multi-criteria evaluation in GIS," *International Journal of Geographical Information Science*, vol. 14, no. 2, pp. 173–184, 2000.

[141] B. N. Joss, R. J. Hall, D. M. Sidders, and T. J. Keddy, "Fuzzy-logic modeling of land suitability for hybrid poplar across the Prairie Provinces of Canada," *Environmental Monitoring and Assessment*, vol. 141, no. 1-3, pp. 79–96, 2008.

[142] J. Oldeland, W. Dorigo, L. Lieckfeld, A. Lucieer, and N. Jürgens, "Combining vegetation indices, constrained ordination and fuzzy classification for mapping semi-natural vegetation units from hyperspectral imagery," *Remote Sensing of Environment*, vol. 114, no. 6, pp. 1155–1166, 2010.

[143] T. Phillips, S. Leyk, H. Rajaram et al., "Modeling moulin distribution on Sermeq Avannarleq glacier using ASTER and WorldView imagery and fuzzy set theory," *Remote Sensing of Environment*, vol. 115, no. 9, pp. 2292–2301, 2011.

[144] A. Al-Faraj, G. E. Meyer, and G. L. Horst, "A crop water stress index for tall fescue (*Festuca arundinacea* Schreb.) irrigation decision-making—a fuzzy logic method," *Computers and Electronics in Agriculture*, vol. 32, no. 2, pp. 69–84, 2001.

[145] S. J. Thomson, R. M. Peart, and J. W. Mishoe, "Parameter adjustment to a crop model using a sensor-based decision support system," *Transactions of the ASAE*, vol. 36, no. 1, pp. 205–213, 1993.

[146] S. J. Thomson and B. B. Ross, "Model-based irrigation management using a dynamic parameter adjustment method," *Computers and Electronics in Agriculture*, vol. 14, no. 4, pp. 269–290, 1996.

[147] C. C. Yang, S. O. Prasher, J. A. Landry, J. Perret, and H. S. Ramaswamy, "Recognition of weeds with image processing and their use with fuzzy logic for precision farming," *Canadian Agricultural Engineering*, vol. 42, no. 4, pp. 195–200, 2000.

[148] C.-C. Yang, S. O. Prasher, J.-A. Landry, and H. S. Ramaswamy, "Development of an image processing system and a fuzzy algorithm for site-specific herbicide applications," *Precision Agriculture*, vol. 4, no. 1, pp. 5–18, 2003.

[149] Y. Gil, C. Sinfort, S. Guillaume, Y. Brunet, and B. Palagos, "Influence of micrometeorological factors on pesticide loss to the air during vine spraying: data analysis with statistical and fuzzy inference models," *Biosystems Engineering*, vol. 100, no. 2, pp. 184–197, 2008.

[150] Z. Qiu, X. Tong, J. Shen, and Y. Bao, "Irrigation decision-making system based on the fuzzy-control theory and virtual instrument," *Nongye Gongcheng Xuebao/Transactions of the Chinese Society of Agricultural Engineering*, vol. 23, no. 8, pp. 165–169, 2007.

[151] F. Touati, M. Al-Hitmi, K. Benhmed, and R. Tabish, "A fuzzy logic based irrigation system enhanced with wireless data logging applied to the state of Qatar," *Computers and Electronics in Agriculture*, vol. 98, pp. 233–241, 2013.

[152] R. B. Ali, E. Aridhi, M. Abbes, and A. Mami, "Fuzzy logic controller of temperature and humidity inside an agricultural greenhouse," in *2016 7th International Renewable Energy Congress (IREC)*, Hammamet, Tunisia, March 2016.

[153] F. Fourati, "Multiple neural control of a greenhouse," *Neurocomputing*, vol. 139, pp. 138–144, 2014.

[154] M. Taki, Y. Ajabshirchi, S. Faramarz Ranjbar, A. Rohani, and M. Matloobi, "Heat transfer and MLP neural network models to predict inside environment variables and energy lost in a semi-solar greenhouse," *Energy and Buildings*, vol. 110, pp. 314–329, 2016.

[155] F. Lafont and J.-F. Balmat, "Optimized fuzzy control of a greenhouse," *Fuzzy Sets and Systems*, vol. 128, no. 1, pp. 47–59, 2002.

[156] P. Salgado and J. B. Cunha, "Greenhouse climate hierarchical fuzzy modelling," *Control Engineering Practice*, vol. 13, no. 5, pp. 613–628, 2005.

[157] M. Azaza, K. Echaieb, F. Tadeo, E. Fabrizio, A. Iqbal, and A. Mami, "Fuzzy decoupling control of greenhouse climate," *Arabian Journal for Science and Engineering*, vol. 40, no. 9, pp. 2805–2812, 2015.

[158] A. Chouchaine, E. Feki, and A. Mami, "Stabilization using a discrete fuzzy PDC control with PID controllers and pole placement: application to an experimental greenhouse," *Journal of Control Science and Engineering*, vol. 2011, Article ID 537491, 9 pages, 2011.

[159] S. Zeng, H. Hu, L. Xu, and G. Li, "Nonlinear adaptive PID control for greenhouse environment based on RBF network," *Sensors*, vol. 12, no. 5, pp. 5328–5348, 2012.

[160] X. Zhu, D. Li, D. He, J. Wang, D. Ma, and F. Li, "A remote wireless system for water quality online monitoring in intensive fish culture," *Computers and Electronics in Agriculture*, vol. 71, pp. S3–S9, 2010.

[161] S. Mahajan, A. Das, and H. K. Sardana, "Image acquisition techniques for assessment of legume quality," *Trends in Food Science and Technology*, vol. 42, no. 2, pp. 116–133, 2015.

14

Soft Sensor Model based on Improved Elman Neural Network with Variable Data Preprocessing and its Application

Hai-bo Zhu and Yong Zhang

School of Electronics and Information Engineering, University of Science and Technology Liaoning, Anshan 114051, China

Correspondence should be addressed to Yong Zhang; 2550345485@qq.com

Academic Editor: Jesús Lozano

In order to solve the problems of strong coupling, nonlinearity, and complex mechanism in real-world engineering process, building soft sensor with excellent performance and robustness has become the core issue in industrial processes. In this paper, we propose a new soft sensor model based on improved Elman neural network (Elman NN) and introduce variable data preprocessing method to the soft sensor model. The improved Elman NN employs local feedback and feedforward network mechanism through context layer to accurately reflect the dynamic characteristics of the soft sensor model, which has the superiority to approximate delay systems and adaption of time-varying characteristics. The proposed variable data preprocessing method adopts combining Isometric Mapping (ISOMAP) with local linear embedding (LLE), which effectively maintains the neighborhood structure and the global mutual distance of dataset to remove the noises and data redundancy. The soft sensor model based on improved Elman NN with variable data preprocessing method by combining ISOMAP and LLE is applied in practical sintering and pelletizing to estimate the temperature in the rotary kiln calcining process. Comparing several conventional soft sensor model methods, the results indicate that the proposed method has more excellent generalization performance and robustness. Its model prediction accuracy and anti-interference ability have been improved, which provide an effective and promising method for the industrial process application.

1. Introduction

Soft sensor is an inferential prediction virtual technique which adopts easily measured variables to predict the process variables which are difficult to measure directly because of technological and economical limitations or complex environment. The soft sensor tries to establish a regression prediction model between easily measured variables and difficultly measured variables, which is adopted to solve the problem that prevents the measurements from being employed as feedback signals for quality control systems. And the soft sensor methods have been employed more widely in the industrial process and become a major developing trend in both academia and industry [1].

In view of the model prediction in the industrial production process, early scholars put forward some model predictive control, such as generalized predictive control [2, 3], dynamic matrix predictive control [4], and model algorithm control [5]. However, these soft sensor prediction methods suffer from common shortcomings. One is that the prediction control of nonlinear systems cannot be effectively solved. Another is that the stability and robustness of multivariable predictive control algorithm need to be solved and it is very difficult to build accurate principle models for complex processes. At present, with the deep research of soft sensor control theory and the continuous progress of engineering technology, some artificial intelligence and machine learning methods based on data-driven technologies are proposed to solve the problems which are difficult to measure key process and quality variables for soft sensor models, such as artificial neural network (ANN) [6–8], rough set [9], partial least squares [10], support vector machine

(SVM) [11–13], and hybrid methods [14, 15]. Some new works on advanced predictive control based on neural networks are referred to, for example, Ławryńczuk proposed MPC algorithms based on double-layer perceptron neural models [16]. And these data-driven soft sensors have been implemented successfully in many industrial application fields, like chemical and metallurgical engineering industries.

However, the complexity of the industrial model will increase and bring about the problem of dimensionality with variables increasing. In addition, there are strong correlations and considerable redundancy between variable data. And the collected variable data may be contaminated by random noises or abnormal working condition data, which indicates that the process variables for soft sensor are stochastic. If those variable data samples are used directly for establishing the soft sensor model, the performance of the soft sensor model may not be guaranteed, which results in terrible prediction accuracy or poor estimation. Hence, it is significant to discover the general trends of data by latent variable models before soft sensor modeling. And the randomness of variable sample data should be taken into consideration and the redundant variable sample data need to be discarded selectively through the use of appropriate variable selection techniques. The core is to exploit the essential information behind the effective sample data to establish a soft sensor model that has excellent performance and strong robustness. The shortcoming of variable sample data has advanced the development of the dimension reduction methods. So many data preprocessing methods for the soft sensor model have been researched in recent years. Sun et al. [17] presented the principal component analysis (PCA) method to analyze the correlation of input variables, which is effective to reduce the dimensions of input variables from the chemical oxygen demand prediction model based on LS-SVM. Najafi et al. [18] proposed a new method called "Path-Based Isomap" to find the low-dimensional embedding. The low-dimensional embedding is computed by exploiting geodesic path-mapping algorithm. Besides, some manifold learning methods have been proposed including Laplacian eigenmaps [19, 20], locally linear embedding [21], and Isometric Mapping method [22, 23]. However, these linear and nonlinear data preprocessing methods have their own limitations, which can only maintain a single feature of the variable dataset and cannot take into account other valuable characteristics. In order to overcome this problem, a data preprocessing method based on combining Isometric Mapping (ISOMAP) with local linear embedding (LLE) for input variable dataset is proposed. Hence, this paper establishes a soft sensor model based on improved Elman NN with input variable data preprocessing method which integrates ISOMAP with LLE under the kernel framework to estimate the quality variable and makes it have better performance and robustness.

The rest of this paper is organized as follows. Section 2 describes the improved Elman NN and data preprocessing method. Section 3 proposes a soft sensor model based on improved Elman NN with variable data selection method by combining ISOMAP with LLE and describes it in detail. In Section 4, an industrial case study on the prediction of temperature in a rotary kiln calcining process in pellet sintering is given. Finally, conclusions are made in Section 5.

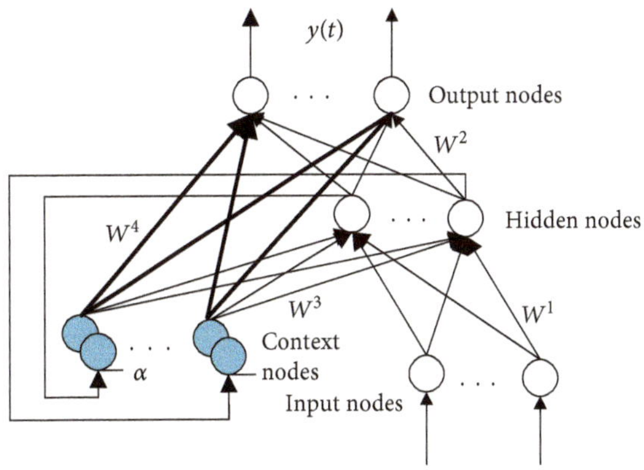

Figure 1: Architecture of improved Elman NN.

2. Background Theory

For convenience, some definitions that will be used throughout this paper are given: NN: neural network; ISOMAP: Isometric Mapping; LLE: local linear embedding.

2.1. Improved Elman Neural Network. Elman neural network (Elman NN) [24–27] is a typical dynamic neural network with recurrent feedback layer based on BP neural network. The Elman NN stores the internal state through the context layer, which has the performance of mapping dynamic characteristics. The self-connected mode makes it sensitive to the value of the historical state and it has better ability to approximate nonlinear mapping than the BP NN. However, this feedback only aims at the hidden layer to adjust the connection weights of the internal feedback, which lacks the ability of global adjustment and dynamic stability. And the learning identification ability has some limitations.

To further improve the dynamic performance and nonlinear approximation capability of Elman NN, the output of context layer as a feedback signal should be fully taken advantage of. Figure 1 shows the architecture of the improved Elman NN that is consisted of an input layer, an output layer, a hidden layer, and a context layer which is connected to hidden layer neurons and output layer neurons.

From Figure 1 we can see that the improved Elman NN adds input signal from context layer nodes to the output layer, which are connected through a novel connection weight matrix $W_{ij}{}^4$ between the context layer and output layer. After the hidden layer process, the output of the hidden layer returns to the context layer, then the intermediate state value of the context layer is transmitted to the output layer and hidden layer through the connection weight matrices $W_{ij}{}^4$ and $W_{ij}{}^3$ separately, which constitutes a feedforward and local feedback network. The feedback loop is formed in the hidden layer and context layer, and the path of feedforward is consisted of the input layer, hidden layer, context layer, and output layer, as shown in Figure 2.

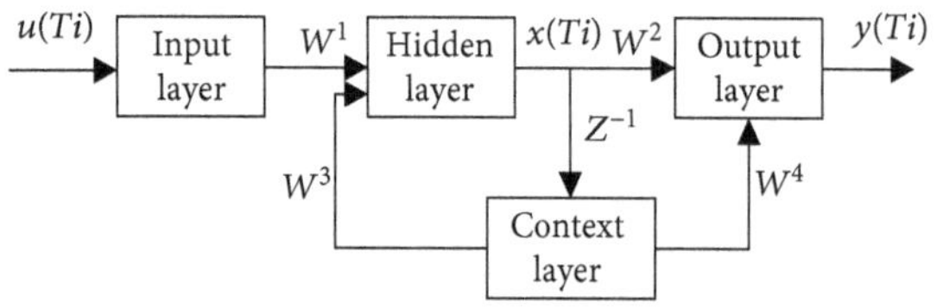

FIGURE 2: Structure block diagram of improved Elman NN.

We define the following symbols: the number of input layer nodes, output layer nodes, and hidden layer nodes is m, n, and l, respectively. The network consisted of input data u, intermediate state oh, and response output y. The activation function of structure from the input layer to the hidden layer and from the hidden layer to the output layer is defined f_θ and g_θ, respectively.

The mathematical model in which the input layer is mapped to the hidden layer through the mapping activation function is

$$\mathrm{oh}_j(k) = f_\theta(u_i(k)) = s\left(\sum_{i=1}^{m} w_{i,j}^1 u_i(k) + \sum_{i=1}^{m} w_{i,j}^3 \mathrm{oc}_j(k) + b_j^h\right), \tag{1}$$

where s represents the *sigmoid* function. Equation (1) is parameterized through $\theta = \{w_{ij}^1, w_{ij}^3, b_j^h\}$, in which w_{ij}^1 is the weight of input layer node i to the hidden layer node j; $w_{ij}^3$is the weight of context layer node i to the hidden layer node j and $b_j^h$is the bias vector of the hidden layer.

The hidden layer oh_j is mapped to the output space, y is the output vector of oh_j, and the activation function is

$$y = g_\theta(\mathrm{oh}_j, \mathrm{oc}_i) = \sum_{i=1}^{n} w_{ij}^2 \mathrm{oh}_j(k) + \sum_{i=1}^{l} w_{ij}^4 \mathrm{oc}_i(k) + b^o. \tag{2}$$

Equation (2) is decided by $\theta' = \{w_{ij}^2, w_{ij}^4, b^o\}$, where w_{ij}^2 is the weight of latent layer node i to the output layer node j; w_{ij}^4 is the connection weight of context layer node i to the output layer node j and b^o represents the bias vector of the output layer.

In the training process, the output of the context layer $\mathrm{oc}_i(k)$ is the sum of output of the hidden layer $\mathrm{oh}_j(k-1)$ and α times output of context layer at a previous time.

$$\mathrm{oc}_i(k) = \mathrm{oh}_j(k-1) + \alpha \cdot \mathrm{oc}_i(k-1). \tag{3}$$

The loss function is defined as the squared error:

$$L(\tilde{y}_d, y) = \|\tilde{y}_d(k) - y(k)\|^2, \tag{4}$$

where $y_d(k)$ is the predicted output value at time k; $y_i(k)$ is the actual output value at time k.

A novel set of parameter vector of the improved Elman NN can be obtained and optimized to minimize the expected average squared error:

$$\begin{aligned}\left(\theta^*, \theta'*\right) &= \underset{\theta,\theta'}{\arg\min}\ \frac{1}{n}\sum_{i=1}^{n} L(\tilde{y}_{d,i}(k), y_i(k)) \\ &= \underset{\theta,\theta'}{\arg\min}\ \frac{1}{n}\sum_{i=1}^{n} L(g_{\theta'}(f_\theta(u_i(k))), y_i(k)).\end{aligned} \tag{5}$$

2.2. Input Variable Data Selection and Dimension Reduction Method. Considering the original input variables of improved Elman NN exist with redundant variable data and abnormal condition variable data, a novel variable data selection methodology by integrating Isometric Mapping (ISOMAP) [28] and local linear embedding (LLE) [29, 30] is proposed to handle measurement uncertainty and determine the dimensionality of the latent variable. The purpose is to combine two manifold learning algorithms effectively under the kernel framework and maintain various characteristics of data set, which establishes a dimensionality reduction method based on manifold learning for high-dimensional nonlinear data sets to effectively solve the network topology and the complex training problems caused by the oversize input vector data. The novel methodology can be divided into three stages. At the first stage, we construct the Mercer kernel matrix K_{ISOMAP} of Isomap. Second, we extract the kernel matrix K_{LLE} of LLE. At the third stage, we obtain the kernel matrix $K_{\mathrm{ISOMAP-LLE}}$ under the kernel framework.

The general steps of the data dimensionality reduction process are the following.

Step 1. Use K neighborhood method or ε radius method to find neighborhood values.

Step 2. Use the shortest path method to calculate the geodesic distance d_{ij} and to form the geodesic distance matrix D^2, in which $D_{ij}^2 = |d_{ij}^2| \in R^{N\times N}$. The matrix $K(D^2)$ is constructed on the basis of approximate geodesic distance matrix D^2.

$$K(D^2) = -\frac{1}{2}HD^2H, \tag{6}$$

where H represents a centralization matrix, $H = I - (1/N)ee^T$ and $e = [1\ 1\ \cdots\ 1]^T \in R^{N\times 1}$.

Step 3. The matrix expression is constructed according to the matrix $K(D^2)$.

$$\begin{bmatrix} 0 & 2K(D^2) \\ -1 & -4K(D) \end{bmatrix}. \tag{7}$$

Step 4. Obtain the maximum eigenvalue c^* of the matrix and construct the Mercer kernel matrix K_{ISOMAP} according to the maximum eigenvalue c^*.

$$K_{\text{ISOMAP}} = K(D^2) + 2cK(D) + \frac{1}{2}c^2 H. \quad (8)$$

In order to ensure that K_{ISOMAP} is a real symmetric semidefinite matrix, we need to satisfy $c \geq c^*$.

Step 5. For each data point of the LLE algorithm $x_i (i = 1, 2, \ldots, N)$, solve its cost function $\varepsilon = \left|x_i - \sum_{j=1}^{k} w_{ij} x_{ij}\right|^2$.

The weight coefficient W_{ij} of data points is determined, and the weight matrix W is constructed. At the same time, a real symmetric semipositive definite matrix M is obtained based on the weight matrix W, and the kernel matrix K_{LLE} of the LLE algorithm is extracted.

$$M = (I - W)^T (I - W), \quad (9)$$

$$K_{\text{LLE}} = \lambda_{\max} I - M, \quad (10)$$

where $\lambda_{\max}$ is the largest eigenvalue of the weight matrix W.

Step 6. Obtain the kernel matrix of $K_{\text{ISOMAP-LLE}}$ by the kernel matrix K_{ISOMAP} of the ISOMAP algorithm and the kernel matrix K_{LLE} of the LLE algorithm. And the kernel matrix regulator β is introduced to adjust the weight of the ISOMAP and LLE algorithms, and a new kernel matrix $K_{\text{ISOMAP-LLE}}$ is constructed based on the kernel function.

$$K_{\text{ISOMAP-LLE}} = K_{\text{ISOMAP}} + (1 - \beta) K_{\text{LLE}}. \quad (11)$$

Step 7. Calculate the eigenvalue matrix P and eigenvector matrix Q of the new kernel matrix $K_{\text{ISOMAP-LLE}}$ and obtain the low-dimensional embedded coordinates Y based on *MDS*:

$$Y = \Lambda^{1/2} Q, \quad (12)$$

where Λ is the diagonal matrix of P.

3. Soft Sensor Model Based on Improved Elman NN with Dimension-Reduction

3.1. Soft Sensor Model Based on Improved Elman NN with Variable Data Preprocessing. Soft sensor model based on improved Elman NN with variable data selection is aimed at exploiting the essential information behind the process data and filter redundant variable data in the soft sensor model. This paper proposes an input variable data selection and dimension-reduction methodology by integrating ISOMAP and LLE manifold learning algorithm, which is effective to solve redundant variable data and abnormal working data of the oversized sample data. This structure of the soft sensor model based on improved Elman NN with variable data selection is shown in Figure 3. The soft sensor model consisted of two parts: an input data selection (ISOMAP and LLE) and a supervised learning (improved Elman NN). In this paper, the variable screening method adopted the manifold learning variable selection method based on the

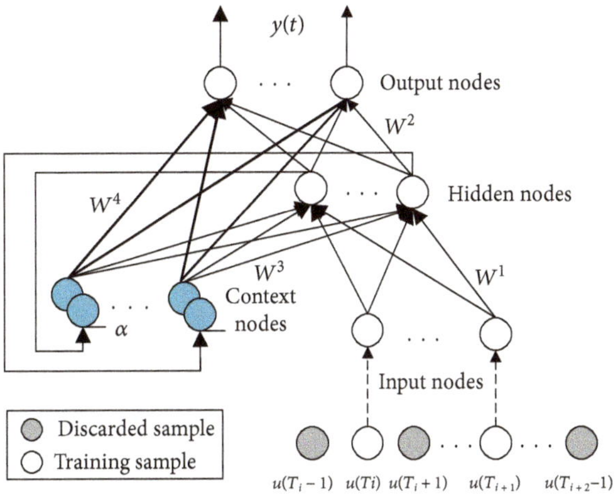

FIGURE 3: The structure of the soft sensor model based on improved Elman NN with dimension-reduction.

fusion of LLE and ISOMAP under the kernel framework, which is mainly reflected in two aspects. (1) The initial input variables should be chosen reasonably to make the "quality" of initial input variables be ensured. (2) The useful information of input variables should not be prematurely eliminated. And adopting supervised criterion trains each layer to obtain the weights which are set as the initial weights instead of the random initial weights.

In Figure 3, W^1, W^2, W^3, and W^4 are the weights of improved Elman NN, respectively, which are adopted as the weights of corresponding layers in the improved Elman NN optimized with respect to a supervised training criterion. The input data of the soft sensor model consisted of training sample data and discarded sample data: the discarded sample data U_{discard}, i.e., noise and redundant values of the primary input variables, and the training sample data (U_{train}, Y), i.e., pairs of input-output sample data without redundant and abnormal working data, are used to train improved Elman NN to have better accuracy for the soft sensor model.

3.2. The Training Based on Supervised Learning. In the training process of improved Elman NN, each layer in the feedforward path is considered as one-layer BP network which can update weight parameters by gradient descent. The representation of ith hidden layer is regarded as the input for the $i + 1$ th context layer which feeds the previous time state of the hidden layer nodes back to the hidden layer. The error is adopted to refresh the weights.

The loss function can be the traditional squared error as in equation (4). Equation (5) can be reformulated by optimizing the parameters to minimize the average squared error. The gradient descent method is used to update parameters of the improved Elman NN and amend the weights $w_{i,j}{}^1$, $w_{i,j}{}^2$, $w_{i,j}{}^3$, and $w_{i,j}{}^4$ to achieve self-adaptive learning. At time k, we can get

$$\Delta w_{ij}^1 = \eta_1 \sum_{k=1}^{T} \sum_{i=1}^{m} \left[\left(\tilde{y}_{d,i}(k) - y_i(k) \right) w_{ij}^2(k) \right] f'(\cdot) \cdot u_i(k), \quad (13)$$

where $u_i(k)$, $i = 1, 2, \ldots, m$, is the signal of input layer nodes. k, $k = 1, 2, \ldots, T$, is the current operation iterations. η_1 is the learning rate of the weights of $w_{i,j}{}^1$.

$$\Delta w_{ij}^2 = \eta_2 \sum_{k=1}^{T}\sum_{i=1}^{n}\left[\left(\tilde{y}_{d,i}(k) - y_i(k)\right)\mathrm{oh}_j(k)\right], \tag{14}$$

where η_2 is the learning rate of the weights of $w_{i,j}{}^2$.

$$\begin{aligned}\Delta w_{ij}^3 &= \eta_3\left(\frac{\partial L}{\partial oh_i(k)}\cdot\frac{\partial oh_i(k)}{\partial w_{ij}^3(k)} + \frac{\partial L}{\partial oc_i(k+1)}\cdot\frac{\partial oc_i(k+1)}{\partial w_{ij}^3(k)}\right)\\ &= \eta_3\left\{\sum_{i=1}^{l}\left[\left(\tilde{y}_{d,i}(k) - y_i(k)\right)w_{ij}^2(k)\right]f'(\cdot)oc_i(k)\right.\\ &\quad\left. + \sum_{i=1}^{l}\left[\left(\tilde{y}_{d,i}(k+1) - y_i(k+1)\right)w_{ij}^4(k+1)\right]f'(\cdot)oc_i(k)\right\},\end{aligned} \tag{15}$$

where η_3 is the learning rate of the weights of $w_{i,j}{}^3$. oc_i is obtained as in equation (3).

$$\Delta w_{ij}^4 = \eta_4 \sum_{k=1}^{T}\sum_{i=1}^{n}\left[\left(\tilde{y}_{d,i}(k) - y_i(k)\right)\mathrm{oc}_i(k)\right], \tag{16}$$

where η_4 is the learning rate of the weights of $w_{i,j}{}^4$.

A new set of weights W_{ij} can be updated by

$$\tilde{w}_{ij}^{k+1} = w_{ij}^k + \Delta w_{ij}^{k+1}. \tag{17}$$

Then, the prediction model of y is obtained by the new weight parameters (θ^*, θ'^*).

$$\begin{aligned}\tilde{y}_{\theta^*,\theta'*} &= g\left(\sum_{i=1}^{n}\tilde{w}_{ij}^2 f\left(\sum_{i=1}^{m}\tilde{w}_{i,j}^1 u_i(k) + \sum_{i=1}^{l}\tilde{w}_{i,j}^3\mathrm{oc}_i(k) + b_j^h\right)\right.\\ &\quad\left. + \sum_{i=1}^{l}\tilde{w}_{ij}^4\mathrm{oc}_i(k) + b^o\right).\end{aligned} \tag{18}$$

3.3. Formulated the Criterion of Evaluating. After the variable selection and training, the prediction value of the soft sensor model based on improved Elman NN as in equation (18) needs be further evaluated. In this paper, the evaluation criterion, the Bayesian information criterion (BIC), is adopted as a measure between the complexity of the soft sensor model and the accuracy.

$$\mathrm{BIC}_{\theta^*,\theta'*} = n\cdot\log_e\left(\frac{1}{2}\sum_{i=1}^{n}\left(\tilde{y}_i - y_i\right)^2\right) + m\cdot\log_e n, \tag{19}$$

where $\tilde{y}_i$ is the prediction value by modified weight parameters as in equation (18). y is the actual value. n represents the observational number of the soft sensor model. m is the number of selected variables through integrating ISOMAP and LLE.

Table 1: Candidate variables of the soft sensor model for temperature prediction in the rotary kiln calcining process.

Variables	Description	Unit
u_1	The speed of the chain grate	m/min
u_2	The speed of the rotary kiln	rin/min
u_3	The speed of the ring refrigerator	m/min
u_4	The temperature of rotary kiln head	°C
u_5	The temperature of rotary kiln tail	°C
u_6	The quantity of coal injection	t
u_7	I section temperature of the cooler	°C
u_8	II section temperature of the cooler	°C
u_9	Material thickness in the chain grate	mm
u_{10}	Preheating I section temperature of the smoke hood in chain grate	°C
u_{11}	Preheating II section temperature of the smoke hood in chain grate	°C

The optimal parameters can be obtained by the minimization of the average BIC values:

$$\left(\theta^*, \theta'^*\right) = \arg\min\left(\frac{1}{N}\sum_{n=1}^{N}\mathrm{BIC}\left(\theta, \theta'\right)\right). \tag{20}$$

The procedure of the soft sensor model based on improved Elman NN with selected variable and dimension-reduction is summarized as follows.

Step 1. Initialization. Train a new improved Elman NN by the training dataset and obtain initial weight parameters of the network.

Step 2. Selection of input variable dataset.

Step 2.1. Initialize variable data preprocessing method based on combining ISOMAP with LLE.

Step 2.2. For the current variable dataset, calculate approximate geodesic distance to obtain the maximum eigenvalue c^* of the matrix and construct the Mercer kernel matrix K_{ISOMAP} by equation (8).

Step 2.3. Obtain the real symmetric semipositive definite matrix M according to cost function ε and extract the kernel matrix K_{LLE} by equation (10).

Step 2.4. Calculate $K_{\mathrm{ISOMAP\text{-}LLE}}$ by equation (11) based on Step 2.2 and Step 2.3.

Step 2.5. Calculate Y by equation (12).

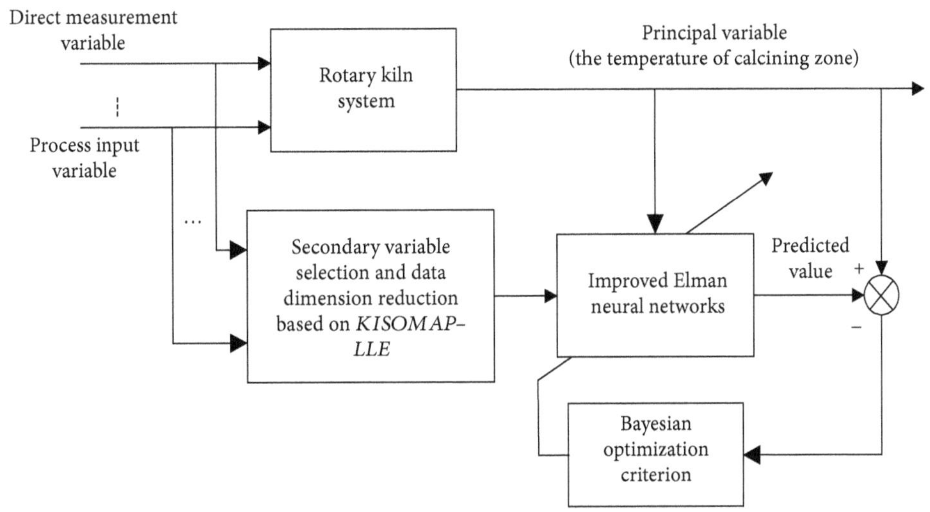

FIGURE 4: The diagram of temperature prediction in the rotary kiln calcining process based on improved Elman NN with variable data preprocessing.

Step 2.6. Obtain the input variable data without redundancy and noises disturbing.

Step 3. Evaluate the parameter optimization by equation (19) and obtain the new parameters by equations (5) and (20).

Step 4. Update the weight parameters by equation (17) and obtain novel neural network by equation (18).

4. Case Study: Application to Temperature Prediction in Rotary Kiln Calcining Process

The rotary kiln is an important part of pellet sintering. Green pellets with a diameter of 9~16 mm are generated in the pellet making system. Then the pellets pass through the kiln tail into the rotary kiln after preheating at the chain grate. In the rotary kiln, the oxidizing roast and consolidation process of pellets is completed, which makes pellets have the physical and chemical properties of the blast furnace burden. To ensure and improve the product quality of pellets, the real-time control of temperature in the rotary kiln calcining process is very important. And the temperature in the rotary kiln calcining process is directly related to production efficiency, energy consumption, and emissions of harmful gas.

By the analysis of the heat balance mechanism and technological characteristics of the rotary kiln calcining process, the temperature in the rotary kiln calcining process is coupled with multiple technological index variables in the grate system and ring cooler system, which makes it very difficult to use the instrument directly to measure the temperature in the rotary kiln calcining process because of poor working conditions or high maintenance costs of hardware sensors. Thus, the soft sensor model based on improved Elman NN with variable data selected and Bayesian optimization is applied to estimate the temperature in the rotary kiln calcining process. For the prediction of the temperature, we use the technological index variables which are correlated with prediction temperature and are easy to measure as the input variables of the soft sensor model in this process, which are listed in Table 1. Besides, a kind of physical and chemical reactions in the rotary kiln calcining process also affect the calcining temperature. A more detailed physico-chemical mechanism of the sintering process is described in literature [31, 32].

4.1. The Soft Sensor Model of Temperature in Rotary Kiln Calcining Process Based on Improved Elman NN with Variable Data Preprocessing. According to the technological characteristics of the rotary kiln and technology analysis, the secondary variables are selected based on the combining of ISOMAP with LLE through kernel function, which can simplify the redundant and abnormal data of the selected variables. The secondary variables are chosen as input of the soft sensor model based on improved Elman NN to predict the temperature in the rotary kiln calcining process. The Bayesian optimization criterion is applied to estimate the accuracy of prediction model and adjust the weights in order to amend the error. The structure diagram of temperature prediction in the rotary kiln calcining process based on improved Elman NN with variable data preprocessing is shown in Figure 4.

4.2. Results and Discussion. For the improved Elman NN training, data samples have been collected and are partitioned into two parts, in which 3000 process datasets of the secondary variables and corresponding temperature in the rotary kiln calcining process are selected for training, and the 1500 process datasets are applied as the testing dataset. At the same time, the data preprocessing method based on the combination of ISOMAP with LLE is used to remove the noise and data redundancy that exist in input data in order to improve prediction performance. Considering that SVM is a state-of-the-art soft sensor model with good generalization ability, the improved Elman NN with variable dataset preprocessing based on ISOMAP and LLE is compared with the soft sensor model based on SVM. The kernel

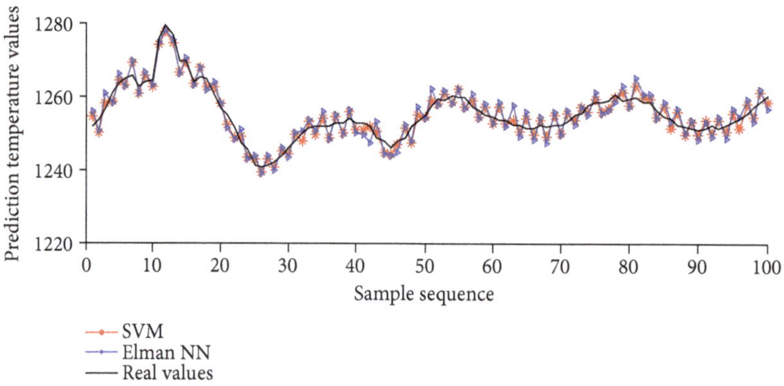

(a) Prediction curve of the kiln burning zone temperature

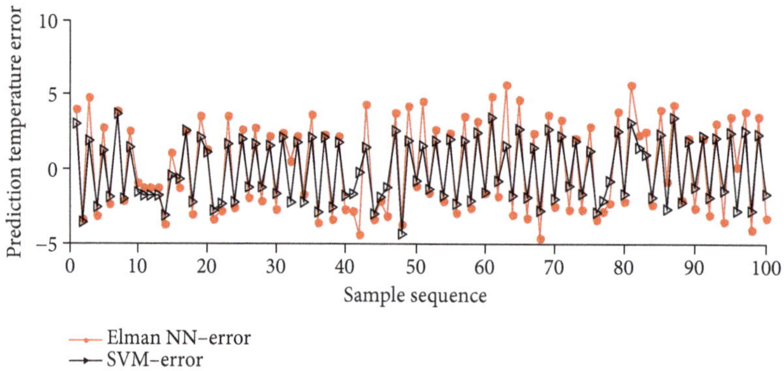

(b) Error curve of the kiln burning zone prediction temperature

FIGURE 5: Comparison of temperature generalization test in rotary kiln between Elman NN and SVM.

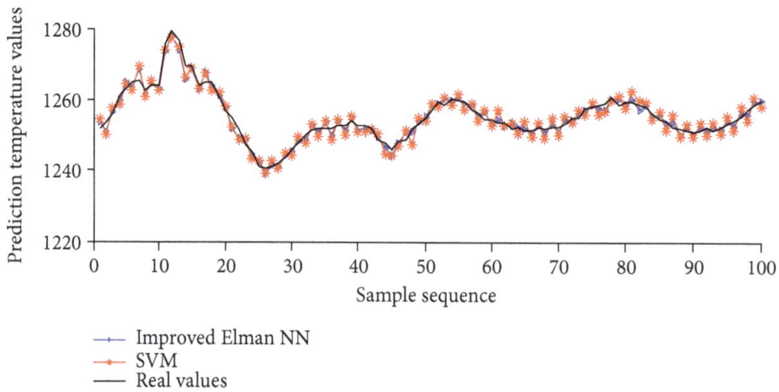

(a) Prediction curve of the kiln burning zone temperature

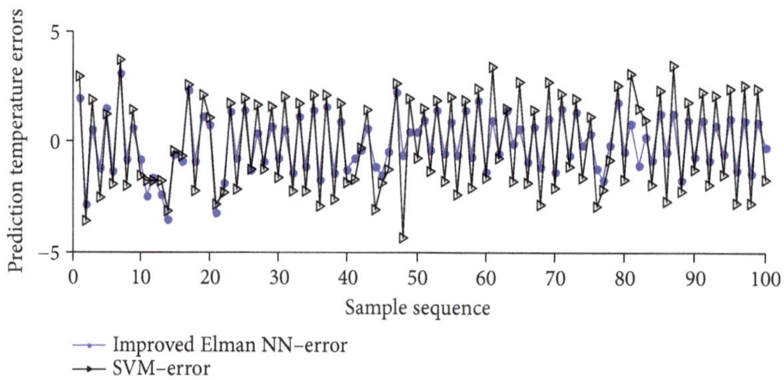

(b) Error curve of the kiln burning zone prediction temperature

FIGURE 6: Comparison of temperature generalization test in rotary kiln between SVM and improved Elman NN.

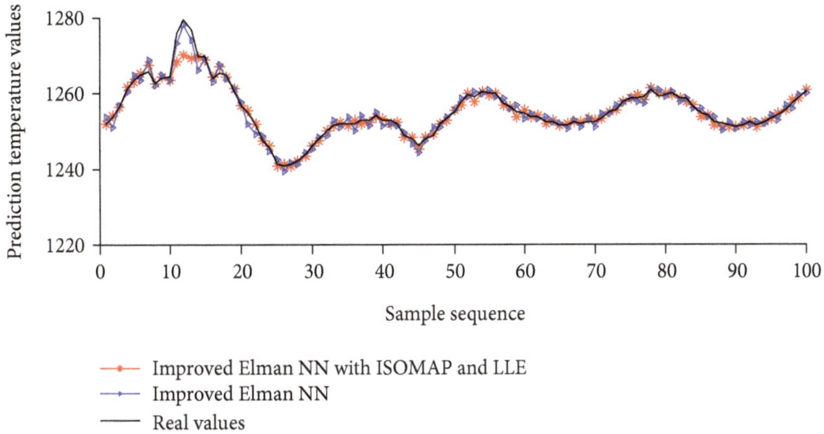

(a) Prediction curve of the kiln burning zone temperature

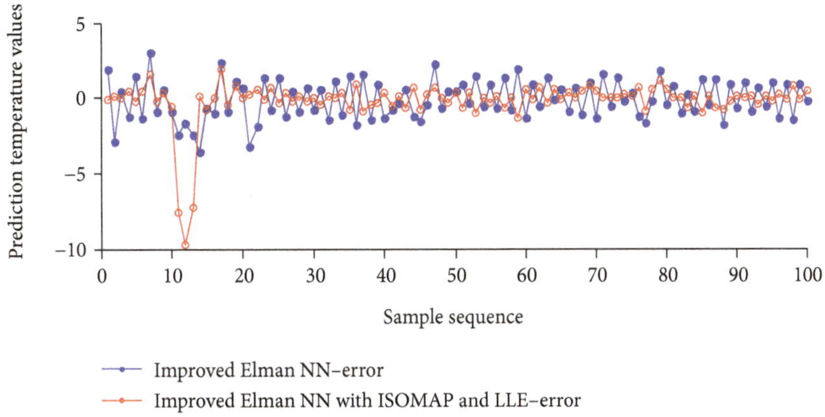

(b) Error curve of the kiln burning zone prediction temperature

FIGURE 7: Comparison of temperature generalization test in rotary kiln between improved Elman NN and improved Elman NN with ISOMAP and LLE.

function of SVM is the radial basis kernel function and the number of support vectors in the SVM is specified as 6. In addition, for a fair comparison, the number of hidden nodes and context nodes is the same in the Elman NN and improved Elman NN. And the number of hidden nodes is set to 36.

In the improved Elman NN with data preprocessing experiments, the gradient descent is employed to train the improved Elman NN. The datasets are chosen by the ISOMAP and LLE method. In addition, the parameters are optimized and decided by comparing validation error curves according to Bayesian criterion. Once the weight parameters are fine-tuned, all training data are employed to train the model again. And the learning rate according to the loss's change is adopted to obtain an excellent performance. To test the prediction performance of the soft sensor model based on different model methods, 100 testing samples are shown in Figures 5–7 because of that the testing datasets are too large to show. To evaluate the performance of the soft sensor model, the root mean square error (*RMSE*), the maximum negative error (*MNE*), and the maximum positive error (*MPE*) are typically used. The values of the testing dataset calculated through the different soft sensor models are compared in Table 2.

TABLE 2: Performance comparison results of prediction temperature under different soft sensor models.

Soft sensor model	RMSE	MNE	MPE
Elman NN	3.0496	−4.582	5.697
SVM	2.1471	−4.332	3.719
Improved Elman NN	1.3003	−3.535	3.061
Improved Elman NN with variable data preprocessing	0.5450	−1.275	1.941

Table 2 shows that improved Elman NN with variable data preprocessing has more excellent learning performance than other soft sensor models. The learning ability of improved Elman NN is improved to some extent and the proposed variable data preprocessing method is effective to remove the redundant data and noises.

Figures 5 and 6 demonstrate that the improved Elman NN has good performance in predicting the temperature in the rotary kiln calcining process. The soft sensor model based on improved Elman NN has much better generalization and learning performance than the soft sensor model based on SVM and conventional Elman NN. The error curves of the SVM prediction, traditional Elman NN prediction, and improved Elman NN prediction which are shown in

Figures 5(b) and 6(b) clearly indicate the difference between the different methods. It can be seen that the prediction error from the improved Elman NN is smaller than that from the SVM and conventional Elman NN.

However, the three methods are limited to handle the noises and data redundancy which exist in input variable data to improve the prediction performance. The soft sensor model based on improved Elman NN with ISOMAP and LLE is able to remove abnormal working condition noise and data redundancy, as shown in Figure 7. It is seen that when several abnormal noise data are added in the early stage, the improved Elman NN with ISOMAP and LLE can effectively avoid disturbance which is from noise. The improved Elman NN prediction is smooth without large fluctuations and shows better robustness than SVM prediction and improved Elman NN prediction.

5. Conclusion

This paper develops a soft sensor model based on improved Elman NN and proposes a data preprocessing method based on ISOMAP and LLE for input variable data selection. The improved Elman NN introduces local feedback and feedforward networks into the Elman NN, which is trained to obtain better generalization and avoid overfitting to a certain extent. The data preprocessing method based on ISOMAP and LLE effectively keeps the neighborhood structure and the global mutual distance of the datasets and better reflects the real character of the original datasets, which is used to remove the abnormal working condition data, noises, and data redundancy in the input variables from improved Elman NN. The soft sensor model based on improved Elman NN with data preprocessing by ISOMAP and LLE is applied to estimate the temperature in the rotary kiln calcining process. The prediction values of the soft sensor model based on improved Elman NN with data preprocessing through ISOMAP and LLE follow the varying trend of the temperature very effectively and have excellent robustness and generalization performance. The great performance illustrates that the soft sensor model based on improved Elman NN with data preprocessing through ISOMAP and LLE provides a powerful and promising method for complex industrial process applications.

Conflicts of Interest

The authors declare that there is no conflict of interests regarding the publication of this paper.

Acknowledgments

This work was supported in part by the Project by National Natural Science Foundation of China under Grant (61473054).

References

[1] W. Yan, D. Tang, and Y. Lin, "A data-driven soft sensor modeling method based on deep learning and its application," *IEEE Transactions on Industrial Electronics*, vol. 64, no. 5, pp. 4237–4245, 2017.

[2] D. W. Clarke and C. Mohtadi, "Properties of generalized predictive control," *Automatica*, vol. 25, no. 6, pp. 859–875, 1989.

[3] P. Albertos and R. Ortega, "On generalized predictive control: two alternative formulations," *Automatica*, vol. 25, no. 5, pp. 753–755, 1989.

[4] C. R. Cutler and B. L. Ramaker, "Dynamic matrix control - computer control algorithm," *Joint Automatic Control Conference*, vol. 17, p. 72, 1980.

[5] J. Richalet, A. Rault, J. L. Testud, and J. Papon, "Model predictive heuristic control: applications to industrial processes," *Automatica*, vol. 14, no. 5, pp. 413–428, 1978.

[6] D. I. Adelani and M. K. Traoré, "Enhancing the reusability and interoperability of artificial neural networks with DEVS modeling and simulation," *International Journal of Modeling, Simulation, and Scientific Computing*, vol. 7, no. 2, article 1650005, 2016.

[7] A. Afram, F. Janabi-Sharifi, A. S. Fung, and K. Raahemifar, "Artificial neural network (ANN) based model predictive control (MPC) and optimization of HVAC systems: a state of the art review and case study of a residential HVAC system," *Energy and Buildings*, vol. 141, pp. 96–113, 2017.

[8] P. Chopra, R. K. Sharma, and M. Kumar, "Prediction of compressive strength of concrete using artificial neural network and genetic programming," *Advances in Materials Science and Engineering*, vol. 2016, Article ID 7648467, 10 pages, 2016.

[9] Q. Cheng, Z. Qi, G. Zhang, Y. Zhao, B. Sun, and P. Gu, "Robust modelling and prediction of thermally induced positional error based on grey rough set theory and neural networks," *International Journal of Advanced Manufacturing Technology*, vol. 83, no. 5–8, pp. 753–764, 2016.

[10] J. Zheng and Z. Song, "Semisupervised learning for probabilistic partial least squares regression model and soft sensor application," *Journal of Process Control*, vol. 64, pp. 123–131, 2018.

[11] K. Y. Bae, H. S. Jang, and D. K. Sung, "Hourly solar irradiance prediction based on support vector machine and its error analysis," *IEEE Transactions on Power Systems*, vol. 32, no. 2, pp. 935–945, 2017.

[12] Z. Ju and H. Gu, "Predicting pupylation sites in prokaryotic proteins using semi-supervised self-training support vector machine algorithm," *Analytical Biochemistry*, vol. 507, pp. 1–6, 2016.

[13] L. Zago, P. Y. Hervé, R. Genuer et al., "Predicting hemispheric dominance for language production in healthy individuals using support vector machine," *Human Brain Mapping*, vol. 38, no. 12, pp. 5871–5889, 2017.

[14] R. Grbić, D. Slišković, and P. Kadlec, "Adaptive soft sensor for online prediction and process monitoring based on a mixture of Gaussian process models," *Computers & Chemical Engineering*, vol. 58, no. 22, pp. 84–97, 2013.

[15] M. Zhang and X. Liu, "A soft sensor based on adaptive fuzzy neural network and support vector regression for industrial melt index prediction," *Chemometrics and Intelligent Laboratory Systems*, vol. 126, no. 126, pp. 83–90, 2013.

[16] M. Ławryńczuk, "Computationally efficient model predictive control algorithms," *Studies in Systems, Decision and Control*, vol. 3, no. 1, pp. 137–149, 2014.

[17] J. Sun, Z. Cheng, R. Yang, and S. Shan, "Soft-sensor modeling for paper mill effluent COD based on PCA-PSO-LSSVM," *Computers & Applied Chemistry*, vol. 34, no. 9, pp. 706–710, 2017.

[18] A. Najafi, A. Joudaki, and E. Fatemizadeh, "Nonlinear dimensionality reduction via path-based isometric mapping," *IEEE Transactions on Pattern Analysis and Machine Intelligence*, vol. 38, no. 7, pp. 1452–1464, 2016.

[19] J. Wang, "Laplacian Eigenmaps," in *Geometric Structure of High-Dimensional Data and Dimensionality Reduction*, pp. 235–247, Springer, 2012.

[20] M. Belkin and P. Niyogi, "Laplacian eigenmaps for dimensionality reduction and data representation," *Neural Computation*, vol. 15, no. 6, pp. 1373–1396, 2014.

[21] Y. Huang, L. Jin, H. S. Zhao, and X. Y. Huang, "Fuzzy neural network and LLE algorithm for forecasting precipitation in tropical cyclones: comparisons with interpolation method by ECMWF and stepwise regression method," *Natural Hazards*, vol. 91, no. 1, pp. 201–220, 2018.

[22] X. Wang, S. Chen, and M. Yao, "Data dimensionality reduction method of semi-supervised isometric mapping based on regularization," *Journal of Electronics & Information Technology*, vol. 33, no. 1, pp. 241–245, 2016.

[23] G. Kanishka and T. I. Eldho, "Watershed classification using Isomap technique and hydrometeorological attributes," *Journal of Hydrologic Engineering*, vol. 22, no. 10, article 04017040, 2017.

[24] G. Ren, Y. Cao, S. Wen, T. Huang, and Z. Zeng, "A modified Elman neural network with a new learning rate scheme," *Neurocomputing*, vol. 286, pp. 11–18, 2018.

[25] K. Kolanowski, A. Świetlicka, R. Kapela, J. Pochmara, and A. Rybarczyk, "Multisensor data fusion using Elman neural networks," *Applied Mathematics and Computation*, vol. 319, pp. 236–244, 2018.

[26] Y. Ge, Y. Huang, D. Hao, G. Li, and H. Li, "An indicated torque estimation method based on the Elman neural network for a turbocharged diesel engine," *Proceedings of the Institution of Mechanical Engineers Part D Journal of Automobile Engineering*, vol. 230, no. 10, pp. 1299–1313, 2016.

[27] A. Wysocki and M. Ławryńczuk, "Elman neural network for modeling and predictive control of delayed dynamic systems," *Archives of Control Sciences*, vol. 26, no. 1, pp. 117–142, 2016.

[28] H. Choi and S. Choi, "Robust kernel Isomap," *Pattern Recognition*, vol. 40, no. 3, pp. 853–862, 2007.

[29] Y. Zhang, Y. Fu, Z. Wang, and L. Feng, "Fault detection based on modified kernel semi-supervised locally linear embedding," *IEEE Access*, vol. 6, pp. 479–487, 2018.

[30] B. Jiang, C. Ding, and B. Luo, "Robust data representation using locally linear embedding guided PCA," *Neurocomputing*, vol. 275, pp. 523–532, 2018.

[31] Y. Qianli and Q. Liyun, "A diagnosis method of rotary kiln lining based on extreme learning machine (ELM) and its application," *Sintering and Pelletizing*, vol. 40, no. 6, pp. 18–22, 2015.

[32] G. Zhang and J. Zhang, "A novel RS-SVM dynamic prediction approach and its application," *Process Automation Instrumentation*, vol. 26, no. 2, pp. 17–20, 2005.

An Indoor Navigation System based on Stereo Camera and Inertial Sensors with Points and Lines

Bo Yang, Xiaosu Xu, Tao Zhang, Yao Li, and Jinwu Tong

Key Laboratory of Micro-Inertial Instrument and Advanced Navigation Technology, Ministry of Education, School of Instrument Science and Engineering, Southeast University, Nanjing 210096, China

Correspondence should be addressed to Xiaosu Xu; xxs@seu.edu.cn

Academic Editor: Frederick Mailly

An indoor navigation system based on stereo camera and inertial sensors with points and lines is proposed to further improve the accuracy and robustness of the navigation system in complex indoor environments. The point and line features, which are fast extracted by ORB method and line segment detector (LSD) method, are both employed in this system to improve its ability to adapt to complex environments. In addition, two different representations of lines are adopted to improve the efficiency of the system. Besides stereo camera, an inertial measurement unit (IMU) is also used in the system to further improve its accuracy and robustness. An estimator is designed to integrate the camera and IMU measurements in a tightly coupled approach. The experimental results show that the performance of the proposed navigation system is better than the point-only VINS and the vision-only navigation system with points and lines.

1. Introduction

Indoor navigation technique, which has been widely applied in the field of mobile robot (e.g., home service robot) or unmanned aerial vehicle (UAV) system [1, 2], has received considerable attention in the past few years. One major challenge of the indoor navigation system is the unavailability of the global position system (GPS) signal in indoor environment. Therefore, many other sensors have been applied in the system such as sonar [3], odometry [4], light detection and ranging (LiDAR) [2], camera [5], and inertial measurement unit (IMU) [6]. With the recent development in vision-based techniques, cameras used as sensors make the vision-based navigation system more and more attractive [7]. Direct method and feature-based method are two main methods for indoor navigation system. While the direct method estimates the motion by minimizing the photometric error [8], we deal with feature-based method in this paper. For conventional feature-based visual navigation system, feature extraction methods are applied to extract the point features from images collected by the camera. After matching these point features between two frames, the pose of the system can be estimated through ego-motion estimation methods (e.g., perspective-n-point (PnP)) or by minimizing the reprojection error [9, 10]. Note that only point features are used in the conventional feature-based visual navigation system. However, in some low-textured scenarios exemplified by man-made or indoor environments, it is hard to extract enough key points by suitable point feature extraction method. As a result, the performance of conventional feature-based method will decrease, which thus implies that other complementary features need to be explored for further performance improvement. In recent years, line feature has received more and more attention since lines or line segments provide significantly more information than points in encoding the structure of the surrounding environment [11]. They are usually abundant in human-made scenarios, which are characterized by regular structures rich in edges and linear shapes [12]. In addition, recent advances in the line segment detection method have made it possible to extract line features fast and accurately. Therefore, in some low-textured environments, the line features can be used in the feature-based visual navigation system to improve the navigation accuracy [11, 13]. In this paper, we choose the stereo camera to acquire images of indoor environments; thereafter, points and lines features are all extracted from these images. In addition, the scale information of the features can be estimated by stereo camera because of the small scale of the indoor environments.

Although the camera can be used efficiently in navigation system, it is not robust to motion blur induced by rapid motion [14] and the vision-based navigation system often fails in rapid motion scenarios. In order to make the system more robust, a common solution is fusing camera with inertial sensors, which leads to the visual-inertial navigation system (VINS) [6, 15]. Inertial sensors usually contain three orthogonal gyroscopes and accelerometers to measure the angular velocity and acceleration of the carrier and estimate the carrier's motion in high frequency. However, it suffers from the accumulated error and relatively large measurement uncertainty at slow motion [16]. It can be found that visual and inertial measurements offer complementary properties which make them particularly suitable for fusion [17], since inertial sensors have a relatively low uncertainty at high velocity, whereas cameras can track features very accurately at low velocity and less accurately with increasing velocity [16]. Consequently, VINS works better than vision-only or pure inertial navigation system. Without loss of generality, the VINS can be classified into two methods, filtering-based method and optimization-based method. Although the accuracy of optimization approach is better than the filtering approach, it is computationally expensive. Considering the extra computational cost in line feature extraction, we adopt the filtering approach which usually uses the extended Kalman filter (EKF) to estimate the pose of the system.

In this paper, we present a visual-inertial navigation system with point and line features for indoor environments. The stereo camera is combined with an inertial sensor in a tightly coupled filtering approach to overcome the defect in single sensor. The point and line features are both used in the proposed system, such that this system can perform well in low-textured environments such as indoor environments. In addition, the robustness of this navigation system can be improved by these two schemes. The performance of this system is also tested in an experiment which uses visual-inertial benchmark dataset.

The rest of the paper is organized as follows. The related work is described simply in Section 2. In Section 3, we present the IMU model based on the inertial measurements and the representations of point and line features. The estimator and the implementation of the algorithm are described in Section 4. Experiments are conducted in Section 5 to demonstrate the performance of the proposed system. Finally, conclusions are drawn in Section 6.

2. Related Work

Compared to the simultaneous localization and mapping (SLAM) system, the visual navigation system does not have loop closures, but in a large part, it follows the SLAM paradigm (especially the front-end of the SLAM). For feature-based method, pioneering work has been carried out in [18]. It presents a monocular camera SLAM system which uses sparse key points and EKF to estimate the camera motion. In [19], it points out that the monocular camera suffers from the "scale drift" problem; however, the stereo camera can efficiently overcome this issue. In [20], it presents a stereo visual SLAM system which uses Harris corner detector to extract point features and EKF to estimate the pose. More recently, the ORB-SLAM has been proposed in [21, 22], which supports all monocular, stereo, and RGB-D cameras. In ORB-SLAM, the ORB point features [23] have been used for the first time and the accuracy of the system is very high. However, the systems mentioned above would reduce their performance in low-textured environment with insufficient point features. Therefore, massive efforts are devoted to the line segment detection method and its application in visual navigation and SLAM system.

A linear-time line segment detector has been proposed in [24, 25] with a high detection accuracy. In [13], a line-based monocular 2D SLAM system has been also presented, which incorporates vertical and horizontal lines to estimate the motion with EKF. In addition, another line-based system has been proposed in [11] to improve the efficiency of the system by adopting two different representations of a line segment. However, the line feature only lends itself to the structured environment, which implies that the system performance is likely to be degraded in complex environment by employing the single line feature. Taking into account the complementarity between point and line features, a combination of both features has beneficial effects on the robustness and accuracy of the navigation system. Recently, visual navigation systems with points and line segments have been proposed in many works [26–28], which reveals superior performance in a wide variety of real-time scenarios. However, the visual-only system has its own limitation and the VINS is developed to overcome it.

The filtering-based VINS is particularly relevant to our work. A vision-aided inertial navigation system based on a multistate constraint Kalman filter has been proposed in [6, 15], presenting a novel system model which does not contain the feature position in state vector. Therefore, the computational overhead of the method is low. However, this system relies on large storage capacity, which makes it unavailable in some applications. A consistent VINS based on EKF has been proposed in [29]. In this paper, the author proposes a new method which improves the consistency of the system based on analyzing the observability of the VINS. An autonomous navigation system combines the stereo camera, and IMU has been proposed in [30], which improves the positioning accuracy by considering both the near and the far point features. However, all these methods mentioned above use point features only. Line-based VINS has been proposed in [31, 32], which use straight lines as features. These systems exhibit superior reconstruction performance against point-based navigation system in line-rich environment. However, in our work, we mainly focus on the VINS combining both point and line features.

3. IMU Model and Feature Representation

In this section, we will describe the IMU model and the camera model which contain the point and line features. To begin with, we define the reference coordinate frames which would be applied in the rest of the paper.

The navigation coordinate frame $\{N\}$ (also can be referred as the world coordinate frame) is chosen as the East-North-Up (ENU) coordinate frame in this paper (the origin of this coordinate frame is fixed in the first frame of the system, and the x-, y-, and z-axes are aligned with the east, north, and up of the system's first frame, resp.). The pose information of the system is all characterized with respect to in this coordinate frame. In addition, the navigation coordinate frame is fixed with the first frame of the system such that the effect of the earth rotation can be ignored (cf. (1) and (2)).

The IMU coordinate frame $\{I\}$ and the camera coordinate frame $\{C\}$ are fixed with the IMU and the stereo camera, respectively. Their origins are located at the center of the IMU and the optical center of the camera. The x-, y-, and z-axes of the IMU coordinate frame are aligned with the right, front, and up of the IMU, respectively. The z-axis of camera coordinate frame points aligns with the optical axis. There are two camera coordinate frames, which are denoted in terms of left camera coordinate frame C_{L} and the right camera coordinate frame C_{R}. The transition matrix between them is obtained though camera calibration, and the C_{L} is chosen as the reference camera coordinate frame $\{C\}$.

In addition, for the sake of simplicity, in Sections 3 and 4, we assume that the IMU coordinate frame $\{I\}$ and the camera coordinate frame $\{C\}$ (left camera coordinate frame) are coincident (in practice, the transition matrix between these two coordinate frames can be extrinsically calibrated following the method proposed in [33]). To avoid confusion, in Sections 3 and 4, we use the other notation $\{B\}$ (body coordinate frame) to represent both coordinate frames (in practice, one of the $\{I\}$ and $\{C\}$ can be selected as the body coordinate frame).

3.1. IMU Model. The angular velocity $\omega_{\mathrm{m}}^{\mathrm{b}}$ and the specific force $f_{\mathrm{m}}^{\mathrm{b}}$ of the system in the body coordinate frame can be measured by the IMU. These measurements include the angular velocity and the acceleration information with noise [34]:

$$\omega_{\mathrm{m}}^{\mathrm{b}} = \omega^{\mathrm{b}} + \varepsilon^{\mathrm{b}} + w_{\mathrm{g}}^{\mathrm{b}}, \tag{1}$$

$$f_{\mathrm{m}}^{\mathrm{b}} = \mathbf{C}_{\mathrm{n}}^{\mathrm{b}}(a^{\mathrm{n}} - \mathbf{g}) + \nabla^{\mathrm{b}} + w_{\mathrm{a}}^{\mathrm{b}}, \tag{2}$$

where ω^{b} denotes the true angular velocity in body coordinate frame, a^{n} denotes the true acceleration in navigation coordinate frame, ε^{b} and ∇^{b} are the constant drifts of the gyroscopes and the accelerometers in body coordinate frame, respectively, $w_{\mathrm{g}}^{\mathrm{b}}$ and $w_{\mathrm{a}}^{\mathrm{b}}$ are the random noises of the gyroscopes and the accelerometers in body coordinate frame, which can be modeled as uncorrelated zero-mean white Gaussian noise, $\mathbf{C}_{\mathrm{n}}^{\mathrm{b}}$ denotes the rotation matrix from the navigation coordinate frame to the body coordinate frame, and $\mathbf{g}$ is the gravity vector.

The estimates of the attitude information $\widehat{Q}$ (described in the form of unit quaternion), the velocity $\widehat{v}^{\mathrm{n}}$, and the position $\widehat{p}^{\mathrm{n}}$ can be updated by the IMU measurements as follows [34]:

$$\begin{aligned} \dot{\widehat{Q}} &= \frac{1}{2} M'\left(\omega_{\mathrm{m}}^{\mathrm{b}} - \widehat{\varepsilon}^{\mathrm{b}}\right)\widehat{Q}, \\ \dot{\widehat{v}}^{\mathrm{n}} &= \widehat{\mathbf{C}}_{\mathrm{b}}^{\mathrm{n}}\left(f_{\mathrm{m}}^{\mathrm{b}} - \widehat{\nabla}^{\mathrm{b}}\right) + \mathbf{g}, \\ \dot{\widehat{p}}^{\mathrm{n}} &= \widehat{v}^{\mathrm{n}}, \end{aligned} \tag{3}$$

where $\widehat{\mathbf{C}}_{\mathrm{b}}^{n}$ denotes the estimated rotation matrix from the body coordinate frame to the navigation coordinate frame, $\widehat{\varepsilon}^{\mathrm{b}}$ and $\widehat{\nabla}^{\mathrm{b}}$ are the estimates of the constant drifts of the gyroscopes and the accelerometers, which can be obtained by the calibration method [35], and

$$M'(\omega) = \begin{bmatrix} 0 & -\omega_x & -\omega_y & -\omega_z \\ \omega_x & 0 & \omega_z & -\omega_y \\ \omega_y & -\omega_z & 0 & \omega_x \\ \omega_z & \omega_y & -\omega_x & 0 \end{bmatrix}. \tag{4}$$

In addition, the unit quaternion Q, the attitudes θ, and the rotation matrixes $\mathbf{C}_{\mathrm{b}}^{\mathrm{n}}$ and $\mathbf{C}_{\mathrm{n}}^{\mathrm{b}}$ can be interchangeable [34].

In addition, ε^{b} and ∇^{b} can be modeled as follows:

$$\begin{aligned} \dot{\varepsilon}^{\mathrm{b}} &= 0, \\ \dot{\nabla}^{\mathrm{b}} &= 0. \end{aligned} \tag{5}$$

According to the analysis above, the IMU state vector can be described as follows:

$$\mathbf{X}_{\mathrm{I}} = \begin{bmatrix} Q^T & v^{\mathrm{n}T} & p^{\mathrm{n}T} & \varepsilon^{\mathrm{b}T} & \nabla^{\mathrm{b}T} \end{bmatrix}^T. \tag{6}$$

In practice, the angular velocity $\omega_{\mathrm{m}}^{\mathrm{b}}$ and the specific force $f_{\mathrm{m}}^{\mathrm{b}}$ are sampled by IMU at discrete times, so (3) should be calculated in a discrete method. Therefore, we assume that the $\omega_{\mathrm{m}}^{\mathrm{b}}$ and $f_{\mathrm{m}}^{\mathrm{b}}$ are constant between the two sampling times; thereafter, (3) can be discretized in an easy way.

3.2. Feature Representation. The point and line features are extracted from the images collected by the stereo camera. In this subsection, we will describe the representations of the point and line features.

3.2.1. Point Features. Owing to small scale of the indoor environments, the depth of the most points in images can be estimated by stereo camera through the baseline between the left and right cameras, so point features can be coded with a Cartesian coordinate frame as follows:

$$P_{\mathrm{p}}^{\mathrm{b}} = [X \quad Y \quad Z]^T, \tag{7}$$

where $P_{\mathrm{p}}^{\mathrm{b}}$ denotes the position of point features in body coordinate frame and it can be estimated from the pixel coordinate value of the point features as follows:

$$\widehat{P}_{\mathrm{p}}^{\mathrm{b}} = \left[\frac{b(u_{\mathrm{L}} - u_0)}{d} \quad \frac{b(v_{\mathrm{L}} - v_0)}{d} \quad \frac{bf}{d}\right]^T, \tag{8}$$

where b is the baseline between the left camera and the right camera, $d = u_{\mathrm{L}} - u_{\mathrm{R}}$ is the disparity, u_{L} and v_{L} are the pixel coordinate value of the point features in the left camera image, u_0 and v_0 are the pixel coordinate value of the optical center in the left camera image, and f is the focal length. In addition, u_0, v_0, and f are the intrinsic parameters of the camera which can be obtained by the camera calibration method [36].

The position of point features in navigation coordinate frame $P_{\mathrm{p}}^{\mathrm{n}}$ is also important in the proposed system, and the estimated position can be computed as follows:

$$\widehat{P}_{\mathrm{p}}^{\mathrm{n}} = \widehat{\mathbf{C}}_{\mathrm{b}}^{\mathrm{n}}\widehat{P}_{\mathrm{p}}^{\mathrm{b}} + \widehat{p}^{\mathrm{n}}. \tag{9}$$

This process is only executed when the point feature is detected in the first time. In this section, for the sake of simplicity, the rotation matrix between IMU coordinate frame and camera coordinate frame $C_{\mathrm{I}}^{\mathrm{C}}$, $C_{\mathrm{C}}^{\mathrm{I}}$ and the relative position between these two coordinate frames are ignored, but in practice, they should be considered.

According to the analysis above, the point feature state vector can be described as follows:

$$\mathbf{X}_{\mathrm{p}} = \left[P_{\mathrm{p},1}^{\mathrm{n}} \quad P_{\mathrm{p},2}^{\mathrm{n}} \quad \cdots \quad P_{\mathrm{p},M}^{\mathrm{n}}\right]^T, \tag{10}$$

where $P_{\mathrm{p},i}^{\mathrm{n}}$ denotes the position of the ith point features in navigation coordinate frame and M is the number of the point features and this number is a variable because of the different images collected by the camera.

3.2.2. Line Features. Different from the points represented by $(X, Y, Z)^T$ in three-dimensional spaces, lines are parameterized in 4 DOFs. In this paper, we introduce two different representations of lines, the Plücker coordinate frame representation and the orthonormal representation [11, 37]. The Plücker coordinate frame representation is initialized when the line features are detected in the first time, and the orthonormal representation is used in the estimator.

The Plücker coordinate frame representation is determined by two points on the line. Assuming the homogeneous coordinate frames of these two points are $P_{X1} = (x_1, y_1, z_1, w_1)^T$ and $P_{X2} = (x_2, y_2, z_2, w_2)^T$, the Plücker matrix $\mathbf{L}$ can be determined as follows:

$$\mathbf{L} = P_{X2}P_{X1}^T - P_{X1}P_{X2}^T. \tag{11}$$

The Plücker matrix $\mathbf{L}$ is a 4×4 antisymmetric matrix in which the diagonal elements are zero and there are six nonzero elements in this matrix. The Plücker coordinate frame is a 6×1 vector composed by these six nonzero elements as below:

$$\mathscr{L} = \begin{bmatrix} \tilde{P}_{X1} \times \tilde{P}_{X2} \\ w_1\tilde{P}_{X2} - w_2\tilde{P}_{X1} \end{bmatrix} = \begin{bmatrix} n \\ v \end{bmatrix}, \tag{12}$$

where $\tilde{P}_{X1}$ and $\tilde{P}_{X2}$ are the 3×1 inhomogeneous coordinate frames of the two points. n is orthogonal to the interpretation plane containing the line and the origin, and v is the direction of the line. The relationship between the Plücker matrix $\mathbf{L}$ and the Plücker coordinate frame $\mathscr{L}$ is expressed as

$$\mathbf{L} = \begin{bmatrix} \lfloor n \times \rfloor & v \\ -v^T & 0 \end{bmatrix}, \tag{13}$$

where $\lfloor \cdot \times \rfloor$ denotes the skew-symmetric cross-product matrix with a vector and the $\lfloor n \times \rfloor$ is defined as

$$\lfloor n \times \rfloor = \begin{bmatrix} 0 & -n_3 & n_2 \\ n_3 & 0 & -n_1 \\ -n_2 & n_1 & 0 \end{bmatrix}. \tag{14}$$

The transformation of the Plücker coordinate frame in different reference coordinate frames (e.g., navigation coordinate frame and body coordinate frame) can be described as follows [38]:

$$\begin{gathered} \begin{bmatrix} n^{\mathrm{b}} \\ v^{\mathrm{b}} \end{bmatrix} = \mathscr{H}_{\mathrm{n}}^{\mathrm{b}} \begin{bmatrix} n^{\mathrm{n}} \\ v^{\mathrm{n}} \end{bmatrix}, \\ \mathscr{H}_{\mathrm{n}}^{\mathrm{b}} = \begin{bmatrix} \mathbf{C}_{\mathrm{n}}^{\mathrm{b}} & \lfloor \mathbf{t}_{\mathrm{n}}^{\mathrm{b}} \times \rfloor \mathbf{C}_{\mathrm{n}}^{\mathrm{b}} \\ 0 & \mathbf{C}_{\mathrm{n}}^{\mathrm{b}} \end{bmatrix}, \end{gathered} \tag{15}$$

$$\begin{gathered} \begin{bmatrix} n^{\mathrm{n}} \\ v^{\mathrm{n}} \end{bmatrix} = \mathscr{H}_{\mathrm{b}}^{\mathrm{n}} \begin{bmatrix} n^{\mathrm{b}} \\ v^{\mathrm{b}} \end{bmatrix}, \\ \mathscr{H}_{\mathrm{b}}^{\mathrm{n}} = \begin{bmatrix} \mathbf{C}_{\mathrm{b}}^{\mathrm{n}} & \lfloor \mathbf{t}_{\mathrm{b}}^{\mathrm{n}} \times \rfloor \mathbf{C}_{\mathrm{b}}^{\mathrm{n}} \\ 0 & \mathbf{C}_{\mathrm{b}}^{\mathrm{n}} \end{bmatrix}, \end{gathered} \tag{16}$$

where $\mathbf{t}_{\mathrm{n}}^{\mathrm{b}}$ denotes the translation vector from navigation coordinate frame to the body coordinate frame.

The 3D lines represented by Plücker coordinate frames in body coordinate frame can also be projected onto the image plane as follows:

$$\begin{gathered} l^{\mathrm{b}} = \mathscr{K} n^{\mathrm{b}}, \\ \mathscr{K} = \begin{bmatrix} f_v & 0 & 0 \\ 0 & f_u & 0 \\ -f_v c_u & f_u c_v & f_u f_v \end{bmatrix}, \end{gathered} \tag{17}$$

where l^{b} is the 2D line in the image plane represented by pixel coordinate frame. It can be found that only n^{b} is used in this projection process.

According to the analysis above, six elements are used in the Plücker coordinate frame representation while the line's DOFs are four. If using this representation in estimator, superfluous elements will cost more computation. In addition, the orthogonal constraint between the n and v, that is, $n^T v = 0$, will decrease the numerical stability, and thus, we need to use another representation method in the estimator.

The orthonormal representation uses minimum 4 parameters to represent a line, and this representation can be derived from the Plücker coordinate frame representation [37].

We define a matrix $\mathbf{S} = [n \mid v]$ and decompose this matrix by QR decomposition as

$$\mathrm{QR}(\mathbf{S}) = \mathbf{U}\begin{bmatrix} w_1 & 0 \\ 0 & w_2 \\ 0 & 0 \end{bmatrix} \tag{18}$$

and set

$$\mathbf{W} = \begin{bmatrix} w_1 & -w_2 \\ w_2 & w_1 \end{bmatrix}. \tag{19}$$

The 3D line can be represented by $(\mathbf{U}, \mathbf{W})$ where $\mathbf{U} \in SO(3), \mathbf{W} \in SO(2)$. This means the $\mathbf{U}$ and $\mathbf{W}$ are three- and two-dimensional rotation matrices, respectively. Thus, the matrix $\mathbf{U}$ can be updated accordingly by a vector containing 3 parameters such as Euler angles $\theta_l = \begin{bmatrix} \theta_{l,1} & \theta_{l,2} & \theta_{l,3} \end{bmatrix}^T$ and $\mathbf{W}$ can be updated by a scalar parameter $\theta_l \in (0, \pi/2)$ as follows:

$$\begin{aligned} \mathbf{U}^* &= \mathbf{U}\mathbf{R}(\theta_l), \\ \mathbf{R}(\theta_l) &= \mathbf{R}_x(\theta_{l,1}), \mathbf{R}_y(\theta_{l,2}), \mathbf{R}_z(\theta_{l,3}), \\ \mathbf{W}^* &= \begin{bmatrix} \cos\theta_l & -\sin\theta_l \\ \sin\theta_l & \cos\theta_l \end{bmatrix}\mathbf{W}, \end{aligned} \tag{20}$$

where $\mathbf{R}_x(\theta_1), \mathbf{R}_y(\theta_2)$, and $\mathbf{R}_z(\theta_3)$ denote the three-dimensional rotation matrices.

We can define a 4×1 increment vector $\boldsymbol{\sigma} = [\theta_l^T, \theta_l]^T$ which can be used in the estimator presented in the next section to represent the error of line features. The detailed discussions will be provided in Section 4.

In addition, the Plücker coordinate frame can be transferred from the orthonormal representation as follows:

$$\mathscr{L} = \begin{bmatrix} w_1 u_1 \\ w_2 u_2 \end{bmatrix}, \tag{21}$$

where u_i denotes the ith column of $\mathbf{U}$.

4. Estimator Establishment and Algorithm Implementation

In this section, an EKF estimator is presented. The IMU measurements along with the point and line features mentioned in Section 3 are combined based on this estimator in a tightly coupled approach. In addition, we will elaborate the feature detection methods, the feature initialization, and some other processing steps.

4.1. Error-State Equation. In our work, the EKF estimator uses the error model, which is defined as follows:

$$\mathbf{X}^T = \begin{bmatrix} \boldsymbol{\delta}\mathbf{X}_\mathrm{I}^T & \boldsymbol{\delta}\mathbf{X}_\mathrm{p}^T & \boldsymbol{\sigma}_\mathrm{l}^T \end{bmatrix}, \tag{22}$$

where $\boldsymbol{\delta}\mathbf{X}_\mathrm{I}^T$, $\boldsymbol{\delta}\mathbf{X}_\mathrm{p}^T$, and $\boldsymbol{\sigma}_\mathrm{l}^T$ are the error-state vectors of IMU, point features, and line features, respectively.

The IMU error-state vector can be defined as

$$\boldsymbol{\delta}\mathbf{X}_\mathrm{I}^T = \begin{bmatrix} \delta\theta_\mathrm{I}^T & \delta v^{\mathrm{n}T} & \delta p^{\mathrm{n}T} & \delta\varepsilon^{\mathrm{b}T} & \delta\nabla^{\mathrm{b}T} \end{bmatrix}, \tag{23}$$

where δv^n, δp^n, $\delta\varepsilon^\mathrm{b}$, and $\delta\nabla^\mathrm{b}$ are the velocity error, the position error, the error of gyroscope constant drift, and the error of accelerometer constant drift, respectively. The aforementioned error is formulated as $\delta x = x - \hat{x}$ where x is the true value and $\hat{x}$ is the estimated value. $\delta\theta_\mathrm{I}$ is the attitude error computed from the error quaternion δQ which is defined as

$$\delta Q \approx \begin{bmatrix} 1 \frac{\theta_\mathrm{I}^T}{1} \end{bmatrix}^T. \tag{24}$$

The relationship between the true quaternion Q and the estimated quaternion $\hat{Q}$ is defined as $Q = \hat{Q} \otimes \delta Q$, where $\otimes$ is the quaternion multiplication [34].

Similarly, the error-state vector of point features can be defined as

$$\boldsymbol{\delta}\mathbf{X}_\mathrm{p}^T = \begin{bmatrix} \boldsymbol{\delta}\mathbf{P}_{\mathrm{p},1}^\mathrm{n} & \boldsymbol{\delta}\mathbf{P}_{\mathrm{p},2}^\mathrm{n} & \cdots & \boldsymbol{\delta}\mathbf{P}_{\mathrm{p},M}^\mathrm{n} \end{bmatrix}, \tag{25}$$

where $\delta P_{\mathrm{p},i}^\mathrm{n}$ is the position error of the ith point features in navigation coordinate frame which is defined as $\delta P_{\mathrm{p},i}^\mathrm{n} = P_{\mathrm{p},i}^\mathrm{n} - \hat{P}_{\mathrm{p},i}^\mathrm{n}$. In addition, $\boldsymbol{\delta}\dot{\mathbf{X}}_\mathrm{p} = 0$.

The error-state vector of the line features can be defined as follows:

$$\boldsymbol{\sigma}_\mathrm{l}^T = \begin{bmatrix} \boldsymbol{\sigma}_1^{\mathrm{n}T} & \boldsymbol{\sigma}_2^{\mathrm{n}T} & \cdots & \boldsymbol{\sigma}_N^{\mathrm{n}T} \end{bmatrix}, \tag{26}$$

where $\boldsymbol{\sigma}_j^\mathrm{n}$ denotes the error vector of the jth line features in navigation coordinate frame and N is the number of the line features. It is worth to mention that N is a variable because of the different images collected by the camera. In addition, $\dot{\boldsymbol{\sigma}}_\mathrm{l} = 0$.

According to the analysis above, the error-state equation can be established as follows:

$$\begin{bmatrix} \delta\dot{\mathbf{X}}_{\mathrm{I}} \\ \delta\dot{\mathbf{X}}_{\mathrm{P}} \\ \dot{\sigma}_{\mathrm{l}} \end{bmatrix} = F \begin{bmatrix} \delta\mathbf{X}_{\mathrm{I}} \\ \delta\mathbf{X}_{\mathrm{P}} \\ \sigma_{\mathrm{l}} \end{bmatrix} + w,$$

$$F = \begin{bmatrix} F_{\mathrm{I}} & 0_{15\times 3M} & 0_{15\times 4N} \\ 0_{3M\times 15} & 0_{3M\times 3M} & 0_{3M\times 4N} \\ 0_{4N\times 15} & 0_{4N\times 3M} & 0_{4N\times 4N} \end{bmatrix}, \quad (27)$$

$$w = \begin{bmatrix} w_{\mathrm{I}} \\ 0_{3M\times 4N} \\ 0_{3M\times 4N} \end{bmatrix},$$

where w_{I} is the process noise assumed to be uncorrelated white zero-mean Gaussian noise the covariance matrix of which is Q_w and

$$F_{\mathrm{I}} = \begin{bmatrix} -\left\lfloor \left(\omega_{\mathrm{m}}^{\mathrm{b}} - \varepsilon^{\mathrm{b}}\right) \times \right\rfloor & 0_{3\times 3} & 0_{3\times 3} & -I_3 & 0_{3\times 3} \\ -\mathbf{C}_{\mathrm{b}}^{\mathrm{n}} \left\lfloor \left(f_{\mathrm{m}}^{\mathrm{b}} - \nabla^{\mathrm{b}}\right) \times \right\rfloor & 0_{3\times 3} & 0_{3\times 3} & 0_{3\times 3} & -\mathbf{C}_{\mathrm{b}}^{\mathrm{n}} \\ 0_{3\times 3} & I_3 & 0_{3\times 3} & 0_{3\times 3} & 0_{3\times 3} \\ 0_{6\times 3} & 0_{6\times 3} & 0_{6\times 3} & 0_{6\times 3} & 0_{6\times 3} \end{bmatrix}. \quad (28)$$

Owing to the estimator is executed in the discrete time, and the differential equation needs to be transformed into the discrete equation as follows:

$$\mathbf{X}_k = \Phi_{k,k-1}\mathbf{X}_{k-1} + w_{k-1}, \qquad \Phi_{k,k-1} \approx I + \mathrm{FT}, \quad (29)$$

where T is the sampling period and k denotes the kth frame.

In addition, in any environment, the stereo camera is used to capture a series of fixed images. According to the point and line feature extraction methods, for a series of fixed images, the extracted points and lines are also fixed and we assume that the matched features are all true positive, and thus, there is no random noise for these features. Actually, they only contain the measurement noise the differential of which equal to zero. Therefore, in (27), there is no noise of point and line features.

4.2. Measurement Equation. The features' reprojection errors in the image plane are selected as the measurement. For the sake of illustration, we present the reprojection errors by considering the case where a single point feature and a single line feature are present.

We first define $P_{\mathrm{p},i}^{\mathrm{n}} = [X_i \quad Y_i \quad Z_i]^T$ as the ith point feature which can be observed by the camera in the kth frame. The reprojection error of the point feature $z_{\mathrm{p},i}$ can be defined as follows:

$$\widehat{P}_{\mathrm{p},i}^{\mathrm{b}} = \widehat{\mathbf{C}}_{\mathrm{n}}^{\mathrm{b}}\left(P_{\mathrm{p},i}^{\mathrm{n}} - \widehat{p}_{\mathrm{k}}^{\mathrm{n}}\right) = \left[X_i^{\mathrm{b}} \quad Y_i^{\mathrm{b}} \quad Z_i^{\mathrm{b}}\right]^T,$$

$$\widehat{P}_{\mathrm{p},i}^{\,\prime} = \begin{bmatrix} \widehat{u}_i \\ \widehat{v}_i \end{bmatrix} = \begin{bmatrix} f_u \dfrac{X_i^{\mathrm{b}}}{Z_i^{\mathrm{b}}} + u_0 \\ f_v \dfrac{Y_i^{\mathrm{b}}}{Z_i^{\mathrm{b}}} + v_0 \end{bmatrix}, \quad (30)$$

$$z_{\mathrm{p},i} = P_{\mathrm{p},i}^{\,\prime} - \widehat{P}_{\mathrm{p},i}^{\,\prime} = \begin{bmatrix} u_i - \widehat{u}_i \\ v_i - \widehat{v}_i \end{bmatrix},$$

where $\widehat{p}_k^{\mathrm{n}}$ is the system's position in the kth frame, $\widehat{P}_{\mathrm{p},i}^{\mathrm{b}}$ is the point feature's estimated position in the body coordinate frame, $\widehat{P}_{\mathrm{p},i}^{\,\prime}$ is the point feature's reprojection position in the image plane, and $(\widehat{u}_i, \widehat{v}_i)$ is its pixel value. $P_{\mathrm{p},i}^{\,\prime}$ is the point feature's observation value in the image plane, and (u_i, v_i) is its pixel value.

We define $\mathscr{L}_j^{\mathrm{n}} = \left[n_j^{\mathrm{n}\,T} v_j^{\mathrm{n}\,T}\right]^T$ as the jth line feature which can be observed by the camera in the kth frame. The reprojection error of the line feature $z_{\mathrm{l},j}$ can be defined as follows:

$$\widehat{\mathscr{L}}_j^{\mathrm{b}} = \begin{bmatrix} \widehat{n}_j^{\mathrm{b}} \\ \widehat{v}_j^{\mathrm{b}} \end{bmatrix} = \begin{bmatrix} \widehat{\mathbf{C}}_{\mathrm{n}}^{\mathrm{b}} & \left\lfloor \widehat{\mathbf{t}}_{\mathrm{n}}^{\mathrm{b}} \times \right\rfloor \widehat{\mathbf{C}}_{\mathrm{n}}^{\mathrm{b}} \\ 0 & \widehat{\mathbf{C}}_{\mathrm{n}}^{\mathrm{b}} \end{bmatrix} \mathscr{L}_j^{\mathrm{n}}, \quad (31)$$

$$\vec{l}_j^{\mathrm{b}} = \begin{bmatrix} f_v & 0 & 0 \\ 0 & f_u & 0 \\ -f_v c_u & f_u c_v & f_u f_v \end{bmatrix} \widehat{n}_j^{\mathrm{b}} = \begin{bmatrix} \vec{l}_{j,1}^{\mathrm{b}} \\ \vec{l}_{j,2}^{\mathrm{b}} \\ \vec{l}_{j,3}^{\mathrm{b}} \end{bmatrix}, \quad (32)$$

$$z_{\mathrm{l},j} = \begin{bmatrix} \dfrac{x_{j,s}^T \cdot \vec{l}_j^{\mathrm{b}}}{\sqrt{\left(\vec{l}_{j,1}^{\mathrm{b}}\right)^2 + \left(\vec{l}_{j,2}^{\mathrm{b}}\right)^2}} & \dfrac{x_{j,e}^T \cdot \vec{l}_j^{\mathrm{b}}}{\sqrt{\left(\vec{l}_{j,1}^{\mathrm{b}}\right)^2 + \left(\vec{l}_{j,2}^{\mathrm{b}}\right)^2}} \end{bmatrix}^T, \quad (33)$$

where $\widehat{\mathscr{L}}_j^{\mathrm{b}}$ is the line feature's estimated position in the body coordinate frame, $\vec{l}_j^{\mathrm{b}}$ is the point feature's reprojection position in the image plane. In addition, when the line feature is observed by the camera, we can obtain the pixel value of the start-point and end-point of the line feature. In (33), $x_{j,s}^T$ and $x_{j,e}^T$ denote the start-point and end-point of the line feature's observation value in the image plane. Note that $x_{j,s}^T$ and $x_{j,e}^T$ are represented by the homogeneous coordinate frames which are 3×1 vectors (the first two elements are the pixel value, and the third element is 1).

According to the analysis above, the measurement vector can be selected as follows:

$$\mathbf{Z}^T = \left[z_{\mathrm{p},1}, z_{\mathrm{p},2}, \dots, z_{\mathrm{p},M}, z_{\mathrm{l},1}, z_{\mathrm{l},2}, \dots, z_{\mathrm{l},N}\right]. \quad (34)$$

The error-state equation can be established as follows:

$$\mathbf{Z} = \mathbf{H}\mathbf{X} + \eta, \quad (35)$$

where η is the measurement noise assumed to be uncorrelated white zero-mean Gaussian process the covariance matrix of which is $\mathbf{R}_\eta$ and $\mathbf{H}$ is the measurement Jacobian matrix formulated as:

$$\mathbf{H} = \begin{bmatrix} \mathbf{H}_{\mathrm{p},1}^T & \cdots & \mathbf{H}_{\mathrm{p},M}^T & \mathbf{H}_{\mathrm{l},1}^T & \cdots & \mathbf{H}_{\mathrm{l},N}^T \end{bmatrix}^T, \quad (36)$$

where the individual matrixes $\mathbf{H}_{\mathrm{p},i}$ and $\mathbf{H}_{\mathrm{l},j}$ can be computed as

$$\begin{aligned}
\mathbf{H}_{\mathrm{p},i} &= J_{z,i}\left[\left\lfloor \mathbf{C}_\mathrm{n}^\mathrm{b}\left(P_{\mathrm{p},i}^\mathrm{n} - p^\mathrm{n}\right)\times\right\rfloor \quad 0_{3\times3} \quad -\mathbf{C}_\mathrm{n}^\mathrm{b} \quad 0_{3\times6} \quad \cdots \quad \mathbf{C}_\mathrm{n}^\mathrm{b} \quad \cdots\right],\\
\mathbf{H}_{\mathrm{l},j} &= J_{\xi,j}\begin{bmatrix} -\left\lfloor \mathbf{C}_\mathrm{n}^\mathrm{b} n^\mathrm{n}\times\right\rfloor - \left\lfloor\left\lfloor \mathbf{t}_\mathrm{n}^\mathrm{b}\times\right\rfloor \mathbf{C}_\mathrm{n}^\mathrm{b} v^\mathrm{n}\times\right\rfloor & 0_{3\times3} & -\left\lfloor \mathbf{C}_\mathrm{n}^\mathrm{b} v^\mathrm{n}\times\right\rfloor & 0_{3\times6} & \cdots & J_{\delta,j} & \cdots \\ -\left\lfloor \mathbf{C}_\mathrm{n}^\mathrm{b} v^\mathrm{n}\times\right\rfloor & 0_{3\times3} & 0_{3\times3} & 0_{3\times6} & & & \end{bmatrix},\\
J_{z,i} &= \begin{bmatrix} \dfrac{f_u}{Z_i^\mathrm{b}} & 0 & \dfrac{-f_u X_i^\mathrm{b}}{\left(Z_i^\mathrm{b}\right)^2} \\ 0 & \dfrac{f_v}{Z_i^\mathrm{b}} & \dfrac{-f_v Y_i^\mathrm{b}}{\left(Z_i^\mathrm{b}\right)^2} \end{bmatrix},\\
J_{\delta,j} &= \mathscr{H}_\mathrm{n}^\mathrm{b}\begin{bmatrix} -\left\lfloor w_{j,1}u_{j,1}\times\right\rfloor & -w_{j,2}u_{j,1} \\ -\left\lfloor w_{j,2}u_{j,2}\times\right\rfloor & -w_{j,1}u_{j,2} \end{bmatrix},\\
J_{\xi,j} &= \frac{1}{l_\mathrm{n}}\begin{bmatrix} u_{j,1} - \dfrac{l_{j,1}^\mathrm{b} e_{j,1}}{l_\mathrm{n}^2} & v_{j,1} - \dfrac{l_{j,2}^\mathrm{b} e_{j,1}}{l_\mathrm{n}^2} & 1 \\ u_{j,2} - \dfrac{l_{j,1}^\mathrm{b} e_{j,2}}{l_\mathrm{n}^2} & v_{j,2} - \dfrac{l_{j,2}^\mathrm{b} e_{j,2}}{l_\mathrm{n}^2} & 1 \end{bmatrix}\left[\mathscr{K} 0_{3\times3}\right],
\end{aligned} \quad (37)$$

where $e_{j,1} = x_{j,s}^T l_j^\mathrm{b}$, $e_{j,2} = x_{j,e}^T l_j^\mathrm{b}$, $l_\mathrm{n} = \sqrt{(l_{j,1}^\mathrm{b})^2 + (l_{j,2}^\mathrm{b})^2}$, $x_{j,s}^T = \begin{bmatrix} u_{j,1} & v_{j,1} & 1 \end{bmatrix}$, and $x_{j,e}^T = \begin{bmatrix} u_{j,2} & v_{j,2} & 1 \end{bmatrix}$.

The measurement equation can also be written as the discrete form:

$$\mathbf{Z}_k = \mathbf{H}_k\mathbf{X}_k + \eta_k. \quad (38)$$

4.3. Estimator and Algorithm Implementation. According to the analysis in the previous section, it can be found that the error-state equation and the measurement equation are nonlinear. In order to deal with the linearization error, we use the EKF estimator to fuse the inertial measurement and the images observed by the stereo camera. As presented in the previous section, these nonlinear functions have been linearized.

Based on the linearized error-state and measurement equations, the EKF estimator can be executed with the following two steps:

(i) Predict

$$\begin{aligned}
\mathbf{X}_{k+1,k} &= \Phi_{k+1,k}\mathbf{X}_k,\\
\mathbf{P}_{k+1,k} &= \Phi_{k+1,k}\mathbf{P}_k\Phi_{k+1,k}^T + \mathbf{Q}_{w,k},
\end{aligned} \quad (39)$$

where $\mathbf{P}_k$ is the covariance matrix of the state vector.

(ii) Update

$$\begin{aligned}
\mathbf{K}_k &= \boldsymbol{P}_{k+1,k}\mathbf{H}_k^T\left(\mathbf{H}_k\mathbf{P}_{k+1,k}\mathbf{H}_k^T + \mathbf{R}_{\eta,k}\right)^{-1},\\
\mathbf{X}_{k+1,k+1} &= \mathbf{X}_{k+1,k} + \mathbf{K}_k(\mathbf{Z}_k - \mathbf{H}_k\mathbf{X}_{k+1,k}),\\
\mathbf{P}_{k+1,k+1} &= (I - \mathbf{K}_k\mathbf{H}_k)\mathbf{P}_{k+1,k},
\end{aligned} \quad (40)$$

where $\mathbf{K}_k$ is the Kalman gain matrix.

The covariance matrices $\mathbf{P}_k$, $\mathbf{Q}_w$, and $\mathbf{R}_\eta$ in the above equations are empirically initialized based on the characteristics of the sensors. Specifically, $\mathbf{Q}_w$ and $\mathbf{R}_\eta$ are initialized, respectively, associated with the random noise of the inertial sensors and the scales of the feature space. $\mathbf{P}_k$ is initialized concerning the constant drifts of the inertial sensors and the scales of the feature space.

Finally, we summarize the whole algorithm. In this system, static base alignment is achieved for obtaining the initial attitude angle by following the method in [30]. Subsequently, analogous to the inertial navigation system [34], the IMU measurement is used to update the pose and the velocity of the system by solving (3).

On the other hand, the images are collected by the stereo camera; after performing the feature detection and initialization which will be discussed in the next subsection, the error-state and measurement equations can be established by (22), (23), (24), (25), (26), (27), (28), (29), (30), (31), (32), (33), (34), (35), (36), (37), and (38) with the IMU measurement and the update result of the inertial navigation. In the last step, the EKF estimator is executed by (39) and (40). In addition, the covariance matrix of the EKF should be augmented when new point or line features are observed.

4.4. Feature Detection and Initialization. In terms of the feature detection, the ORB method [23] is used to extract point features from the images. This method deals with the orientation problem of the FAST method and generates a very fast binary descriptor based on BRIEF to describe the detected point features. Therefore, this method can detect and describe the point features in a fast, efficient, and accurate manner. In [21, 22], the ORB method was used in the visual SLAM system, which resulted in an excellent performance.

On the other hand, the LSD method [24, 25] is used to extract the line features from the images. Its mechanism is to merge pixels with the similar gradient direction. This method can extract the lines with subpixel accuracy in linear time. In addition, the detected lines are described by a binary descriptor, line band descriptor (LBD) [39], which allows accurate matching between two frames or two cameras (left and right).

The nearest-neighbor method is used to match the features between the two frames based on the descriptor of features. In order to reduce the false matches, the random sample consensus (RANSAC) method is used to filter out the outliers for both point features and line features.

Once a new feature is detected and described, we next perform binocular matching. For a point feature in the left image, its counterpart in the right image can be acquired along the parallel epipolar line in the right image. The Hamming distance is used to measure the similarity of point features with a predefined threshold. For the line feature, we exploit the matching method proposed in [40].

The unsuccessfully matched features will be discarded. After matching the point and line features between the two images, the positions of the point features and the two endpoints of the line features can be obtained by (8). The representations of line features in the 3D space can be computed by (12). The point features and lines in the navigation coordinate frame can be computed by (9) and (16). With the above-mentioned feature initialization, the initial positions of the features in the navigation system are obtained and are thus used to compute the reprojection error in the measurement equation. In addition, after executing one filtering process, the error of the features' positions in the navigation coordinate frame can be estimated after feature filtering. Therefore, these positions can be updated in the filtering processes and become progressively accurate.

5. Experiments and Results

In this section, experiments are conducted to test the effectiveness of the proposed indoor navigation system. We assume our system is tailored for any structured indoor environment where the geometric scale information can be calculated by the stereo camera and the object textures can be characterized by either point or line features. We use EuRoC dataset [41] for evaluating our approach and use root-mean-square error (RMSE) for performance measure. In the comparative study, we compare the proposed navigation system (PL-VINS), with the StVO-PL [26] and the system based on multistate constraint Kalman filter (MSCKF) [15]. The StVO-PL is a state-of-the-art visual navigation system using stereo camera with points and lines, while the MSCKF is an EKF-based visual-inertial navigation system with high accuracy. In implementation, we directly use the publicly available source code for both StVO-PL and MSCKF methods. For the sake of consistency, we repeat all the competing approaches ten times on all the sequences and the final experimental results are obtained by averaging the accuracy scores.

5.1. Experiment Description. The EuRoC dataset contains 11 sequences which are categorized into two types both recorded in three indoor environments. The first type of the dataset contains the first five sequences which are recorded in a machine hall (MH). The second type of the dataset contains the remaining six sequences recorded in Vicon room 1 (V1) and Vicon room 2 (V2) with an approximate size of $8\,\mathrm{m} \times 8.4\,\mathrm{m} \times 4\,\mathrm{m}$. All eleven sequences include the stereo images and the IMU measurements containing the gyroscopes, accelerometers data along with the ground truth. More specifically, the stereo images are recorded by a stereo camera, MT9V034 with $2 \times 20\,\mathrm{Hz}$ while the IMU measurements are recorded by a MEMS inertial sensor, ADIS16448 with 200 Hz. Besides, the ground truths are provided by Leica MS50 and VICON. All these equipment are mounted on a micro air vehicle (MAV) which flies in the indoor environment to record the data. More details about the EuRoC dataset can be found in [41]. In addition, these 11 sequences provide varying challenges relevant to the speed of the MAV, illumination, texture, and so forth. In some sequences, especially for the sequence Vicon room 2 03, the environment is low-textured with few visual contents.

According to the description of the EuRoC dataset, it can be found that these sequences are suitable for testing the proposed indoor navigation system because of the visual-inertial data and the low-textured indoor environments. In addition, to the best of our knowledge, this dataset is the only dataset which provides visual-inertial data and ground truth in indoor environments.

In order to test the algorithms quantitatively, the transformation error matrix between two coordinate frames can be computed as follows:

$$\boldsymbol{\delta}\mathbf{T} = \widehat{\mathbf{T}}^{-1}\mathbf{T}, \quad \mathbf{T} = \begin{bmatrix} \mathbf{C}_{\mathrm{n}}^{\mathrm{b}} & \mathbf{t}_{\mathrm{n}}^{\mathrm{b}} \\ 0_{1\times 3} & 1 \end{bmatrix}, \tag{41}$$

where $\mathbf{T}$ is the transformation matrix between two coordinate frames provided by the ground truth, $\widehat{\mathbf{T}}$ is the estimation transformation matrix from the navigation system, and $\boldsymbol{\delta}\mathbf{T}$ is the error matrix. The translation errors and the rotation errors can be obtained from this error matrix.

The RMSE of translation and rotation can also be computed for comparison of different systems:

$$\begin{aligned} \mathrm{RMSE(Transal.)} &= \left(\frac{1}{m}\sum_{i=1}^{m} \|\mathrm{trans}(\boldsymbol{\delta}\mathbf{T}_i)\|^2\right)^{1/2}, \\ \mathrm{RMSE(Rot.)} &= \left(\frac{1}{m}\sum_{i=1}^{m} \|\mathrm{Rot}(\boldsymbol{\delta}\mathbf{T}_i)\|^2\right)^{1/2}, \end{aligned} \tag{42}$$

where m is the number of the frames, $\mathrm{trans}(\boldsymbol{\delta}\mathbf{T}_i)$ is the 3×1 translation error vector, and $\mathrm{Rot}(\boldsymbol{\delta}\mathbf{T}_i)$ is the 3×1 rotation error vector which are represented by the Euler angles.

5.2. Results and Discussion

5.2.1. Translation Errors. Extensive evaluations on 11 sequences are carried out by making use of three different navigation systems mentioned above. First, we analyze the translation errors of the dataset. Figures 1 and 2 show the translation errors of different methods on sequences machine hall 04 and Vicon room 2 02. Table 1 gives the relative x, y, z, and whole translation RMSE errors of all 11 sequences. In Table 1, the "T" column means the whole translation RMSE errors. "MH," "V1," and "V2," respectively, indicate the "machine hall," the "Vicon room 1," and the "Vicon room 2." "MH01" implies the sequence 01 in machine hall, and another four sequences named from "MH02" to "MH05" are also used in our evaluations. Besides, we also use the respective three sequences in V1 and V2 in our experiments, namely, "V1 01," "V1 02," "V1 03," "V2 01," "V2 02," and "V2 03." In addition, in order to demonstrate the robot movement in some environment, we give the ground truth trajectories and the trajectories estimated by the proposed method as shown in Figure 3.

It can be seen from Figures 1 and 2 and Table 1 that the proposed navigation system reports higher accuracy than MSCKF and StVO-PL in terms of translation errors. The StVO-PL achieves the lowest positioning accuracy with the largest x, y, z, and whole translation RMSE errors. For most of the sequences, the translation RMSE errors of the proposed method are better than those of the MSCKF. In addition, the proposed navigation system is more robust than the other two systems. The error curves of the

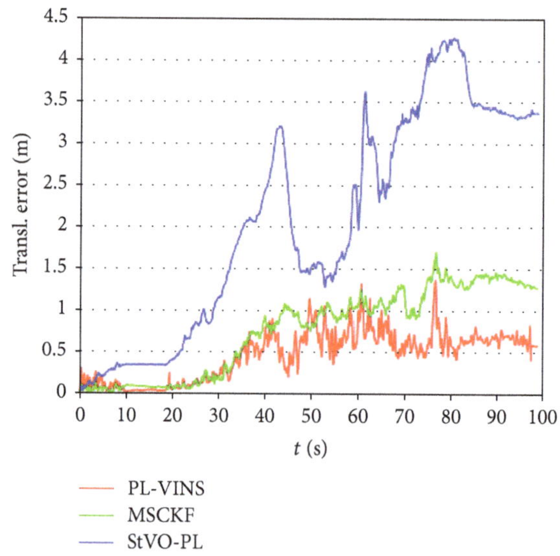

FIGURE 1: Translation errors of machine hall 04.

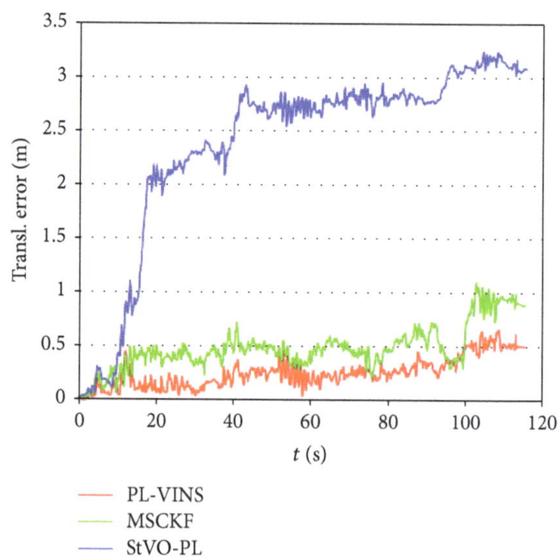

FIGURE 2: Translation errors of Vicon room 2 02.

proposed method are smoother than those of the StVO-PL. On sequence Vicon room 2 03, both the StVO-PL and MSCKF underperform while the proposed system exhibits superior performance.

It is observed that in terms of the sequences with substantial point features generated, point-based VINS performs well such that the line features provide limited performance improvements. In some cases, particularly, the measurement noise of the line features and the incorrect matches could lead to a slight worse performance such as the results in sequence machine hall 02. It is also shown that our approach outperforms MSCKF in the case when the features are insufficient. Thus, in specific cases, the performance of the proposed system is likely to be compromised by the measurement noise and the false matches; however, most of the results

TABLE 1: Translation errors (RMSE) in meters of the EuRoC dataset.

	VINS (point and lines) (m)				MSCKF (m)				StVO-PL (m)			
	x	y	z	T	x	y	z	T	x	y	z	T
MH 01	0.105	0.112	0.116	0.193	0.352	0.139	0.071	0.386	0.813	0.271	0.388	0.941
MH 02	0.082	0.168	0.108	0.216	0.134	0.126	0.078	0.201	0.929	0.251	0.202	0.984
MH 03	0.306	0.137	0.172	0.377	0.174	0.491	0.124	0.535	1.669	1.069	0.271	2.001
MH 04	0.323	0.322	0.307	0.551	0.629	0.665	0.114	0.923	0.865	1.365	1.853	2.458
MH 05	0.376	0.335	0.209	0.546	0.569	0.326	0.645	0.920	0.672	2.232	0.677	2.423
V1 01	0.286	0.101	0.109	0.261	0.797	1.004	0.717	0.322	0.797	1.003	0.718	1.468
V1 02	0.081	0.085	0.175	0.211	0.181	0.146	0.055	0.239	0.250	0.379	0.392	0.599
V1 03	0.202	0.247	0.134	0.347	0.291	0.357	0.396	0.608	0.264	1.823	0.453	1.897
V2 01	0.098	0.171	0.045	0.202	0.206	0.214	0.238	0.381	0.971	0.388	0.962	1.421
V2 02	0.096	0.246	0.120	0.289	0.412	0.203	0.264	0.531	1.519	0.968	1.766	2.522
V2 03	0.529	0.649	0.567	1.011	—	—	—	—	—	—	—	—

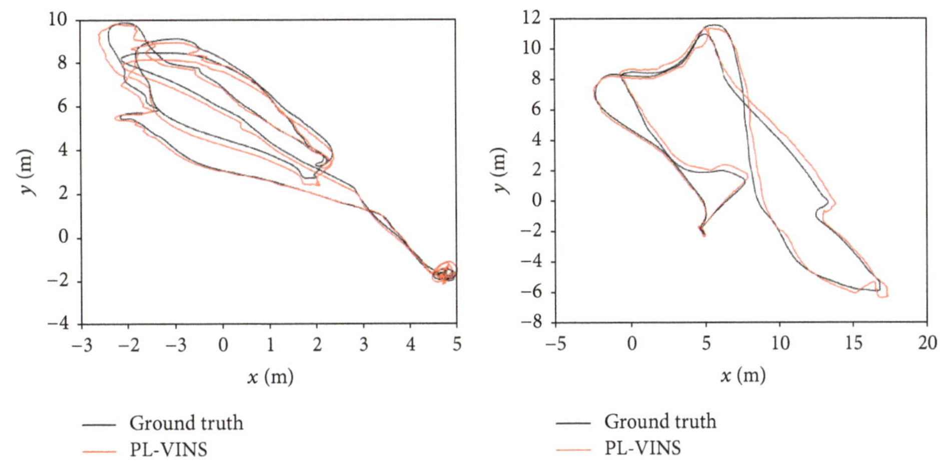

FIGURE 3: (a) Ground truth, the trajectories estimated by our algorithm on MH 02 sequence, which is viewed from the gravity direction. (b) Ground truth, the trajectories estimated by our algorithm on MH 05 sequence, which is viewed from the gravity direction.

significantly demonstrate the advantage of our system against other competing system.

It is also observed from figures that the error of the StVO-PL increases over time. Since StVO-PL is a vision-based navigation system, serious accumulated error is produced, which significantly impairs the performance of the system. By contrast, the proposed method and the MSCKF are visual-inertial navigation system, which enables reducing the error accumulation. Besides, both of them achieve very high positioning accuracy. Furthermore, compared to the MSCKF, the proposed method utilizes both the line features and point features. The line features abundant in indoor or man-made environments can use edges and linear shapes in environments. Thus, in some complex low-textured environments such as sequence Vicon room 2 03, the proposed system can still extract enough features to estimate the motion, while other systems such as MSCKF suffer from the lack of point features, which contributes to better positioning accuracy and the robustness of the proposed system.

5.2.2. Rotation Errors. In this subsection, we analyze the rotation errors of the EuRoC dataset. Similarly, as for sequences machine hall 05 and Vicon room 1 03, the rotation errors of three systems along with time are shown in Figures 4 and 5. The relative x, y, z, and whole rotation RMSE errors of all 11 sequences are shown in Table 2. In Table 2, the "R" column denotes the whole rotation RMSE errors.

The results of the rotation is similar to the results of the translation. It can be seen from the two figures and the table that the rotation accuracy of the StVO-PL is much worse than the other two systems with large errors of roll, pitch, and yaw angles. Despite the smooth motion in some sequences, the StVO-PL system does not perform well, due to the accumulated error of the visual navigation system. As for the other two navigation systems, the proposed navigation system has the relatively small rotation errors. However, in some sequences, the rotation errors of these two systems are very close, probably due to the smooth movement. For the challenging sequences with sudden angle changes, the

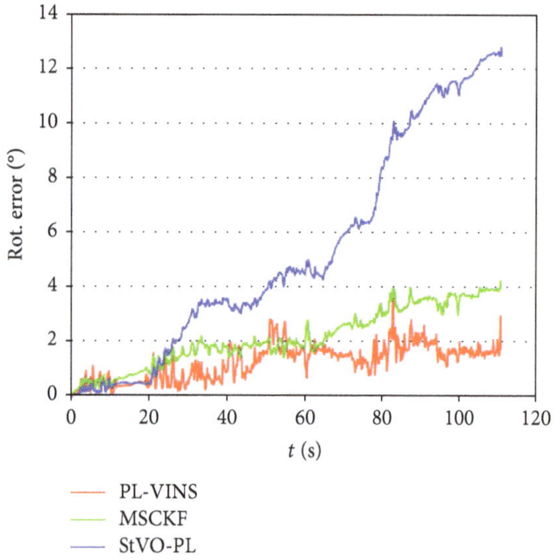

FIGURE 4: Rotation errors of the machine hall 05.

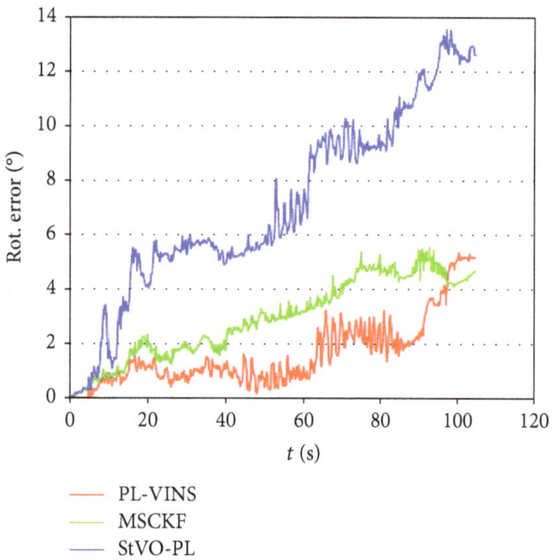

FIGURE 5: Rotation errors of the Vicon room 1 03.

proposed navigation system performs better than the MSCKF. In addition, the robustness of the proposed navigation is also better than other two systems. Both StVO-PL and MSCKF fail for the sequence Vicon room 2 03 which has many low-textured scenes, but the proposed navigation system executes successfully. As analogous to the translation errors, our method reports higher rotation errors than the MSCKF in some scenario, which can be caused by the measurement noise and false matches of line features.

Finally, it can be observed from Tables 1 and 2 that our approach has been evaluated on all the eleven video clips which provides different challenges, and the experimental results demonstrate that our method beats the other competing techniques on ten video clips, which considerably suggests the consistent superiority of our method.

To summarize, the proposed navigation system performs much better than the visual-only navigation system with points and lines and better than the point-only VINS. The reason is that the inertial sensors can aid the visual sensors to improve the performance of the navigation system. What is more, the datasets used in our work are all recorded in man-made environments which contain abundant line features and in some sequences; these environments contain some low-textured senses which lack enough point features. Therefore, the application of line feature can improve the performance of the system to a certain degree.

5.2.3. Qualitative Evaluations of Both Features in Different Environment. Figures 6 and 7 qualitatively illustrate the matched point and line features between two frames in the dark environment and in the environment with some curves.

It is shown in Figure 6 that it is difficult to extract substantial point features and line features due to insufficient visual contents in the dark environment. To be specific, the point features are distributed intensively while the line patterns are scattered sparsely. Despite the significant variance in visual distribution, combining point and line features sufficiently encode the most image textures, which indicates the strong correlation between the two features.

It is observed in Figure 7 that line features can be extracted on the smooth surfaces or bar-shaped objects except some curves with high curvature. By contrast, the point features can be extracted on these curves with high curvature yet fail to be mined on the smooth surfaces or bar-shaped objects. Based on this observation, we argue that there exists substantial complementarity between these two features. Therefore, fusing the point and line features significantly contributes to the performance improvements.

In addition, sufficient correct matches are observed from these two figures, which thus guarantees the estimation accuracy. Subsequently, the estimated position errors of both features can be derived and revised accordingly. Therefore, the accuracy of these positions will be progressively accurate.

6. Conclusions

In this paper, we present an indoor navigation system combining stereo camera and inertial sensors with point and line features. The stereo images and the IMU measurements are fused by the EKF estimator for more accurate and robust pose estimation. Furthermore, the point and line features are all extracted from the stereo images. This scheme can further improve the robustness and accuracy of the system and makes it perform well in complex indoor environment such as low-texture or human-made scenes. Experiments based on eleven sequences provided by the EuRoC dataset are conducted to test the proposed system. The results show that the proposed navigation system performs better than the point-only VINS or the vision-only navigation system with points and lines in indoor environment, especially in more complex and challenging scenarios. So, the proposed navigation system can be applied to the complex indoor environment with high accuracy and robustness.

Table 2: Rotation errors (RMSE) in degrees of the EuRoC dataset.

	VINS (point and lines) (°)				MSCKF (°)				StVO-PL (°)			
	Roll	Pitch	Yaw	R	Roll	Pitch	Yaw	R	Roll	Pitch	Yaw	R
MH 01	0.852	0.442	0.768	1.229	0.419	0.250	1.383	1.467	2.452	1.825	4.021	5.051
MH 02	1.349	1.169	1.338	2.231	1.049	2.377	0.676	2.685	1.973	1.636	4.569	5.239
MH 03	0.961	0.507	0.525	1.207	1.084	0.681	1.683	2.071	2.525	4.439	3.963	5.954
MH 04	0.918	1.612	1.734	1.271	0.314	0.869	1.970	2.170	4.936	1.423	3.985	6.503
MH 05	1.658	1.121	1.988	1.411	0.452	0.966	2.179	2.426	1.081	2.621	6.056	6.687
V1 01	1.326	1.267	2.193	2.869	2.234	2.322	0.908	3.348	2.744	2.644	3.130	4.931
V1 02	0.374	0.877	1.301	1.613	0.553	0.371	1.761	1.882	3.138	3.646	3.771	6.113
V1 03	1.544	0.517	1.362	2.123	1.179	0.549	3.041	3.307	2.841	4.004	6.178	7.892
V2 01	0.613	0.489	0.879	1.178	0.631	0.518	0.682	1.064	3.413	2.742	2.808	5.202
V2 02	0.888	0.841	1.138	1.671	0.599	0.697	1.867	2.081	2.096	2.391	5.184	6.081
V2 03	2.087	2.661	2.878	4.441	—	—	—	—	—	—	—	—

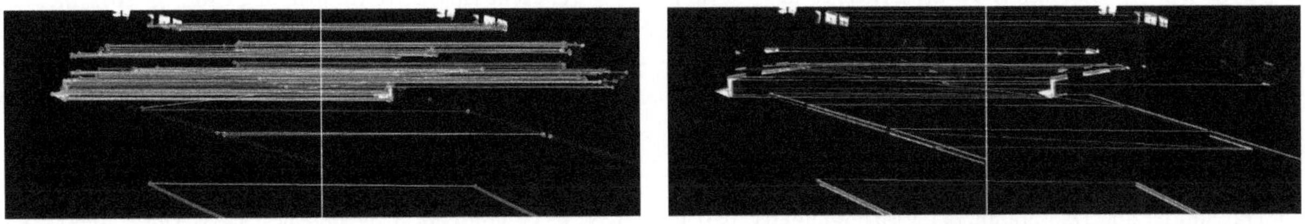

(a) (b)

Figure 6: (a) Matched point features between two frames in the dark environment. (b) Matched line features between two frames in the dark environment.

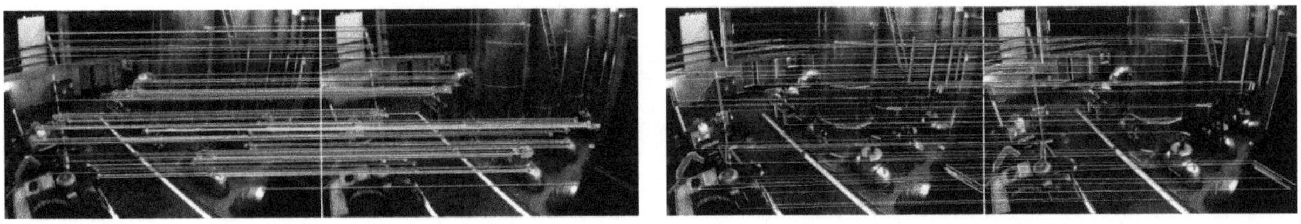

(a) (b)

Figure 7: (a) Matched point features between two frames when there are some curves in environment. (b) Matched line features between two frames when there are some curves in environment.

In addition, the MEMS inertial sensors and the camera are all the low-cost and passive sensors, so, this system can be used in most environments and suitable for the low-cost autonomous navigation system. On the other hand, this system does not include the loop closure detection, so, its accuracy may be no higher than the visual-inertial SLAM system. In addition, we do not consider the features which cannot obtain the scale information by stereo camera because of the far distance, so, the performance in outdoor environment will be worse. In future work, we will further improve this system in these two fields.

Conflicts of Interest

The authors declare no conflict of interest.

Acknowledgments

This study is supported in part by the National Natural Science Foundation of China (Grant nos. 61473085, 51775110, and 51375088), Fundamental Research Funds for the Central Universities (2242015R30031), and key laboratory fund of the Ministry of Public Security based on large data structure (2015DSJSYS002).

References

[1] M. Jung and J.-B. Song, "Robust mapping and localization in indoor environments," *Intelligent Service Robotics*, vol. 10, no. 1, pp. 55–66, 2017.

[2] G. A. Kumar, A. K. Patil, R. Patil, S. S. Park, and Y. H. Chai, "A LiDAR and IMU integrated indoor navigation system for

UAVs and its application in real-time pipeline classification," *Sensors*, vol. 17, no. 6, p. 1268, 2017.

[3] M. W. M. G. Dissanayake, P. Newman, S. Clark, H. F. Durrant-Whyte, and M. Csorba, "A solution to the simultaneous localization and map building (SLAM) problem," *IEEE Transactions on Robotics*, vol. 17, no. 3, pp. 229–241, 2001.

[4] W. Chen and T. Zhang, "An indoor mobile robot navigation technique using odometry and electronic compass," *International Journal of Advanced Robotic Systems*, vol. 14, no. 3, article 1729881417711643, 2017.

[5] A. J. Davison, I. D. Reid, N. D. Molton, and O. Stasse, "MonoSLAM: real-time single camera SLAM," *IEEE Transactions on Pattern Analysis and Machine Intelligence*, vol. 29, no. 6, pp. 1052–1067, 2007.

[6] A. I. Mourikis and S. Roumeliotis, "A multi-state constraint Kalman filter for vision-aided inertial navigation," in *2007 IEEE International Conference on Robotics and Automation*, pp. 3565–3572, Roma, Italy, April 2007.

[7] A. Ben-Afia, V. Gay-Bellile, A.-C. Escher et al., "Review and classification of vision-based localisation techniques in unknown environments," *IET Radar, Sonar & Navigation*, vol. 8, no. 9, pp. 1059–1072, 2014.

[8] J. Engel, V. Koltun, and D. Cremers, "Direct sparse odometry," *IEEE Transactions on Pattern Analysis and Machine Intelligence*, vol. 40, no. 3, pp. 611–625, 2018.

[9] F. Fraundorfer and D. Scaramuzza, "Visual odometry: part I: the first 30 years and fundamentals," *IEEE Robotics and Automation Magazine*, vol. 18, no. 4, pp. 80–92, 2011.

[10] F. Fraundorfer and D. Scaramuzza, "Visual odometry : part II: matching, robustness, optimization, and applications," *IEEE Robotics and Automation Magazine*, vol. 19, no. 2, pp. 78–90, 2012.

[11] G. Zhang, J. H. Lee, J. Lim, and I. H. Suh, "Building a 3-D line-based map using stereo SLAM," *IEEE Transactions on Robotics*, vol. 31, no. 6, pp. 1364–1377, 2015.

[12] R. Gomez-Ojeda, J. Briales, and J. Gonzalez-Jimenez, "PL-SVO: semi-direct monocular visual odometry by combining points and line segments," in *2016 IEEE/RSJ International Conference on Intelligent Robots and Systems (IROS)*, pp. 4211–4216, Daejeon, South Korea, October 2016.

[13] G. Zhang and I. Suh, "Building a partial 3D line-based map using a monocular SLAM," in *2011 IEEE International Conference on Robotics and Automation*, pp. 1497–1502, Shanghai, China, May 2011.

[14] L. Clement, V. Peretroukhin, J. Lambert, and J. Kelly, "The battle for filter supremacy: a comparative study of the multi-state constraint Kalman filter and the sliding window filter," in *2015 12th Conference on Computer and Robot Vision (CRV)*, pp. 23–30, Halifax, NS, Canada, June 2015.

[15] M. Li and A. I. Mourikis, "High-precision, consistent EKF-based visual-inertial odometry," *International Journal of Robotics Research*, vol. 32, no. 6, pp. 690–711, 2013.

[16] P. Corke, J. Lobo, and J. Dias, "An introduction to inertial and visual sensing," *The International Journal of Robotics Research*, vol. 26, no. 6, pp. 519–535, 2007.

[17] S. Leutenegger, S. Lynen, M. Bosse, R. Siegwart, and P. Furgale, "Keyframe-based visual–inertial odometry using nonlinear optimization," *The International Journal of Robotics Research*, vol. 34, no. 3, pp. 314–334, 2015.

[18] A. J. Davison, "Real-time simultaneous localisation and mapping with a single camera," in *2003. Proceedings. Ninth IEEE International Conference on Computer Vision*, pp. 1403–1410, Nice, France, October 2003.

[19] L. M. Paz, P. Pinies, J. D. Tardos, and J. Neira, "Large-scale 6-DOF SLAM with stereo-in-hand," *IEEE Transactions on Robotics*, vol. 24, no. 5, pp. 946–957, 2008.

[20] A. J. Davison and D. W. Murray, "Simultaneous localization and map-building using active vision," *IEEE Transactions on Pattern Analysis and Machine Intelligence*, vol. 24, no. 7, pp. 865–880, 2002.

[21] R. Mur-Artal, J. M. M. Montiel, and J. D. Tardos, "ORB-SLAM: a versatile and accurate monocular SLAM system," *IEEE Transactions on Robotics*, vol. 31, no. 5, pp. 1147–1163, 2015.

[22] R. Mur-Artal and J. D. Tardós, "ORB-SLAM2: an open-source SLAM system for monocular, stereo, and RGB-D cameras," *IEEE Transactions on Robotics*, vol. 33, no. 5, pp. 1255–1262, 2017.

[23] E. Rublee, V. Rabaud, K. Konolige, and G. Bradski, "ORB: an efficient alternative to SIFT or SURF," in *2011 International Conference on Computer Vision*, pp. 2564–2571, Barcelona, Spain, November 2011.

[24] R. G. von Gioi, J. Jakubowicz, J.-M. Morel, and G. Randall, "LSD: a fast line segment detector with a false detection control," *IEEE Transactions on Pattern Analysis and Machine Intelligence*, vol. 32, no. 4, pp. 722–732, 2010.

[25] R. G. von Gioi, J. Jakubowicz, J.-M. Morel, and G. Randall, "LSD: a line segment detector," *Image Processing on Line*, vol. 2, pp. 35–55, 2012.

[26] R. Gomez-Ojeda and J. Gonzalez-Jimenez, "Robust stereo visual odometry through a probabilistic combination of points and line segments," in *2016 IEEE International Conference on Robotics and Automation (ICRA)*, pp. 2521–2526, Stockholm, Sweden, May 2016.

[27] A. Pumarola, A. Vakhitov, A. Agudo, A. Sanfeliu, and F. Moreno-Noguer, "PL-SLAM: real-time monocular visual SLAM with points and lines," in *2017 IEEE International Conference on Robotics and Automation (ICRA)*, pp. 4503–4508, Singapore, Singapore, May 2017.

[28] R. Gomez-Ojeda, D. Zuñiga-Noël, F.-A. Moreno, D. Scaramuzza, and J. Gonzalez-Jimenez, "PL-SLAM: A stereo SLAM system through the combination of points and line segments," 2017, http://arxiv.org/abs/1705.09479.

[29] G. Huang, M. Kaess, and J. Leonard, "Towards consistent visual-inertial navigation," in *2014 IEEE International Conference on Robotics and Automation (ICRA)*, pp. 4926–4933, Hong Kong, China, May 2014.

[30] Z. Xian, X. Hu, and J. Lian, "Fusing stereo camera and low-cost inertial measurement unit for autonomous navigation in a tightly-coupled approach," *Journal of Navigation*, vol. 68, no. 3, pp. 434–452, 2015.

[31] D. Kottas and S. Roumeliotis, "Efficient and consistent vision-aided inertial navigation using line observations," in *2013 IEEE International Conference on Robotics and Automation*, pp. 1540–1547, Karlsruhe, Germany, May 2013.

[32] H. Yu and A. Mourikis, "Vision-aided inertial navigation with line features and a rolling-shutter camera," in *2013 IEEE International Conference on Robotics and Automation (ICRA)*, pp. 892–899, Karlsruhe, Germany, September 2015.

[33] C. Guo and I. Roumeliotis, "IMU-RGBD camera 3D pose estimation and extrinsic calibration: observability analysis

and consistency improvement," in *2013 IEEE International Conference on Robotics and Automation*, pp. 2935–2942, Karlsruhe, Germany, May 2013.

[34] D. H. Titterton and J. L. Weston, *Strapdown Inertial Navigation Technology*, Lavenham Press Ltd, London, UK, 2nd edition, 2004.

[35] J. Rohac, M. Sipos, and J. Simanek, "Calibration of low-cost triaxial inertial sensors," *IEEE Instrumentation & Measurement Magazine*, vol. 18, no. 6, pp. 32–38, 2015.

[36] Z. Zhang, "A flexible new technique for camera calibration," *IEEE Transactions on Pattern Analysis and Machine Intelligence*, vol. 22, no. 11, pp. 1330–1334, 2000.

[37] A. Bartoli and P. Sturm, "Structure-from-motion using lines: representation, triangulation and bundle adjustment," *Computer Vision and Image Understanding*, vol. 100, no. 3, pp. 416–441, 2005.

[38] A. Bartoli and P. Sturm, "The 3D line motion matrix and alignment of line reconstructions," *International Journal of Computer Vision*, vol. 57, no. 3, pp. 159–178, 2004.

[39] L. Zhang and R. Koch, "An efficient and robust line segment matching approach based on LBD descriptor and pairwise geometric consistency," *Journal of Visual Communication and Image Representation*, vol. 24, no. 7, pp. 794–805, 2013.

[40] D.-M. Woo, D.-C. Park, S.-S. Han, and S. Beack, "2D line matching using geometric and intensity data," in *2009. AICI '09. International Conference on Artificial Intelligence and Computational Intelligence*, pp. 99–103, Shanghai, China, November 2009.

[41] M. Burri, J. Nikolic, P. Gohl et al., "The EuRoC micro aerial vehicle datasets," *International Journal of Robotics Research*, vol. 35, no. 10, pp. 1157–1163, 2016.

16

Dynamic Characteristics Analysis of the Six-Axis Force/Torque Sensor

Liyue Fu and Aiguo Song

School of Instrument Science and Engineering, Southeast University, Nanjing, Jiangsu 210096, China

Correspondence should be addressed to Aiguo Song; a.g.song@seu.edu.cn

Academic Editor: Antonio Lazaro

In this study, dynamic characteristics of a robot six-axis wrist force/torque (F/T) sensor with crossbeam elastomer are analyzed by two methods of model identification, a method for simultaneous identification of order and parameters of the model (SIM) and a method based on the differential evolution (DE) algorithm. Firstly, by establishing the simplified mechanical model and finite element (FE) model, respectively, natural frequency of the six-axis wrist F/T sensor is calculated. Secondly, dynamic calibration experiment is conducted. Lastly, two dynamic models of the sensor are identified by SIM and DE methods and the dynamic characteristics of the sensor, such as natural frequency and working band, are further analyzed. Comparing experimental values with the theoretical values, the results show that this sensor has a wide dynamic range with the first natural frequency at more than 1600 Hz, working bands (±5%) are more than 400 Hz, and the step response oscillation is intense. This study can provide a reference for the application of the six-axis F/T sensor in the field of dynamic measurement.

1. Introduction

A six-axis force/torque (F/T) sensor can measure the tangential force terms along X-, Y-, and normal force term along Z-axis (F_X, F_Y, and F_Z) as well as the moment terms about X-, Y-, and Z-axis (M_X, M_Y, and M_Z) simultaneously. Six-axis F/T sensor can be used in robotics, aerospace, automotive, industrial areas [1]. Dynamic force measurement is more and more used in various industries, which make the dynamic performance analysis of six-axis F/T sensors particularly important. In the field of micromachining, multiaxis F/T sensors are used to measure 3D machining forces which involve very high frequencies due to the high spindle speeds used during the process [2]. As research in vibration-based damage identification increases, multiaxis F/T sensors with high dynamic ranges are used in real-time structural health monitoring [3]. With the quick development of the robot field and the needs of special areas, such as precise assembly, contour tracing, and two-hand coordination [4], six-axis F/T sensor has become one of the most important sensors in the field of the intelligent robot. When F/T sensor is used to sense the collision between robot and environment, it is necessary to detect the size and direction of the dynamic collision force [5]. As a detecting element in the force reflection control system, it should respond quickly to the load, namely, having excellent dynamic characteristics [6]. These applications demonstrate the importance of the research on the dynamic characteristics of the F/T sensor. This study arises from the demand for dynamic measurements in various industries.

The acquisition of the dynamic characteristics of the six-axis F/T sensor is based on the dynamic calibration experiment. The dynamic calibration of a six-axis F/T sensor is to obtain the relationship between the input and output of the sensor when performing varied six-axis forces. Essentially, it is a process to acquire dynamic characteristics' indices such as natural frequency, time constant, and damping ratio. Huang verified that the frequency band has the most influence for the sensor and test system and therefore becomes the most important practical index for dynamic calibration [7]. Regular incentives are impulse, step, and sinusoidal forces, and the corresponding calibration methods are called impulse response, step response, and frequency response methods, respectively [8]. Previously, the research on

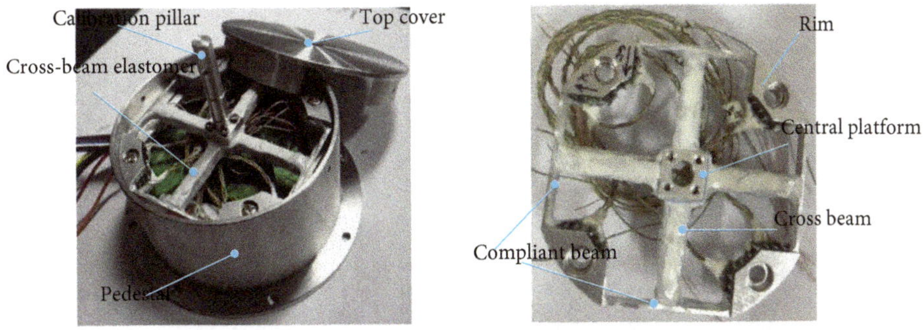

(a) (b)

FIGURE 1: Prototype of the six-axis wrist F/T sensor for the robot. (a) Mechanical composition of the sensor. (b) Crossbeam elastomer.

dynamic characteristics is mainly based on the impulse response method. Xu and Li used a pendulum to exert an impact force along one axis of a multiaxis wrist F/T sensor [9] and developed a dynamic test to compare the dynamic characteristics between the wrist sensor and a JR3® multiaxis F/T sensor [10]. However, only a few channels were utilized, since it is difficult to calibrate the torque, which cannot fully reflect the six-axis wrist F/T sensor's performance. Liu et al. developed a method-based correlation wavelet for measuring the dynamic performance index, but the cantilever was tested, and the robot's multiaxis wrist force sensor was only simulated [11]. In this study, the negative step excitation is constructed by cutting the rope of hanging weight to obtain the step response to carry out the dynamic calibration experiment of the six-axis wrist F/T sensor.

There has been limited literatures on dynamic performance analysis of the multiaxis F/T sensor, but most of them are not comprehensive enough. Song et al. studied dynamic performance of the 2-axis force sensor applied for the force feedback maglev control system by using an impulse stimulation method [12], yet dynamic indices were not attained. Xu and Zhu carried out the dynamic calibration experiment of the six-axis wrist F/T sensor, without building the necessary difference equation model or transfer function model [13]. Ballo et al. employed a six-axis F/T sensor for the frontal impact test on a Hybrid III 50th percentile dummy [14], but the dynamic analysis method was not given. Thus, the purpose of this paper is to analyze the dynamic characteristics of the six-axis F/T sensor comprehensively.

The rest of this paper is organized as follows. Section 2 describes the mechanical structure of the F/T sensor. The mechanical model is established, and the relationship between the natural frequency and weight of the elastomer is obtained in Section 3. In Section 4, the first six natural frequencies and vibration modes are acquired by modeling and modal analysis of the sensor in ANSYS software. In Section 5, dynamic calibration experiment is conducted and the dynamic models are established by model identification. The methods of model identification used in this section are a method for simultaneous identification of order and parameters of the model (SIM) and a method based on the differential evolution (DE) algorithm. Furthermore, the dynamic performance indices of the sensor are achieved. Finally, the concluding remarks follow in Section 6.

2. Sensor Structure and Calibration Method

As shown in Figure 1(a), the prototype of the six-axis wrist F/T sensor for the robot consists of four parts: pedestal, crossbeam elastomer, top cover, and calibration pillar. The sensor's housing is designed as cylindrical structure to connect with the robotic arm and claws. Crossbeam elastomer is the core component of the sensor, and the performance of the sensor mainly relies on it. Figure 1(b) shows the structure of the crossbeam elastomer. It comprises of four crossbeams, four compliant beams, a central platform, and four rims, which are characterized by a compliant beam flexible link at the connection between a crossbeam and two rims. The force or torque is applied to the six-axis F/T sensor through the calibration pillar which transmits the force or torque to the crossbeam elastomer by connecting with the central platform. Figure 2(a) illustrates the loading setup of six-axis forces or torques. The calibration methods of various channels are depicted in Figures 2(b)–2(e).

The excitation force applied to the sensor through the calibration pillar caused deformation of the crossbeam elastomer. Since the compliant beam is very thin, it can be regarded as the flexible link when the applied force is perpendicular to the plane of it and it can be regarded as the ideal rigid body when the applied force is parallel to the plane. Therefore, the force applied to the sensor leads to the bending deformation of two crossbeams whose deformation degree is proportional to the applied force and tension or compression deformation of the other two crossbeams whose deformation degree can be neglected. The resistive strain gauges attached to the bending crossbeams can detect the amplitude of the corresponding force.

3. Theoretical Analysis

3.1. Simplified Statics Model. To get the equivalent stiffness value of the flexible link of the crossbeam elastomer, simplified statics model is established. Figure 3(a) shows the schematic diagram of crossbeam elastomer, where F represents the force applied to the elastomer along Y direction through central platform. Since its width (or height) is much less than the length, the crossbeam can be regarded as a slender beam. When the central platform is subjected to a tensile force along Y direction, bending occurs on crossbeams AB and C

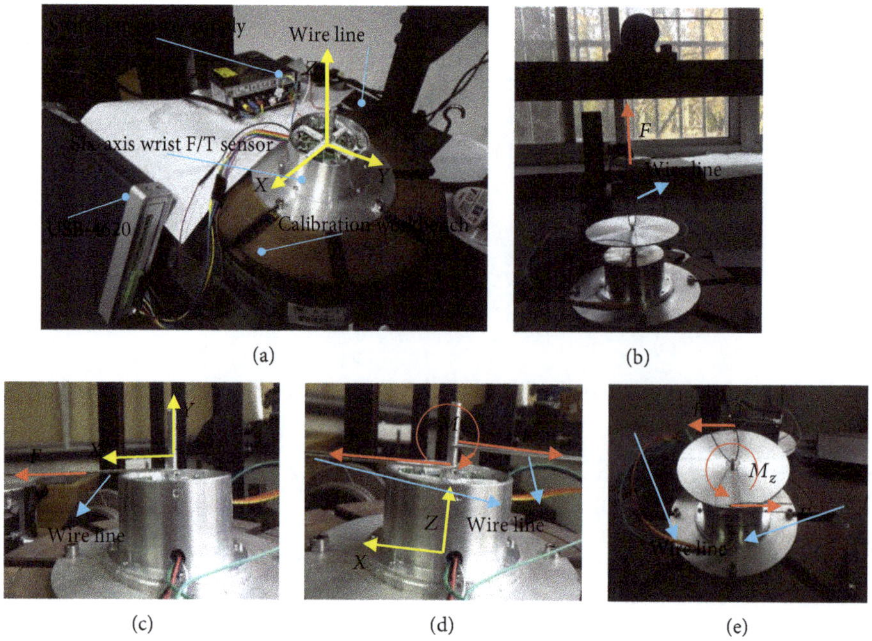

FIGURE 2: Calibration of the six-axis F/T sensor. (a) Calibration experiment setup. (b) Calibration in F_z channel. (c) Calibration in F_x channel. (d) Calibration in M_y channel. (e) Calibration in M_z channel.

D and compliant beams JK and MN. The compliant beams EF and GH are considered as the roller support of the crossbeams AB and CD. As the compliant beams are laminated elements, compliant beams JK and MN are idealized as flexible bodies and crossbeams PQ and RS are idealized as rigid bodies. Thus, the beam AD can be simplified as a simply supported beam with a fixed hinge at one end and a movable hinge at the other, whereas the beams JK and MN as cantilever beams. Figure 3(b) depicts the mechanical model of the whole crossbeam elastomer, where ΔX_e and ΔX_c are the displacement of midpoints of beams AD and JK, respectively, and F_Q and F_C are shear forces on beams AD and JK, respectively. The simplified statics model of the crossbeams AB and CD is shown as Figure 3(c).

According to the principle of the balanced force and moment, the supporting reaction forces of the supports are $F_A = F_B = F_e/2$, where F_e represents the force applied to the beam AD which is component of force F and $F_e = 2F_Q$. The central platform can be regarded as a lumped mass owing to smaller size relative to the crossbeam, that is, the BC segment in Figure 3(d) can be concentrated on a point O. The shear force and bending moment on an arbitrary section of the beam are as follows:

$$F_Q = \begin{cases} \dfrac{F_e}{2}, & 0 \le x \le \dfrac{l}{2}, \\ \dfrac{-F_e}{2}, & \dfrac{l}{2} < x \le l, \end{cases}$$
$$M(x) = \begin{cases} \dfrac{F_e x}{2}, & 0 \le x \le \dfrac{l}{2}, \\ \dfrac{F_e(l-x)}{2}, & \dfrac{l}{2} < x \le l. \end{cases} \tag{1}$$

The deformation of the crossbeam can be expressed by the deflection and rotation angle of a point on this beam. As is shown in Figure 3(d), the point m on this beam is shifted to the point m_1 under the action of the force F_e, the distance between the two points along Y direction is deflection $\omega(x)$, and the angle at which the cross section turns relative to the original position is rotation angle $\varphi(x)$. Both the bending moment and the shear force can cause the deformation of the beam [15].

$$\begin{cases} \omega(x) = v_m(x) + v_q(x), \\ \varphi(x) = \gamma_m(x) + \gamma_q(x), \end{cases} \tag{2}$$

where $v_m(x)$ are $\gamma_m(x)$ are deflection and rotation angle brought on by bending moment, respectively, and $v_q(x)$ and $\gamma_q(x)$ are deflection and rotation angle brought on by shear force, respectively. Based on the differential equation of the deflection and equations (1) and (2), the deflection and rotation angle of a point on this beam are derived [16]:

$$\omega(x) = \begin{cases} \dfrac{F_e}{2EI}\left(\dfrac{x^3}{6} + C_1 x\right) + \dfrac{F_e x}{2KGA} + C_3, & 0 \le x \le \dfrac{l}{2}, \\ \dfrac{F_e l x^2}{4EI} - \dfrac{F_e x^3}{12EI} + C_2 x + C_4 + \dfrac{F_e x}{2KGA}, & \dfrac{l}{2} < x \le l, \end{cases}$$
$$\varphi(x) = \begin{cases} \dfrac{F_e l x^2}{4EI} + C_1, & 0 \le x \le \dfrac{l}{2}, \\ \dfrac{F_e}{2EI}\left(lx - \dfrac{x^2}{2}\right) + C_2, & \dfrac{l}{2} < x \le l, \end{cases} \tag{3}$$

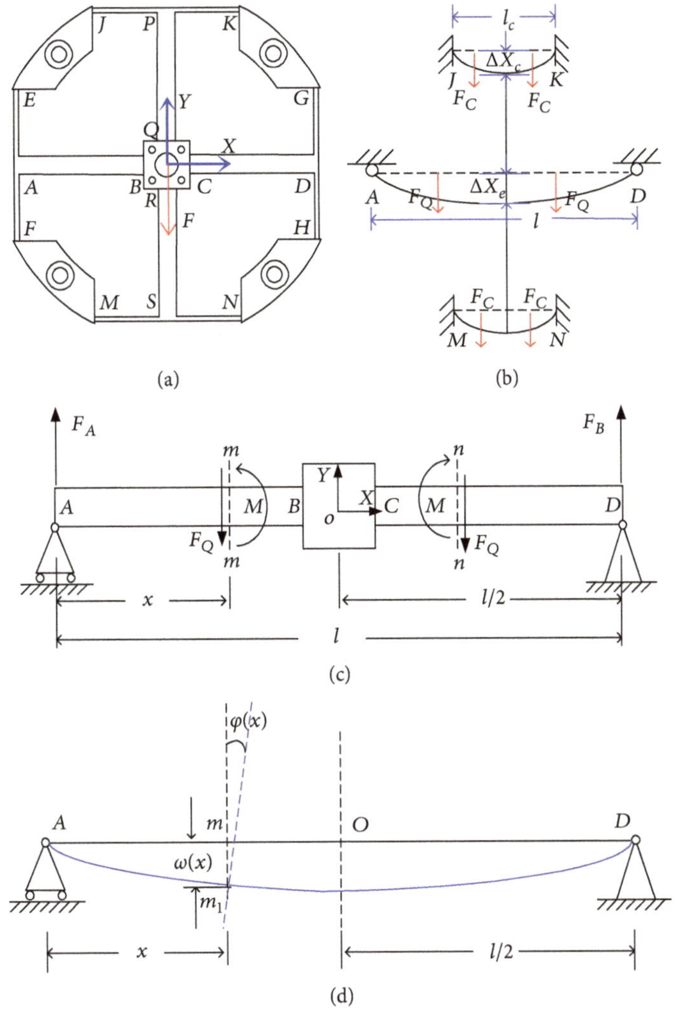

FIGURE 3: The crossbeam elastomer of the six-axis wrist F/T sensor. (a) The schematic diagram of the elastomer. (b) Mechanical model of crossbeam elastomer under F_y. (c) The simplified statics model of the elastomer under F_y. (d) Deformation of the beam AD under force.

where C_1, C_2, C_3, and C_4 are undetermined coefficients whose values can be derived in virtue of boundary and continuous conditions. The final solution of the deflection of the beam is computed as follows:

$$\omega(x) = \begin{cases} \dfrac{F_e l^2 x}{16EI} - \dfrac{F_e x^3}{12EI} + \dfrac{F_e x}{2KGA}, & 0 \le x \le \dfrac{l}{2}, \\ \dfrac{F_e x^3}{12EI} - \dfrac{F_e l x^2}{4EI} + \dfrac{3F_e l^2 x}{16EI} - \dfrac{F_e l^3}{48EI} + \dfrac{F_e (l-x)}{2KGA}, & \dfrac{l}{2} < x \le l, \end{cases} \tag{4}$$

so that the midspan deflection, namely, the displacement of the beam AD, is shown as equation (5). In the same way, the displacement of the beam JK is summarized as equation (6).

$$\omega\left(\frac{l}{2}\right) = \frac{F_e l^3}{48EI} + \frac{F_e l}{4KGA} = \Delta X_e, \tag{5}$$

$$\Delta X_c = \frac{F_G l_c^{\ 3}}{24EI_c} + \frac{2F_G l_c}{KGA_c}, \tag{6}$$

where E and G are the elastic modulus and shear modulus of the material, respectively, $G = E/(1+\mu)$, A and A_c are the sectional area of the beams AD and JK, respectively, and I and I_c are the moment of inertia of the beams AD and JK, respectively. Based on the moment of inertia of a rectangular section, I and I_c can be expressed as equations (7) and (8), respectively.

$$I = \frac{bh^3}{12}, \tag{7}$$

$$I_c = \frac{bh_2^3}{12}. \tag{8}$$

The material of the crossbeam elastomer of the six-axis wrist F/T sensor is aluminium alloy 2024. Parameters of the crossbeam elastomer are shown in Table 1, where b and h denote the width and thickness of crossbeam, respectively, h_2 and l_2 represent the height and length of

Table 1: Parameters of the cr·ossbeam elastomer.

Parameter	b (mm)	h (mm)	l (mm)	h_2 (mm)	l_2 (mm)	Density (kg/m^3)	E (MPa)	μ
Value	3.8	3.8	64	1	32	2820	72,000	0.3

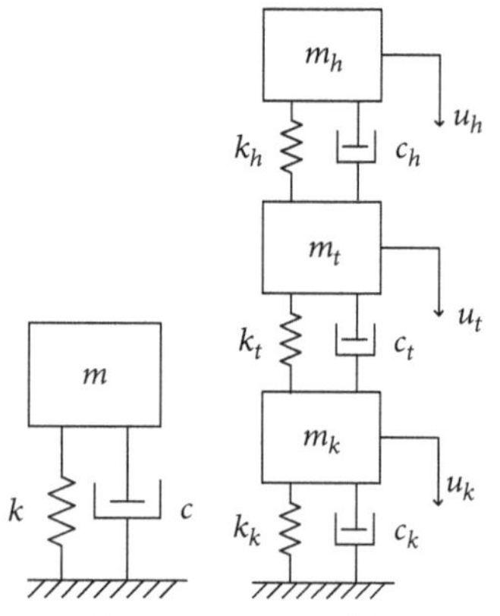

Figure 4: The simplified dynamics models of the six-axis wrist F/T sensor for the robot. (a) A single DOF system. (b) A 3-DOF system.

compliant beam, respectively, and E and μ are the elastic modulus and Poisson's ratio of aluminium alloy 2024 which is used to make of the crossbeam elastomer. Moreover, equation (9) can be obtained according to the geometric characteristic of the deformation.

$$\Delta X_e = \Delta X_c. \tag{9}$$

By substituting equations (5)–(8) to equation (9), the formula is derived as follows:

$$F_Q = 8F_c = 0.4F. \tag{10}$$

Subsequently, the equivalent stiffness coefficient is deducted as follows:

$$k = \frac{F}{\Delta X_e} = \frac{1}{\left(l^3/60EI\right) + \left(l/5KGA\right)}. \tag{11}$$

Substituting the parameters in Table 1 into equation (7), the equivalent stiffness coefficient can be calculated as

$$k = 1.824 \times 10^5. \tag{12}$$

3.2. Simplified Dynamics Model. The six-axis wrist F/T sensor is a multiple input and multiple output (MIMO) system. When the sensor is constrained on a foundation surface, the joint between the foundation surface and the pedestal is a rigid connection, and thus the input-output characteristics in any direction can be simplified as a single degree-of-freedom (DOF) system. When the sensor is mounted on the wrist of the robot, the whole mechanism consists of three components: the pedestal of the sensor, the crossbeam elastomer, and the robotic hand. Every component has one translation/rotation DOF in any direction of the six-axis F/T sensor, so the mechanism is a 3-DOF system. As shown in Figure 4, m, u, k, and c are the mass, displacement of the mass, equivalent stiffness, and damp, respectively, and subscripts h, t, and k are the hand of the robot, the crossbeam elastomer, and pedestal of the sensor, respectively.

Assuming that k_h, k_t, and k_k are all compressed and the displacements of the mass change according to sinusoidal law, the displacements of the masses and the vibration equation are expressed as equations (13) and (14). Then, equation (13) is substituted with equation (14), and equation (15) is obtained.

$$\begin{bmatrix} u_h \\ u_t \\ u_k \end{bmatrix} = \begin{bmatrix} \varphi_h \\ \varphi_t \\ \varphi_k \end{bmatrix} \sin(\omega t + \theta), \tag{13}$$

$$\begin{bmatrix} m_h & 0 & 0 \\ 0 & m_t & 0 \\ 0 & 0 & m_k \end{bmatrix} \begin{bmatrix} \ddot{u}_h \\ \ddot{u}_t \\ \ddot{u}_k \end{bmatrix} + \begin{bmatrix} c_h & -c_k & 0 \\ -c_h & c_h + c_t & 0 \\ 0 & -c & c_t + c_k \end{bmatrix} \begin{bmatrix} \dot{u}_h \\ \dot{u}_t \\ \dot{u}_k \end{bmatrix} + \begin{bmatrix} k_h & -k_h & 0 \\ -k_h & k_h + k_t & -k_t \\ 0 & -k_t & k_t + k_k \end{bmatrix} \begin{bmatrix} u_h \\ u_t \\ u_k \end{bmatrix} = 0, \tag{14}$$

$$Z \begin{bmatrix} \varphi_h \\ \varphi_t \\ \varphi_k \end{bmatrix} \sin(\omega t + \theta) = 0, \tag{15}$$

where Z is mechanical impedance matrix and can be denoted as the following form:

$$Z = \begin{bmatrix} -m_h\omega^2 + j\omega c_h + k_h & -j\omega c_k - k_h & 0 \\ -j\omega c_h - k_h & -m_t\omega^2 + j\omega(c_h + c_t) + k_h + k_t & 0 \\ 0 & -j\omega c_t - k_t & -m_k\omega^2 + j\omega(c_k + c_t) + k_k + k_t \end{bmatrix}. \tag{16}$$

TABLE 2: The results of simplified dynamics analysis of the six-axis F/T sensor.

Channel	Natural frequency (Hz)
F_x	1642
F_y	1642
F_z	1973
M_x	3376
M_y	3376
M_z	4892

The eigenvalue of the matrix Z is the natural frequency of the system. Considering that the joint between robotic hand and the elastomer and the joint between robotic arm and the pedestal are rigid connections, the 3-DOF system can be simplified to a single-DOF system as shown in Figure 4(a). Moreover, the mass is $m = m_h + m_t$, and thus, the natural frequency of the system is resolved in equation (17). Obviously, in case of considering the equivalent mass of the robotic hand, mounted resonance frequency of the six-axis wrist F/T sensor is diminished. In fact, the change of gripping workpiece mass, the mass, shape, and mass distribution of end effector will affect the dynamic characteristics of the sensor. Before testing and evaluating the dynamic properties of the wrist force sensor, it is necessary to determine the state of the sensor and the environment in which it is used. And the unified basis and meaningful results can be obtained under a uniform load.

$$\omega = \sqrt{\frac{k}{m_h + m_t}}. \tag{17}$$

In view of the equivalent stiffness coefficient and the mass of the crossbeam elastomer and calibration pillar (4.9×10^{-2} kg), the natural frequency of the six-axis wrist F/T sensor in F_y direction can be calculated to be 1642 Hz. Similarly, the natural frequencies of the sensor's other channels can be solved as shown in Table 2.

4. Modal Analysis

Vibration modal is one of the intrinsic characteristics of elastic structures. The modal analysis of the elastomer is carried out in ANSYS software, and the inherent vibration properties of the six-axis wrist F/T sensor are studied, including the undamped natural frequency, damped ratio, and vibration modal. The modal extraction method is set to block lanczos, and the number of extraction is 6. The results are shown in Table 3, and six vibration modes are shown in Figures 5(a)–5(f).

TABLE 3: The results of modal analysis of the six-axis F/T sensor.

Set	Natural frequency (Hz)	Vibration mode
1	1660	Translation along X-axis
2	1661	Translation along Y-axis
3	2025	Translation along Z-axis
4	3577	Rotation around X-axis
5	3578	Rotation around Y-axis
6	5094	Rotation around Z-axis

5. Dynamic Calibration Experiment and Model Identification

5.1. Dynamic Calibration Experiment. The step response method is used to calibrate the six-axis F/T sensor. The negative step excitation is constructed by cutting the rope of hanging weight to obtain the step response to carry out the dynamic calibration experiment of the six-axis wrist F/T sensor. One end of the rope is hung on the calibration pillar, and the other end suspended the weight by a pulley. The rope is cut off at a given time, thus a negative step excitation is applied to the six-axis F/T sensor. The cotton rope, fishing line, and ordinary plastic rope are, respectively, used to suspend weights during calibration experiment. Figure 6 shows the three ropes. The test results show that the fishing line has the best effect among the three ropes; hence, the fishing line with diameter of 0.4 mm is used in the dynamic calibration experiment. The step responses of the sensor have been acquired via a high-speed Smacq® USB-4620 data acquisition card and transmitted to the host computer. The sampling frequency of this data acquisition card can be changed by software and set to 100 kHz in the dynamic calibration experiment. Figure 2(a) presents the experimental facilities.

The method of constructing step excitation by cutting the rope of hanging weight during dynamic calibration experiment is simple, but the step rise time is long. Force excitation can be generated in other ways, such as dynamic pressure generator based on a double-acting pneumatic actuator [17].

5.2. Model Identification Based on the SIM Method. Based on the acquired input and output data, the dynamic models are established by model identification. The methods of model identification used in this paper are the SIM method and DE method. The SIM method is a method for simultaneous identification of order and parameters of the model. This method can directly calculate the minimum value of the index function corresponding to all the models of interest, and then the order is determined. Compared with the least square method, this approach has some important features, such as less calculation, simpler program, and more flexible applications. Meanwhile, the problem of numerical "morbid" caused by ill-conditioned regular equations in the least square method is avoided.

Let the differential equation of the six-axis F/T sensor in F_x channel is described as follows:

$$B(d^{-1})u(k) - A(d^{-1})y(k) = \varepsilon(k), \tag{18}$$

where $u(k)$ and $y(k)$ are the input and output observations of the sensor, respectively, d^{-1} is a backward shift operator, $\varepsilon(k)$

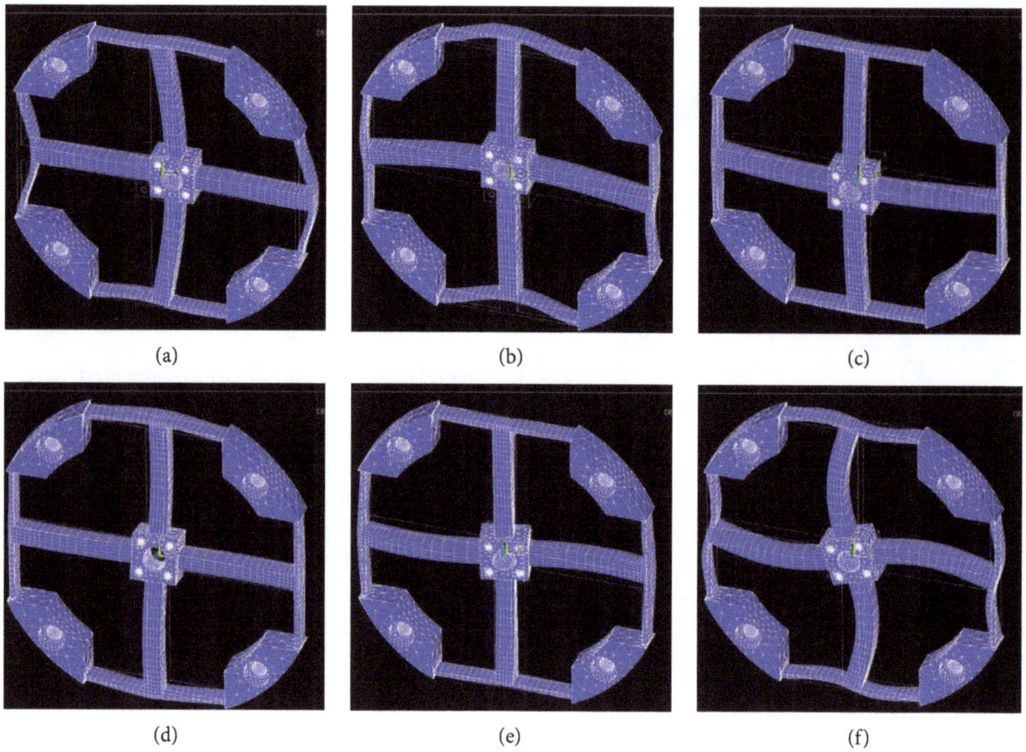

FIGURE 5: The first six-order vibration mode of the six-axis wrist F/T sensor for the robot. (a) The first-order mode. (b) The second-order mode. (c) The third-order mode. (d) The fourth-order mode. (e) The fifth-order mode. (f) The sixth-order mode.

FIGURE 6: The three ropes used in dynamic calibration experiment.

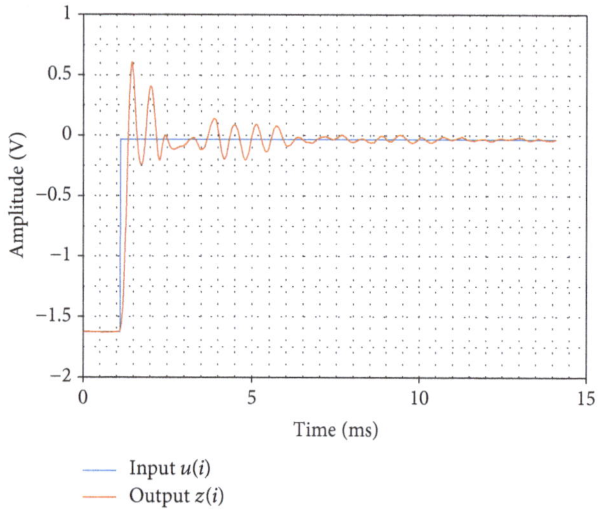

FIGURE 7: The input and output signals of the sensor in F_x channel in dynamic calibration experiment.

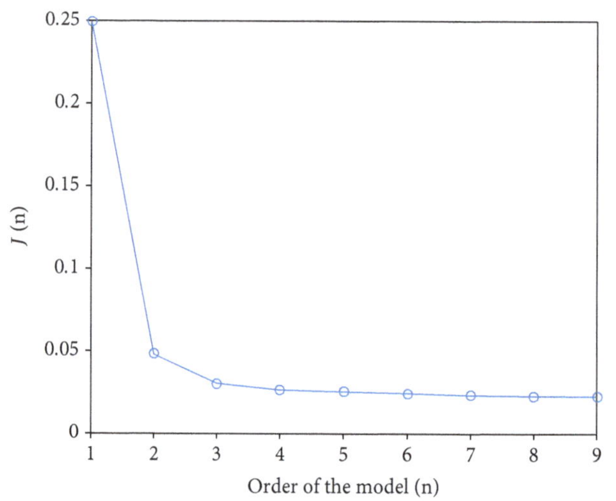

FIGURE 8: The relationship curve between the index function and model order.

is the fitting error, $A(d^{-1}) = 1 + a_1 d^{-1} + a_2 d^{-2} + \cdots + a_n d^{-n}$, and $B(d^{-1}) = b_0 + b_1 d^{-1} + b_2 d^{-2} + \cdots + b_n d^{-n}$.

The negative step excitation constructed and step response acquired in experiment are shown in Figure 7. The relationship curve between the index function, described as the sum of squared residuals, and model order is depicted in Figure 8. It can be seen that the changes after $J(3)$ is very slow, and thus the model order is 3. Then, the model parameter vector can be expressed as equation (19) by the SIM

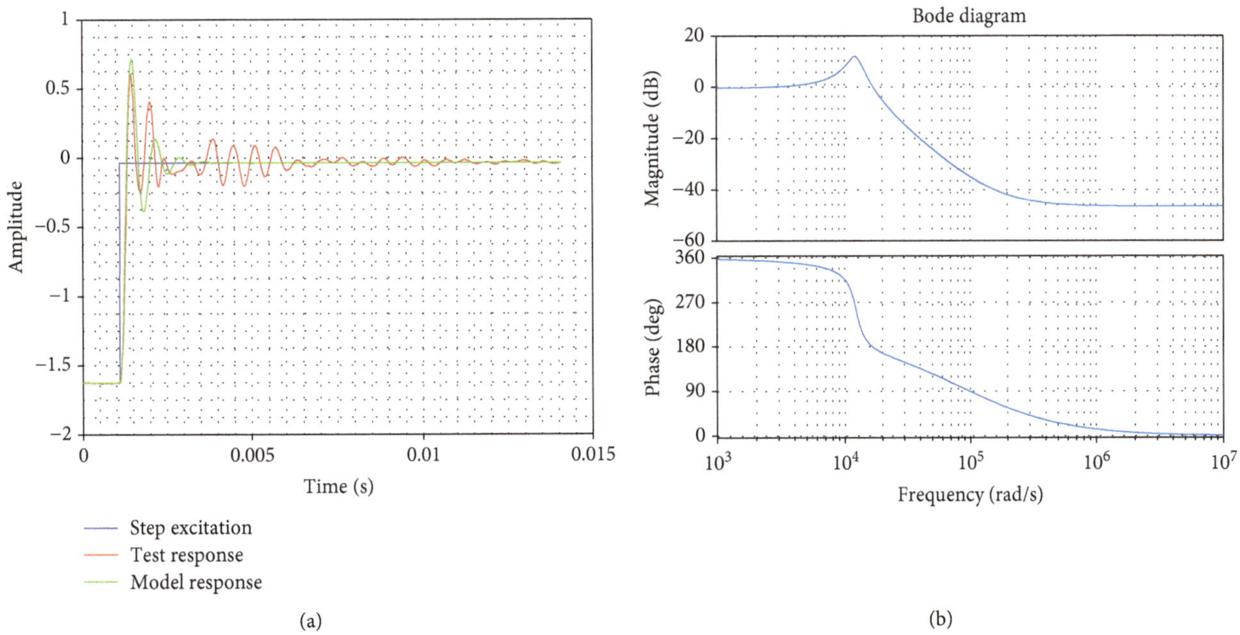

FIGURE 9: The model response and bode diagram of the sensor. (a) The model response in comparison with test response. (b) Bode diagram of the model.

method model identification. The difference equation and transfer function of the sensor are shown in equations (20) and (21), respectively.

$$\theta(n) = [0.0083\ 0.5841\ 0.0047\ -0.2608\ -0.0011\ -1.3076\ 0.0037]^T, \tag{19}$$

$$\begin{aligned} &y(n) - 1.3076y(n-1) - 0.2608y(n-2) + 0.5841y(n-3) \\ &\quad = 0.0037x(n) - 0.0011x(n-1) + 0.0047x(n-2) \\ &\qquad + 0.0083x(n-3), \end{aligned} \tag{20}$$

$$G(s) = \frac{(8.2e-4)s^3 + 4430s^2 - (5.36e8)s + 8.532e13}{s^3 + (8.643e5)s^2 + (5.494e9)s + 8.587e13}. \tag{21}$$

The negative step excitation constructed is used to get the model's response. Figure 9(a) shows that the calculated values of the model fit well with the experimental response curve, and this model can be used. Bode diagram of this model is depicted in Figure 9(b).

From the bode diagram and transfer function, we can see that the natural frequency of the model is 1720 Hz and the working bands of amplitude error within ±10% and within ±5% are 560 Hz and 414 Hz, respectively. Compared with the aforementioned theoretical analysis and modal analysis, this model identified by the SIM method is correct.

So the dynamic models from other channels can be also identified by the SIM method. The dynamic performance indices of the sensor can be obtained as shown in Table 4.

TABLE 4: The dynamic performance indices of the model identified by the SIM method.

Index	F_x	M_y	M_z	F_z
Natural frequency (Hz)	1720	3530	5060	2020
Working band (±10%) (Hz)	560	997	1622	652
Working band (±5%) (Hz)	414	693	1294	475

5.3. Model Identification Based on the DE Algorithm. Differential evolution (DE) algorithm is a kind of the optimization algorithm based on swarm intelligence theory, which is often used to solve the problem of optimization search in complex real space. The special memory ability of the DE algorithm enables it to dynamically track the current search situation to adjust its search strategy. It has strong global convergence ability and robustness and does not need the characteristics' information of the problem. The parameter identification of the complex system can be realized.

The basic idea of the DE algorithm is that starting from a random initial population, a new individual is generated by summing weighted vector difference between any two individuals in the population and the third individual within the certain rules. By comparing to a predetermined individual, if the fitness of the new individual is better, the new individual will replace the old one in the next generation; otherwise the old remains. Through continuous iterative computation, we retain good individuals, eliminate inferior individuals, and guide the search process to approach the optimal solution.

Assumed to be a three-order model, the discrete transfer function is expressed as equation (22) and parameters to be identified can be written as follows: $\widehat{\theta} =$

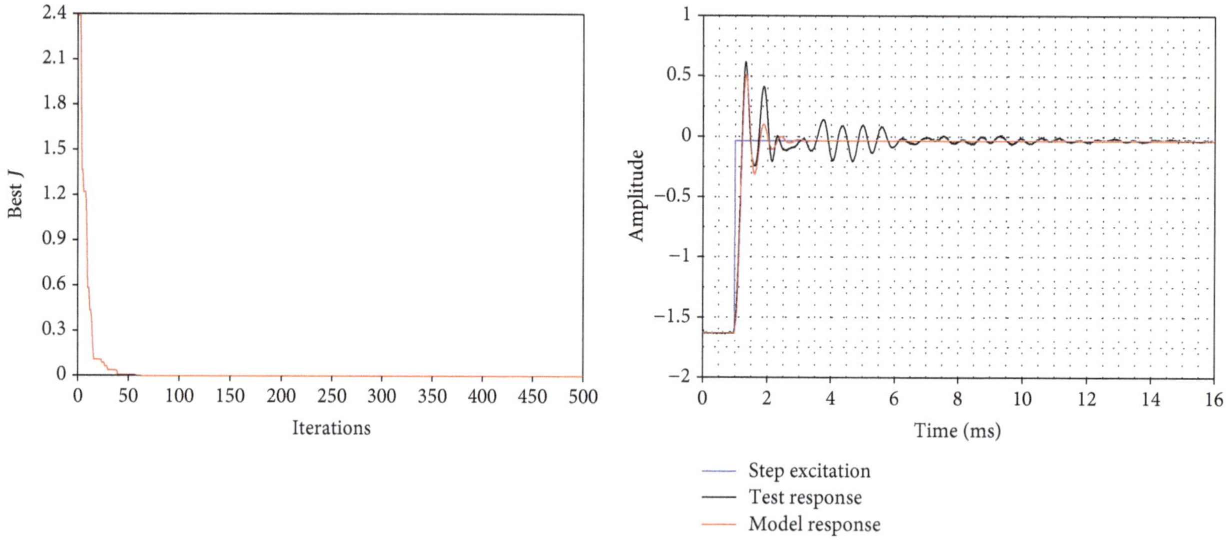

(a) (b)

FIGURE 10: The dynamic model identification of the six-axis wrist F/T sensor in F_x channel. (a) The fitness-iteration relational curve. (b) The response of the model obtained by DE algorithm identification.

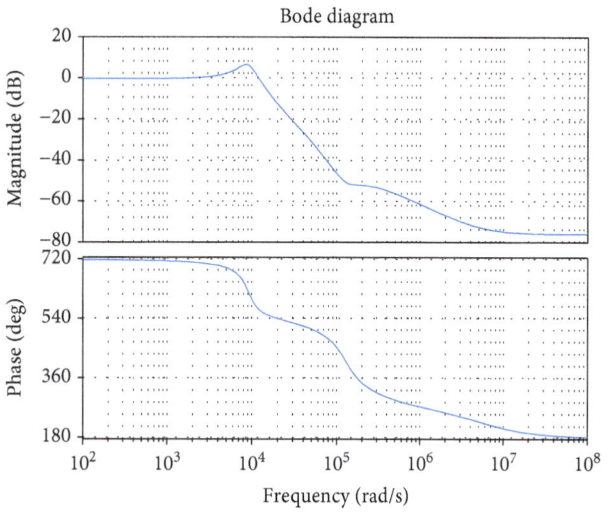

FIGURE 11: The bode diagram of the model identified by the DE algorithm in the sensor's F_x channel.

$[b_3 \quad a_3 \quad b_2 \quad a_2 \quad b_1 \quad a_1]$. The ranges of these parameters are $[-5, 5]$. In the DE algorithm, the mutation factor is 0.8, the cross factor is 0.7, the size of samples is 100, and the maximum of iterations is 500. After 60 steps of iteration, the best sample can be obtained as equation (23). Figures 10(a) and 10(b) depict the fitness-iteration relational curve and the model response, respectively.

$$H(z) = \frac{b_3 z^{-3} + b_2 z^{-2} + b_1 z^{-1}}{a_3 z^{-3} + a_2 z^{-2} + a_1 z^{-1} + 1}, \tag{22}$$

$$\hat{\theta} = [0.0063 \quad -0.8597 \quad -0.01281 \quad 2.702 \quad 0.00798 \quad -2.841]. \tag{23}$$

In the light of the transfer function identified by the DE algorithm, the bode diagram can be depicted as shown in Figure 11.

From Figure 11 and equation (23), it can be obtained that the natural frequency is 1680 Hz and the working bands of amplitude error within ±10% and within ±5% are 530 Hz and 395 Hz, respectively. These results agree with the model identified by the SIM method.

In the same way, the model identification results of M_y, M_z, and F_z channels are shown as Figure 12. The transfer functions of the three channels are identified as equations (24)–(26). The dynamic performance indices are shown as Table 5.

$$H(z) = \frac{-0.01755z^{-1} + 0.04019z^{-2} - 0.02001z^{-3}}{1 - 2.797z^{-1} + 2.615z^{-2} - 0.816z^{-3}}, \tag{24}$$

$$H(z) = \frac{-0.00606z^{-1} + 0.009565z^{-2} - 0.003314z^{-3}}{1 - 1.009z^{-1} - 0.9739z^{-2} + 0.9831z^{-3}}, \tag{25}$$

$$H(z) = \frac{-0.0007735z^{-1} - 0.002627z^{-2} + 0.005051z^{-3}}{1 - 1.053z^{-1} - 0.8864z^{-2} + 0.941z^{-3}}. \tag{26}$$

Combined with the aforementioned models, the dynamic performance indices are integrated into Table 6. It can be seen in the table that the analysis results of these models are consistent and the first natural frequency is more than 1600 Hz.

6. Conclusions

In this study, a systematic theoretical analysis, modal analysis, dynamic calibration experiment, and model identification are carried out for testing and verifying the six-axis

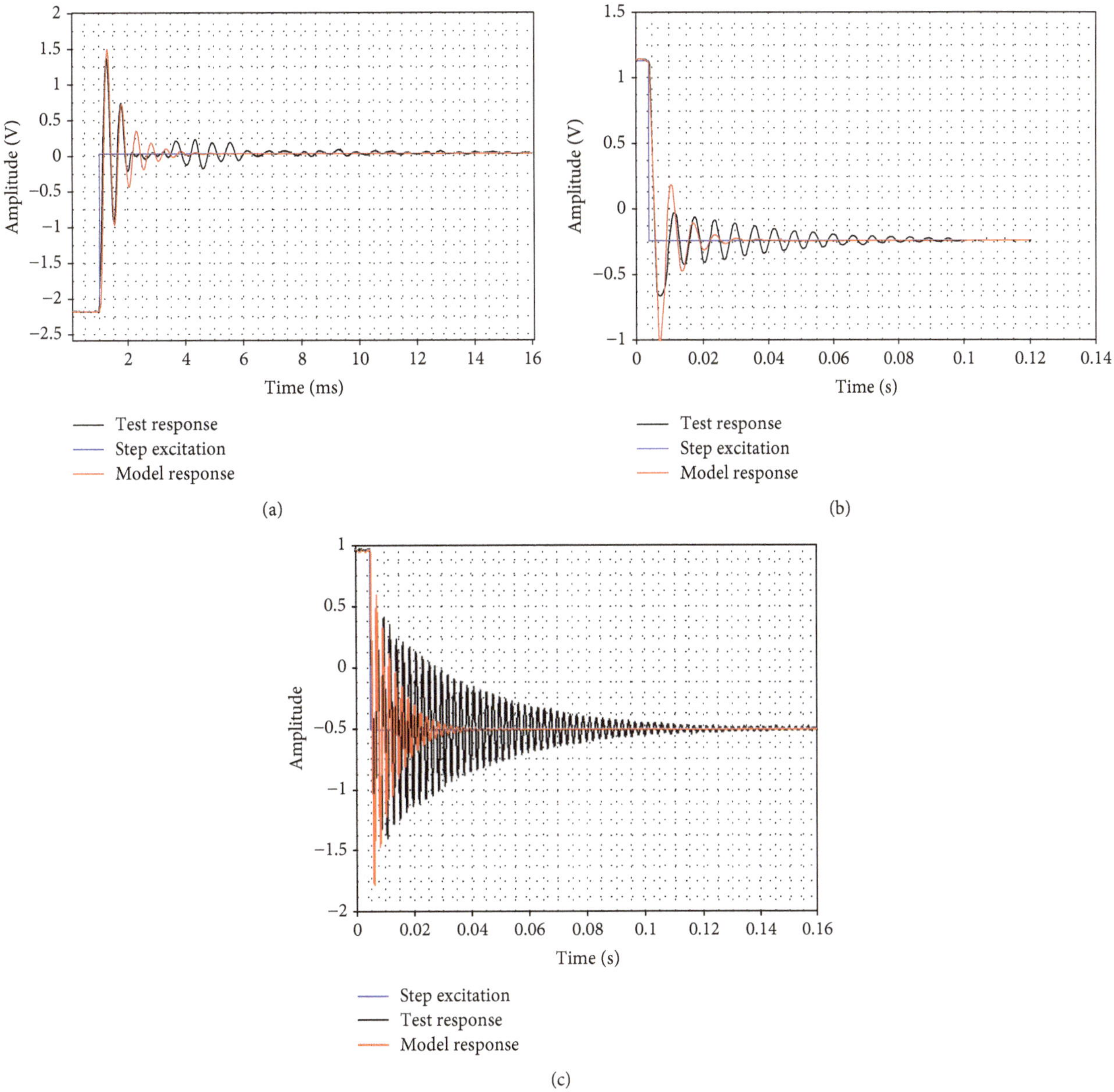

FIGURE 12: The model's identification results of other channels. (a) M_y channel. (b) M_z channel. (c) F_z channel.

TABLE 5: The dynamic performance indices of the model identified by the DE algorithm.

Index	F_x	M_y	M_z	F_z
Natural frequency (Hz)	1680	3460	4910	1986
Working band (±10%) (Hz)	530	980	1610	648
Working band (±5%) (Hz)	405	678	1285	466

wrist F/T sensor. The SIM method and DE algorithm are used to identify the dynamic models. From Tables 2–6, the maximum difference among the aforementioned four models within 5.6% and the dynamic performance indices obtained by the identified models are consistent with the results from the theoretical analysis and thus showing the identified models are reasonable. The results from the dynamic analysis for the six-axis wrist F/T sensor provide a theoretical basis for the application of the sensor in the dynamic working environment.

The results show that this sensor has a wide dynamic range with the first resonant frequency at more than 1600 Hz, working bands (±5%) are more than 400 Hz, and the step response oscillation is intense. To widen its working band and reduce the overshoot, it is necessary to add a compensation link to the sensor which will be the future work. In this dynamic calibration experiment, each channel is individually excited and only this channel's responses are acquired simultaneously, so the coupling is weak and there are no decoupling in this experiment. However, the

TABLE 6: The natural frequencies derived from various models.

Model	F_x	M_y	M_z	F_z
Simplified dynamics model	1642 Hz	3376 Hz	4892 Hz	1973 Hz
Finite element model	1660 Hz	3578 Hz	5094 Hz	2025 Hz
The model identified by the SIM method	1720 Hz	3530 Hz	5060 Hz	2020 Hz
The model identified by the DE algorithm	1680 Hz	3460 Hz	4910 Hz	1986 Hz
Maximum difference among the above models	4.5%	5.6%	4.0%	2.6%

coupling is serious in actual dynamic applications, and dynamic decoupling of the six-axis F/T sensor will also be the future work.

Conflicts of Interest

The authors declare that there is no conflict of interest regarding the publication of this paper.

Acknowledgments

This work was supported by the National Key R&D Program of China (no. 2016YFB1001301), National Nature Science Foundation of China (no. U1713210), and Suzhou Science and Technology Planning Project (no. SGC201653).

References

[1] Y. Zhao, C. Zhang, D. Zhang, Z. Shi, and T. Zhao, "Mathematical model and calibration experiment of a large measurement range flexible joints 6-UPUR six-axis force sensor," *Sensors*, vol. 16, no. 8, p. 1271, 2016.

[2] E. Korkmaz, B. A. Gozen, B. Bediz, and O. B. Ozdoganlar, "High-frequency compensation of dynamic distortions in micromachining force measurements," *Procedia Manufacturing*, vol. 1, pp. 534–545, 2015.

[3] P. Q. Khang, N. M. Dung, N. B. Khiem, H.-P. Phan, K. Matsumoto, and I. Shimoyama, "Multi-axis force sensor with dynamic range up to ultrasonic," in *2014 IEEE 27th International Conference on Micro Electro Mechanical Systems (MEMS)*, San Francisco, CA, U.S.A, January 2014.

[4] E. Kyrkjebø, "Inertial human motion estimation for physical human-robot interaction using an interaction velocity update to reduce drift," in *Companion of the 2018 ACM/IEEE International Conference on Human-Robot Interaction - HRI '18*, pp. 163-164, Chicago, IL, USA, March 2018.

[5] Y.-Q. Leng, Z.-C. Chen, X. He, Y. Zhang, and W. Zhang, "Collision sensing using force/torque sensor," *Journal of Sensors*, vol. 2016, Article ID 6291216, 10 pages, 2016.

[6] R. Osypiuk, J. Piskorowski, and D. Kubus, "A method of improving the dynamic response of 3D force/torque sensors," *Mechanical Systems and Signal Processing*, vol. 68-69, pp. 366–377, 2016.

[7] J. Q. Huang, *Measurement System Dynamics and Application*, National Defense Industry Press, 1st edition, 2013.

[8] G. Qin, "Research on a new robot wrist force sensor," M.S. Thesis, Southeast University, Nanjing, China, 2004.

[9] K.-J. Xu and C. Li, "Dynamic decoupling and compensating methods of multi-axis force sensors," *IEEE Transactions on Instrumentation and Measurement*, vol. 49, no. 5, pp. 935–941, 2000.

[10] K. Li, B. Pan, Y. Fu, and S. Wang, "Experimental study of static and dynamic characteristics of a miniature 6-axis force and torque sensor," in *2015 IEEE International Conference on Information and Automation*, Lijiang, China, August 2015.

[11] Z. Liu, Y. Wang, E. Chen, L. Wen, and Y. Ge, "A method for measuring dynamic performance index of robot's multi-axis wrist force sensor," in *2005 IEEE International Conference on Information Acquisition*, Hong Kong, China, June–July 2005.

[12] F. Song, L. Deng, and B. Song, "Experimental research on dynamic performances of 2-dimensional force sensor applied for force feedback magnetic levitation control system," in *Proceeding of the 11th World Congress on Intelligent Control and Automation*, pp. 136–141, Shenyang, China, June-July 2014.

[13] K. J. Xu and Z. Zhu, "Experimental modelling of six-axis wrist force sensor based on step responses," *ROBOT*, vol. 22, pp. 251–255, 2000.

[14] F. Ballo, M. Gobbi, G. Mastinu, and G. Previati, "A six axis load cell for the analysis of the dynamic impact response of a hybrid III dummy," *Measurement*, vol. 90, pp. 309–317, 2016.

[15] J. Ma and A. Song, "Fast estimation of strains for cross-beams six-axis force/torque sensors by mechanical modeling," *Sensors*, vol. 13, no. 5, pp. 6669–6686, 2013.

[16] W. Ke, F. Du, and X. Zhang, "Algorithm and experiments of six-dimensional force/torque dynamic measurements based on a Stewart platform," *Chinese Journal of Aeronautics*, vol. 29, no. 6, pp. 1840–1851, 2016.

[17] A. Svete, M. Štefe, A. Maček, J. Kutin, and I. Bajsić, "Dynamic pressure generator for dynamic calibrations at different average pressures based on a double-acting pneumatic actuator," *Sensors and Actuators A: Physical*, vol. 247, pp. 136–143, 2016.

17

A Novel Bioinspired Multiobjective Optimization Algorithm for Designing Wireless Sensor Networks in the Internet of Things

Jun Huang,[1] Liqian Xu,[1] Cong-cong Xing,[2] and Qiang Duan[3]

[1]*School of Communication and Information Engineering, Chongqing University of Posts and Telecommunications, Chongqing 400065, China*
[2]*Math and Computer Science Department, Nicholls State University, Thibodaux, LA 70310, USA*
[3]*Information Science and Technology Department, The Pennsylvania State University, Abington, PA 19001, USA*

Correspondence should be addressed to Jun Huang; xiaoniuadmin@gmail.com

Academic Editor: Anand Paul

The design of wireless sensor networks (WSNs) in the Internet of Things (IoT) faces many new challenges that must be addressed through an optimization of multiple design objectives. Therefore, multiobjective optimization is an important research topic in this field. In this paper, we develop a new efficient multiobjective optimization algorithm based on the chaotic ant swarm (CAS). Unlike the ant colony optimization (ACO) algorithm, CAS takes advantage of both the chaotic behavior of a single ant and the self-organization behavior of the ant colony. We first describe the CAS and its nonlinear dynamic model and then extend it to a multiobjective optimizer. Specifically, we first adopt the concepts of "nondominated sorting" and "crowding distance" to allow the algorithm to obtain the true or near optimum. Next, we redefine the rule of "neighbor" selection for each individual (ant) to enable the algorithm to converge and to distribute the solutions evenly. Also, we collect the current best individuals within each generation and employ the "archive-based" approach to expedite the convergence of the algorithm. The numerical experiments show that the proposed algorithm outperforms two leading algorithms on most well-known test instances in terms of Generational Distance, Error Ratio, and Spacing.

1. Introduction

The Internet of Things (IoT) is an emerging paradigm for information collection, communications, process, and application that is rapidly gaining ground in a wide variety of fields, including manufacture, transportation, healthcare, environment surveillance, and many other areas. The basic idea of this new networking and computing paradigm is the pervasive presence of a variety of objects (things), such as sensors, actuators, mobile devices, and RFID tags, which are able to interact with each other and communicate with the Internet infrastructure. Wireless sensor networks (WSNs) form an indispensable ingredient of the IoT for data collections and transmissions, thus having a significant impact on the overall service performance of the IoT. Therefore, optimization of the WSN design plays a crucial role in constructing a high performance IoT for meeting the requirements of various applications.

Designing a large scale WSN for supporting the IoT faces many challenges. These challenges are mainly caused by the limited resources available in WSNs, including battery lifetime, computing capacity, and memory space at sensor nodes, and also by dynamic network topology especially in ad hoc sensor networks. On the other hand, due to the wide spectrum of applications that may be supported by the IoT in the future, WSN design faces various design goals that often conflict with each other, for example, short delay, high throughput, minimal energy consumption, and low cost. Network design in such application scenarios must consider multiple factors in order to achieve trade-offs among multiple objectives to achieve optimal overperformance for the entire network. Therefore, multiobjective optimization algorithms (MOAs) are fundamentally important to WSN design in the IoT development.

MOAs may be employed to address various key problems in WSN design. For example, the sensor node placement problem involves tradeoffs between multiple objectives such as maximum coverage and minimum energy consumption. When sensor nodes are organized in clusters, the selection of cluster heads is also a challenging issue that must be tackled with a multiobjective optimization vision. Routing strategy is another fairly important problem in WSN design [1], which may have a direct impact on multiple aspects of network performance, including data transmission delay, throughput, and lifetime of individual nodes as well as the entire network. Mobile agents are often used in WSNs to visit a sequence of sensors and fuse impotent data. Optimal agent routes also often need to meet multiple objectives such as minimizing total path delay, loss, and energy consumption as well [2]. These multiobjective problems are all challenging problems that need complex algorithms. On the other hand, special features of WSNs, such as limited computing power, storage space, and battery capacity in particular, bring in new requirements on time and space efficiency of the MOAs that can be applied in such an environment. This motivates research on devising new algorithms for multiobjective optimizations.

During the last twenty years, many MOAs have been proposed [3, 4]. In particular, the multiobjective evolutionary algorithms have been extensively studied [5–14]. For instance, the nondominated sorting genetic algorithm II (NSGA-II) [12] and the strength Pareto evolutionary algorithm 2 (SPEA2) [10] are the most favorable ones used in the field of engineering, economics, and business management [15–17]. On the other hand, some other MOAs have also been proposed with the development of bioinspired heuristics [18–21]. The representative one in this line of research is the multiobjective particle swarm optimization (MOPSO), which is a multiobjective optimizer extended by particle swarm optimization (PSO). In addition, Coello et al. [21] showed that MOPSO can achieve a better performance than both NSGA-II and SPEA2 with some well-known testing instances.

Inspired by a biological experiment of ants' chaotic behavior, a chaotic ant swarm (CAS) optimization algorithm that models ants' chaotic behaviors mathematically has been given in [22]. Specifically, CAS is a derivative-free method; the individuals of the ant colony exchange information and benefit from either their own experiences or the experiences of other individuals, while exploring promising areas of the search space. The model and mechanism of CAS are different from those of ant colony optimization (ACO) [23]. The CAS algorithm has been successfully applied to various areas such as fuzzy system identification [24], economic dispatch [25], computation of multiple global optima [26], data clustering, and parameter identification [27].

In this paper, we mainly focus on extending the CAS algorithm to a multiobjective optimizer. The major contributions of this paper are summarized as follows.

(i) We employ the basic concepts of "nondominated sorting" and "crowding distance" referred to by NSGA-II to allow the proposed algorithm to obtain the true or near Pareto optima.

(ii) We redefine the rule of "neighbor" selection for each individual (ant) to enable the algorithm to converge and to distribute the solutions evenly. Also, we allow CAS to collect the current best individuals within each generation and we employ the "archive-based" approach to speed up the convergence of the algorithm.

(iii) We conduct thorough comparisons between CAS and two leading multiobjective optimization algorithms (MOPSO and NSGA-II) over representative benchmarks. The results show that the proposed algorithm outperforms MOPSO and NSGA-II on most of testing instances in terms of *Generational Distance, Error Ratio*, and *Spacing*.

The remainder of this paper is organized as follows. Section 2 presents notations and basic concepts of multiobjective optimization. Section 3 briefly describes the CAS algorithm. The multiobjective version of CAS and its related analysis are proposed in Section 4. Section 5 shows the experimental results and compares the average performance of the proposed algorithm against that of both MOPSO and NSGA-II with benchmark testing problems. Section 6 draws the conclusion of this paper.

2. Notations and Basic Concepts

In this section, we introduce the basic concepts of multiobjective optimization. Terminology is consistent with that in [12, 13, 21].

Definition 1 (global minimum). Given a function $f : \Omega \subseteq R^n \rightarrow R$, $\Omega \neq \Theta$ for $\vec{x} \in \Omega$ the value $f^* \equiv f(\overrightarrow{x^*}) > -\infty$ is called global minimum if and only if

$$\forall \vec{x} \in \Omega : f\left(\overrightarrow{x^*}\right) \leq f(\vec{x}). \quad (1)$$

In this case, $\overrightarrow{x^*}$ is the minimum solution, f is the objective function, the set Ω is the feasible region ($\Omega \in S$), Θ is the empty set, and S represents the whole search space.

Definition 2 (multiobjective optimization problem). The multiobjective optimization problem is to minimize $F(\vec{x}) = (f_1(\vec{x}), \ldots, f_m(\vec{x}))$ subject to $\vec{x} \in \Omega$, where Ω is the decision (variable) space, R^m is the objective space, and $F : \Omega \rightarrow R^m$ consists of m real-valued objective functions.

Definition 3 (dominate, Pareto set, Pareto front). Let $u = (u_1, \ldots, u_m)$, $v = (v_1, \ldots, v_m) \in R^m$ be two vectors; u is said to *dominate* v if $u_i \leq v_i$ for all $i = 1, \ldots, m$, and $u \neq v$. A point $\overrightarrow{x^*} \in \Omega$ is called (globally) Pareto optimal if there is no $\vec{x} \in \Omega$ such that $F(\vec{x})$ dominates $F(\overrightarrow{x^*})$. The set of all Pareto optimal points, denoted by PS, is called the *Pareto set*. The set of all Pareto objective vectors, $\text{PF} = \{F(\vec{x}) \in R^m, \vec{x} \in \text{PS}\}$, is called the *Pareto front*.

3. Chaotic Ant Swarm Algorithm

The phenomenon and behaviors of social insect societies such as ant colonies have fascinated many researchers in recent years. What particularly strikes the occasional observers as well as the scientists is the high degree of societal organization of ants; that is, the ant colonies can accomplish complex tasks in spite of the limited capability of individual ants. Through information exchange, the global self-organization behavior of social ants emerges. Such emergent behavior was reported in Leptothorax ant colonies which was related to the chaotic dynamics. Global oscillations of colony activity were also reported together with the observation that individual ant behavior can be characterized by low-dimensional strange attractors [28]. Meanwhile, the study of ant chaotic behaviors and of their self-organizing capacities has greatly interested computer scientists in developing models for distributed organizations which are useful in solving difficult optimization and distributed control problems. In the following, we will give the detailed chaotic ant swarm algorithm based on some biological observations and investigations on the chaotic and self-organizing behaviors of ants.

3.1. Chaotic and Self-Organizing Behaviors of Ants. The chaotic behavior of insect was first presented by Cole in his experimental studies on activity cycles of Leptothorax ants [28]. Cole used a solid-state automatic digital camera to examine the dynamical behaviors of isolated individual ants and ant colony. He finally found some interesting results, "the ant behavior may be chaotic. The attractor of the movement activity of single, isolated Leptothorax longispinosus ants has a small, non-integer dimension characteristic of low-dimensional chaos. The activity of entire colonies of ants yields an integer dimension that is consistent with periodicity in activity." He even speculated that, "The existence of chaos animal behavior can have several important implications. Variation in the temporal component of individual behavior may not be due simply to chance variations in the stochastic world, but to deterministic processes that depend on initial conditions." Inspired by Cole's observation, we give a nonlinear dynamic model called CAS to describe the chaotic and self-organizing behaviors of ants.

3.2. Nonlinear Dynamic Model of Chaotic Ant Swarm. ACO algorithms as well as CAS algorithm are both inspired by ant behaviors, but the fundamental principle of these two algorithms are quite different. ACO algorithms, which are based on probability theory, explain how ants find the shortest paths between food sources and their nests by disseminating pheromone. However, ACO algorithms do not consider any chaotic behaviors of single ants. Chaotic ant swarm algorithm, on the other hand, is based on chaotic search strategy and self-organizing of ant colony. We now give the nonlinear dynamic model of CAS as follows.

Consider an ant colony including n ants, which located in the decision (search) space Ω and they try to minimize a function $F : \Omega \rightarrow R^m$. The position of each ant i is denoted by $s_i = (z_{i1}, \ldots, z_{iL})$, where $i = 1, 2, \ldots, n$.

In the initial stage, we employ the chaotic system $z(t+1) = z(t)e^{[3-\psi z(t)]}$ in our equation to perform chaotic search. The system obtained from $z(t+1) = z(t)e^{\mu[1-z(t)]}$ is described by Sole et al. in [29] (for more details please see our previous work [22–28, 30]). Each ant is influenced by its current position, the best position so far, its neighbors and the organization process of the swarm. The individual ant's chaotic behavior is adjusted by introducing a successive decrement of the organization variable $y_i(t)$. Such an organization variable can lead the individual to moving to the new site then acquiring the best fitness. In addition, to achieve the goals of the information exchange between individuals and the movements to the new site aiming on the best fitness, we further introduce an quantity $[p_{id}(t-1) - z_{id}(t-1)]$. The term p_{id} is selected based on the fitness theory which has been substantially developed in optimization theory. Thus, we obtain the following detailed dynamical optimization system of the chaotic ant swarm:

$$\begin{aligned} y_i(t) &= y_i(t-1)^{(1+r_i)}, \\ z_{id}(t) &= z_{id}(t-1)\, e^{[1-e^{-ay_i(t)}][3-\psi_d z_{id}(t-1)]} \\ &\quad + [p_{id}(t-1) - z_{id}(t-1)]\, e^{-2ay_i(t)+b}, \end{aligned} \tag{2}$$

where a is a sufficiently large positive constant and can be selected as $a = 200$, b is a constant and $0 \le b \le 2/3$, ψ_d determines the selection of the search range of the dth element of the variable in search space, $r_i \ge 0$ is termed as the organization factor of ant i, $y_i(0) = 0.999$, and $z_{id}(t)$ is the current state of the dth dimension of the individual ant i, where $d = 1, 2, \ldots, L$. $p_{id}(t-1)$ is the best position found by the ith ant and its neighbors within $t-1$ steps, $y_i(t)$ is the current state of the organization variable, t means the current time step, and $t-1$ is the previous step.

$y_i(t)$ and r_i control the convergence of (2); $y_i(t)$ may converge faster if r_i gets larger. The organization factor $r_i = 0$ means the ant swarm is not organized. Under this condition, $e^{-ay_i(t)}$ and $e^{-2ay_i(t)+b}$ are approximately equal to zero, and (2) can be rewritten as $z_{id}(t) = z_{id}(t-1)e^{[3-\psi_d z_{id}(t-1)]}$. If r_i is very large, then the total time of "chaotic" search is small and the system converges very quickly; however, the desired optima (or near optima) cannot be obtained in this case. On the other hand, if r_i is small, then the time of chaotic search is large, the system converges slowly, and the running time will be longer. The value of r_i is thus chosen between $0 \le r_i \le 0.5$. To denote a different r_i of each ant, for example, we can set $r_i = 0.1 + 0.2\,\mathrm{rand}()$, where rand( ) is a uniformly distributed random number in the interval $[0, 1]$. After the chaotic search, $y_i(t)$ approximately equals zero and the convergence of (2) will be mainly determined by $z_{id}(t) = z_{id}(t-1) + \exp(b)[p_{id}(t-1) - z_{id}(t-1)]$. Under this condition, when $0 < b < \ln(2)$, the state $z_{id}(t)$ of (2) will converge to $p_{id}(t)$.

We found that the above model of chaotic ant swarm equation (2) searches for optima in constrained positive or negative intervals. Specifically, if $\psi_d > 0$, (2) can be used to realize the search process in the intervals in which all $z_{id}(t) \ge 0$, and if $\psi_d < 0$, (2) can be used to realize the search process

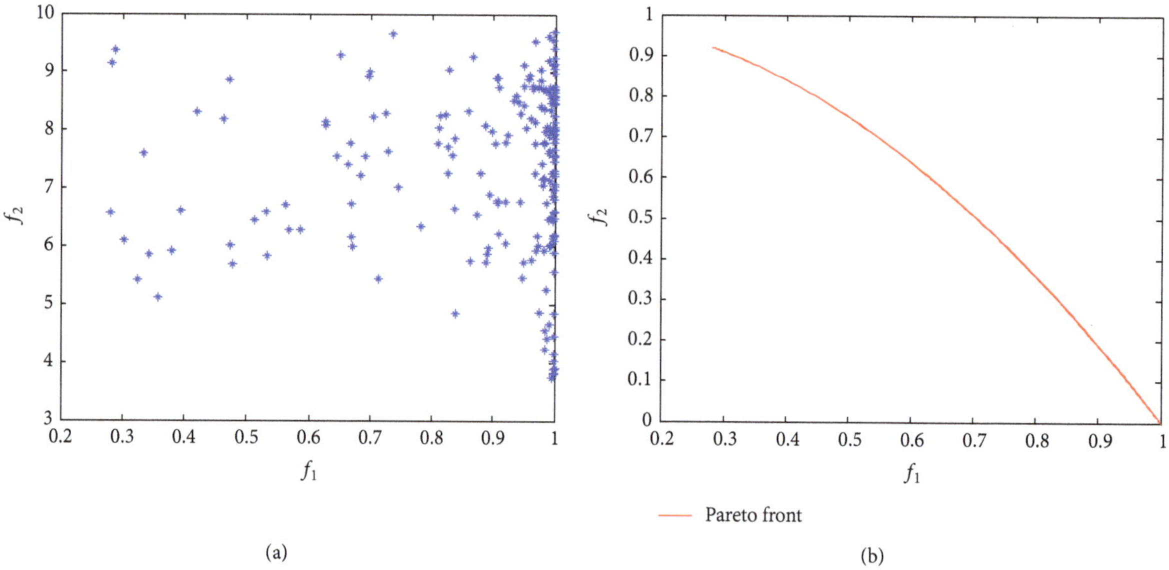

FIGURE 1: Solutions (a) generated by the algorithm and the Pareto front (b) for problem ZDT6.

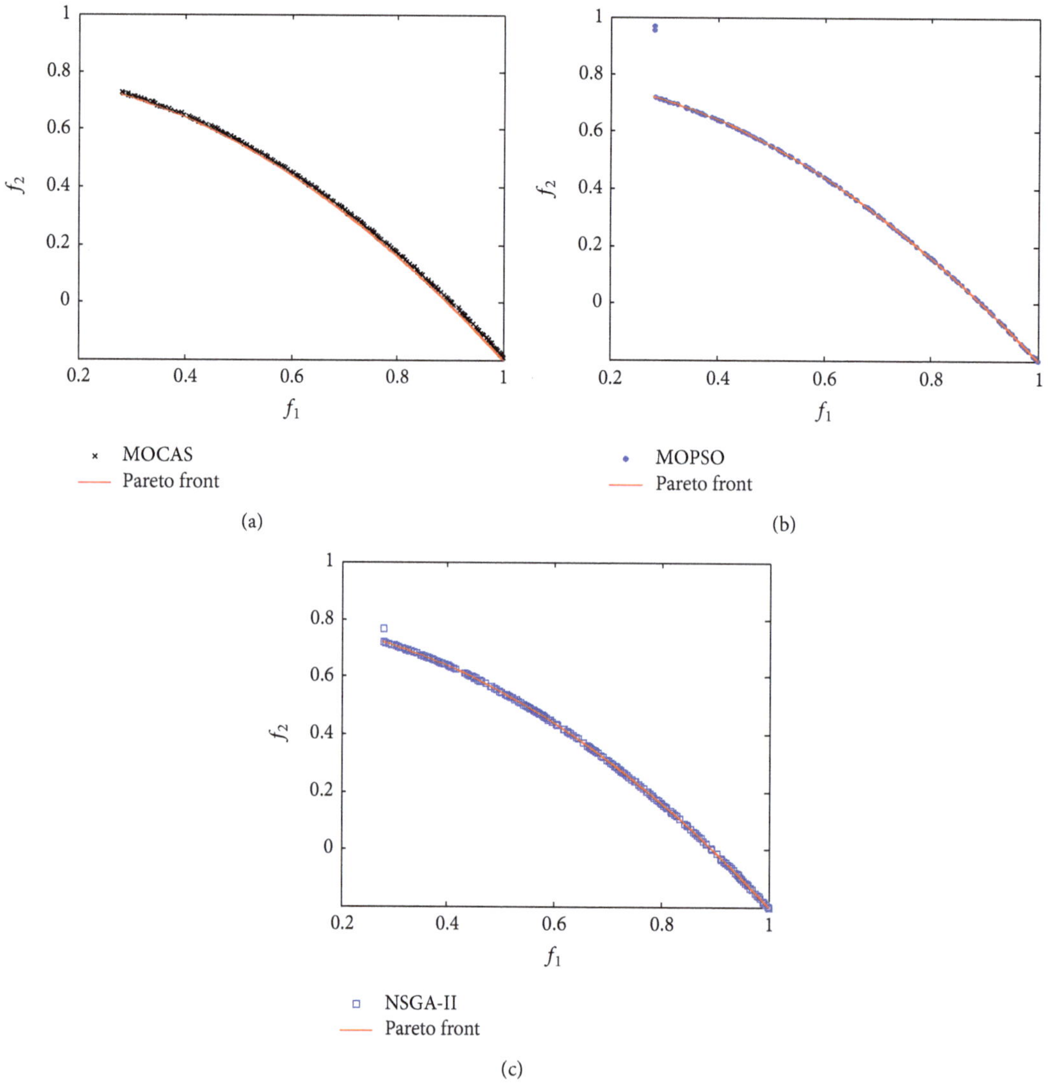

FIGURE 2: Pareto front generated by the MOCAS (a), MOPSO (b), and NSGA-II (c) for P1.

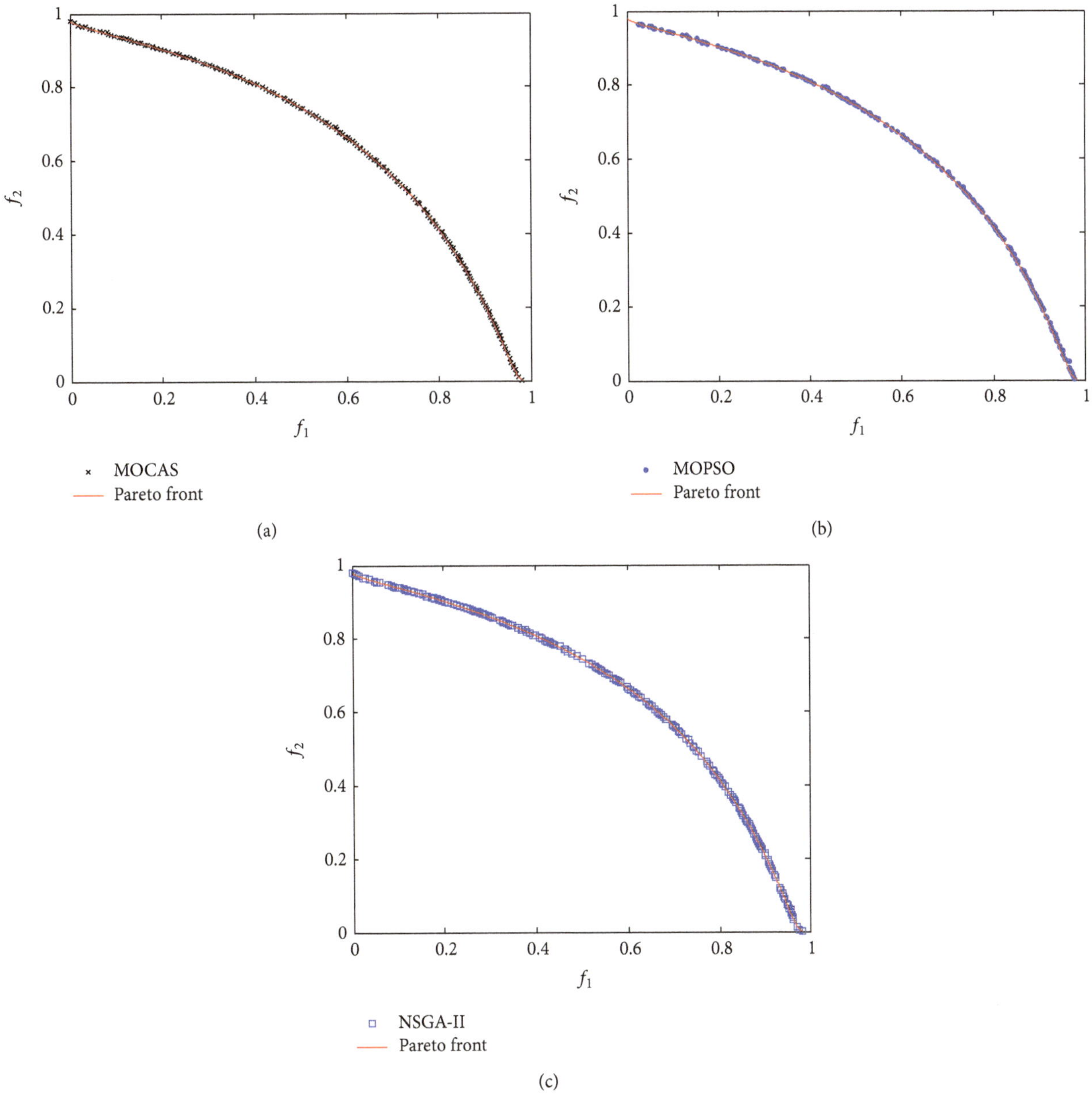

FIGURE 3: Pareto front generated by the MOCAS (a), MOPSO (b), and NSGA-II (c) for P2.

in the intervals for all $z_{id}(t) \leq 0$. We therefore introduce $V_i(7.5/\psi_d)$ and give the following version of CAS model:

$$\begin{aligned} y_i(t) &= y_i(t-1)^{(1+r_i)}, \\ z_{id}(t) &= \left[z_{id}(t-1) + V_i \frac{7.5}{\psi_d} \right] e^{[1-e^{-a y_i(t)}][3-\psi_d z_{id}(t-1)]} \\ &\quad + \left[p_{id}(t-1) - z_{id}(t-1) \right] e^{-2a y_i(t)+b} - V_i \frac{7.5}{\psi_d}, \end{aligned} \tag{3}$$

where $0 \leq V_i \leq 1$ determines the search region of ant i and offers the advantage that ants could search diverse regions of the problem space. If $V_i = 1/2$, it means that the chaotic attractor of ant i moves a half to the negative orientation compared to the chaotic attractor of (2). The values of V_i should be suitably selected according to the actual optimization problems. We call (3) the general algorithmic model of chaotic ant swarm. In this model we can select the initial position of an individual ant as $z_{id}(0) = 7.5/\psi_d(1 - V_i)\,\text{rand}(\,)$, where $\psi_d > 0$. More details about CAS are given in [22].

4. Multiobjective Chaotic Ant Swarm Algorithm

4.1. From Single-Objective to Multiobjective. Equation (3) enables the CAS algorithm to obtain the optima or near optima with "satisfactory" precision when handling single-objective optimization problems [22–28, 30]. In a naïve

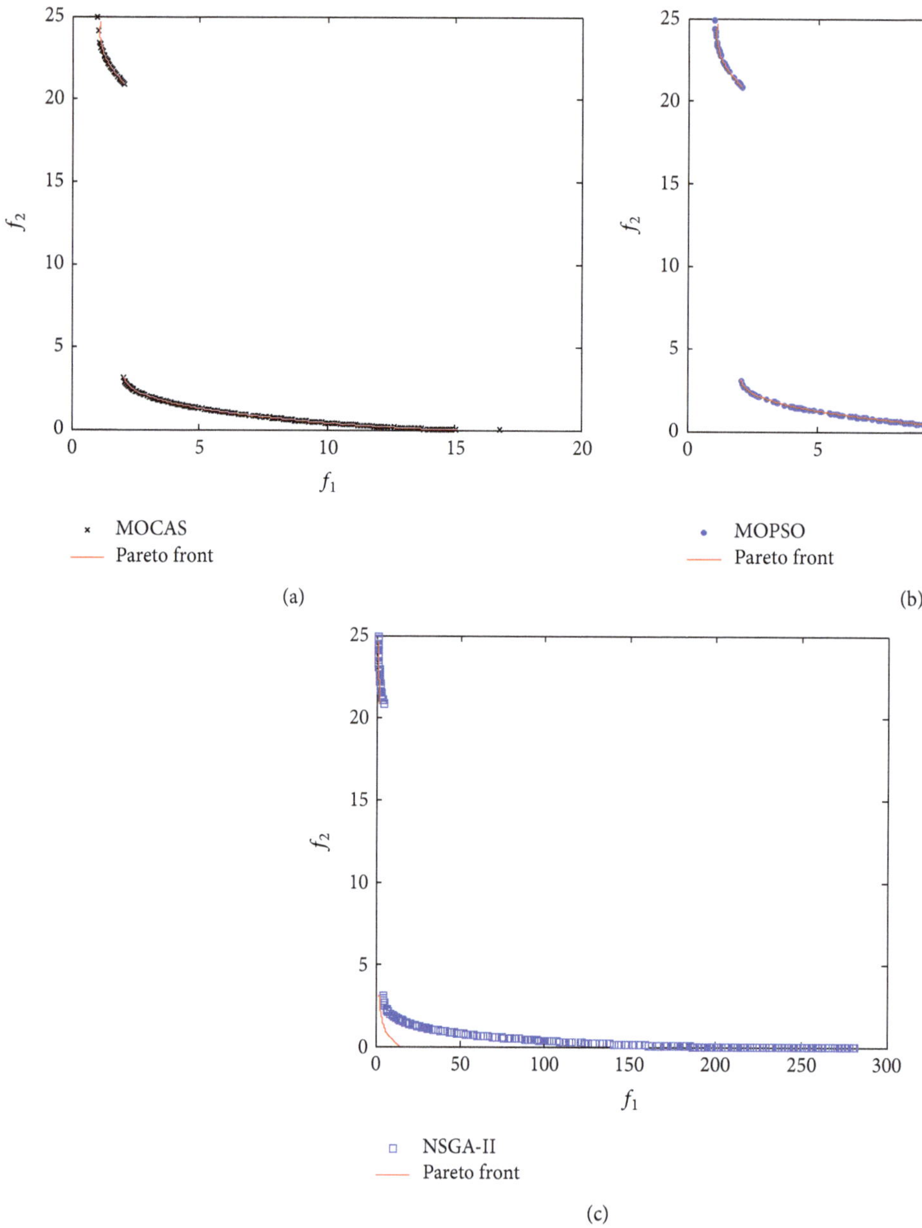

FIGURE 4: Pareto front generated by the MOCAS (a), MOPSO (b), and NSGA-II (c) for P3.

way, we directly extend the CAS to multiobjective CAS by employing the basic concepts of "nondominated sorting" and "crowding distance" [12]. Unfortunately, the extended CAS algorithm cannot achieve our desired goal. Figure 1 presents the obtained result of such an algorithm when solving a well-known multiobjective optimization problem ZDT6 [12]:

$$f_1(x) = 1 - \exp(-4x_1)\sin^6(6\pi x_1),$$
$$f_2(x) = g(x)\left[1 - \left(\frac{f_1(x)}{g(x)}\right)^2\right], \tag{4}$$

where

$$g(x) = 1 + 9\left(\sum_{i=2}^{n}\frac{x_i}{n-1}\right)^{0.25} \tag{5}$$

subject to $0 \le x_i \le 1, i = 1, \ldots, 10$.

As shown in Figure 1, the obtained results are far from the Pareto set; that is, this naively extended algorithm fails to explore the optima. This failure motivates us to reconsider the behaviors of both individual ants and ant colony.

Practically, in the real world, an individual ant performs the chaotic search at the initial time. If a pioneer ant has found a better position, the neighboring ants will follow him

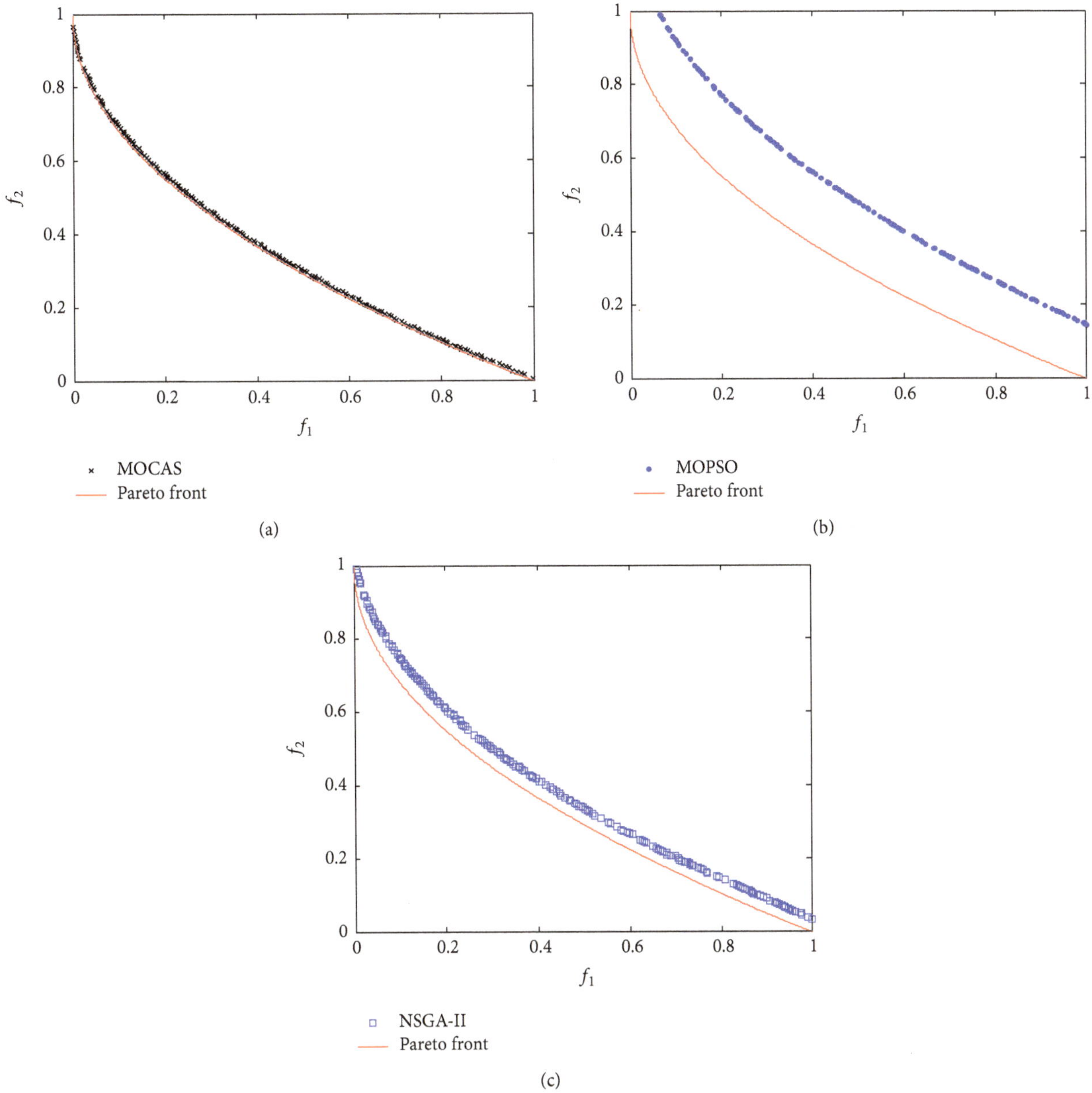

FIGURE 5: Pareto front generated by the MOCAS (a), MOPSO (b), and NSGA-II (c) for P4.

by exchanging information. With the search being evolved, all of the ants will eventually reach the source of food. This is the basic idea behind the single-objective CAS algorithm. However, the organization of ant colony for multiobjective optimization is quite different. In the single-objective optimization, the algorithm could find the optima based on the assumption that the "better" spatial position of a neighbor leads to a "better" object value. But this assumption does not hold in the multiobjective optimization. The neighbor should be redefined as the individual ant in the current first front to enable the algorithm to converge. On the other hand, individual ants that are in the current first front would travel to each other via information exchange. For example, if ant i and ant j are two neighboring ants in the current front, then ant i would update its position to that of ant j by exchanging information with ant j since ant i supposes ant j is much better than itself. In the same way, ant j will update its position to that of ant i.

4.2. Multiobjective Chaotic Ant Swarm (MOCAS) Algorithm. Besides the differences mentioned above, the archive-based approach in the MOCAS is also employed to store nondominated solutions generated in the past. The use of redefined concept of neighbor selection and the historical record of previously found nondominated vectors allow the MOCAS algorithm to converge to the globally nondominated solutions with fast speed.

The detailed steps of the MOCAS algorithm are given in Algorithm 1.

```
Step 1. Initialize the main population (ant colony) P;
        for each individual in P
          initialize its position, search range, organization factor and other variables;
        initialize the archive S = Φ;
Step 2. Evaluate each individual in the population;
Step 3. Classify the individuals in P into dominated individuals D and non-dominated individuals ND;
        for each individual p in P
          flag = 0;
          for each individual q in P
            if p is dominated by q
              D = D ∪ {q};
            else
              D = D ∪ {p};
              flag = 1;
          if flag == 0
            ND = ND ∪ {p};
Step 4. Calculate crowding distance for ND;
        Initialize the distance to be zero for all individuals in ND (nd denotes the size of ND);
        for each objective m
          sort the individuals in ND based on objective m;
        assign infinite distance to boundary values for each individual in ND, i.e. ND(d_1) = ∞ and ND(d_nd) = ∞;
        for i = 2 to (nd − 1)
          ND(d_i) = ND(d_i) + (ND(i + 1)·m − ND(i − 1)·m) / (f_m^max − f_m^min);
        /*ND (i)·m is the value of the mth objective function of the ith individual in P*/
Step 5. Update the archive S;
        S' = S ∪ ND;
        classify S' into dominated individuals D' and non-dominated individuals ND';
        S = ND';
Step 6. Generate new population by (3) and re-defined concept of neighbor selection;
Step 7. if terminate is true
          Output the population;
      else
          goto step 2;
```

ALGORITHM 1: MOCAS.

Another desired goal of MOCAS is to distribute the obtained solutions on the Pareto front evenly. In the proposed MOCAS, we use the crowding distance mechanism to guide the algorithm to generate the well-distributed Pareto front. Specifically, in Step 6, the algorithm first sorts the *ND* according to the crowding distance. When the individual with the largest value of crowding distance selects the neighbor, he will determinately select the one with the smallest value of crowding distance in *ND*. This idea is also inspired by the ant behavior in the real world. When ants found several sources of food, the ant in the crowded source would move toward the uncrowded regions. We will validate the effectiveness of this idea in the next section.

Now we analyze the time complexity of one iteration of the entire algorithm. Step 1 initializes the whole algorithm involving ant colony and some related parameters. It takes $O(N \cdot (D + M))$ time, where N is the population size, D is the dimension of decision variables, and M is number of objectives.

Step 2 steps into the main loop and evaluates objectives for each individual; this procedure takes $O(NM)$ time. Step 3 classifies the population and takes $O(MN^2)$ in the worst case. Step 4 calculates the crowding distance for individuals in the current front; this step takes $O(2MN\log(2N))$. Step 5 updates the archive by nondominated sorting, so it has the same time complexity of Step 3 $O(MN^2)$. Step 6 generates a new population by (3); this will take $O(2MN\log(2N))$. Hence, the time complexity of one iteration of MOCAS in the worst case is $O(MN^2)$.

5. Numeric Simulations

In this section, we test the MOCAS algorithm with several well-known problems and we compare its performance against that of the state-of-the-art multiobjective optimization algorithms, NSGA-II (the code is available at http://www.iitk.ac.in/kangal/codes.shtml) and MOPSO (the code is available at http://www.cs.cinvestav.mx/~EVOCINV/software.html). Another important algorithm SPEA2 is not compared here since its performance is worse than that of NSGA-II in most cases [12]. We use the typical parameter settings for these three algorithms except that the population size and the number of generations are all set to 200 and

TABLE 1: Testing functions used in this study.

Instance	Objective functions	Variable boundaries	Optimal Solutions
P1	$f_1(x) = 1 - e^{-4x_1}\sin^6(6\pi x_1)$ $f_2(x) = g(x)\left(1 - \left(\frac{f_1(x)}{g(x)}\right)^2\right)$ $g(x) = 1 + 9\left(\sum_{i=2}^{6}\frac{x_i}{4}\right)^2$	$[0, 1]$	$x_1 \in [0, 1]$ $x_i = 0$ $i = 2, \ldots, 6$
P2	$f_1(x) = 1 - \exp\left(-\sum_{i=1}^{3} x_i - \left(\frac{1}{\sqrt{3}}\right)^2\right)$ $f_2(x) = 1 - \exp\left(-\sum_{i=1}^{3} x_i + \left(\frac{1}{\sqrt{3}}\right)^2\right)$	$[-4, 4]$	$x_1 = x_2 = x_3 \in \left[-\frac{1}{\sqrt{3}}, \frac{1}{\sqrt{3}}\right]$
P3	$f_1(x) = \left[1 + (A_1 - B_1)^2 + (A_2 - B_2)^2\right]$ $f_2(x) = \left[(x_1 + 3)^2 + (x_2 + 1)^2\right]$ $A_1 = 0.5\sin 1 - 2\cos 1 + \sin 2 - 1.5\cos 2$ $A_2 = 1.5\sin 1 - \cos 1 + 2\sin 2 - 0.5\cos 2$ $B_1 = 0.5\sin x_1 - 2\cos x_1 + \sin x_2 - 1.5\cos x_2$ $B_2 = 1.5\sin x_1 - \cos x_1 + 2\sin x_2 - 0.5\cos x_2$	$[-\pi, \pi]$	$x_1 = -\pi$ $x_2 \in \left[-\frac{4\pi}{3}, \frac{\pi}{4}\right]$
P4	$f_1(x) = x_1$ $f_2(x) = g(x)\left(1 - \sqrt{\frac{x_1}{g(x)}}\right)$ $g(x) = 1 + \frac{9\left(\sum_{i=2}^{n} x_i\right)}{n-1}$	$[0, 1]$ $n = 30$	$x_1 \in [0, 1]$ $x_i = 0$ $i = 2, \ldots, 10$
P5	$f_1(x) = x_1$ $f_2(x) = g(x)\left(1 - \left(\frac{x_1}{g(x)}\right)^2\right)$ $g(x) = 1 + \frac{9\left(\sum_{i=2}^{n} x_i\right)}{n-1}$	$[0, 1]$ $n = 30$	$x_1 \in [0, 1]$ $x_i = 0$ $i = 2, \ldots, 10$
P6	$f_1(x) = x_1$ $f_2(x) = g(x)\left[1 - \sqrt{\frac{x_1}{g(x)}} - \left(\frac{x_1}{g(x)}\right)\sin(10\pi x_1)\right]$ $g(x) = 1 + \frac{9\left(\sum_{i=2}^{n} x_i\right)}{n-1}$	$[0, 1]$ $n = 30$	$x_1 \in [0, 1]$ $x_i = 0$ $i = 2, \ldots, 10$
P7	$f_1(x) = 1 - \exp(-4x_1)\sin^6(6\pi x_1)$ $f_2(x) = g(x)\left[1 - \left(\frac{f_1(x)}{g(x)}\right)^2\right]$ $g(x) = 1 + 9\left(\frac{\sum_{i=2}^{n} x_i}{n-1}\right)^{0.25}$	$[0, 1]$ $n = 10$	$x_1 \in [0, 1]$ $x_i = 0$ $i = 2, \ldots, 10$
P8	$f_1(x) = x^2$ $f_2(x) = (x - 2)^2$	$[-10^3, 10^3]$	$x \in [0, 2]$

2000, respectively. We run each test instance 30 times to statistically evaluate the performance of three algorithms.

5.1. Simulation Results. The testing problems are listed in Table 1, which are the most frequently used testing instances in multiobjective optimizations. All problems have two objective functions. None of these problems have any constraints. Table 1 also describes the number of variables, their boundaries, and the optima solutions for each problem.

Figures 2–9 depict the nondominated solutions produced by the MOCAS, MOPSO, and NSGA-II algorithms, respectively, when they are applied to each of the problems in Table 1. The red curve in these figures represents the true Pareto front for the corresponding testing problem. It can be seen and intuitively observed from these figures that our proposed MOCAS algorithm either (1) clearly outperforms both MOPSO and NSGA-II algorithms, or (2) outperforms one of the two but performs roughly the same with the other, or (3) performs roughly the same with both of them, for all testing cases. This immediately suggests an average superiority of the MOCAS algorithm over the MOPSO and NSGA-II algorithms.

Specifically, Figures 5 and 6 show that the output of the MOCAS algorithm converges to the true Pareto front for testing problems P4 and P5 while neither the MOPSO algorithm nor the NSGA-II algorithm is able to do so for either of the two problems. Figure 4 shows the comparison on a noncontinuous testing function. The results demonstrate

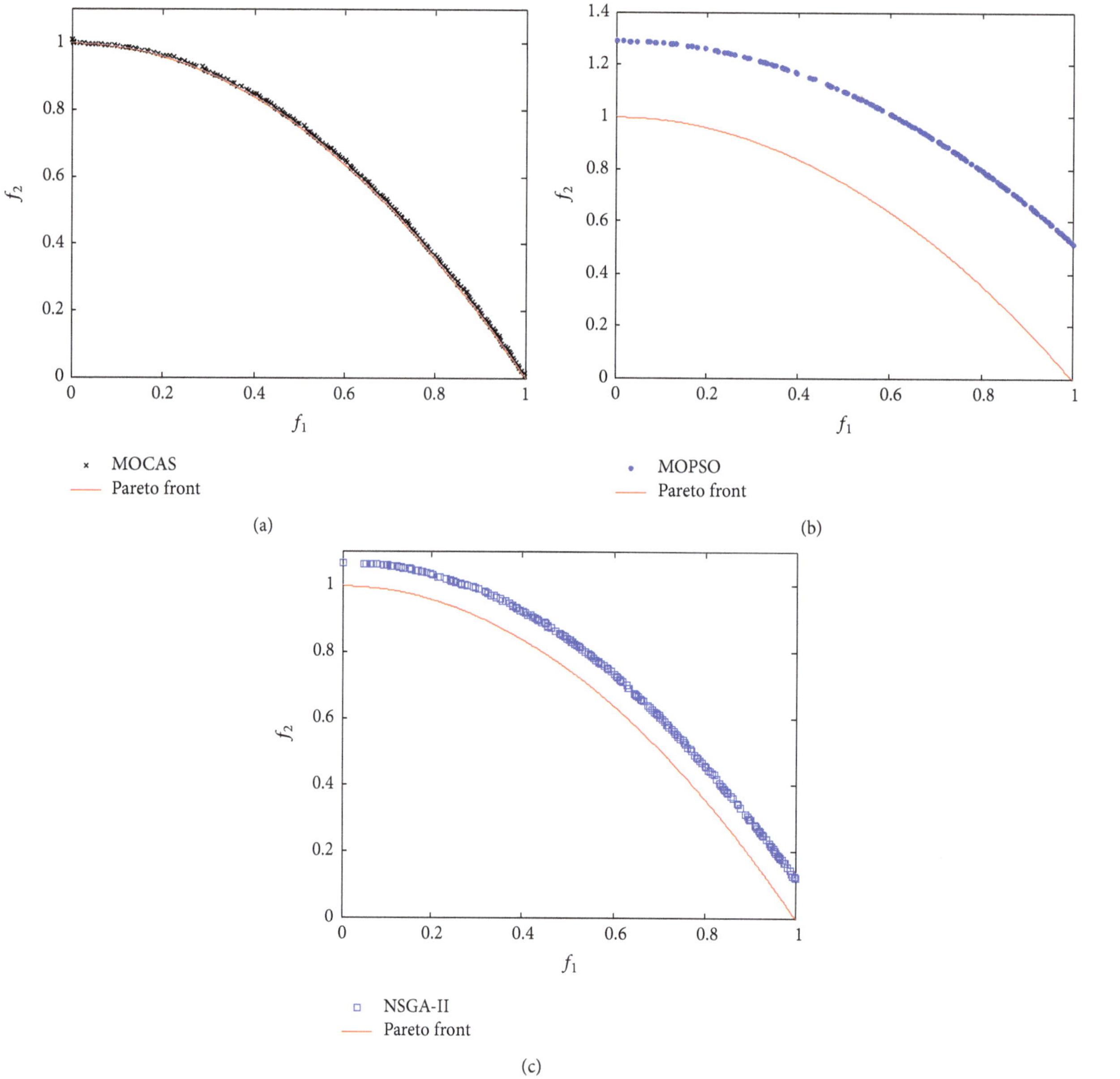

FIGURE 6: Pareto front generated by the MOCAS (a), MOPSO (b), and NSGA-II (c) for P5.

that the performance of the MOCAS algorithm is roughly equivalent to that of MOPSO but superior to that of NSGA-II for testing problem P3, and Figure 7 demonstrates that the performance of the MOCAS algorithm is roughly equivalent to that of NSGA-II but superior to that of MOPSO for testing problem P6. Figures 2, 3, 8, and 9 all suggest that the performance of the MOCAS algorithm is closely equivalent to that of both MOPSO and NSGA-II for testing problems P1, P2, P7, and P8, respectively. Quantitatively detailed performance analysis is given in the next subsection to show the precise comparison of three algorithms.

5.2. Performance Evaluations. In order to further evaluate the performance of these three algorithms quantitatively, we adopt the following three performance metrics as suggested in [21, 31–33].

5.2.1. Generational Distance (GD). This metric measures how far the obtained nondominated solutions are from the true Pareto front. Its definition is

$$\mathrm{GD} = \frac{\sqrt{\sum_{i=1}^{n} d_i^2}}{n}, \tag{6}$$

where n is the number of obtained nondominated solutions and d_i is the Euclidean distance (measured in objective space) between the ith nondominated solution and the Pareto front.

5.2.2. Error Ratio (ER). This metric was originally proposed by Van Veldhuizen [32] to measure the percentage of obtained nondominated solutions that are not on the Pareto front:

$$\mathrm{ER} = \frac{\sum_{i=1}^{n} e_i}{n}, \tag{7}$$

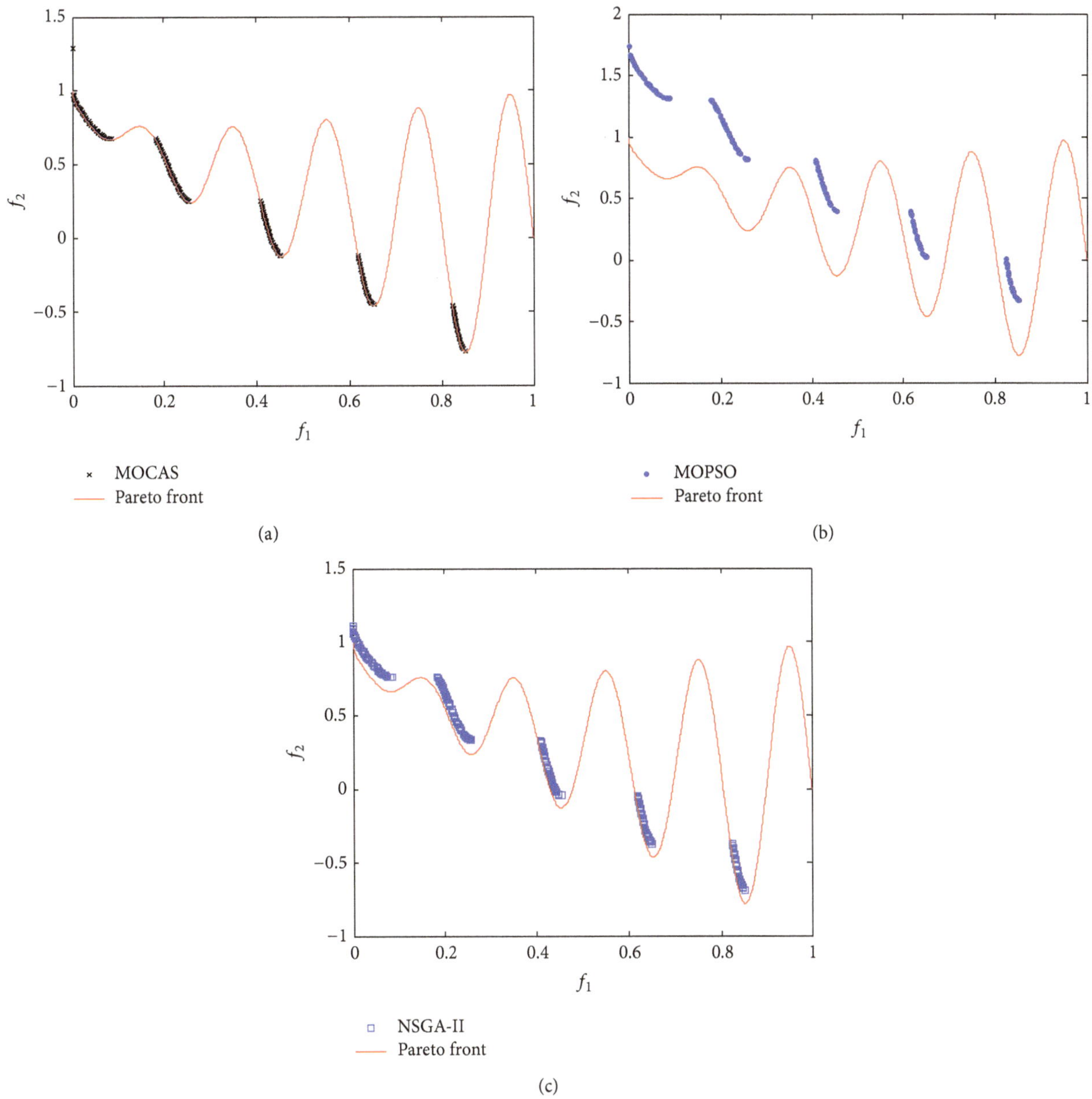

FIGURE 7: Pareto front generated by the MOCAS (a), MOPSO (b), and NSGA-II (c) for P6.

where n is the number of obtained nondominated solutions. $e_i = 0$ means the ith solution is a member of the Pareto set, and $e_i = 1$, otherwise.

5.2.3. Spacing (SP). This metric measures the extent of spread of obtained nondominated solutions. It is defined as

$$\mathrm{SP} = \sqrt{\frac{1}{n-1}\sum_{i=1}^{n}\left(\bar{d} - d_i\right)^2}, \quad (8)$$

where $d_i = \min_j[|f_1^i(\vec{x}) - f_1^j(\vec{x})| + |f_2^i(\vec{x}) - f_2^j(\vec{x})|]$, $i, j = 1, \ldots, n$, $\bar{d}$ is the mean of all d_i's, and n is the number of obtained nondominated solutions.

It should be noted that metrics GD and ER describe the first goal of multiobjective optimization algorithms as mentioned previously in the design of MOCAS, while SP reflects the second goal. In the following, we present the comparison for three algorithms with respect to above performance metrics.

Tables 2, 3, 4, 5, 6, 7, 8, and 9 show the comparison of the results from the three algorithms with respect to GD, ER, and SP, where the boldfaced numbers indicate the best average CD (or ER or SP) across the three algorithms. It can be seen that the average performance of MOCAS is the best when the three algorithms run the test problems P2, P3, P4, P5, P6, and P8. For testing problems P1 and P7, however, the tables show that the NSGA-II has the best GD and ER, but the MOCAS has the best SP. By zooming in

TABLE 2: Comparison results of three metrics for P1.

	MOCAS			MOPSO			NSGA-II		
	GD	ER	SP	GD	ER	SP	GD	ER	SP
Best	0.003346	0.915	0.006114	0.021315	0.03	0.045803	0.006018	0.025	2.03*E* − 07
Worst	0.057707	0.995	0.186609	0.054298	0.206522	0.815469	0.060076	0.09	0.342528
Average	0.024415	0.976	**0.082727**	0.034578	0.069526	0.344084	**0.023784**	**0.054**	0.094153
Median	0.022859	0.985	0.086083	0.035318	0.055	0.322122	0.021513	0.045	0.044569
Std. dev.	0.019502	0.02481	0.060023	0.009498	0.050566	0.275979	0.016115	0.025473	0.118375

TABLE 3: Comparison results of three metrics for P2.

	MOCAS			MOPSO			NSGA-II		
	GD	ER	SP	GD	ER	SP	GD	ER	SP
Best	5.66*E* − 05	0	0.002721	8.72*E* − 05	0	3.55*E* − 03	9.32*E* − 05	0	0.003052
Worst	7.07*E* − 05	0.005	0.003334	0.000141	0.025	0.004423	0.000129	0.02	0.003885
Average	**6.45E − 05**	**0.0005**	**0.003032**	0.000123	0.0065	0.003828	0.000107	0.009	0.003314
Median	6.57*E* − 05	0	0.003015	0.00013	0.005	0.003812	0.000107	0.0125	0.003282
Std. dev.	4.66*E* − 06	0.001581	0.000207	1.83*E* − 05	0.007835	2.62*E* − 04	1.05*E* − 05	0.008097	0.000226

TABLE 4: Comparison results of three metrics for P3.

	MOCAS			MOPSO			NSGA-II		
	GD	ER	SP	GD	ER	SP	GD	ER	SP
Best	0.008367	0.295	0.056603	0.017705	0.825	0.131108	8.94454	1	0.598082
Worst	0.019111	0.38	0.229375	0.036684	0.86	0.29519	9.73034	1	0.753096
Average	**0.01012**	**0.331429**	**0.110787**	0.025308	0.8405	0.187466	9.34286	1	0.672726
Median	0.00917	0.33	0.064751	0.023319	0.84	0.177388	9.37408	1	0.665563
Std. dev.	0.003223	0.029681	0.079856	0.00636	0.010659	0.049485	0.22547	0	0.055491

TABLE 5: Comparison results of three metrics for P4.

	MOCAS			MOPSO			NSGA-II		
	GD	ER	SP	GD	ER	SP	GD	ER	SP
Best	0.001762	0.52	0.003063	0.010027	1	0.004461	0.000405	0.985	0.003226
Worst	0.003037	0.75	0.003842	0.016015	1	0.005487	0.015986	1	0.213161
Average	**0.002539**	**0.623**	**0.003379**	0.011864	1	0.005067	0.004408	0.997	0.039783
Median	0.002517	0.6125	0.003244	0.011309	1	0.005196	0.000546	1	0.006887
Std. dev.	0.000372	0.082131	0.000288	0.002002	0	0.000348	0.006263	0.00483	0.068458

TABLE 6: Comparison results of three metrics for P5.

	MOCAS			MOPSO			NSGA-II		
	GD	ER	SP	GD	ER	SP	GD	ER	SP
Best	0.002383	0.31	2.18*E* − 10	0.018673	1	0.00268	0.00032	1	0.003371
Worst	0.004709	0.58	0.005356	0.06841	1	0.313675	0.028573	1	0.004965
Average	**0.003966**	**0.4565**	**0.003019**	0.035906	1	0.064442	0.005302	1	0.004118
Median	0.004256	0.455	0.003241	0.032347	1	0.014006	0.001079	1	0.004122
Std. dev.	0.000718	0.082329	0.001462	0.016319	0	0.113717	0.009635	0	0.00042

TABLE 7: Comparison results of three metrics for P6.

	MOCAS			MOPSO			NSGA-II		
	GD	ER	SP	GD	ER	SP	GD	ER	SP
Best	0.000875	0.01	0.00271	0.018633	1	0.003987	0.000275	0.905	0.005761
Worst	0.001535	0.035	0.003566	0.031005	1	0.210964	0.015983	1	0.006931
Average	**0.001174**	**0.022**	**0.003229**	0.024214	1	0.060328	0.00436	0.97	0.006255
Median	0.001165	0.02	0.003323	0.024216	1	0.022439	0.001551	0.9925	0.006259
Std. dev.	0.00022	0.008563	0.000248	0.004111	0	0.078422	0.005767	0.037565	0.000365

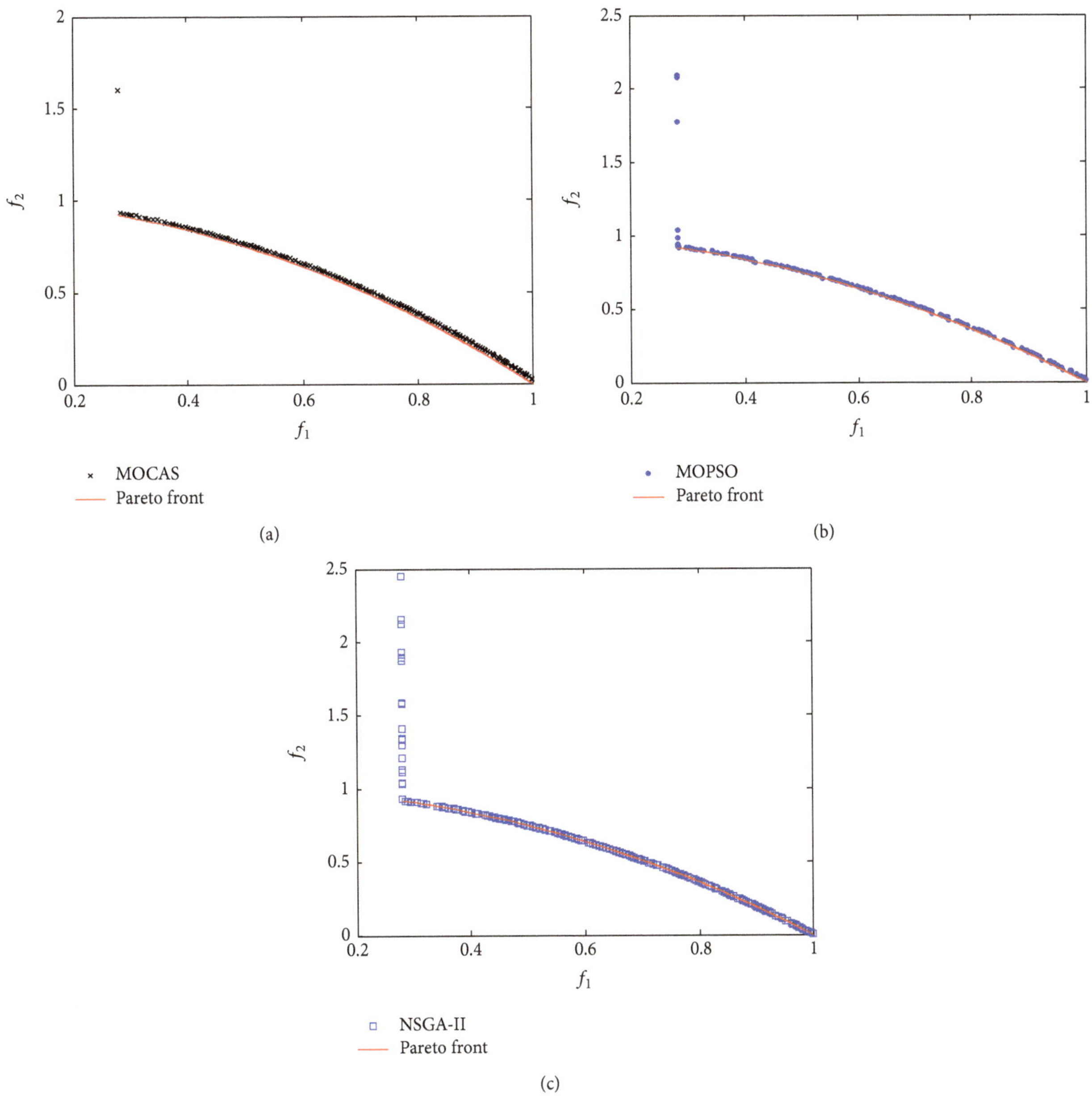

FIGURE 8: Pareto front generated by the MOCAS (a), MOPSO (b), and NSGA-II (c) for P7.

TABLE 8: Comparison results of three metrics for P7.

	MOCAS			MOPSO			NSGA-II		
	GD	ER	SP	GD	ER	SP	GD	ER	SP
Best	0.013159	1	0.002741	0.012502	0.636364	0.038502	0.003501	0.04	0.004511
Worst	0.03909	1	0.197246	0.343949	1	0.451114	0.053784	0.11	0.320704
Average	0.027409	1	**0.104**	0.211105	0.849299	0.16362	**0.016799**	**0.072**	0.135646
Median	0.031508	1	0.09552	0.230836	0.908974	0.123775	0.009905	0.0725	0.095691
Std. dev.	0.009926	0	0.071008	0.128875	0.128953	0.129505	0.015584	0.02429	0.135664

the corresponding figures (Figures 2 and 8) and inspecting the details, we see that the MOCAS would not converge to the true Pareto front, but it has a fairly good distribution of obtained solutions. This observation matches the quantitative performance evaluations.

6. Conclusion

We proposed in this paper a novel bioinspired multiobjective optimization algorithm named MOCAS for designing wireless sensor networks in the Internet of Things by extending

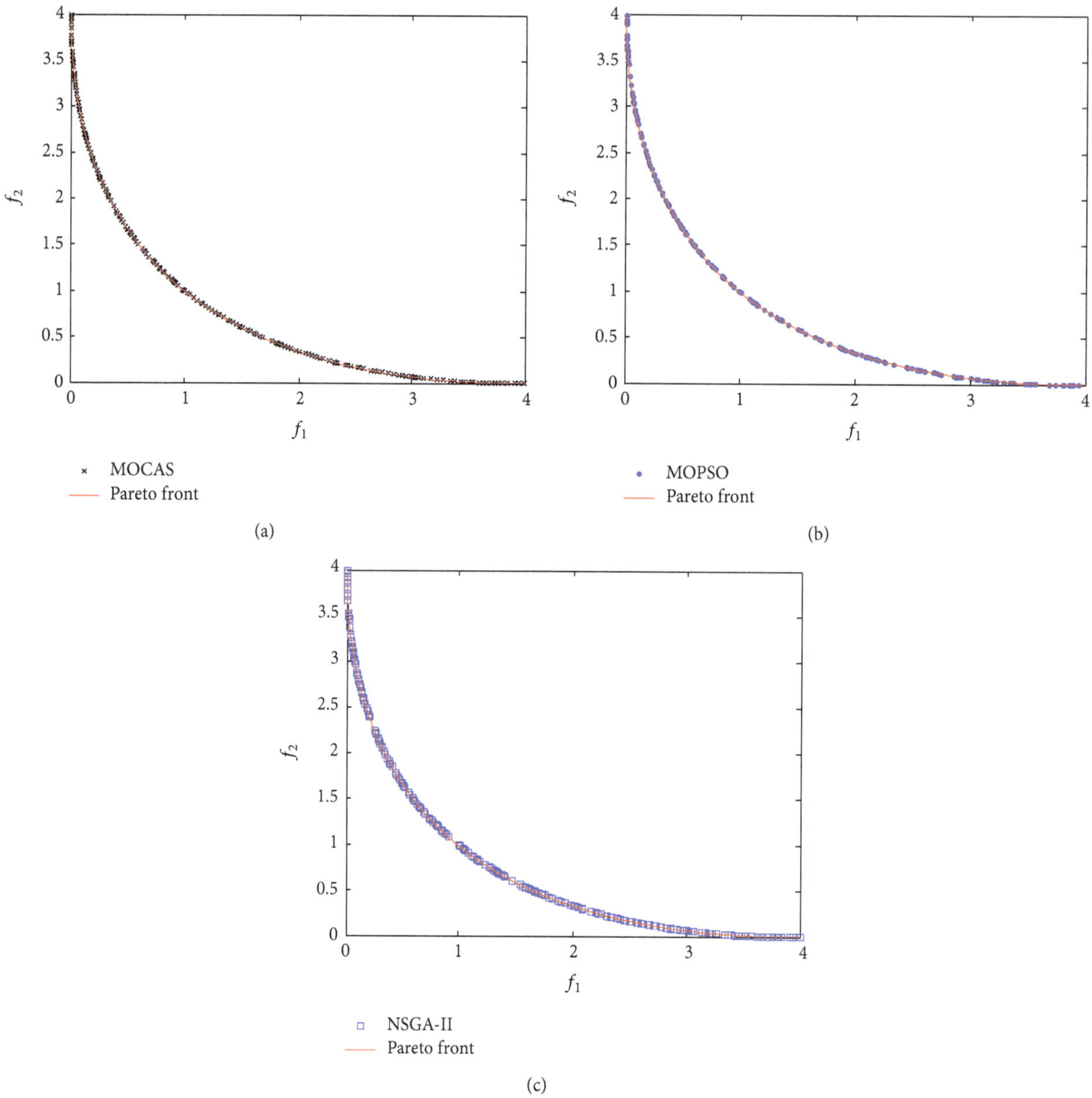

FIGURE 9: Pareto front generated by the MOCAS (a), MOPSO (b), and NSGA-II (c) for P8.

TABLE 9: Comparison results of three metrics for P8.

	MOCAS			MOPSO			NSGA-II		
	GD	ER	SP	GD	ER	SP	GD	ER	SP
Best	0	0	0.012175	0	0	0.017289	0	0	0.016588
Worst	0	0	0.016525	0	0	0.023579	0	0	0.020291
Average	0	0	**0.014496**	0	0	0.020478	0	0	0.017275
Median	0	0	0.014371	0	0	0.020433	0	0	0.01685
Std. dev.	0	0	0.001377	0	0	0.001774	0	0	0.001093

previous work on chaotic ant swarms. A straightforward extension from the single-objective optimization CAS model to the multiobjective optimization CAS model was initially attempted by introducing the "nondominated sorting" and "crowding distance" notions into the single-objective optimization CAS model. This approach however turns out to be a failure since its outcomes are not even close to the true Pareto front. We then redefined the concept of "neighbors" and "neighbor-selecting" rules and incorporated an "archive-based" approach into the algorithm allowing the resulting MOCAS algorithm to converge fast to the true Pareto front with an evenly distributed set of solutions. By testing MOCAS on some well-known multiobjective optimization problems and comparing the results with those produced by the state-of-the-art peer algorithms MOPSO and NSGA-II, we have seen that the proposed MOCAS algorithm outperforms the other two algorithms on average, which evidently shows the competitiveness of the MOCAS algorithm in dealing with multiobjective optimization problems.

Since the variables of MOCAS are bounded by symmetric intervals in this paper, one issue that we would like to explore in the future is how to set the parameter for the algorithm to cover a wider range of multiobjective optimization problems. Another interesting issue that we also would like to study is the further extension of MOCAS to handle constrained multiobjective optimization problems.

Conflict of Interests

The authors declare that there is no conflict of interests regarding the publication of this paper.

References

[1] S. Okdem and D. Karaboga, "Routing in wireless sensor networks using ant colony optimization," in *Proceedings of the 1st NASA/ESA Conference on Adaptive Hardware and Systems (AHS '06)*, pp. 401–404, 2006.

[2] T. Camilo, C. Carreto, J. S. Silva et al., "An energy-efficient ant-based routing algorithm for wireless sensor networks," in *Ant Colony Optimization and Swarm Intelligence*, pp. 49–59, Springer, Berlin, Germany, 2006.

[3] F. Y. Cheng and X. S. Li, "Generalized center method for multi-objective engineering optimization," *Engineering Optimization*, vol. 31, no. 5, pp. 641–661, 1999.

[4] T. Marler, *Multi-Objective Optimization: Concepts and Methods for Engineering*, VDM Publishing, Saarbrücken, Germany, 2009.

[5] M. Fonseca and P. J. Fleming, "An overview of evolutionary algorithms in multiobjective optimization," *Evolutionary Computation*, vol. 3, no. 1, pp. 1–16, 1995.

[6] K. Deb, *Multi-Objective Optimization Using Evolutionary Algorithms*, John Wiley & Sons, Chichester, UK, 2001.

[7] A. C. Coello, "20 years of evolutionary multi-objective optimization: what has been done and what remains to be done," in *Computational Intelligence: Principles and Practice*, G. Y. Yen and D. B. Fogel, Eds., chapter 4, pp. 73–88, IEEE Computational Intelligence Society, New York, NY, USA, 2006.

[8] J. D. Schaffer, *Some experiments in machine learning using vector evaluated genetic algorithms [Ph.D. dissertation]*, Vanderbilt University, Nashville, Tenn, USA, 1984.

[9] J. D. Knowles and D. W. Corne, "Approximating the non-dominated front using the Pareto archived evolution strategy," *Evolutionary computation*, vol. 8, no. 2, pp. 149–172, 2000.

[10] E. Zitzler, M. Laumanns, and L. Thiele, "SPEA2: improving the strength pareto evolutionary algorithm," Tech. Rep. 103, Swiss Federal Institute of Technology, Zürich, Switzerland, 2001.

[11] A. Coello Coello and G. T. Pulido, "Multiobjective optimization using a micro-genetic algorithm," in *Proceedings of the Genetic and Evolutionary Computation Conference (GECCO '01)*, L. Spector, E. D. Goodman, A. Wu et al., Eds., pp. 274–282, San Francisco, Calif, USA, 2001.

[12] K. Deb, A. Pratap, S. Agarwal, and T. Meyarivan, "A fast and elitist multiobjective genetic algorithm: NSGA-II," *IEEE Transactions on Evolutionary Computation*, vol. 6, no. 2, pp. 182–197, 2002.

[13] H. Li and Q. Zhang, "Multiobjective optimization problems with complicated pareto sets, MOEA/ D and NSGA-II," *IEEE Transactions on Evolutionary Computation*, vol. 13, no. 2, pp. 284–302, 2009.

[14] B. Soylu and M. Köksalan, "A favorable weight-based evolutionary algorithm for multiple criteria problems," *IEEE Transactions on Evolutionary Computation*, vol. 14, no. 2, pp. 191–205, 2010.

[15] J. Huang and Y. Liu, "MOEAQ: a QoS-aware multicast routing algorithm for MANET," *Expert Systems with Applications*, vol. 37, no. 2, pp. 1391–1399, 2010.

[16] M. A. Abido, "Multiobjective evolutionary algorithms for electric power dispatch problem," *IEEE Transactions on Evolutionary Computation*, vol. 10, no. 3, pp. 315–329, 2006.

[17] T. Siegfried, S. Bleuler, M. Laumanns, E. Zitzler, and W. Kinzelbach, "Multiobjective groundwater management using evolutionary algorithms," *IEEE Transactions on Evolutionary Computation*, vol. 13, no. 2, pp. 229–242, 2009.

[18] M. P. Hansen, "Tabu search for multiobjective optimization," in *Proceedings of the 13th International Conference Multiple Criteria Decision Making (MCDM '97)*, pp. 1–16, Cape Town, South Africa, 1997.

[19] E. L. Ulungu, P. H. Teghem, and D. Tuyttens, "MOSA method: a tool for solving multiobjective combinatorial optimization problems," *Journal of Multi-Criteria Decision Analysis*, vol. 8, pp. 221–236, 1999.

[20] A. Jaszkiewicz, "On the computational efficiency of multiple objective metaheuristics. The knapsack problem case study," *European Journal of Operational Research*, vol. 158, no. 2, pp. 418–433, 2004.

[21] C. A. C. Coello, G. T. Pulido, and M. S. Lechuga, "Handling multiple objectives with particle swarm optimization," *IEEE Transactions on Evolutionary Computation*, vol. 8, no. 3, pp. 256–279, 2004.

[22] L. Li, Y. Yang, H. Peng, and X. Wang, "An optimization method inspired by "chaotic" ant behavior," *International Journal of Bifurcation and Chaos in Applied Sciences and Engineering*, vol. 16, no. 8, pp. 2351–2364, 2006.

[23] M. Dorigo, V. Maniezzo, and A. Colorni, "Ant system: optimization by a colony of cooperating agents," *IEEE Transactions on Systems, Man, and Cybernetics, Part B: Cybernetics*, vol. 26, no. 1, pp. 29–41, 1996.

[24] L. Li, Y. Yang, and H. Peng, "Fuzzy system identification via chaotic ant swarm," *Chaos, Solitons and Fractals*, vol. 41, no. 1, pp. 401–409, 2009.

[25] J. Cai, X. Ma, L. Li, Y. Yang, H. Peng, and X. Wang, "Chaotic ant swarm optimization to economic dispatch," *Electric Power Systems Research*, vol. 77, no. 10, pp. 1373–1380, 2007.

[26] L. Li, Y. Yang, and H. Peng, "Computation of multiple global optima through chaotic ant swarm," *Chaos, Solitons and Fractals*, vol. 40, no. 3, pp. 1399–1407, 2009.

[27] L. Li, Y. Yang, H. Peng, and X. Wang, "Parameters identification of chaotic systems via chaotic ant swarm," *Chaos, Solitons and Fractals*, vol. 28, no. 5, pp. 1204–1211, 2006.

[28] B. J. Cole, "Is animal behaviour chaotic? Evidence from the activity of ants," *Proceedings of the Royal Society B: Biological Sciences*, vol. 244, no. 1311, pp. 253–259, 1991.

[29] R. V. Sole, O. Miramontes, and B. C. Goodwin, "Oscillations and chaos in ant societies," *Journal of Theoretical Biology*, vol. 161, no. 3, pp. 343–357, 1993.

[30] H. Peng, L. Li, Y. Yang, and F. Liu, "Parameter estimation of dynamical systems via a chaotic ant swarm," *Physical Review E: Statistical, Nonlinear, and Soft Matter Physics*, vol. 81, no. 1, Article ID 016207, 2010.

[31] D. A. van Veldhuizen and G. B. Lamont, "Multiobjective evolutionary algorithm research: a history and analysis," Tech. Rep. TR-98-03, Department of Electrical and Computer Engineering, Graduate School of Engineering, Air Force Institute of Technology, Wright-Patterson AFB, Ohio, USA, 1998.

[32] D. A. Van Veldhuizen, *Multiobjective evolutionary algorithms: Classifications, analyzes, and new innovations [Ph.D. thesis]*, Department of Electrical and Computer Engineering, Graduate School of Engineering, Air Force Institute of Technology, Wright-Patterson AFB, Wright-Patterson Air Force Base, Ohio, ISA, 1999.

[33] J. R. Schott, *Fault tolerant design using single and multicriteria genetic algorithm optimization [M.S. thesis]*, Department of Aeronautics and Astronautics, Massachusetts Institute of Technology, Cambridge, Mass, USA, May 1995.

18

A Portable Random Key Predistribution Scheme for Distributed Sensor Network

Shihui Zheng, Yuan Tian, Lei Jin, and Yu Yang

Information Security Center, National Engineering Laboratory for Disaster Backup and Recovery, Beijing University of Posts and Telecommunications, Beijing 100876, China

Correspondence should be addressed to Shihui Zheng; shihuizh@gmail.com

Academic Editor: Christos Riziotis

A distributed sensor network (DSN) can be deployed to collect information for military or civilian applications. However, due to the characteristics of DSNs such as limited power, key distribution for a distributed sensor network is complex. In this paper, a neighbor-based path key establishing method and a seed-based algorithm are put forward to improve the original random key pre-distribution scheme. The new scheme is portable because it is independent of the routing protocol. Moreover, the connectivity of the entire network also approaches 1. In particular, the new scheme can keep high connectivity by setting a small amount of redundancy in parameter values when the number of neighbors drops because of the node dormancy or death. The resilience against node capture in our scheme is not lower than that in the l-path scheme and the basic schemes when the number of hops in a path is larger than 5, and the simulation result shows that the efficiency of our scheme is also slightly higher.

1. Introduction

A lot of secure key distribution techniques, for instance, the public cryptography, cannot be applied in Wireless sensor networks (WSNs) due to the characteristics of WSNs like large scale network, lacking of trusted infrastructure, and physical constraints to energy and memory. A naive solution is that a single master key is used in all communications which is too vulnerable under the node capture attack. Matsumoto and Imai [1] defined key predistribution (KPD) which is the method of distribution of keys onto nodes before deployment. Therefore, the nodes build up the network using their secret keys after deployment, that is, when they reach their target position. Blom [2] (see also [3]) presented λ-secure key predistribution schemes. The tradeoff is that, unlike the $(N - 1)$ pairwise-key scheme, those schemes are no longer perfect against node capture. Instead, the λ-secure KPD schemes [2–7] have perfect security when the number of captured nodes is less than λ. However, once more than λ nodes are controlled by a specific attacker, all keys in the entire network are compromised. Deployment knowledge based KPD schemes [8–10] utilized deployment knowledge to improve the connectivity and the resilience against node capture. However, these solutions are not quite viable since the location of each node may be unknown before deployment. Local center based scheme [11–13] is another type of general KPD techniques, assuming that the network consists of a local trusted infrastructure. Some trusted and powerful nodes distribute keys to sensor nodes around them. However, it is not accessible in a random deployment and dynamic network. In WSNs, random key predistribution (RKPD) schemes [14–21] are widely accepted due to its simplicity, low overhead, scalability, and high global connectivity.

1.1. Related Work. In 2002, Eschenauer and Gligor [14] proposed a random key predistribution (RKPD) scheme, often called basic scheme, using giant component theory. The basic scheme includes 3 phases: (1) key predistribution, (2) shared key discovery, and (3) path key establishment. The first phase is executed before the deployment of nodes. A controller generates a large key pool, and each node draws keys (to form a key ring) out of the pool without replacement. The second phase may occur at the neighbor discovery phase. Two neighbor nodes (two nodes connected physically) try to find a common key identifier in their key rings. If such a key identifier is found, these nodes are logically connected and the key corresponding to the same key identifier is called a shared key. Then, the two nodes separately add a record including the other party's identifier and the shared keys' identifier to a list, called a neighbor list. The last phase occurs just before the information transmission. A pair of nodes x and y (including nodes which are only physically connected with x) try to find a path $(x, w_1), (w_1, w_2), \ldots, (w_{t-1}, w_n), (w_t, y)$, where (i, j) represents that sensor nodes i and j are logically connected neighbors. Afterwards, x and y can distribute a key on the path, and the key is called a path key. The basic scheme can be parameterized to meet the demands that the connected probability of the entire network closes to 1.

Later, Chan et al. [15] gave a q-composite scheme in which any two logically connected nodes need to share q keys instead of only one to establish a path key. An l-path scheme was also presented in [15], in which the path key is broken up into l nuggets, and these nuggets are passed to y along l disjoint paths. But the problem of discovering multiple disjoint paths is computationally hard, and too much overhead may be incurred in this process. Zhu et al. [18] relaxed the requirement where a single physical path is used as long as the nuggets of the path key are transmitted through multiple logically disjoint paths. However, the cost of discovering multiple logically disjoint paths is also expensive.

Gu et al. [20] pointed that the performance of the basic scheme is satisfactory only in highly dense sensor networks, where the average number of physical neighbors per node is more than 20. They proposed a methodology called network decoupling and designed a new protocol to establish the path key between physical neighbors with the help of a node proxy. Li et al. [21] also proposed a multihop proxies random key predistribution scheme. In their schemes, each node constructs a local logical graph and a local physical graph before the path-key establishment, which incurs a large amount of storage, computation, and communication overload. In addition, the physical graph of WSNs changes all the time when the node moves or dies. Thus, if the local graph is not updated in time, the path key establishment will fail.

1.2. Our Contributions. Generally, a key establishment scheme has nothing to do with the routing protocol. However, the path in lots of RKPD schemes, such as the basic scheme, is set up through a routing protocol. That is, a customized routing protocol needs to be used along with those key distribution schemes, which will seriously affect the portability of those RKPD schemes. Moreover, the number of hops of the path may increase because the adjacent nodes in the path must be logically connected (see Section 6). In fact, the total amount of computation cost for deciding whether a shared key exists in key rings of two adjacent nodes is also very huge because the routing process may involve many nodes (see Section 6).

A new method to establish the path key with the help of neighbors is presented in this paper. The basic steps of Phase 3 are changed, and a path that the adjacent nodes only need to be physically connected can be achieved by original routing protocols [22]. To reduce the communication overload, a seed-based method is used to chose key ring for each node upon the input of node's identifier. Although the seed-based method has been mentioned in paper [18, 19], we firstly construct a deterministic algorithm **G**, with which the times that each key is selected by nodes approximate the average value. In addition, the connectivity of the new scheme is also higher if suitable parameters are chosen. Specifically, when the number of neighbors is less than the predefined out-degree due to the node dormancy or death, the probability that the entire network is connected still approaches 1 by setting a small amount of redundancy in parameters. The probability that a link is compromised when nodes are captured in our scheme is equal to that of the basic scheme, and it is higher than that of the q-composite scheme and that of the l-path scheme. However, the probability that a path key is compromised in the establishment phase is lower than that of l-path schemes when the number of hops is more than 4. Our scheme is less competitive in terms of the computation and communication complexity analysis. However, the execution time of a path key establishment for our scheme is slightly less than that of the basic scheme in the simulation. The reason is that, the complexity does not include the extra traffic and calculations that are caused by the customized routing protocol in the basic scheme.

2. The New RKPD Scheme

In the rest of this paper, N represents the number of nodes in the entire wireless sensor network, and each node has a node identifier i, $(i = 1, 2, \ldots, N)$. L is the number of keys in the key pool, and each key also has a key identifier k, $(k = 1, 2, \ldots, L)$. n is the average number of nodes in single-hop communication range, called node density. And s is the number of keys in the key ring for each node. In addition, t expresses the number of hops of a path between the source node x and the destination node y.

2.1. The Scheme. A new random key predistribution scheme is described in this section. The scheme also includes three phases: (1) key predistribution, (2) shared key discovery, and (3) path key establishment.

Phase 1 (key predistribution). Before the deployment of nodes, for each node, a control center (CC) randomly chooses a key ring and loads it into the node (the flow chart is shown in Figure 1).

Step 1. The CC randomly generates L keys and assigns a unique key identifier k $(k \in [1, L])$ to each key_k; those keys and the corresponding identifiers compose a key pool.

Step 2. The CC chooses a deterministic algorithm **G** to decide the key identifiers allocated to each node on the input of the node's identifier.

Step 3. For each node i $(i = 1, \ldots, N)$, the CC inputs its identifier i into **G** and output s distinct values between 1 and L, denoted by $\{T_1, T_2, \ldots, T_s\}$. At last, the CC draws s keys whose key identifiers are $\{T_l \mid l = 1, \ldots, s\}$. Those keys $\{\text{key}_{T_l} \mid l = 1, \ldots, s\}$ and the corresponding key identifiers

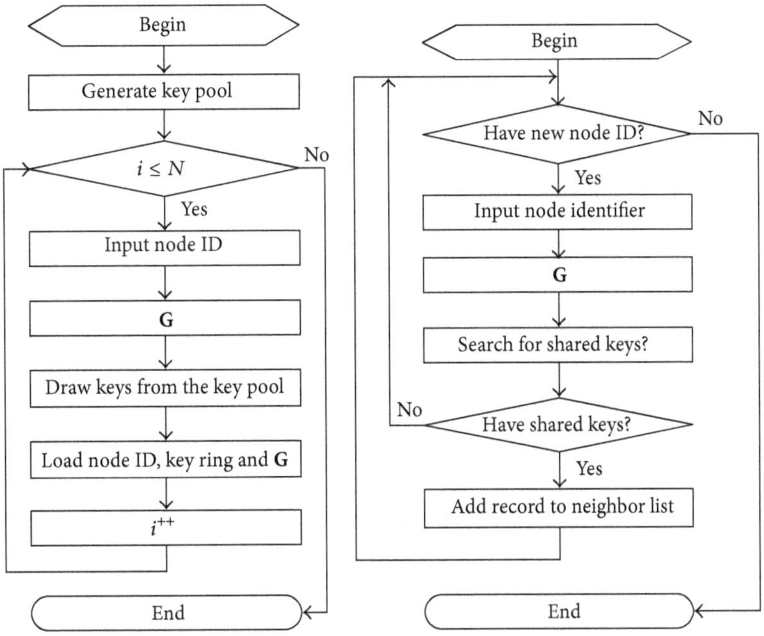

(a) Key predistribution (b) Shared key discovery

FIGURE 1: The operation process of Phases 1 and 2.

$\{T_l \mid l = 1, \ldots, s\}$ compose a key ring which is loaded into node i. Also the algorithm **G** is loaded into each node.

Phase 2 (shared key discovery). After the deployment of nodes, each node creates its own neighbor list.

Step 1. Each node i ($i = 1, \ldots, N$) broadcasts its identifier and records the received identifiers, denoted by $\{i_r \mid r = 1, 2, \ldots\}$.

Step 2. For each node i_r, node i runs the procedure **G** and generates the key identifier set of nodes i_r. If there is a common key identifier in such set and its own key ring, they are logically connected. Then node i adds a record involving the node identifier i_r and the same key identifier to its neighbor list.

Phase 3 (path key establishment). Node x wants to establish a path key with node y. If they are in wireless communication range and have a shared key, that is, they are on each other's neighbor list, the shared key can be used as the path key. Else, x randomly generates a path key pk and encrypts pk with some key in y's key ring. The ciphertext, denoted by $CT = E(pk)$, together with the identifier of the encryption key is sent to y on a physical connected path founded by a routing protocol [22]. Finally, y finds the encryption key corresponding to the received key identifier and decrypts CT to obtain pk. The following procedure explains how x can find an encryption key which belongs to y's key ring. Figure 2 shows the process of source node x and Figure 3 shows the process of destination node y and x's neighbors z.

Step 1. x inputs y's node identifier i_y into the procedure **G** to generate y's key identifier set and then searches for a shared key identifier between y's key identifier set and its own's in its key ring using binary search method. If a same key identifier k_{xy} is found, x looks up k_{xy} in its own key ring and obtains the encryption key. Else, if all the key identifiers are different, x goes to step 2.

Step 2. x broadcasts y's identifier i_y.

Step 3. x's neighbor z_j which receives the identifier looks i_y up in its neighbor list. If x's identifier i_x is not on the list, z_j stops.

Step 4. z_j inputs y's identifier i_y into the procedure **G** to generate y's key identifiers and then tries to find a shared key identifier between y's key identifiers and its own's. If z_j does not find a same key identifier, it stops. Otherwise, z_j sends "1" to x for confirmation.

Step 5. x chooses z^* from neighbors which have responded and obtains the shard $\text{key}_{k_{z^*x}}$ through a neighbor list query and a key ring query. Then, x encrypts pk with the shared key and sends the temp ciphertext CT_{temp} and the key identifier k_{z^*x} to z^*.

Step 6. z^* searches its key ring for the shared $\text{key}_{k_{z^*x}}$ and decrypts CT_{temp} to obtain pk. After that, z^* searches its key ring for the shared $\text{key}_{k_{z^*y}}$ and encrypts pk with $\text{key}_{k_{z^*y}}$. At last, z^* sends the final ciphertext CT and the encryption key identifier k_{z^*y} to x.

Step 7. x forwards CT and k_{z^*y} to y.

2.2. The Algorithm **G**. The deterministic algorithm **G** is used to decide the key ring allocated to each node. Specifically, for

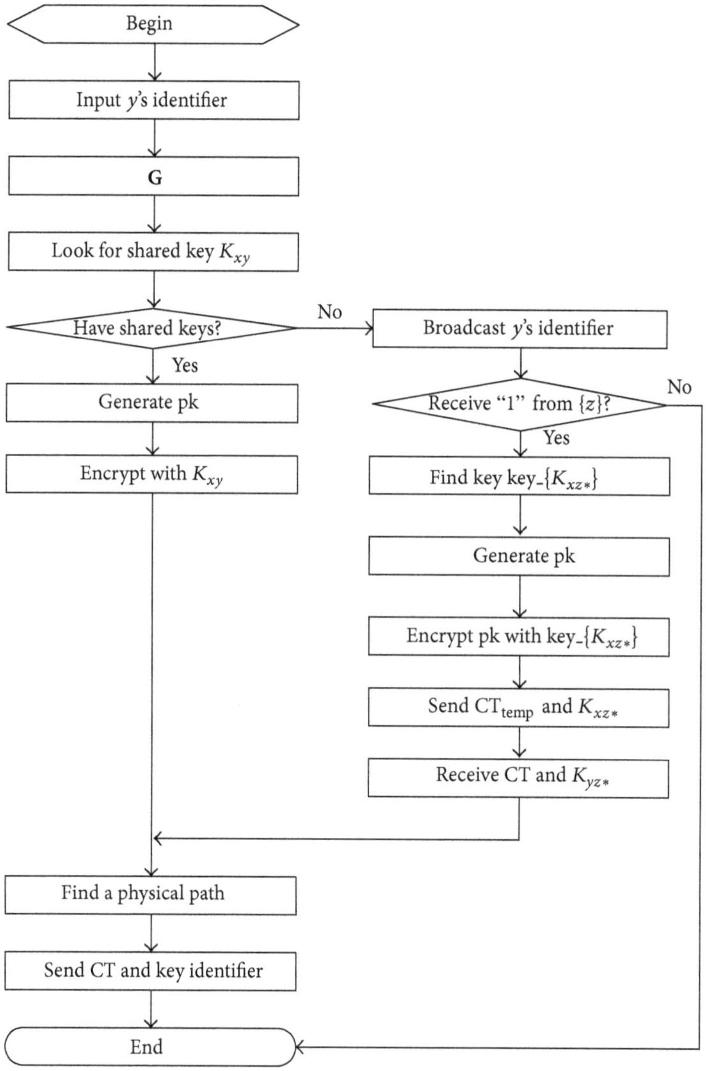

FIGURE 2: The operation process of node x in Phase 3.

each node with a unique node identifier, the algorithm generates s distinct integers between 1 and L using pseudorandom number generator upon the input of a node identifier. These s integers are the identifiers of the keys for this node. In this paper, the algorithm **G** is based on a bit-oriented linear feedback shift registers (LFSR). The content of the LFSR is denoted by $u_0, u_1, \ldots, u_{\alpha-1}$. The feedback polynomial of the LFSR, $f(x)$ is a primitive polynomial of degree α.

2.2.1. Initialization. Before any bit stream is generated, the register must be initialized with the node identifier. Let the bits of the node identifier k be denoted by k_j, $0 \leq j < \lceil \log_2 N \rceil$. The initialization phase is done as follows. First, load the LFSR with the node identifier bits, $u_j = k_j$, $0 \leq j < \lceil \log_2 N \rceil$, and then the remaining bits of the LFSR are filled with "1"s.

2.2.2. Key Identifier Set Generation. Create a one-dimensional array T of length s and initialize the array with "0"s. After the following steps, output the array which is fulfilled with key identifiers for a node.

For each $j \in [0, s-1]$

(1) the LFSR shift to producing bit stream;

(2) once every $\lceil \log_2 L \rceil$ bits $v_\beta, v_{\beta+1}, v_{\beta+2}, \ldots, v_{\beta+\lceil \log_2 L \rceil}$ have been generated, compute value $v = 2^{\lceil \log_2 L \rceil - 1} \cdot v_{\beta+\lceil \log_2 L \rceil} + \cdots + 2 \cdot v_{\beta+1} + v_\beta$;

(3) if $v > L$ or $v = T[\gamma]$ $(\gamma = 0, \ldots, j)$, go to (1);

(4) else, $T[j] = v$;

(5) sort the sequence $T[0], T[1], \ldots, T[j]$ using standard insertion sort method

end for.

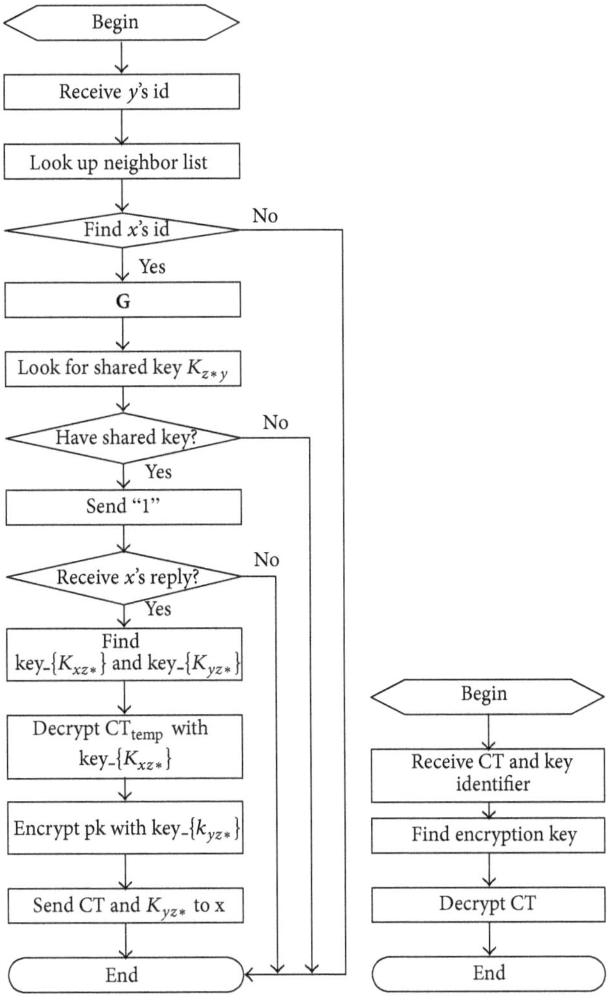

(a) The process of neighbor z (b) The process of y

FIGURE 3: The operation process of nodes z and y in Phase 3.

2.2.3. Design Criteria

(1) The Number of Nodes, In Which Each Key in the Key Pool is Allocated to, Closes to sN/L. In random sampling without replacement, each member of a population has an equal chance of being included in the sample. Hence, each key in the key pool has a probability of s/L to be chosen by a node. Since the network size is N, each key should be statistically allocated to sN/L nodes. In practice, however, using computational algorithms can only produce long sequences which seems to be random because the outputs are in fact determined by a shorter initial value, called a seed. So the generators are regarded as pseudorandom number generators. As a result, the number of nodes that each key is selected by can only close to sN/L in reality.

A binary sequence generated by a linear feedback shift which registers with a primitive feedback polynomial has balance property, run property, and correlation property. Therefore, any truncation of the sequence can be regarded as a pseudorandom number. The experimental result (see Figure 5) also shows that the times in which each key identifier is selected approximate to sN/L.

(2) $\alpha \geq \max(\lceil \log_2 N \rceil, \lceil \log_2 L \rceil)$. First of all, $\alpha > \lceil \log_2 N \rceil$ ensures that the register can be initialized to distinct states according to distinct node identifiers. Moreover, if $\alpha \leq \lceil \log_2 L \rceil$, the key identifiers (in binary format) that have a run of length e ($e > \alpha$) will never be selected by **G** (refer to run test of LFSR for the proof [23]).

(3) $\alpha < 3 \cdot \lceil \log_2 L \rceil$. This condition guarantees that a long subsequence of "0" will not appear at the start of the sequence, since most of those bits will be discarded in step 4 of the key identifier set generation procedure.

(4) $2^{\alpha} - 1 \gg \lceil \log_2 L \rceil \cdot \lceil \log_2 N \rceil$. This condition ensures that the key identifier set for a node can be generated in one cycle of the sequence produced by the LFSR.

Example 1. Suppose $N = 1000$, $L = 100$, and $s = 12$. According to the design criteria, we choose $\alpha = 12$ and the

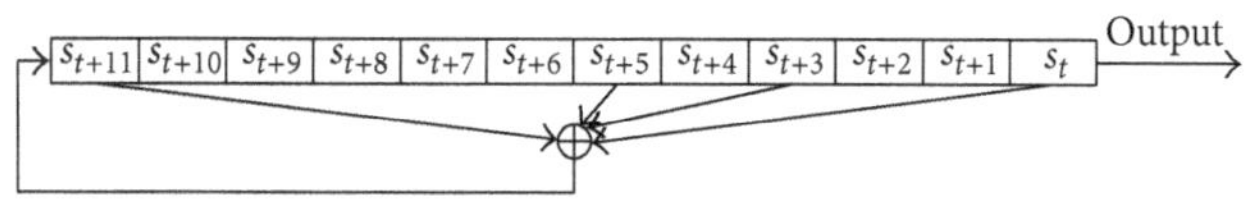

Figure 4: The structure of the LFSR.

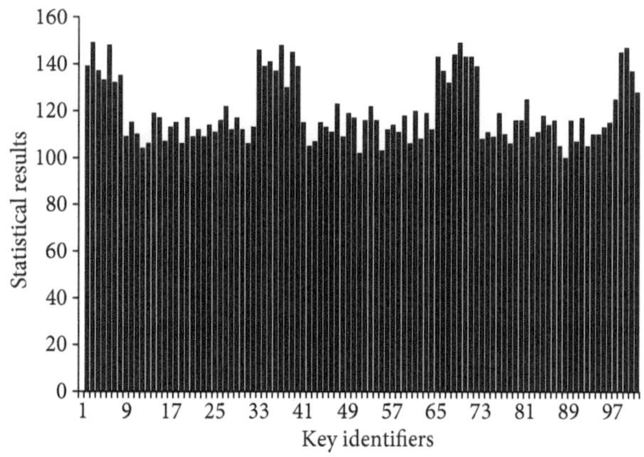

Figure 5: The times of each key identifier being chosen.

feedback polynomial of the LFSR $f(x) = x^{12} + x^6 + x^4 + x + 1$ (see also Figure 4). Obviously, α meets the conditions (2), (3), and (4).

The average times that each key is drawn by nodes are $sN/L = 12000/100 = 120$, and the actual statistical results are shown in Figure 5. It can be seen that most values are between 100 and 140, which close to 120.

3. Connectivity

The connected probability p refers to the probability that any two nodes in a distributed sensor network can establish a path key successfully. Next, we will prove that the connected probability p of the new RKPD scheme is

$$p = 1 - \frac{C_L^s \cdot C_{L-s}^s}{(C_L^s)^2} \cdot \left(1 - \frac{\sum_{i=1}^{s-1}\sum_{j=1}^{s-i} C_s^i \cdot C_s^j \cdot C_{L-2s}^{s-i-j}}{C_L^s}\right)^{n-1}. \quad (1)$$

As mentioned in Phase 3, a path key can be securely sent to y as long as an encryption key can be found, and the probability of which will be discussed in two situations.

Case 1. x finds a shared key k_{xy} with y in step 1 of Phase 3, and this key is used as the encryption key.

Each key ring of size s is randomly drawn out of the key pool, so the probability that two nodes do not share any key is $\overline{P_{xy}} = C_L^s \cdot C_{L-s}^s/(C_L^s)^2$. In other words, the first key ring is picked at random and the second key ring is drawn out of the remaining $L-s$ unused keys in the pool. Therefore, at least one shared key exists with probability $P_{xy} = 1 - (C_L^s \cdot C_{L-s}^s/(C_L^s)^2)$.

Case 2. x cannot find any shared key with y in step 1 of Phase 3, but a logically connected neighbor z^* which has a shared key k_{yz^*} with y can be found. Hence, that shared key will be used as the encryption key.

A neighbor of node x, denoted by z, has both a shared key with x and a shared key with y which means that i $(1 \le i \le s-1)$ keys of z's key ring are taken from x's ring and j $(1 \le j \le s-i)$ keys are taken from y's ring. So the total number of the possible key rings of z is $\sum_{i=1}^{s-1}\sum_{j=1}^{s-i} C_s^i \cdot C_s^j \cdot C_{L-2s}^{s-i-j}$. Therefore, such a neighbor exists with probability $P_{xyz} = \sum_{i=1}^{s-1}\sum_{j=1}^{s-i} C_s^i \cdot C_s^j \cdot C_{L-2s}^{s-i-j}/C_L^s$.

In addition, the node density is n, so there are $n-1$ neighbors in average. Therefore, the probability that at least one neighbor can help x encrypt pk in step 5 and step 6 is $P_{xyz,n} = 1 - (1 - P_{xyz})^{n-1}$. In sum, the connected probability is

$$\begin{aligned} p &= \Pr[\text{Case 1}] + \Pr\left[\text{Case 2} \mid \overline{\text{Case 1}}\right] \\ &= P_{xy} + \overline{P_{xy}} \cdot P_{xyz,n} = \left(1 - \frac{C_L^s \cdot C_{L-s}^s}{(C_L^s)^2}\right) \\ &\quad + \frac{C_L^s \cdot C_{L-s}^s}{(C_L^s)^2} \cdot \left[1 - \left(1 - \frac{\sum_{i=1}^{s-1}\sum_{j=1}^{s-i} C_s^i \cdot C_s^j \cdot C_{L-2s}^{s-i-j}}{C_L^s}\right)^{n-1}\right] \\ &= 1 - \frac{C_L^s \cdot C_{L-s}^s}{(C_L^s)^2} \cdot \left(1 - \frac{\sum_{i=1}^{s-1}\sum_{j=1}^{s-i} C_s^i \cdot C_s^j \cdot C_{L-2s}^{s-i-j}}{C_L^s}\right)^{n-1}. \end{aligned} \quad (2)$$

According to (2), the system parameters can be decided. For instance, let $p = 0.9999$ while the parameters are listed in Table 1. The first row is the size of key pool and the first column is the node density. Each entry of the table is the theoretical values of the key ring size.

Table 1: System parameters in the new scheme when $p = 0.9999$.

n/L	100	1000	10000	100000
10	12	38	119	380
20	10	30	97	310
30	9	27	86	270
40	8	25	77	250
50	8	23	72	230
60	7	22	69	220

Let the total number of nodes in the network be $N = 1000$ and let the connected probability be $p = 0.9999$. Figure 6 shows how the key ring size s varies along with the node density n.

In the figures of this section, symbol (O) denotes the new scheme, symbol (B) denotes the basic scheme, and symbol (Q) denotes the q-composite scheme. It can be seen that the theoretical value of s in the basic scheme and the q-composite scheme increases sharply when n closes to 20. The reason is that the predefined out-degree is $d = 20$. In other words, if the

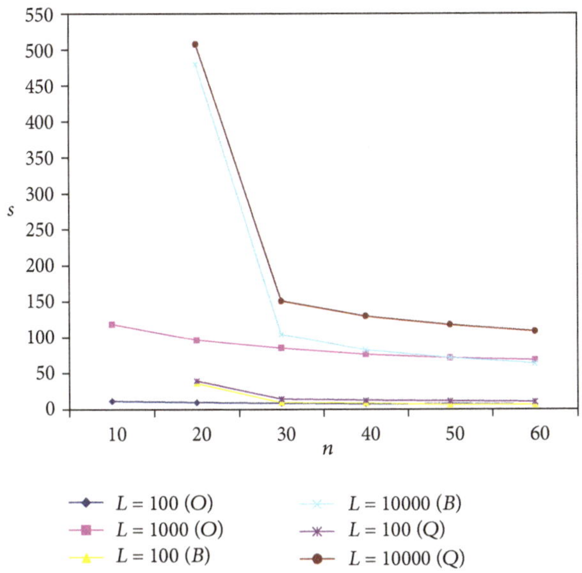

FIGURE 6: Changes of the key ring size.

node density n is equal to or less than d, the connectivity will be seriously affected. Since the out-degree cannot be modified when nodes are deployed, the size of the key ring must be very large to keep high connectivity when nodes sleep or die. As a result, the amount of storage per node in the basic scheme and the q-composite scheme will increase. However, the ring size in the new scheme increases smoothly as the node density decreases. Thus, by adding a small amount of redundancy to the key ring size, the network will work well even in the case in which the number of active nodes is less than d. For example, let $s = 12$ when $L = 100$.

4. Security

Resilience in WSNs refers to the resistance of key distribution schemes against node capture. When sensor nodes are deployed in hostile areas (e.g., battle surveillance), an adversary can mount a physical attack on a sensor node and recover secret information from its memory. So we are interested in the question: for any two nodes x and y which have not been captured by the attacker, what is the probability that the attacker can eavesdrop their communications using the subset of the key pools that was recovered from the nodes captured. That is, the shared key of two physically connected neighbors (also called a link) is in the compromised key set or a path key is compromised in path key establishment.

Theorem 2. *The probability that any secure link in the shared-key discovery phase between two uncompromised nodes is compromised when m nodes have been captured is* $\sum_{\tau=s}^{\min(ms,L)}[C_L^\tau \cdot (\tau/L)\cdot(((C_\tau^s)^m - \sum_{\lambda=1}^{\tau-s}(-1)^{\lambda+1}C_\tau^\lambda(C_{\tau-\lambda}^s)^m)/(C_L^s)^m)]$.

Proof. Let $I \subset \{1,2,\ldots,L\}$ be a subset of the key identifers of the key pool, where $s \le |I| \le \min(ms, L)$. And let C_I be the event in which key with identifier $k_j \in I$ is compromised after m nodes are captured by an adversary. The probability $\Pr[C_I]$ is $((C_{|I|}^s)^m - \sum_{\lambda=1}^{|I|-s}(-1)^{\lambda+1}C_{|I|}^\lambda(C_{|I|-\lambda}^s)^m)/(C_L^s)^m$ (please refer to paper [24] for the proof).

Let C be the event in which the link between the two nodes is compromised, so $\Pr[C] = \sum_I \Pr[C \mid C_I]\cdot\Pr[C_I]$. If I is fixed, a link in our scheme is compromised if and only if the shared key's identifier of the two nodes is in I. Hence, $\Pr[C \mid C_I] = |I|/L$.

From the above two aspects, the probability of compromising a link in the shared-key discovery is

$$
\begin{aligned}
\Pr[C] &= \sum_I \Pr[C \mid C_I]\cdot\Pr[C_I] \\
&= \sum_I\left[\frac{|I|}{L}\cdot\left(\frac{(C_{|I|}^s)^m - \sum_{\lambda=1}^{|I|-s}(-1)^{\lambda+1}C_{|I|}^\lambda(C_{|I|-\lambda}^s)^m}{(C_L^s)^m}\right)\right] \\
&= \sum_{\tau=s}^{\min(ms,L)}\left[C_L^\tau\cdot\frac{\tau}{L}\cdot\left(\frac{(C_\tau^s)^m - \sum_{\lambda=1}^{\tau-s}(-1)^{\lambda+1}C_\tau^\lambda(C_{\tau-\lambda}^s)^m}{(C_L^s)^m}\right)\right].
\end{aligned}
\tag{3}
$$

The probability that a link is compromised in our scheme is equal to that in the basic scheme, and it is higher than that of the q-composite scheme. □

Theorem 3. *The probability that a path key between two nodes is compromised in the path key establishment phase when m nodes have been captured is* $\sum_{\tau=s}^{\min(ms,L)}[C_L^\tau \cdot (P_{xy}\cdot(\tau/L) + (1-P_{xy})\cdot(2\cdot(\tau/L)-(\tau/L)^2))\cdot(((C_\tau^s)^m - \sum_{\lambda=1}^{\tau-s}(-1)^{\lambda+1}C_\tau^\lambda(C_{\tau-\lambda}^s)^m)/(C_L^s)^m)]$.

Proof. In path key establishment, let B be the event in which a path key between the two nodes is compromised in the path key establishment phase when m nodes have been captured. Similarly, $\Pr[B] = \sum_I \Pr[B \mid C_I]\cdot\Pr[C_I]$.

In our scheme, a path key is encrypted with one shared $\text{key}_{k_{xy}}$, named direct encryption, with probability P_{xy} (see Figure 7) and is encrypted with two shared keys ($\text{key}_{k_{xz^*}}$, $\text{key}_{k_{yz^*}}$), named indirect encryption, with probability $1 - P_{xy}$ (see Figure 8). We already know that a shared key is compromised if and only if the shared key's identifier is in I. So in the direct encryption, the path key is compromised with probability $|I|/L$. And in the indirect encryption, the probability that the path key is compromised is the probability of compromised one shared key plus the probability of compromised another shared key minus the probability of compromising both shared keys. □

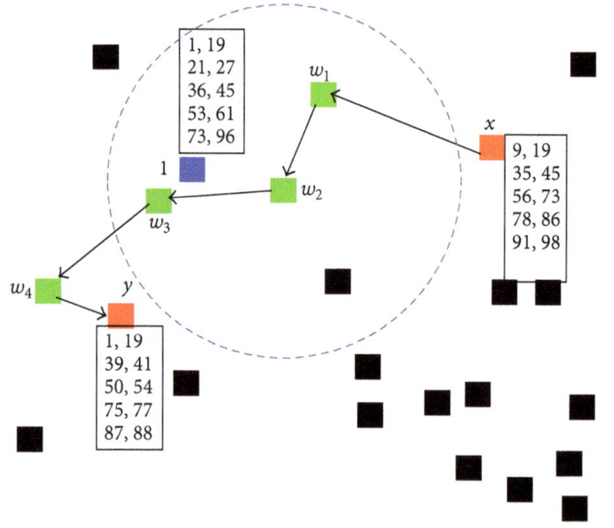

FIGURE 7: The path key is compromised in the direct encryption. [1]x is source node and y is destination node. The attacker captures a node (the blue node) whose key ring includes key identifier 19. So the path key between x and y may be eavesdropped at anywhere near the path (x-w_1-w_2-w_3-w_4-y), for example, at the link (w_2-w_3).

So the probability that the path key is compromised is $\Pr[B \mid C_I] = P_{xy}\cdot(|I|/L)+(1-P_{xy})\cdot(2\cdot(|I|/L)-(|I|/L)^2)$ when I is fixed. To sum up, in our scheme, the probability that the path key is compromised is

$$
\begin{aligned}
\Pr[B] &= \sum_I \Pr[B \mid C_I]\cdot\Pr[C_I] \\
&= \sum_I \left[\left(P_{xy}\cdot\frac{|I|}{L}+\left(1-P_{xy}\right)\cdot\left(2\cdot\frac{|I|}{L}-\left(\frac{|I|}{L}\right)^2\right)\right)\right. \\
&\qquad \left.\cdot\left(\frac{\left(C_{|I|}^{s}\right)^m-\sum_{\lambda=1}^{|I|-s}(-1)^{\lambda+1}C_{|I|}^{\lambda}\left(C_{|I|-\lambda}^{s}\right)^m}{\left(C_L^s\right)^m}\right)\right] \\
&= \sum_{\tau=s}^{\min(ms,L)}\left[C_L^{\tau}\cdot\left(P_{xy}\cdot\frac{\tau}{L}+\left(1-P_{xy}\right)\right.\right. \\
&\qquad \left.\cdot\left(2\cdot\frac{\tau}{L}-\left(\frac{\tau}{L}\right)^2\right)\right) \\
&\qquad \left.\cdot\left(\frac{\left(C_{\tau}^{s}\right)^m-\sum_{\lambda=1}^{\tau-s}(-1)^{\lambda+1}C_{\tau}^{\lambda}\left(C_{\tau-\lambda}^{s}\right)^m}{\left(C_L^s\right)^m}\right)\right].
\end{aligned}
\tag{4}
$$

In the basic scheme [14] and the q-composite [15] scheme, the path key is established on a multihop path (mainly refers to multiple links using different encryption key), so the path key is compromised when at least one hop is compromised (see Figure 9). Assuming that one hop (a link) is compromised with probability $\Pr[C \mid C_I]$ when m nodes are captured, the probability that all t-hops are secure is $(1-\Pr[C \mid C_I])^t$. As a result, the probability that the path key is compromised is $\Pr[B] = \sum_{\tau=s}^{\min(ms,L)}[C_L^{\tau}\cdot(1-(1-\Pr[C \mid C_I])^t)\cdot(((C_{\tau}^{s})^m-\sum_{\lambda=1}^{\tau-s}(-1)^{\lambda+1}C_{\tau}^{\lambda}(C_{\tau-\lambda}^{s})^m)/(C_L^s)^m)]$.

As a result, in the basic scheme, the probability that the path key is compromised is $\Pr[B] = \sum_{\tau=s}^{\min(ms,L)}[C_L^{\tau}\cdot(1-(1-(\tau/L))^t)\cdot(((C_{\tau}^{s})^m-\sum_{\lambda=1}^{\tau-s}(-1)^{\lambda+1}C_{\tau}^{\lambda}(C_{\tau-\lambda}^{s})^m)/(C_L^s)^m)]$. And in the q-composite scheme, the probability that the path key is compromised is $\Pr[B] = \sum_{\tau=s}^{\min(ms,L)}[C_L^{\tau}\cdot(1-(1-\sum_{j=q}^{s}(C_{\tau}^{j}/C_L^{j})\cdot(p(j)/p'))^t)\cdot(((C_{\tau}^{s})^m-\sum_{\lambda=1}^{\tau-s}(-1)^{\lambda+1}C_{\tau}^{\lambda}(C_{\tau-\lambda}^{s})^m)/(C_L^s)^m)]$, where $p(i) = (C_L^i)\cdot(C_{L-i}^{2(s-i)})\cdot(C_{2(s-i)}^{s-i})/(C_L^s)^2$ and $p' = d/n$. Please refer to paper [24] for the proof of $\Pr[C \mid C_I]$ in the basic scheme and the q-composite scheme.

In the l-path scheme [15], a path key is compromised if and only if l disjointed path is all compromised, that is, $\Pr[C \mid C_I] = \prod_{i=1}^{l}(1-(1-\Pr[C \mid C_I])^{t_i})$, where t_i is the number of hops on the ith path. So the probability that the path key is compromised is $\Pr[B] = \sum_{\tau=s}^{\min(ms,L)}[C_L^{\tau}\cdot\prod_{i=1}^{l}(1-(1-(\tau/L))^{t_i})\cdot(((C_{\tau}^{s})^m-\sum_{\lambda=1}^{\tau-s}(-1)^{\lambda+1}C_{\tau}^{\lambda}(C_{\tau-\lambda}^{s})^m)/(C_L^s)^m)]$.

The probability that a path key will be compromised in the establishment phase when $m = 1$ is listed in Table 2 ($L = 100$, $s = 10$, and $q = 2$). It can be seen that the probability in our scheme will be lower than that of the l-path scheme when the number of hops is larger than 4.

5. Complexity

In this section, the storage, computation, and communication complexity of the new scheme are presented.

(1) Storage Complexity. Same as the basic scheme, each node in the new scheme stores a key ring and a neighbor list. There are s key-identifier/key pairs in the key ring. Let |key| represent the length of a key, so the key ring takes ($s\cdot|\text{key}| + s\cdot\lceil\log_2 L\rceil$) bits of storage space. On the other hand, the probability that two neighbor nodes are logically connected is also P_{xy}, and the average number of neighbors for a node is $n-1$. Thus, there are about $[P_{xy}\cdot(n-1)]$ records on the neighbor list. Each record includes a node identifier and a key identifier. Therefore, the neighbor list uses $[P_{xy}\cdot(n-1)\cdot(\lceil\log_2 N\rceil+\lceil\log_2 L\rceil)]$ bits memory space. Consequently, the total storage complexity of the new scheme is $O(s\cdot|\text{key}|+s\cdot\lceil\log_2 L\rceil+P_{xy}\cdot(n-1)\cdot(\lceil\log_2 N\rceil+\lceil\log_2 L\rceil))$.

(2) Communication Complexity. Let |CT| be the length of a ciphertext. If x and y share a key, the main message transmitted on the channel is the ciphertext and the key identifier of the encryption key. So, the communication complexity is $O((t+1)\cdot(\lceil\log_2 L\rceil+|\text{CT}|))$, where t represents the number of hops in a path.

If x and y do not share any keys, the process of searching for an encryption key (step 2 to step 6) will bring extra traffic. y's identifier with length $\lceil\log_2 N\rceil$ is sent in step 2, and a key identifier and a ciphertext are transmitted both in step 5 and in step 6. The probability that a neighbor z has shared keys both with x and with y is $P_{xy}\cdot P_{xy}$, so about $[(n-1)\cdot P_{xy}\cdot P_{xy}]$ bits confirmation message will be sent to x

TABLE 2: The probability that a path key is compromised.

t	1	2	3	4	5	6
Our scheme	0.129743	0.129743	0.129743	0.129743	0.129743	0.129743
Basic scheme	0.100000	0.190000	0.271000	0.343900	0.409510	0.468559
q-composite scheme	0.003741	0.007469	0.011183	0.014882	0.018568	0.022240
l-path scheme	0.010000	0.036100	0.073441	0.0118267	0.167698	0.219548

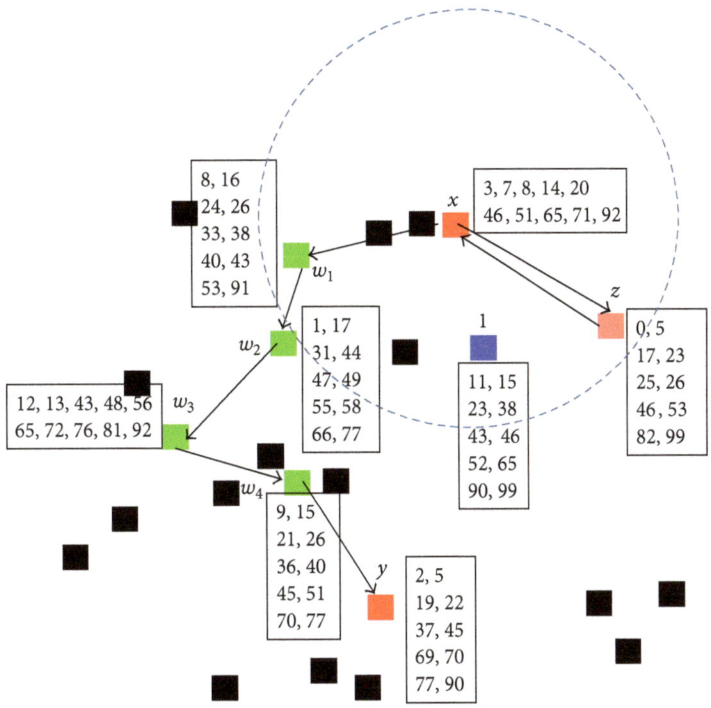

FIGURE 8: The path key is compromised in the indirect encryption. [2]x is source node and y is destination node. The attacker captures a node (the blue node) whose key ring includes key identifier 46. So the path key between x and y may be eavesdropped only near the link (x-z).

in step 4. Hence, the extra traffic is $O(\lceil \log_2 N \rceil + 2(\lceil \log_2 L \rceil + |\text{CT}|) + [(n-1) \cdot P_{xy} \cdot P_{xy}])$.

And because the probability that x and y do not share any keys is $\overline{P_{xy}}$, the average communication complexity is $O((t+1) \cdot (\lceil \log_2 L \rceil + |\text{CT}|) + \overline{P_{xy}} \cdot (\lceil \log_2 N \rceil + 2(\lceil \log_2 L \rceil + |\text{CT}|) + [(n-1) \cdot P_{xy} \cdot P_{xy}]))$.

The communication complexity of the new scheme is slightly higher than that of the basic scheme. Figure 10 shows that the difference of the communication complexity between the two schemes can be almost negligible.

(3) Computation Complexity. Let $|G|$ denote the time complexity of random sequence generation such as LFSR, and let $|E|$ represent the time complexity of an encryption (or decryption) operation.

The amount of calculation of sorting a (key identifier) set of size s is $O(s \cdot (s-1)/2 \cdot \lceil \log_2 L \rceil)$, such as bubble sort algorithm or insertion sort algorithm. And the calculation amount of comparing two ordered sets of size s to find a common key identifier (binary search), called key identifier search, is $O(s \cdot \lceil \log_2 s \rceil \cdot \lceil \log_2 L \rceil)$. According to a key identifier, the amount of calculation of looking up the corresponding key in a key ring (binary search), denoted by key ring query, is $O(\lceil \log_2 s \rceil \cdot \lceil \log_2 L \rceil)$.

The average length of the neighbor list is $[(n-1) \cdot P_{xy}]$ and a node identifier has $\lceil \log_2 N \rceil$ bits, so the computation amount of deciding whether a node identifier is on a neighbor list, denoted by neighbor list query, will be $O([(n-1) \cdot P_{xy}] \cdot \lceil \log_2 N \rceil)$.

Case 1. If x and y have a shared key, only x and y are involved in establishing a path key regardless of the receiving and forwarding operations of nodes in the path. x executes a random sequence generation, a key identifier search, a key ring query, and a path key encryption in step 1, so the computation complexity of x is $O(|G| + s \cdot (s-1)/2 \cdot \lceil \log_2 L \rceil + s \cdot \lceil \log_2 s \rceil \cdot \lceil \log_2 L \rceil + |E|)$.

Accordingly, y executes a key ring query and a decryption; hence, the computation complexity of y is $O(\lceil \log_2 s \rceil \cdot \lceil \log_2 L \rceil + |E|)$.

Case 2. If x and y do not have any shared key, x's neighbors are involved in finding the encryption key. $n-1$ neighbors may run neighbor list query (step 3), so the complexity is $O((n-1) \cdot [(n-1) \cdot P_{xy}] \cdot \lceil \log_2 N \rceil)$. About $[(n-1) \cdot P_{xy}]$ neighbors will run step 4 which includes a random

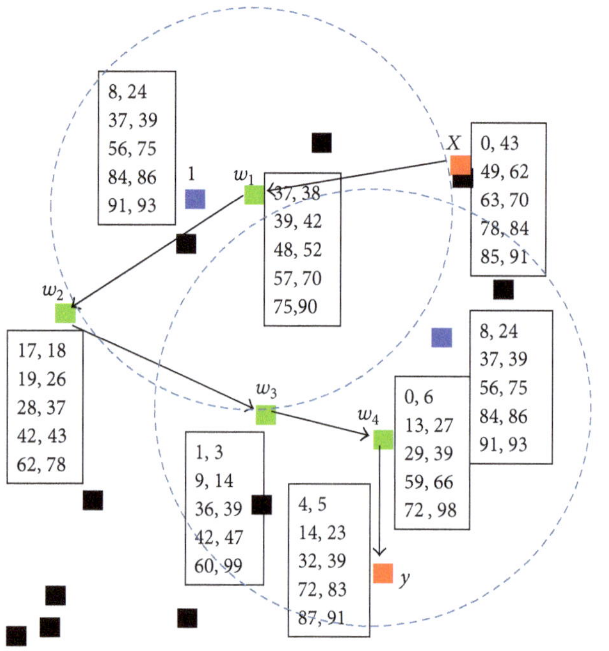

FIGURE 9: The path key is compromised in the basic scheme. [3]x is source node and y is destination node. The attacker captures a node (the blue node) whose key ring includes key identifiers 37 and 39. So the path key between x and y may be compromised at the link (w_1-w_2) which share the key 37, and it may also be compromised at the link (w_3-w_4) which share the key 39.

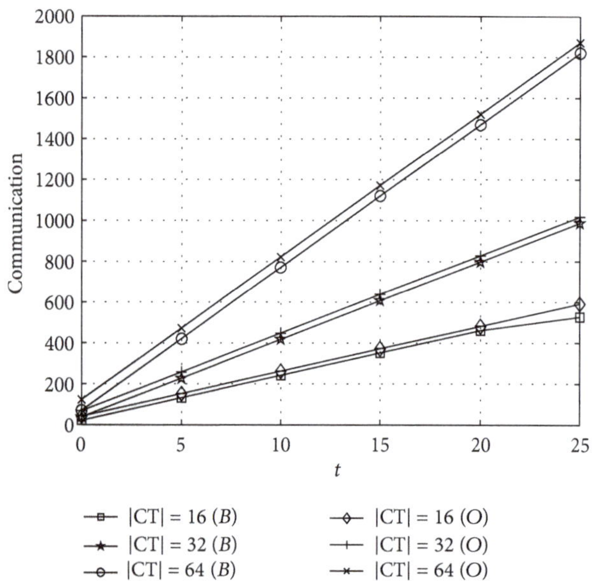

FIGURE 10: The comparison of the communication complexity.

sequence generation and a key identifier search; therefore, the computation amount is $O(\lceil (n-1)\cdot P_{xy}\rceil \cdot (|G| + s\cdot (s-1)/2\cdot \lceil \log_2 L\rceil + s\cdot \lceil \log_2 s\rceil \cdot \lceil \log_2 L\rceil))$.

In step 5, x runs a neighbor list query and a key ring query to find the shared key with z^* and encrypts the ciphertext with the shared key. The total computation complexity is $O(\lceil (n-1)\cdot P_{xy}\rceil \cdot \lceil \log_2 N\rceil + \lceil \log_2 s\rceil \cdot \lceil \log_2 L\rceil + |E|)$.

In step 6, z^* executes key ring query twice, encryption once, and decryption once. So the calculation amount is $O(2\lceil \log_2 s\rceil \cdot \lceil \log_2 L\rceil + 2|E|)$.

Considering the above two aspects, the computational complexity is

$$\begin{aligned}
\text{Comp} &= \{\text{Comp of } x\} + \{\text{Comp of } y\} \\
&\quad + \{\text{Comp of neighbors}\} \\
&= \Big\{ O\Big(|G| + s\cdot \frac{(s-1)}{2}\cdot \lceil \log_2 L\rceil \\
&\quad + s\cdot \lceil \log_2 s\rceil \cdot \lceil \log_2 L\rceil + |E|\Big) \\
&\quad + \overline{P_{xy}}\cdot \Big(\lceil (n-1)\cdot P_{xy}\rceil \cdot \lceil \log_2 N\rceil \\
&\quad + \lceil \log_2 s\rceil \cdot \lceil \log_2 L\rceil \Big)\Big\} \\
&\quad + \{\lceil \log_2 s\rceil \cdot \lceil \log_2 L\rceil + |E|\} \\
&\quad + \overline{P_{xy}}\cdot \Big\{\big((n-1)\cdot \lceil (n-1)\cdot P_{xy}\rceil \cdot \lceil \log_2 N\rceil\big) \\
&\quad + \Big(\lceil (n-1)\cdot P_{xy}\rceil \\
&\quad \cdot \Big(|G| + s\cdot \frac{(s-1)}{2}\cdot \lceil \log_2 L\rceil \\
&\quad + s\cdot \lceil \log_2 s\rceil \cdot \lceil \log_2 L\rceil \Big)\Big) \\
&\quad + (2\lceil \log_2 s\rceil \cdot \lceil \log_2 L\rceil + 2|E|)\Big\} \\
&= O\Big(\big(\lceil (n-1)\cdot P_{xy}\rceil \cdot \overline{P_{xy}} + 1\big)\cdot |G| \\
&\quad + \big(\lceil (n-1)\cdot P_{xy}\rceil \cdot \overline{P_{xy}} + 1\big) \\
&\quad \cdot \Big(s\cdot \frac{(s-1)}{2}\cdot \lceil \log_2 L\rceil\Big) \\
&\quad + \big(s + \lceil (n-1)\cdot P_{xy}\rceil \cdot \overline{P_{xy}}\cdot s + 3\overline{P_{xy}} + 1\big) \\
&\quad \cdot (\lceil \log_2 s\rceil \cdot \lceil \log_2 L\rceil) + 2\big(\overline{P_{xy}} + 1\big)\cdot |E| \\
&\quad + n\cdot \lceil (n-1)\cdot P_{xy}\rceil \cdot \overline{P_{xy}}\cdot \lceil \log_2 N\rceil\Big).
\end{aligned} \tag{5}$$

In Phase 2 of the new scheme, the algorithm **G** is used to save on traffic at the cost of increasing the computation amount. So the total traffic in this stage is about $O(N\cdot s\cdot \lceil \log_2 L\rceil)$, and the cost of computation is about $O(N\cdot (|G| + s\cdot (s-1)/2\cdot \lceil \log_2 L\rceil + s\cdot \lceil \log_2 s\rceil \cdot \lceil \log_2 L\rceil))$. Finally, the complexity of the basic scheme and the new scheme is listed in Table 3.

The efficiency (in Phase 3) of the basic scheme and the new scheme is shown in Figures 11 and 10. In those figures, symbol (O) denotes the new scheme and symbol (B) denotes the basic scheme.

From (5), the computational complexity of the new scheme is concerned with L, s, and $|E|$. Besides that, the basic scheme varies with the number of hops t, which is shown in

TABLE 3: The complexity of the basic scheme and the new scheme.

	Basic scheme	New scheme
Storage	$O\left(s \cdot \lvert\text{key}\rvert + s \cdot \lceil \log_2 L \rceil + P_{xy} \cdot (n-1) \cdot (\lceil \log_2 N \rceil + \lceil \log_2 L \rceil)\right)$	$O\left(s \cdot \lvert\text{key}\rvert + s \cdot \lceil \log_2 L \rceil + P_{xy} \cdot (n-1) \cdot (\lceil \log_2 N \rceil + \lceil \log_2 L \rceil)\right)$
Communication (Phase 2)	$O(N \cdot s \cdot \lceil \log_2 L \rceil + N \cdot \lceil \log_2 N \rceil)$	$O(N \cdot s \cdot \lceil \log_2 L \rceil)$
Computation (Phase 2)	$O(N \cdot s \cdot \lceil \log_2 s \rceil \cdot \lceil \log_2 L \rceil)$	$O\left(N \cdot \left(\lvert G \rvert + s \cdot \frac{s-1}{2} \cdot \lceil \log_2 L \rceil + s \cdot \lceil \log_2 s \rceil \cdot \lceil \log_2 L \rceil\right)\right)$
Communication (Phase 3)	$O((t+1) \cdot (\lceil \log_2 L \rceil + \lvert \text{CT} \rvert))$	$O\left((t+1) \cdot (\lceil \log_2 L \rceil + \lvert \text{CT} \rvert) + \overline{P_{xy}} \cdot \left(\lceil \log_2 N \rceil + 2(\lceil \log_2 L \rceil + \lvert \text{CT} \rvert) + \left[(n-1) \cdot P_{xy} \cdot P_{xy}\right]\right)\right)$
Computation (Phase 3)	$O\left((t+1) \cdot 2\left((n-1) \cdot P_{xy} \cdot \lceil \log_2 N \rceil + \lceil \log_2 s \rceil \cdot \lceil \log_2 L \rceil + \lvert E \rvert\right)\right)$	Computation in formula (5)

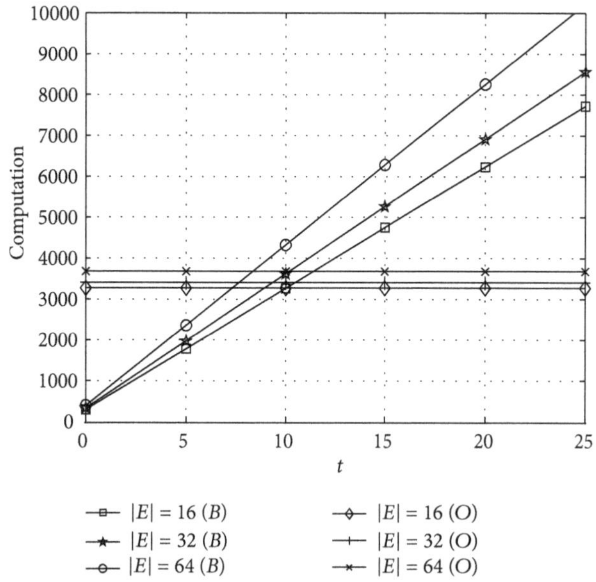

FIGURE 11: Comparisons of the computation complexity.

Figure 11 ($L = 100$ and $s = 10$). It can be easily noticed that if t is small, the basic scheme performs faster than the new scheme. However, the computation complexity of the basic scheme grows rapidly with the increase of the number of hops t. When t is more than 10, the complexity of the new scheme is lower than that of the basic one.

It is worth noting that the computation complexity of the basic scheme in Table 3 does not include the extra computation cost of finding logically connected neighbors in routing phase, which will be estimated in the next section.

6. Simulations

In this section, we use computer to simulate the basic scheme and our scheme. In the basic scheme, every hop of the path is not only physically but also logically connected. We try to estimate the amount of computation on deciding whether two adjacent nodes are logically connected (neighbor list query) in routing process. On the other hand, we prove that the path length between two nodes will increase in the basic scheme. At last, the execution time of the path key establishment of the new scheme and the basic scheme is tested. The simulation environment is as follows:

the operating system: ubuntu10.04,

CPU: Intel Core2 Quad Q8400 2.66 GHz,

RAM: 4.00 GB,

communication bandwidth: 4 KB/s,

routing protocol: AODV.

In those simulations, parameters are set as $N = 1000$, $L = 100$, $s = 10$, and $n = 28$.

(1) Computation Amount. As previously metioned, the basic scheme will spend lots of extra computation on neighbor list query in the routing process. Table 4 shows the average number of nodes which execute the neighbor list query in 208 random routing trails. Moreover, the hop limit is 30.

TABLE 4: The number of nodes involved in routing process of the basic scheme.

Number of hops	Number of trails	Average number of nodes
1–5	19	606
6–10	37	552
11–15	36	548
16–20	28	612
21–25	36	570
26–30	20	656

Obviously, more than half nodes are involved in a routing process. The computation costs on a neighbor list query are $O([(n-1) \cdot P_{xy}])$, so the total computation complexity is $O(N/2 \cdot [(n-1) \cdot P_{xy}])$. The theoretical value is more than 90000 when parameters are the same as those of routing trails.

(2) Path Length. In fact, the path length t of the basic scheme may be longer than that of the new scheme since the routing protocol is affected by the basic scheme. Consequently, the communication complexity of the basic scheme will be more than the communication amount shown in Figure 10. We randomly chose two nodes which are physically connected neighbors and then conducted 10240 routing trails to find a path in which two adjacent nodes are logically connected. The length of the path in each trail was counted, and the count results are shown in Table 5.

Obviously, the probability that the path length increases ($t > 0$) because of the influence of the basic scheme is at least $1/2$.

It should be pointed out that parameters in those simulations are small, for instance, $L = 100$ and $s = 10$, because we assume that the network has a lot of simple sensor nodes which have limited power. In fact, if parameters increase, $L = 10000$ for example, the above conclusions also hold.

(3) Execution Time of the Path Key Establishment. We simulate the path key establishment process on the computer, in which the encryption algorithm is DES. Figure 12 shows the average time spent on path key establishment.

The x-coordinate in Figure 12 is the physical hops of two nodes (x, y). As mentioned above, the real number of hops in the path key establishment of the basic scheme may be larger than that. "Direct" means that x and y share a key K_{xy}. The path key is encrypted with this key, and then pk is sent to y. "Indirect" means that x transmits the pk to y with the help of a neighbor z^*. It can be seen in the figure that our scheme is slightly more efficient than the basic scheme.

7. Conclusions

The random key predistribution scheme is a good solution to distributed sensor networks, which reduces the energy

TABLE 5: The growth of the number of hops.

	$t = 0$	$t = 1$	$t = 2$	$t = 3$	$t \geq 4$
Number of trails	4777	3428	1441	456	138
Percent	46.65	33.48	14.07	4.45	1.35

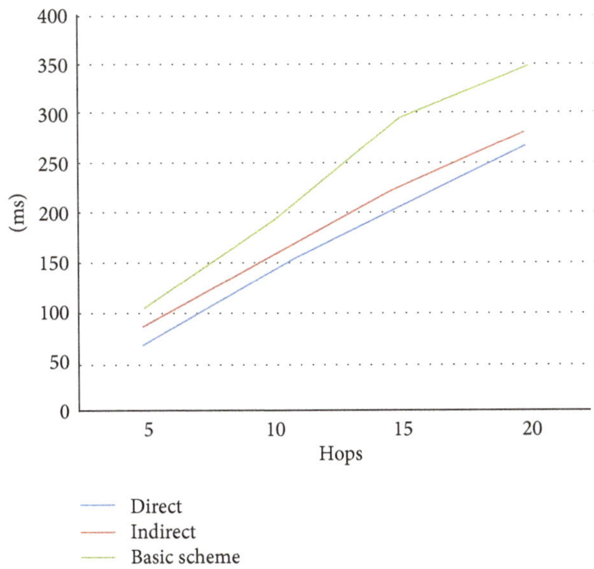

FIGURE 12: The simulation result of path key establishment.

consumption of nodes though losing a little connectivity and security of the scheme. In this paper, a new RKPD scheme which is independent of the routing protocol is given. The new scheme can be parameterized to meet the appropriate level of performance and keep the stability of the connectivity in a certain range by bringing small redundant in parameters.

Moreover, an algorithm with the input of node's identifier is constructed to reduce the communication overload. However, the traffic of the new scheme in the path key establishment phase is still higher than that of the basic scheme. Since Pottic and Kaiser [25] revealed that the energy of transmitting 1 K bits over 100 meters could be used to execute 3×10^6 instructions, the communication overload of the new scheme should be further reduced. A intuitive solution is that we do not generate a new path key. Instead, if x and y have a shared key k_{xy}, the key is used as the path key. Else, the key$_{k_{yz^*}}$ that z^* shares with y will be sent to x (covered by key$_{k_{xz^*}}$) by z^*. In this way, x only transmits the key identifier k_{xy} or k_{yz^*} to y. However, the above solution may incur new security flaw which is the same as the scheme in [17]. For example, a captured node may be cloned and redeployed in WSNs, and then they send fake message (as step 1 in the path key establishment) to its neighbors and defraud fresh (not compromised) keys in the key pool.

Conflict of Interests

The authors declare that there is no conflict of interests regarding the publication of this paper.

Acknowledgments

This work is supported by National Natural Science Foundation of China (nos. 61370194 and 61202082) and the Fundamental Research Funds for the Central Universities (nos. BUPT2012RC0219 and BUPT2013RC0311).

References

[1] T. Matsumoto and H. Imai, "On the KEY PREDISTRIBUTION SYSTEM: a practical solution to the key distribution problem," in *Advances in Cryptology—CRYPTO'87*, vol. 293 of *Lecture Notes in Computer Science*, pp. 185–193, Santa Barbara, Calif, USA, August 1987.

[2] R. Blom, "An optimal class of symmetric key generation systems," in *Proceedings of the EUROCRYPT 84 Workshop on Advances in Cryptology*, pp. 335–338, Paris, France, April 1984.

[3] C. Blundo, A. D. Santis, A. Herzberg, S. Kutten, U. Vaccaro, and M. Yung, "Perfectly secure key distribution for dynamic conferences," in *Advances in Cryptology —CRYPTO' 92*, vol. 740 of *Lecture Notes in Computer Science*, pp. 471–486, 1993.

[4] W. Du, J. Deng, Y. S. Han, P. K. Varshney, J. Katz, and A. Khalili, "A pairwise key predistribution scheme for wireless sensor networks," *ACM Transactions on Information and System Security*, vol. 8, no. 2, pp. 228–258, 2005.

[5] D. Liu and P. Ning, "Establishing pairwise keys in distributed sensor networks," in *Proceedings of the 10th ACM Conference on Computer and Communications Security*, pp. 52–61, October 2003.

[6] D. Chakrabarti, S. Maitra, and B. Roy, "A key pre-distribution scheme for wireless sensor networks: merging blocks in combinatorial design," *International Journal of Information Security*, vol. 5, no. 2, pp. 105–114, 2006.

[7] S. Mitra, S. Mukhopadhyay, and R. Dutta, "Unconditionally-secure key pre-distribution for triangular grid based wireless sensor network," *Journal of Applied Mathematics and Computing*, vol. 44, no. 1-2, pp. 229–249, 2014.

[8] M. F. Younis, K. Ghumman, and M. Eltoweissy, "Location-aware combinatorial key management scheme for clustered sensor networks," *IEEE Transactions on Parallel and Distributed Systems*, vol. 17, no. 8, pp. 865–882, 2006.

[9] D. Liu and P. Ning, "Improving key predistribution with deployment knowledge in static sensor networks," *ACM Transactions on Sensor Networks*, vol. 1, no. 2, pp. 204–239, 2005.

[10] D. W. Carman, "New directions in sensor network key management," *International Journal of Distributed Sensor Networks*, vol. 1, no. 1, pp. 3–15, 2005.

[11] A. Perrig, R. Szewczyk, V. Wen, D. Culler, and J. D. Tygar, "SPINS: security protocols for sensor networks," in *Proceedings of the 7th Annual ACM/IEEE International Conference on Mobile Computing and Networking (MobiCom '01)*, pp. 189–199, July 2001.

[12] S. Hussain, F. Kausar, and A. Masood, "An efficient key distribution scheme for heterogeneous sensor networks," in *Proceedings of the International Wireless Communications and Mobile*

Computing Conference (IWCMC '07), pp. 388–392, August 2007.

[13] A. Rasheed and R. Mahapatra, "Key predistribution schemes for establishing pairwise keys with a mobile sink in sensor networks," *IEEE Transactions on Parallel and Distributed Systems*, vol. 23, no. 1, pp. 176–184, 2011.

[14] L. Eschenauer and V. D. Gligor, "A key-management scheme for distributed sensor networks," in *Proceedings of the 9th ACM Conference on Computer and Communications Security (CCS '02)*, pp. 41–47, November 2002.

[15] H. Chan, A. Perrig, and D. Song, "Random key predistribution schemes for sensor networks," in *Proceedings of the IEEE Symposium on Research in Security And Privacy*, pp. 197–213, May 2003.

[16] T. Shan and C. Liu, "Enhancing the key pre-distribution scheme on wireless sensor networks," in *Proceedings of the IEEE Asia-Pacific Services Computing Conference*, pp. 1127–1131, 2008.

[17] C.-F. Law, K.-S. Hung, and Y.-K. Kwok, "A novel key redistribution scheme for wireless sensor networks," in *Proceedings of the IEEE International Conference on Communications (ICC '07)*, pp. 3437–3442, June 2007.

[18] S. Zhu, S. Xu, S. Setia, and S. Jajodia, "Establishing pair-wise keys for secure communication in ad hoc networks: a probabilistic approach," in *Proceedings of the 11th IEEE International Conference on Network Protocols (ICNP '03)*, vol. 1, pp. 326–335, 2003.

[19] R. Di Pietro, L. V. Mancini, and A. Mei, "Random key-assignment for secure wireless sensor networks," in *Proceedings of the 1st ACM Workshop on Security of Ad Hoc and Sensor networks (SASN '03)*, pp. 62–71, October 2003.

[20] W. Gu, X. Bai, S. Chellappan, and D. Xuan, "Network decoupling for secure communications in wireless sensor networks," in *Proceedings of the 14th IEEE International Workshop on Quality of Service (IWQoS '06)*, pp. 188–198, June 2006.

[21] W.-S. Li, T.-S. Su, and W.-S. Hsieh, "Multi-neighbor random key pre-distribution: a probabilistic analysis," *IEEE Communications Letters*, vol. 13, no. 5, pp. 306–308, 2009.

[22] M. Abazeed, N. Faisal, S. Zubair, and A. Ali, "Routing protocols for wireless multimedia sensor network: a survey," *Journal of Sensors*, vol. 2013, Article ID 469824, 11 pages, 2013.

[23] http://en.wikipedia.org/wiki/Linear_feedback_shift_register.

[24] D. H. Yum and P. J. Lee, "Exact formulae for resilience in random key predistribution schemes," *IEEE Transactions on Wireless Communications*, vol. 11, no. 5, pp. 1638–1642, 2012.

[25] G. Pottic and W. Kaiser, "Wireless sensor networks," *Communications of the ACM*, vol. 43, pp. S1–S8, 2000.

19

Fault Line Selection Method of Small Current to Ground System Based on Atomic Sparse Decomposition and Extreme Learning Machine

Xiaowei Wang,[1] Yanfang Wei,[1] Zhihui Zeng,[1] Yaxiao Hou,[2] Jie Gao,[1] and Xiangxiang Wei[1]

[1]*School of Electrical Engineering and Automation, Henan Polytechnic University, Jiaozuo 454000, China*
[2]*School of Electrical Engineering and Automation, Harbin Institute of Technology, Harbin 150001, China*

Correspondence should be addressed to Xiaowei Wang; proceedings@126.com

Academic Editor: Sergiu Dan Stan

This paper proposed a fault line voting selection method based on atomic sparse decomposition (ASD) and extreme learning machine (ELM). Firstly, it adopted ASD algorithm to decompose zero sequence current of every feeder line at first two cycles and selected the first four atoms to construct main component atom library, fundamental atom library, and transient characteristic atom libraries 1 and 2, respectively. And it used information entropy theory to calculate the atom libraries; the measure values of information entropy are got. It constructed four ELM networks to train and test atom sample and then obtained every network accuracy. At last, it combined the ELM network output and confidence degree to vote and then compared the vote number to achieve fault line selection (FLS). Simulation experiment illustrated that the method accuracy is 100%, it is not affected by fault distance and transition resistance, and it has strong ability of antinoise interference.

1. Introduction

For small current to ground system, FLS study focus is fault line identified when single phase to ground fault occurred; at this moment, the fault current is weaker, and Petersen coil to ground mode also has the features. Therefore, the conventional method with using current amplitude size and phase information is difficult to obtain satisfactory results.

In recent years, modern signal processing technology used FLS to get fault characteristic information, such as wavelet transform [1], *S* transform [2], mathematical morphology [3], Hilbert-Huang transform (HHT) [4], Prony algorithm [5], and Hough transform [6]. Besides, the common method of FLS criterion had artificial neural networks [7], support vector machines [8], and Bayesian classification [9].

Zero sequence current was decomposed by wavelet transform and calculated wavelet modulus maxima to determine arrival time of traveling wave's head and then compared the amplitude and polarity of every feeder line at this time to achieve FLS [1]. It used *S* transform to get modulus value and phase angle of every frequency range and compared modulus value and phase angle to obtain characteristic frequency and voting mechanism, respectively; the experiments indicated that the method could not only judge the fault line accurately but also obtain the FLS confidence degree [2]. Paper [10] used *S* transform to get transient fault feature, and, based on the frequency point of transient maximum energy, it chose characteristic frequency sequence; therefore, the criterion with relative entropy values of multiple combination modes determined the fault section. Paper [3] proposed a novel method which was based on mathematical morphology; the method included two aspects: one used morphological filters to preprocess the data and removed the noise impact for FLS at the maximum extent and the other adopted morphological operators to detect the denoised signal with mutant aspect to judge the fault line. Paper [4] calculated transient instantaneous power by Hilbert-Huang transform (HHT) and got fault direction well; the method took advantage of transient high-frequency component at lower sampling rate. It tried to divide the zero sequence current signal into several segments to ensure good continuity and smaller mutation at every subsegment, and Prony algorithm was applied to choose transient dominant component with maximum energy principle and then calculated the relative entropy and voted by preliminary vote and k values check to judge the fault line [5]. Hough transform was adopted to construct whole mutant direction angle which indicated overall trend of zero sequence current at initial stage, and FLS was achieved by distinguishing the direction angle [6]. Paper [7] replaced ordinary neurons with rough neurons and fuzzy neurons to identify 10 kinds of fault type; the method improved the training speed and reduced training samples and fault identification accuracy was enhanced. It used correlation coefficient of zero sequence voltage and charge as characteristic input to construct FLS process which was based on transient zero sequence Q-U features; the method adopted support vector machine algorithm with small sample [8]. For incomplete information of fault diagnosis, [9] adopted evidence uncertainty reasoning and compared abnormal events to reduce computation amount.

This paper proposed a novel FLS method which was based on combination of ASD and ELM. Firstly, it used atomic sparse algorithm to decompose zero sequence current of every feeder line and extracted the first four atoms to construct fault sample library, respectively; besides, it calculated information entropy measure of every library. Then, it trained the ELM network to improve network output accuracy. At last, fault voting was adopted to vote every feeder line and compared the values, and then the fault line was judged. Simulation results showed that the accuracy rate of proposed method is 100% and had strong ability of antinoise interference.

The remaining of this paper is organized as follows. In Section 2, we analyzed the physical characteristics of zero sequence current. In Sections 3 and 4, the theory of time-frequency atom decomposition and ELM work principle are presented, respectively. In Section 5, test signals analysis is given in the paper. In Section 6, we chose the characteristic atoms of zero sequence current. In Section 7, the FLS methods are proposed. In Section 8, example analysis is applied to verify the proposed method. In Section 9, we discussed the applicability of the method. In Section 10, the paper is completed with conclusions and future directions.

2. Physical Characteristics Analysis

Transient zero sequence circuit of single phase to ground fault is shown in Figure 1, where C_0 and L_0 are zero sequence capacitance and inductance, respectively, R_g is transition resistance of grounding point, R_p and L_p are equivalent resistance and inductance of arc suppression coil, and $e(t)$ is zero sequence voltage.

When the fault occurred in compensation network, Figure 1, the transient zero sequence current flows through

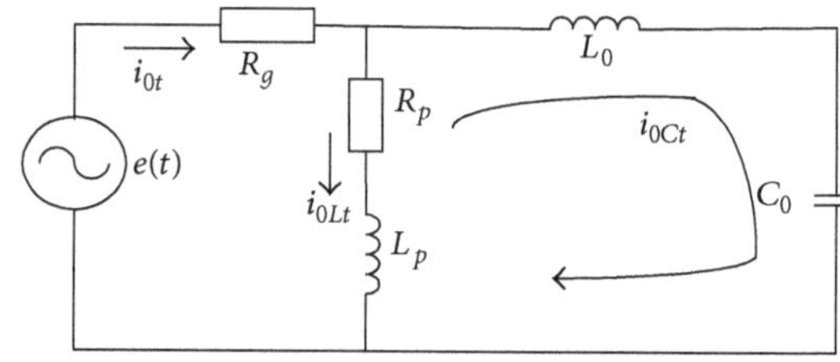

FIGURE 1: Transient zero sequence equivalent circuit of single phase to ground.

fault location [11, 12]; the calculation is shown in the following formula:

$$\begin{aligned} i_{0t} &= i_{0Lt} + i_{0Ct} \\ &= I_{Lm}\cos\varphi e^{-t/\tau_L} + I_{Cm} \\ &\quad \times \left(\frac{\omega_f}{\omega}\sin\varphi\sin\omega t - \cos\varphi\cos\omega_f t\right)e^{-\delta t}, \end{aligned} \tag{1}$$

where i_{0Lt} and i_{0Ct} are inductance component and capacitance component of transient zero sequence current, I_{Lm} and I_{Cm} are initial value of inductance current and capacitance current, respectively, ω is angular frequency, ω_f and δ are oscillation angle frequency and attenuation coefficient of transient zero sequence current capacitance component, respectively, τ_L is decay time constant of inductance current, and φ is initial phase of fault line.

ω_f and δ calculations are shown in the following formula, respectively:

$$\omega_f = \sqrt{\left|\frac{1}{L_0 C_0} - \left(\frac{R_g}{2L_0}\right)^2\right|}, \tag{2}$$

$$\delta = \frac{R_g}{2L_0}. \tag{3}$$

Transient zero sequence current is comprised of sinusoidal function from formula (1), and its waveform has attenuation characteristics. It can be seen from formula (2) and (3) that oscillation angle frequency ω_f is influenced by L_0, C_0, and R_g; attenuation coefficient δ is also influenced by L_0 and R_g; when transition resistance R_g increased, ω_f decreased and δ increased; it reflected that wave oscillation trend of zero sequence current becomes slow and the attenuation time become fast, and then transient process will end soon and into steady state. Therefore, if it could extract accurately transient component to achieve FLS exactly at the large resistance to ground fault, it will be an important index to test the applicability of the FLS methods.

Figure 2 is zero sequence current of actual distribution network when overhead line 1 caused fault; it can be seen from Figure 2 that, whether the overhead line, cable line, or hybrid line, its zero sequence current has oscillation attenuation characteristics.

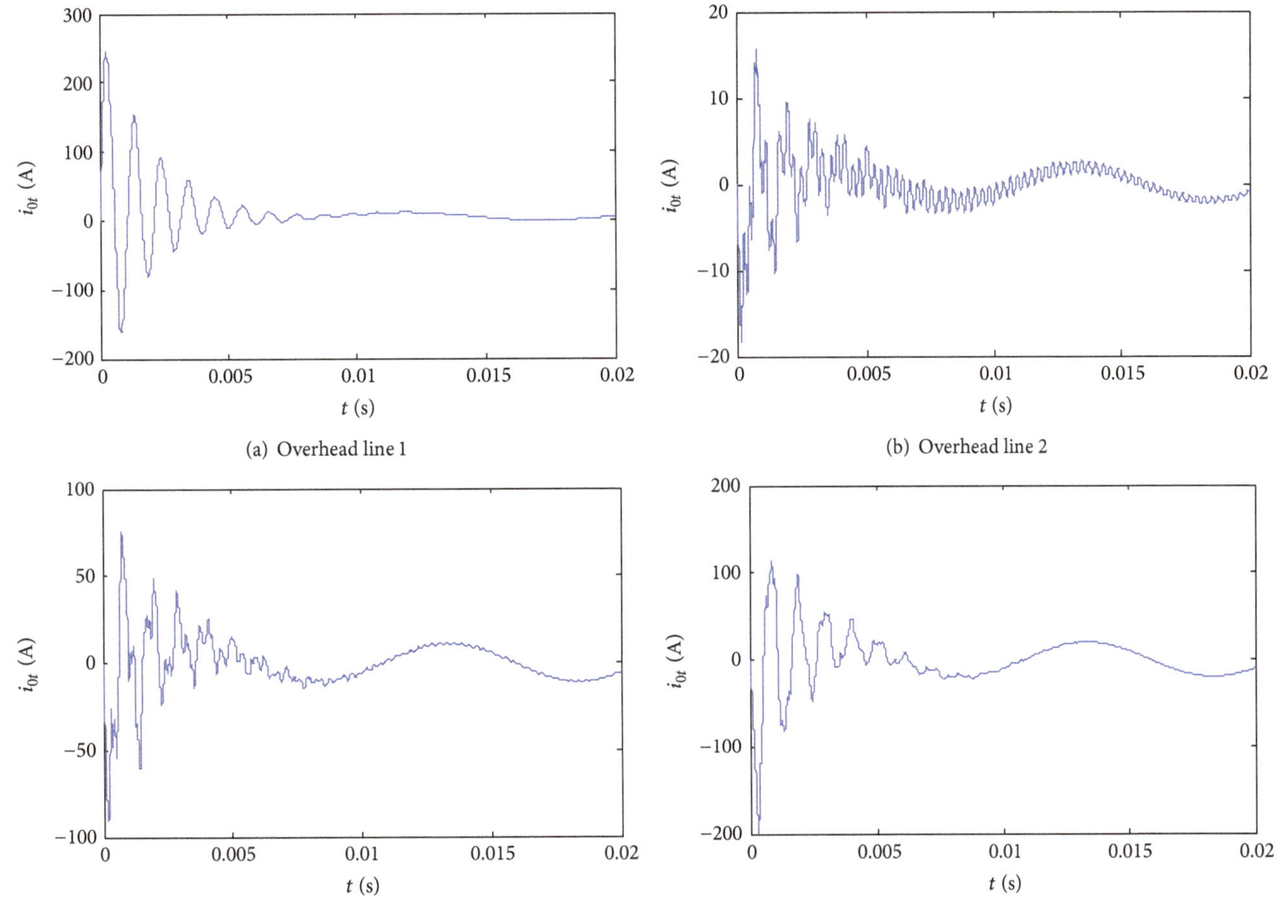

(a) Overhead line 1

(b) Overhead line 2

(c) Hybrid line

(d) Cable line

FIGURE 2: Transient zero sequence current.

3. Time-Frequency Atom Decomposition Theory

3.1. Decomposition Methods. For continuous signal $f(t) \in H$, H is Hilbert space and transformed $f(t)$ into $f(n)$; its process is discretization [13–15]. Defined atom dictionary $D = (g_r)_{r\in\Gamma}$, Γ is group of parameters r, $\|g_r\| = 1$. Choose the atoms to match the signal $f(n)$ from atom dictionary D, that is, the maximum inner product between $f(n)$ and all atoms. $g_{(r_0)}(n)$ meet the following formula:

$$\left|\left\langle f(n), g_{(r_0)}(n)\right\rangle\right| = \sup_{\gamma\in\Gamma}\left|\left\langle f(n), g_r\right\rangle\right|. \quad (4)$$

The signal could be decomposed by the best matching atom $g_{(r_0)}(n)$ component and the residual signal $Rf(n)$, and the calculation expression is shown in the following formula:

$$f(n) = \left\langle f(n), g_{(r_0)}(n)\right\rangle g_{(r_0)}(n) + Rf(n). \quad (5)$$

In (5), $Rf(n)$ approached along $g_{(r_0)}(n)$ direction. Obviously, $g_{(r_0)}(n)$ and $Rf(n)$ were orthogonal; therefore, the following formula was got:

$$\|f(n)\|^2 = \left|\left\langle f(n), g_{(r_0)}(n)\right\rangle\right|^2 + \|Rf(n)\|^2. \quad (6)$$

Because atom dictionaries are over completeness, the optimal solution could be turned to suboptimal solution; that is, choose the approximation atom to a certain extent. The calculation is shown in

$$\left|\left\langle f(n), g_{(r_0)}(n)\right\rangle\right| = \alpha \sup_{\gamma\in\Gamma}\left|\left\langle f(n), g_r\right\rangle\right|. \quad (7)$$

In formula (7), $0 \le \alpha \le 1$, decompose $Rf(n)$ and chose the best matching atom $g_{(r_1)}(n)$ from atom dictionary, make $R^0 f(n) = f(n)$ iterative k times; the k time residual component $R^k f(n)$ could be expressed as

$$R^k f(n) = \left\langle R^k f(n), g_{(r_k)}(n)\right\rangle g_{(r_k)}(n) + R^{k+1} f(n). \quad (8)$$

The signal $f(n)$ decomposed m times, and its expression is shown in

$$f(n) = \sum_{k=0}^{m-1}\left\langle R^k f(n), g_{(r_k)}(n)\right\rangle g_{(r_k)}(n) + R^m f(n). \quad (9)$$

Therefore, signal energy $\|f(n)\|^2$ could be expressed in

$$\|f(n)\|^2 = \sum_{k=0}^{m-1}\left|\left\langle R^k f(n), g_{(r_k)}(n)\right\rangle g_{(r_k)}(n)\right|^2 + \|R^m f(n)\|^2. \quad (10)$$

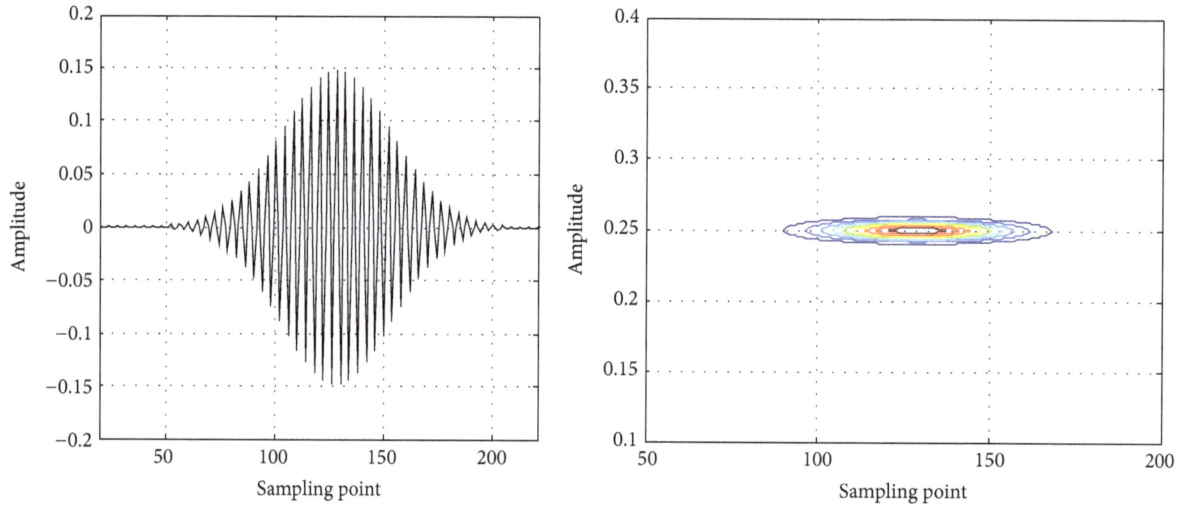

(a) Time-frequency diagram

(b) Wigner-Ville distribution

Figure 3: Gabor atoms time-frequency diagram and Wigner-Ville distribution. $s = 2^6$, $u = 128$, $\xi = \pi/2$.

In (10), $g_{(r_k)}(n)$ meet the formula

$$\left|\left\langle R^k f(n), g_{(r_k)}(n)\right\rangle\right| = \alpha \sup_{\gamma\in\Gamma}\left|\left\langle R^k f(n), g_r\right\rangle\right|. \quad (11)$$

If it decomposed m times to meet the accuracy, then stop the decomposition. Because the residual component $R^m f(n)$ tends to 0, $f(n)$ could be expressed by chosen atoms. It was shown in

$$f(n) = \sum_{k=0}^{m-1}\left\langle R^k f(n), g_{(r_k)}(n)\right\rangle g_{(r_k)}(n). \quad (12)$$

The similarity degree C_m between original signal $f(n)$ and constructed signal $f_m(n)$ is shown in

$$C_m = \frac{\langle f(n), f_m(n)\rangle}{\|f(n)\|\cdot\|f_m(n)\|}. \quad (13)$$

Because $\|g_r\| = 1$, calculated Wigner-Ville distribution of formula (12), it could get

$$\begin{aligned}Wf(n,l) &= \sum_{k=0}^{M-1}\left|\left\langle R^k f(n), g_{r_k}(n)\right\rangle\right|^2 Wg_{r_k}(n,l) \\ &+ \sum_{k=0}^{M-1}\sum_{m=0,m\neq k}^{M-1}\left\langle R^k f(n), g_{r_k}(n)\right\rangle \overline{\left\langle R^m f(n), g_{r_m}(n)\right\rangle} \\ &\cdot W\left[g_{r_k}(n), g_{r_m}(n)\right](n,l).\end{aligned} \quad (14)$$

In (14), $Wg_{r_k}(n,l)$ is Wigner-Ville distribution of atom $g_{r_k}(n)$ and l is discretization frequency variable. The last item in (14) is cross terms of every atom. Mallat eliminated the cross terms and got the energy distribution in

$$Ef(n,l) = \sum_{k=0}^{M-1}\left|\left\langle R^k f(n), g_{(r_k)}(n)\right\rangle\right|^2 Wg_{(r_k)}(n,l). \quad (15)$$

In (15), $|\langle R^k f(n), g_{(r_k)}(n)\rangle|^2$ is energy intensity and $Ef(n,l)$ is density function of $f(n)$ energy distribution.

3.2. Gabor Atoms. Gabor atoms are constructed by Gauss energy function with telescopic, translation, and modulation transform, and its expression is shown in the following formula:

$$g_r(t) = \frac{1}{\sqrt{s}}g\left(\frac{t-u}{s}\right)e^{j\xi t}. \quad (16)$$

The expression of corresponding real Gabor atom is shown in the following formula:

$$g_{(r,\phi)}(t) = \frac{1}{\sqrt{s}}g\left(\frac{t-u}{s}\right)\cos(\xi t + \phi). \quad (17)$$

In (17), $g(t)$ is standard Gauss signal and is equal to $2^{1/4}e^{-\pi t^2}$, s is scale parameter, $1/\sqrt{s}$ is atom normalization parameter, and u, ξ, and ϕ are parameters of time shift, frequency modulation, and phase.

Parameter r is equal to (s, u, ξ), and its discretization treatment is $r = (a^j, pa^j\Delta u, ka^{-j}\Delta\xi)$, where $0 < j \le \log_2 N$, $0 \le p \le N2^{-j+1}$, $0 \le k \le 2^{j+1}$, and N is sampling points, where $a = 2$, $\Delta u = 0.5$, and $\Delta\xi = \pi$. ϕ is discrete as $\phi = v\cdot\pi/6$, where $0 \le v \le 12$, v is integer.

Single atom time-frequency diagram and Wigner-Ville distribution of Gabor atom are shown in Figure 3.

It can be known that Gabor atom has the best time-frequency aggregation in Figure 3 and utilized sparse signal representation to fully reveal signal time-frequency characteristics. The deficiency of Gabor atom is that atom frequency

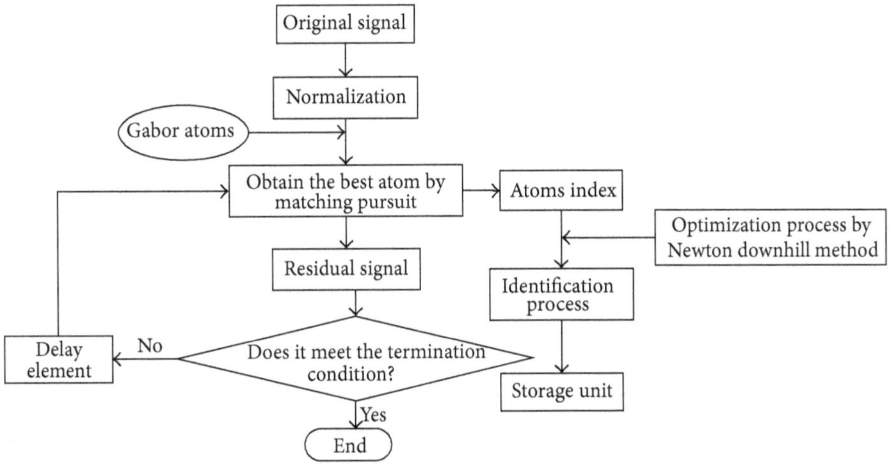

FIGURE 4: Atomic sparse decomposition process.

is not changed with time, and the division way of time-frequency plane belongs to lattice segmentation. Compare Figures 2 and 3; it is known that the similarity degree between Gabor atom waveforms and zero sequence currents is higher and, therefore, adopts matching pursuit way to match; it could not only accurately extract the fault feature components but also save a large number of calculation time. Hence, it used Gabor atoms to extract the fault features in the paper.

4. ELM Work Principle

Extreme learning machine is a novel feed-forward neural network [16–18], which is assumed to have N training sample $\{(\mathbf{x}_k, t_k)\}_{k=1}^{N}$; its expression is shown in (16):

$$o_k = \boldsymbol{\omega}^T f(\mathbf{W}_{\text{in}}\mathbf{x}_\mathbf{k} + \mathbf{b}), \quad k = 1, 2, \ldots, N. \tag{18}$$

In (18), $\mathbf{x}_\mathbf{k}$, $\mathbf{b}$, and o_k are input vector, hidden layer bias, and network output, respectively, $\mathbf{W}_{\text{in}}$ is input weight linked input node and hidden layer node, $\boldsymbol{\omega}$ is output weight linked hidden layer node and output node, f is hidden layer activation function and its form is Sigmoid function generally, and N is sample number.

At the beginning of training, $\mathbf{W}_{\text{in}}$ and $\mathbf{b}$ randomly generated and remain unchanged; it only needs to train output weight $\boldsymbol{\omega}$. Assume that feed-forward neural network approached training sample with zero error; that is, $\sum_{k=1}^{N}\|o_k - t_k\| = 0$. Therefore, $\mathbf{W}_{\text{in}}$, $\mathbf{b}$, and $\boldsymbol{\omega}$ meet the formula

$$\boldsymbol{\omega}^T f(\mathbf{W}_{\text{in}}\mathbf{x}_\mathbf{k} + \mathbf{b}) = t_k, \quad k = 1, 2, \ldots, N. \tag{19}$$

Formula (19) is written to matrix form; that is, $\mathbf{H}\boldsymbol{\omega} = \mathbf{T}$, where

$$\mathbf{H} = \begin{bmatrix} f(\mathbf{W}_{\text{in}}\mathbf{x}_1 + \mathbf{b}_1) & \cdots & f(\mathbf{W}_{\text{in}}\mathbf{x}_1 + \mathbf{b}_m) \\ \vdots & \ddots & \vdots \\ f(\mathbf{W}_{\text{in}}\mathbf{x}_N + \mathbf{b}_1) & \cdots & f(\mathbf{W}_{\text{in}}\mathbf{x}_N + \mathbf{b}_m) \end{bmatrix}_{N\times m}. \tag{20}$$

In (20), $\mathbf{H}$ and m are output matrix and node number of hidden layer, respectively, and $\mathbf{T}$ is expected output vector and it could be expressed as $\mathbf{T} = [t_1, t_2, \ldots, t_N]^T$. If the activation function of hidden layer is infinitely differentiable and the number of hidden layer node met the relationship $m \le N$, it could approach training sample with small training error. $\boldsymbol{\omega}$ value is calculated by pseudoinverse algorithm generally [19].

The training process of single-hidden layer feed-forward neural network (SLFN) is equivalent to calculating least squares solution of linear system; it is shown in

$$\|\mathbf{H}\widehat{\boldsymbol{\omega}} - \mathbf{T}\| = \min_{\omega} \|\mathbf{H}\boldsymbol{\omega} - \mathbf{T}\|. \tag{21}$$

In (21), $\widehat{\boldsymbol{\omega}}$ is least squares solution of minimum norm of $\mathbf{H}\boldsymbol{\omega} = \mathbf{T}$. $\mathbf{H}^+$ is generalized inverse of hidden layer output matrix $\mathbf{H}$. For feed-forward neural network, smaller weights have stronger generalization ability. For all least squares solution of equation $\mathbf{H}\boldsymbol{\omega} = \mathbf{T}$, $\widehat{\boldsymbol{\omega}}$ has the smallest norm number; that is, $\|\widehat{\boldsymbol{\omega}}\| = \|\mathbf{H}^+\mathbf{T}\| \le \|\boldsymbol{\omega}\|$, where

$$\forall \boldsymbol{\omega} \in \left\{\boldsymbol{\omega} \mid \|\mathbf{H}\boldsymbol{\omega} - \mathbf{T}\| \le \|\mathbf{H}\mathbf{z} - \mathbf{T}\|,\ \forall \mathbf{z} \in R^{N\times m}\right\}. \tag{22}$$

From (22), not only can ELM achieve the minimum training error but it also has stronger generalization ability than the traditional gradient descent algorithm. There is only one $\mathbf{H}^+$ for the generalized inverse $\mathbf{H}^+$of matrix $\mathbf{H}$, so the $\widehat{\boldsymbol{\omega}}$ value is unique.

5. Test Signal Analysis

Given the test signal, there are three frequency components which have different time scale intervals; the calculation is as follows:

$$s(t) = \begin{cases} 9e^{-t}\sin(150\pi t) & 0 \le t \le 0.2\,\text{s} \\ 3.8e^{-t}\sin(100\pi t) & \\ \quad +2.8e^{-0.5t}\sin(70\pi t) & 0.2\,\text{s} < t < 0.5\,\text{s}. \end{cases} \tag{23}$$

We added 20 db Gauss white noise to the signal and verified anti-interference ability of atom decomposition method. For original signal, it should be normalized the signal is divided by its Euclid norm [20]. The decomposition process was shown in Figure 4.

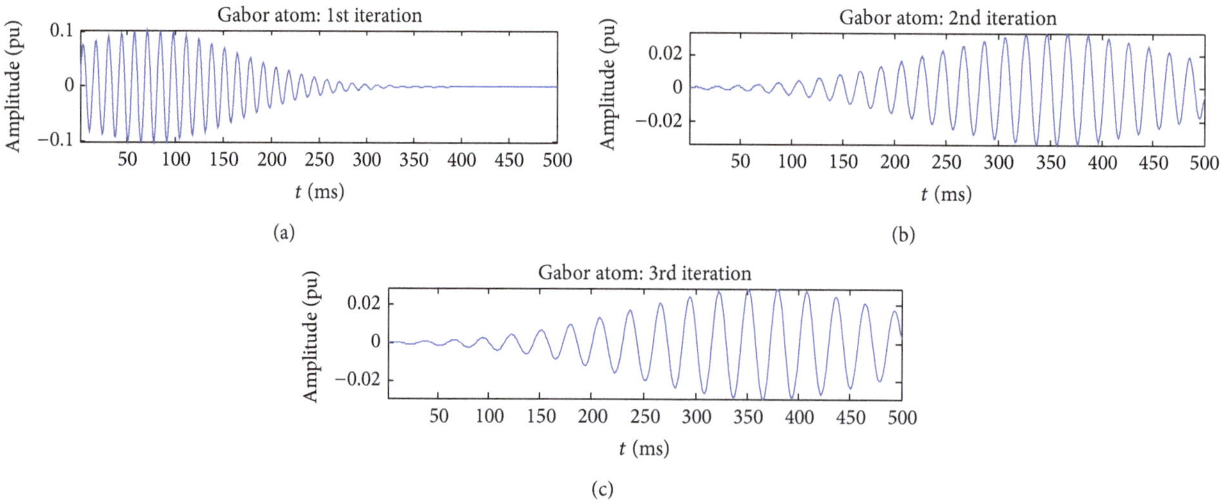

FIGURE 5: The first, second, and third atoms.

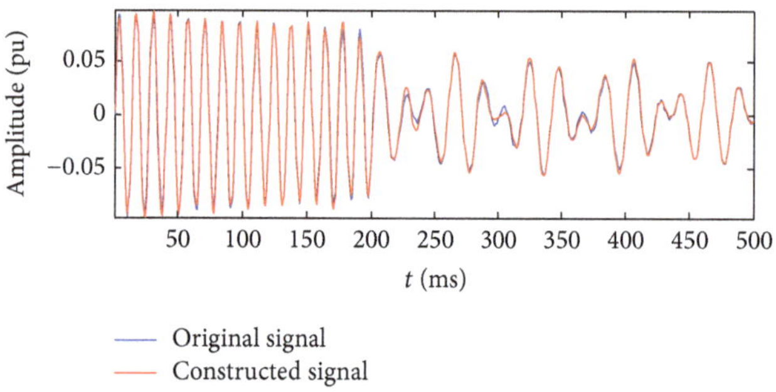

FIGURE 6: Comparison of original signal and constructed signal by 20 iterations.

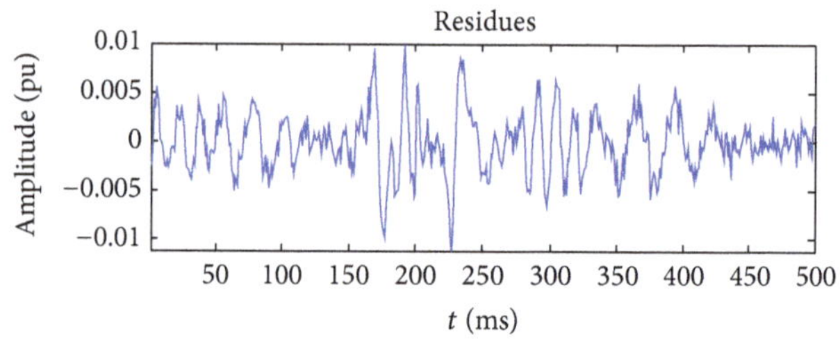

FIGURE 7: Residual component decomposed by 20 iterations.

Identification method: frequency center F is equaled to $f_s\xi/2\pi$, f_s is sampling frequency, start time is $T_s = u - s/2$, end time is $T_e = u+s/2$, ϕ is phase angle, and the amplitude is equaled to atom normalization amplitude to multiply actual energy value which is Euclid norm.

Set $f_s = 1\,\text{kHz}$; simulation time and sampling points areequaled to 0.5 s and 500, respectively. Before the decomposition, the normalization equation is shown in

$$s_g(n) = \frac{s(n)}{\|s(n)\|}. \tag{24}$$

In (24), $s(n)$ is discrete signal of $s(t)$, $s_g(n)$ is normalization expression of $s(n)$, $\|s(n)\|$ is Euclid norm, and its value is 92.7915.

Set iteration number as 20, and decompose $s_g(n)$ by atomic algorithm. Figure 5 shows atom 1, atom 2, and atom 3 generated by iteration, respectively, and all atoms in Figure 5 are normalized results. Comparison of original signal and constructed signal is shown in Figure 6; it indicated that the difference of two signals is smaller by 20 iteration numbers. The residual component amplitude is only 10^{-3} in Figure 7 and further shows that the accuracy is satisfied atomic decomposition requirement.

The parameters of atom 1, atom 2, and atom 3 are shown in Table 1. Notably, every atom decomposed by atomic algorithm does not have physical meaning; it just indicated distribution characteristics of time scales. Hence, the atomic parameters in Table 1 are needed for identification processing and we transformed it to indicate local features of original signal; it is shown in Table 2.

We observed the amplitude, frequency, and phase value of atoms in Table 2; it could know that the atoms 1, 2, and 3 represent $9e^{-t}\sin(150\pi t)$, $3.8e^{-t}\sin(100\pi t)$, and $2.8e^{-0.5t}\sin(70\pi t)$ of original signal, respectively, by identification processing. For atom 1, because the actual end time is 200 ms and the calculated end time is 187.7268 ms, deviation is 12.2732 ms. But, for atoms 2 and 3, comparing the calculated start time with actual start time, the deviations are 1.4297 ms and 7.8642 ms, respectively; the difference is smaller.

Time-frequency analysis by 20 iteration number decomposition is shown in Figure 8; we can see that the atoms could indicate local characteristics of nonstationary test signals accurately, including frequency segment, time interval, and frequency components energy; compared to the traditional FFT spectrum, cross interference and noise interference can be suppressed effectively, and the calculation accuracy is also higher than FFT.

TABLE 1: Characteristic parameters of every atom.

Atoms	Scale s/pu	Location shift u/pu	Frequency shift w/Hz	Phase/rad	Atom amplitude
1	227.6520	72.9008	0.4709	−2.0125	0.0903
2	308.3047	352.7226	0.3150	−2.2285	0.0380
3	316.8729	366.3006	0.2200	−1.8173	0.0281

TABLE 2: Local characteristic parameters of test signal.

Atoms	Amplitude	Frequency/Hz	Phase/rad	Start time/ms	End time/ms
1	9.0112	74.9841	−2.0125	—	186.7268
2	3.7921	50.1592	−2.2285	198.5703	—
3	2.8041	35.0319	−1.8173	207.8642	—

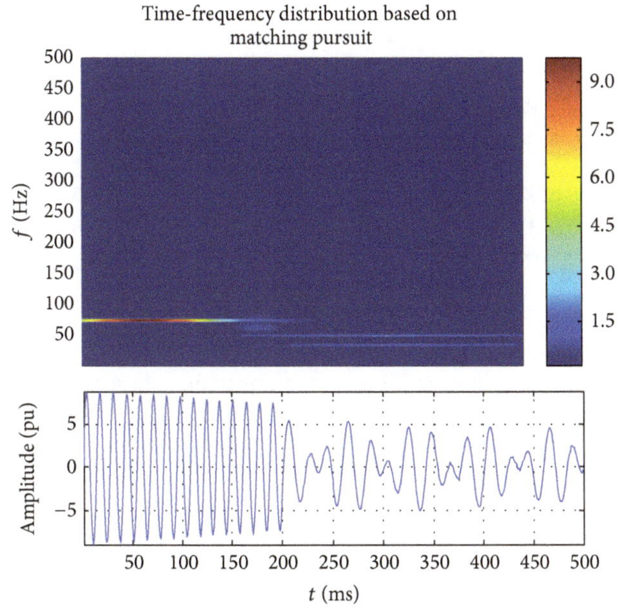

FIGURE 8: Time-frequency representation of test signal.

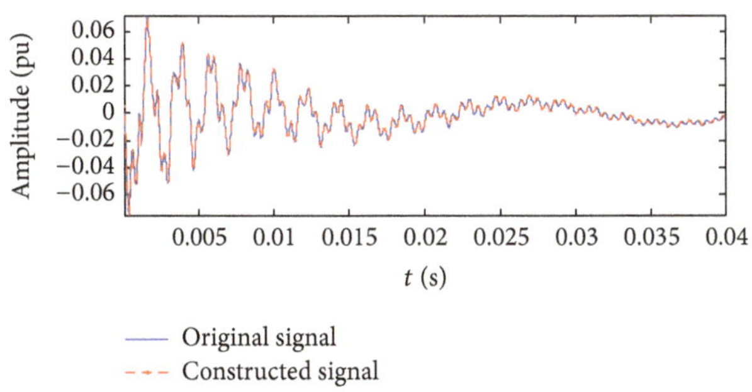

FIGURE 9: Fitting waveform of overhead line S_2.

6. Choose Characteristic Atom of Zero Sequence Current

To verify ASD algorithm extracting fault feature components ability in distribution network, it gives the feeder fault as example; by ASD, set the iteration number as 4; hence, the zero sequence current fitting waveform of overhead line is shown in Figure 9.

The similarity between constructed signal and original overhead line S_2 signal is higher by 4 iterations, and the fitting accuracy meets the requirements. The first four atoms' waveform and specific parameters are shown in Figure 10 and Tables 3 and 4, respectively.

Combined with Figure 10 and Table 4, atom 1 waveform shows oscillation attenuation trend; both waveform and energy value have higher similarity with original signal, and it indicated major information of original signal; therefore, atom 1 is defined as main components atom, and its frequency value is equal to 472.9299 Hz. Atom 2 is defined as fundamental atom, and its frequency value is 50.9554 Hz. Atom 3 and atom 4 all show oscillation attenuation trend, but their frequency values are different from atom 1: they are equal to 1778.6624 Hz and 1176.7516 Hz, respectively, and all the high frequency components; therefore, it is defined as transient characteristic atoms 1 and 2, respectively.

Zero sequence current fitting waveform of cable line S_4 is shown in Figure 11, the first four atoms are shown in Figure 12, and the parameters of every atom are shown in Tables 5 and 6. Combining Figure 12 and Table 6, it is known that atom 1 of cable line is main components atom, and its frequency is equal to 472.9299 Hz. Therefore, the frequencies of fundamental atom and transient characteristic atoms 1 and 2 are equal to 49.3631 Hz, 1175.1592 Hz, and 2366.2420 Hz, respectively. Comparing Figures 10 and 12, it got different characteristic of transient characteristic atoms between overhead line and cable line.

Comparing Figures 10(c), 10(d) and Figures 12(c), 12(d) respectively, for cable line, oscillation attenuation trend of transient characteristic atoms is more obvious; its oscillation process is shorter than overhead line. Because the capacitance to ground value of cable line is larger than overhead line in actual distribution network, it got different oscillation process of fault current.

7. Fault Line Selection Methods

7.1. Atom Dictionary Measure Based on Information Entropy Index. Information entropy can measure uncertain degree of event; the information entropy value is larger; it indicated that

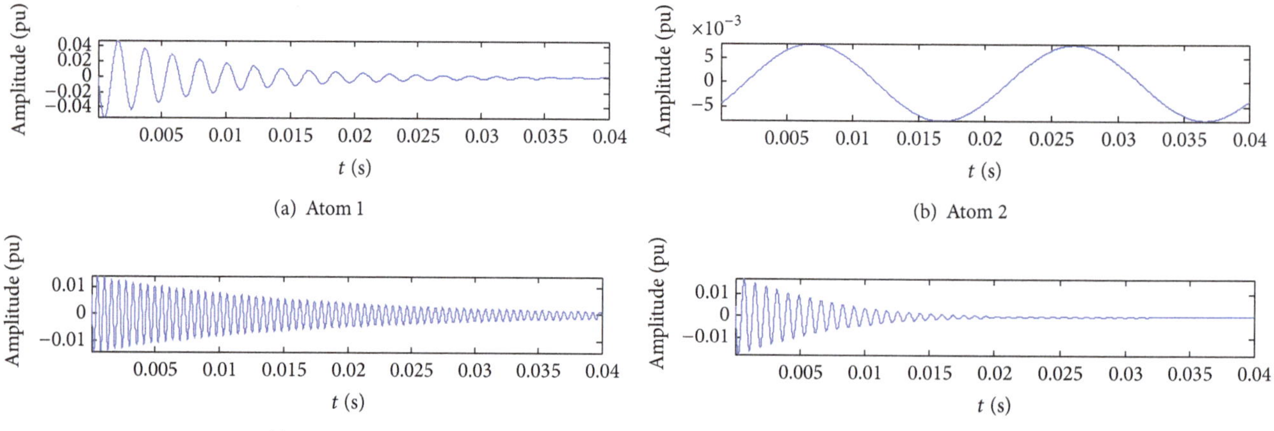

(a) Atom 1 (b) Atom 2 (c) Atom 3 (d) Atom 4

FIGURE 10: Zero sequence current characteristic atom of overhead line S_2.

TABLE 3: Every atom characteristic parameters of overhead line S_2.

Atoms	Scale s/pu	Location shift u/pu	Frequency shift w/Hz	Phase/rad	Atom amplitude
1	3.941×10^4	-2.7094×10^5	0.0297	1.5419	0.0385
2	1.15×10^6	-2.6135×10^6	0.0032	−2.1710	0.0048
3	4.1957×10^4	-1.6524×10^5	0.1117	0.9501	0.0122
4	2.97×10^3	-1.7133×10^3	0.0739	1.2609	0.0135

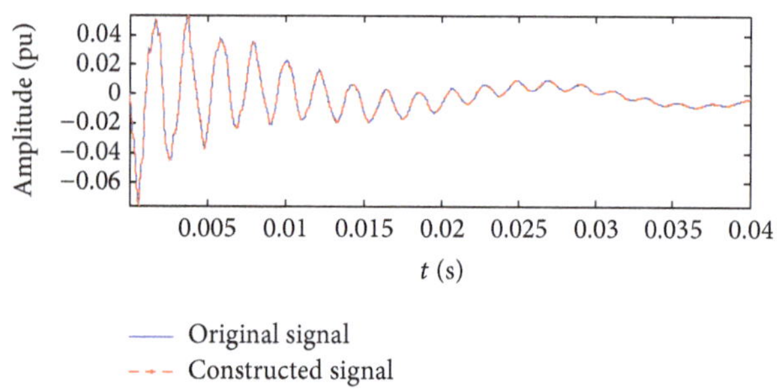

FIGURE 11: Fitting waveform of cable line S_4.

uncertain degree is larger; that is, random characteristics of event are stronger; therefore, the credibility applied to fault diagnosis is lower [21, 22]. According to the characteristics of single phase to ground fault, one fault feature is more reliable, the fault difference of fault line and healthy line is larger, and its information entropy value is smaller; it indicated that certain characteristics of FLS result based on the fault characteristics are larger. Therefore, it used information entropy to measure uncertain characteristics of every feature. To evaluate certain degree of sample library by atoms, it adopted information entropy theory to calculate in the paper; the details are as follows.

Firstly, calculate the ratio of atom library and atom library sum; it is shown in

$$\lambda(k) = \frac{L(k)}{\sum_{k=0}^{N} L(k)}. \tag{25}$$

In (25), $L(k)$ is atom library; it can be main component atom library, fundamental atom library, and transient characteristic atom library. $\lambda(k)$ is probability reflected every line fault, and the calculation formula of information entropy is shown in

$$H = -\sum_{k=0}^{N} \lambda(k) \ln \lambda(k). \tag{26}$$

In (26), information entropy reflected characteristics information content of samples, and the value is larger, it indicated that sample in the atom library has more uncertainty; hence, the fault characteristic component is less, and the credibility is lower. On the contrary, the credibility of atom library is higher.

Figures 13(a), 13(b), 13(c), and 13(d) are information entropy value of main components atom, fundamental atom, transient characteristic atoms 1 and 2; it can be seen from Figure 13 that information entropy values of most atoms are smaller and reflected certainty of the sample is stronger, and it applied to FLS which has more credibility; however, the entropy values of some samples are larger; it reflected that certainty of the samples is weak, and the credibility is lower. To evaluate credibility of every atom, the statistical method is adopted to measure the information entropy; the details are as follows.

Step 1. We selected the maximum entropy value of atom libraries 1, 2, 3, and 4, expressed as $H_{1\max}$, $H_{2\max}$, $H_{3\max}$, and $H_{4\max}$, compared the four values, and determined the maximum entropy value $H_{\max}$.

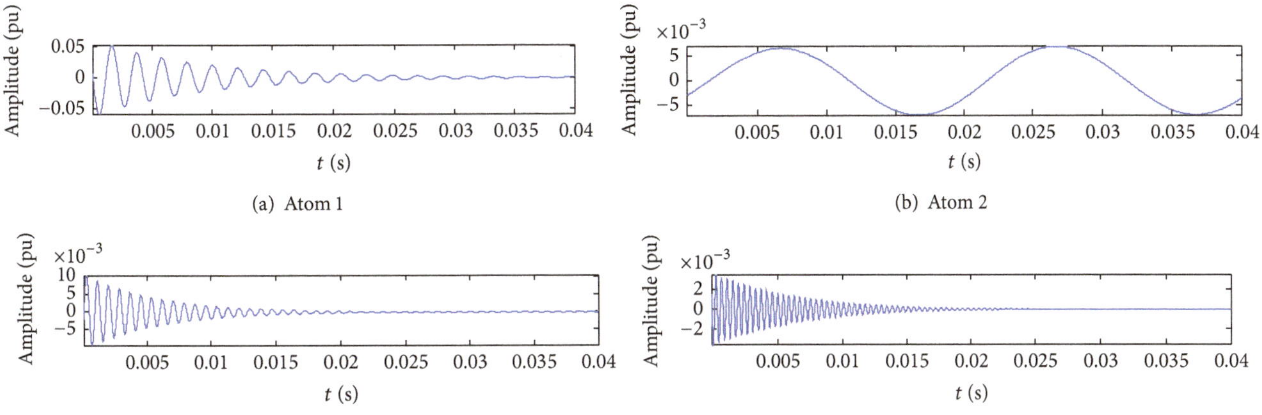

(a) Atom 1
(b) Atom 2
(c) Atom 3
(d) Atom 4

FIGURE 12: Zero sequence current characteristic atom of cable line S_4.

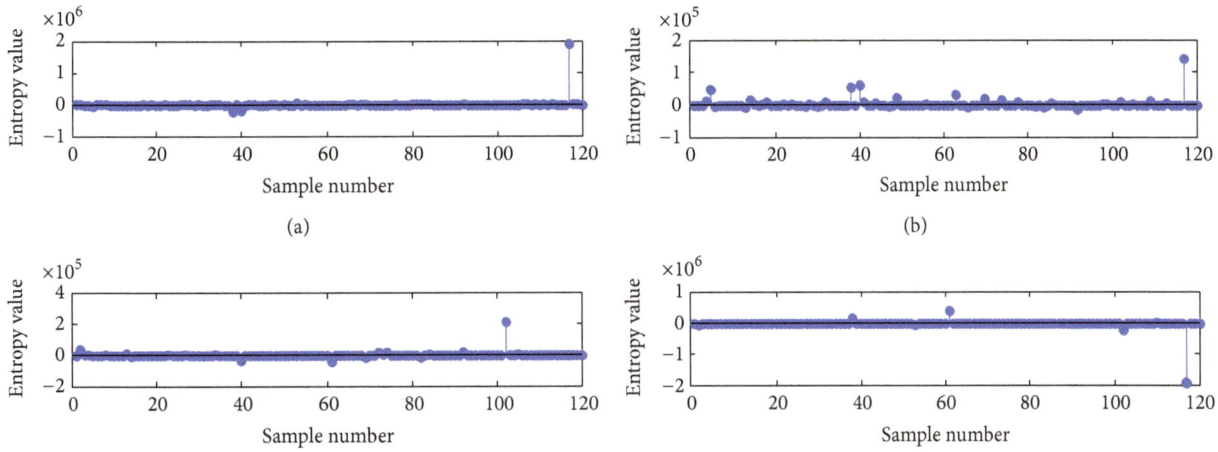

(a)
(b)
(c)
(d)

FIGURE 13: Information entropy value of every atom library.

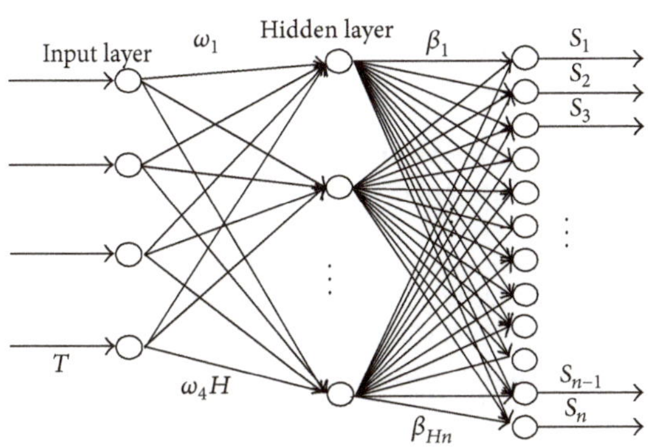

FIGURE 14: ELM network topology structure.

Step 2. We calculated $H/H_{\max}$ and then counted sample number of every atom library which is less than μ (in the paper $\mu = 0.01$).

Step 3. We took sample number by Step 2 to divide total number of samples and got information entropy measure of atom library.

The information entropy measure value can evaluate data credibility of every atom library with FLS, the measure value is smaller, and it indicated that the sample uncertainty is smaller and the certainty is larger; therefore, the credibility with FLS is higher. On the contrary, the value is larger, indicated certainty is smaller, and the credibility is lower.

7.2. Confidence Degree of Fault Line Selection. Judgment results did not add additional constraints in the past FLS method; it only required to show the fault line symbol, and the symbol output results have the following disadvantages.

(1) It can not reflect significant degree of fault feature. When the fault occurred, if fault feature is obvious, the FLS is very reliable; on the contrary, the fault feature is weak, and the results may be wrong, but the difference is hard to reflect in symbol FLS method.

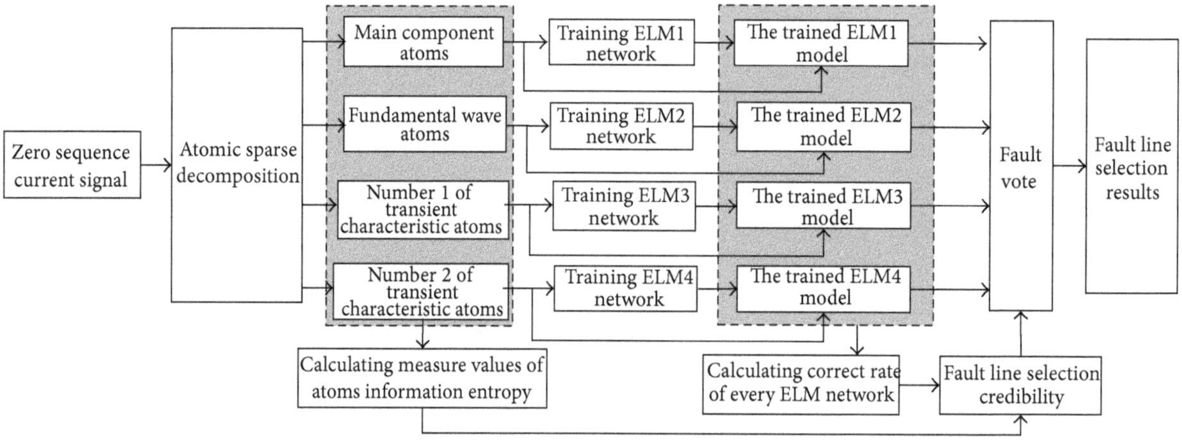

FIGURE 15: Basic framework of fault voting.

TABLE 4: Zero sequence current local characteristic parameters of overhead line S_2.

Atoms	Amplitude	Frequency shift/Hz	Phase/rad	Start time/ms	End time/ms
1	3.8654	472.9299	1.5419	10.2568	—
2	0.0485	50.9554	−2.1710	15.5897	—
3	1.2352	1778.6624	0.9501	8.5679	—
4	1.3446	1176.7516	1.2609	9.5879	22.5625

(2) It can not provide fault indication information of other lines.

(3) It is not conducive to use multiple criteria comprehensively. When using multiple criteria to select fault line, it is not viable to vote results of several criteria simply [23–25].

This paper proposed a novel FLS method based on atom library fusion ideas; its purpose is not to give FLS results by every criterion simply; it quantitatively measured fault symptom degree of every line by every atom characteristic and then trained the ELM to make decision. Finally, it adopted vote to get the results.

It has given concept of FLS confidence degree in the paper; the confidence degree is defined as real variables that is used to measure atom samples certainty and ELM training accuracy; its scope is $[0, \infty)$. The confidence degree value of atom library is larger; it indicated that vote weight of the atom library is larger. The calculation is shown in

$$\begin{aligned} &\text{Confidence degree} \\ &\quad = \text{information entropy measure of atom library} \\ &\qquad \times \text{ELM network accuracy}. \end{aligned} \tag{27}$$

7.3. Fault Line Selection Model of ELM. Based on the acquired main component atom, fundamental atom, and transient characteristic atom of group W, the ELM network is trained. There are three steps for the initial judgment of FLS in ELM network.

Step 1. Normalize the input/output training samples of group W, which is limited to $[0, 1]$, and randomly offer the input weights and hidden layer threshold of the input neurons and the τ hidden layer neurons $\omega_\tau = [\omega_{1\tau}, \omega_{2\tau}, \omega_{3\tau}, \omega_{4\tau}]^T$ $(\tau \in H)$.

Step 2. According to the generalized inverse matrix theory of Penrose Moore, the output weights of the network with the least square solution are calculated in an analytical way $\beta_\tau = [\omega_{\tau 1}, \ldots, \beta_{\tau 12}]^T$ and well-trained ELM network is obtained, from which the nonlinear mapping relations between every sample atom and fault conditions in the line can be shown.

Step 3. Given a set of fault atomic sample input data, the initial selection of fault line is presented based on the well-trained ELM network. The accuracy rate of test set is adopted to test the result of the initial selection [26, 27].

Based on the above analysis, the ELM network topology established in this paper is shown in Figure 14.

7.4. Fault Vote Mechanism. According to the theory of information entropy measure and FLS confidence degree, the paper proposed the basic framework as shown in Figure 15.

From Figure 15, the four atoms correspondingly composed atom library as fault training samples and input it to corresponding ELM network to train, and then, according to ELM network output and FLS confidence degree to achieve the fault vote, finally it judged the FLS results [28, 29]. Based on the vote principle of society, it proposed fault vote selection way based on confidence degree; the specific steps are as follows.

Step 1. Firstly, set that every line is healthy line; in other words, assume that there is no fault.

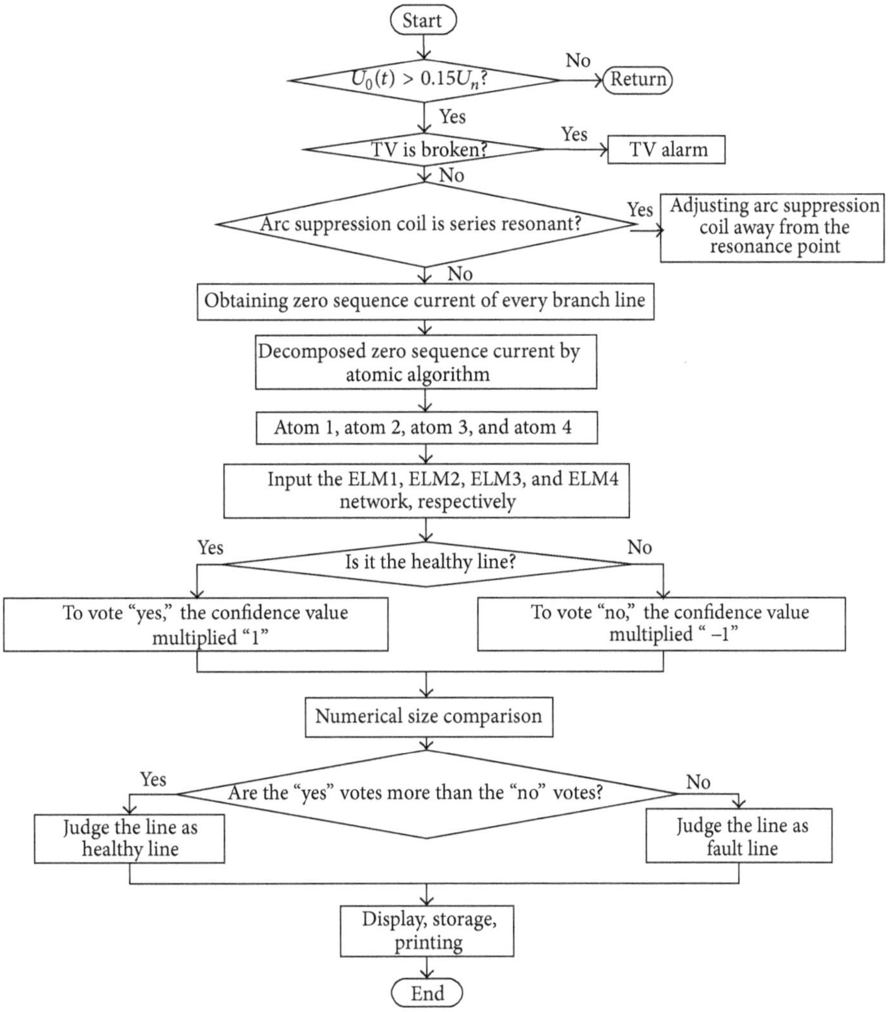

FIGURE 16: Fault line selection process.

TABLE 5: Every atom characteristic parameters of cable line S_4.

Atoms	Scale s/pu	Location shift u/pu	Frequency shift w/Hz	Phase/rad	Atom amplitude
1	3.6216×10^4	-2.2767×10^5	0.0297	1.4440	0.0513
2	1.3019×10^4	2.4196×10^3	0.0031	−2.0734	0.0065
3	2.3151×10^4	-1.5341×10^5	0.0738	−1.8505	0.0123
4	4.6146×10^3	-4.8524×10^3	0.1486	0.9955	0.0043

Step 2. When a line is judged as healthy line by ELM network output, the confidence degree value is multiplied "1," which is consistent with Step 1 assumption, hence voted "agree." On the other hand, when a line is judged as fault line by ELM network output, the confidence degree value is multiplied "−1," which is deviated from Step 1 assumption, hence voted "against."

Step 3. When ELM network judgment is completed, compare the vote number value of "agree" and "against" and then, when "agree" value is larger than "against," judge the line as healthy line; on the contrary, the line is judged as fault line.

The specific process of FLS is shown in Figure 16.

8. Example Analysis

In this paper, the ATP-EMTP is used to simulate a single phase to ground fault, and the simulation model is shown in Figure 17. The parameters of simulation model are as follows [30].

In order to simplify the analysis, the power supply adopts ideal source; therefore, the internal impedance of the source is 0.

Overhead line positive-sequence parameters are $R_1 = 0.17\,\Omega/\text{km}$, $L_1 = 1.2\,\text{mH/km}$, and $C_1 = 9.697\,\text{nF/km}$; zero sequence parameters are $R_0 = 0.23\,\Omega/\text{km}$, $L_0 = 5.48\,\text{mH/km}$, and $C_0 = 6\,\text{nF/km}$.

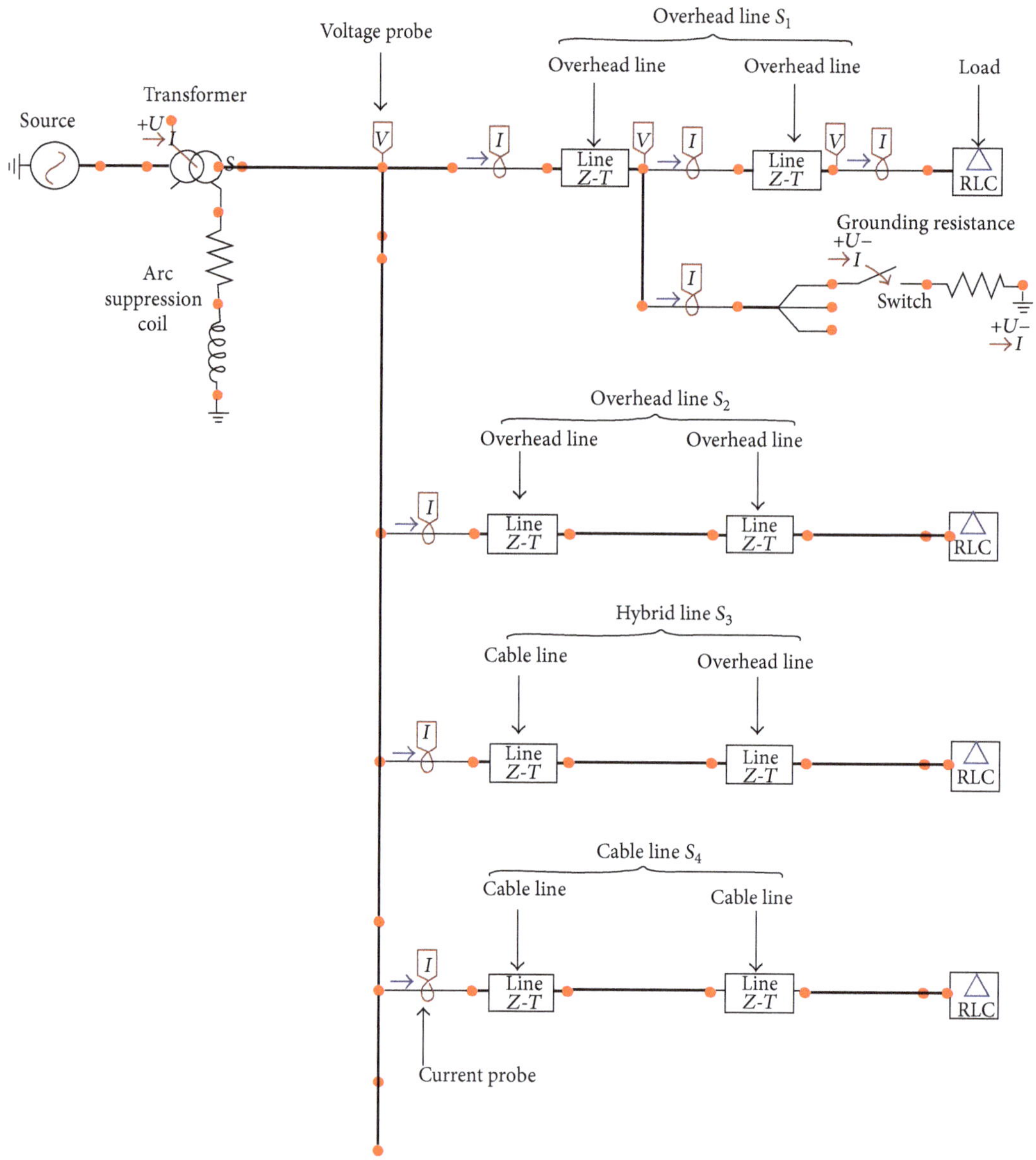

Figure 17: ATP simulation model.

Cable line positive-sequence parameters are R_{11} = 0.193 Ω/km, L_{11} = 0.442 mH/km, and C_{11} = 143 nF/km; zero sequence parameters are R_{00} = 1.93 Ω/km, L_{00} = 5.48 mH/km, and C_{00} = 143 nF/km.

The overhead lines S_1 and S_2 are 13.5 km and 24 km, respectively. Cable line S_4 is 10 km. Hybrid line S_3 is 17 km, where the cable line is 5 km and the overhead line is 12 km.

Transformer is 110/10.5 kV, connection mode indicates that the primary side is triangle connection, and the second side is star connection. Primary resistance is 0.40 Ω, inductance is 12.2 Ω, secondary resistance is 0.006 Ω, inductance is 0.183 Ω, excitation current is 0.672 A, magnetizing flux is 202.2 Wb, magnetic resistance is 400 kΩ. Load: all are delta, Z_L = 400 + j20 Ω. Petersen coil: L_N = 1281.9 mH, R_N = 40.2517 Ω.

Sampling frequency f_s is equal to 10^5 Hz, simulation time is 0.06 s, and the single phase to ground fault occurred at 0.02 s. Based on the simulation model, when the initial phase angle is 0°, the transition resistance is 1 Ω, 10 Ω, 100 Ω, 1000 Ω, or 2000 Ω, respectively; the single phase to ground fault tests are carried out at the points of 5 km and 10 km in line S_1, 9 km and 17 km in line S_3, and 6 km and 10 km in line S_4 with the arc suppression coil to ground (overcompensation is 10%). The zero sequence current signals of four feeder lines, which are chosen from 2 circles after the fault, can be collected for each fault, and the total number is 4 × 5 × 2 × 3 = 120. After the atomic decomposition of these 120 zero sequence current signals, the first 4 atoms of each group are picked out, respectively, to comprise a main component atomic library, a fundamental atomic library, and two transient atomic libraries. Each library contains 120 atomic samples, the first 100 samples of which are taken as the training set and the last 20 samples of which as the test set [31].

According to the ELM theory, when the number of hidden layer neurons equals the number of the training set samples, then, for any W_{in} and b, ELM can approximate to the

TABLE 6: Zero sequence current local characteristic parameters of cable line S_4.

Atoms	Amplitude	Frequency/Hz	Phase/rad	Start time/ms	End time/ms
1	33.6181	472.9299	1.4440	11.5875	—
2	4.2596	49.3631	−2.0734	14.8956	—
3	8.0605	1175.1592	−1.8505	8.6987	21.5614
4	2.8179	2366.2420	0.9955	9.4893	28.4462

TABLE 7: Fault voting result of overhead line S_1 under 0°.

Fault location/km	Transition resistance/Ω	Feeder line style	Atom 1	Atom 2	Atom 3	Atom 4	Voting results	Fault line selection results
10	100	**Overhead line S_1**	**0.9667 × (−1)**	**0.855 × (−1)**	**0.8358 × (−1)**	**0.7866 × (−1)**	**3.4441 > 0**	**Vote S_1 is fault line**
		Overhead line S_2	0.9667 × 1	0.855 × 1	0.8358 × (−1)	0.7866 × (−1)	1.8217 > 1.6224	Vote S_2 is healthy line
		Hybrid line S_3	0.9667 × 1	0.855 × 1	0.8358 × 1	0.7866 × 1	3.4441 > 0	Vote S_3 is healthy line
		Cable line S_4	0.9667 × 1	0.855 × 1	0.8358 × 1	0.7866 × (−1)	2.6575 > 0.7866	Vote S_4 is healthy line
5	1000	**Overhead line S_1**	**0.9667 × (−1)**	**0.855 × (−1)**	**0.8358 × (−1)**	**0.7866 × (−1)**	**3.4441 > 0**	**Vote S_1 is fault line**
		Overhead line S_2	0.9667 × 1	0.855 × 1	0.8358 × 1	0.7866 × 1	3.4441 > 0	Vote S_2 is healthy line
		Hybrid line S_3	0.9667 × 1	0.855 × (−1)	0.8358 × 1	0.7866 × 1	2.5891 > 0.855	Vote S_3 is healthy line
		Cable line S_4	0.9667 × 1	0.855 × 1	0.8358 × 1	0.7866 × (−1)	2.6575 > 0.7866	Vote S_4 is healthy line
10	1000	**Overhead line S_1**	**0.9667 × (−1)**	**0.855 × (−1)**	**0.8358 × (−1)**	**0.7866 × (−1)**	**3.4441 > 0**	**Vote S_1 is fault line**
		Overhead line S_2	0.9667 × 1	0.855 × 1	0.8358 × 1	0.7866 × 1	3.4441 > 0	Vote S_2 is healthy line
		Hybrid line S_3	0.9667 × 1	0.855 × 1	0.8358 × 1	0.7866 × 1	3.4441 > 0	Vote S_3 is healthy line
		Cable line S_4	0.9667 × 1	0.855 × 1	0.8358 × (−1)	0.7866 × 1	2.6083 > 0.8358	Vote S_4 is healthy line
5	2000	**Overhead line S_1**	**0.9667 × (−1)**	**0.855 × (−1)**	**0.8358 × (−1)**	**0.7866 × 1**	**2.6575 > 0.7866**	**Vote S_1 is fault line**
		Overhead line S_2	0.9667 × 1	0.855 × 1	0.8358 × 1	0.7866 × 1	3.4441 > 0	Vote S_2 is healthy line
		Hybrid line S_3	0.9667 × 1	0.855 × 1	0.8358 × 1	0.7866 × 1	3.4441 > 0	Vote S_3 is healthy line
		Cable line S_4	0.9667 × 1	0.855 × (−1)	0.8358 × 1	0.7866 × 1	2.5891 > 0.855	Vote S_4 is healthy line
10	2000	**Overhead line S_1**	**0.9667 × (−1)**	**0.855 × (−1)**	**0.8358 × (−1)**	**0.7866 × (−1)**	**3.4441 > 0**	**Vote S_1 is fault line**
		Overhead line S_2	0.9667 × 1	0.855 × 1	0.8358 × 1	0.7866 × 1	3.4441 > 0	Vote S_2 is healthy line
		Hybrid line S_3	0.9667 × 1	0.855 × 1	0.8358 × (−1)	0.7866 × 1	2.6083 > 0.8358	Vote S_3 is healthy line
		Cable line S_4	0.9667 × 1	0.855 × 1	0.8358 × 1	0.7866 × 1	3.4441 > 0	Vote S_4 is healthy line

TABLE 8: Fault voting result of hybrid line S_3 under 0°.

Fault location/km	Transition resistance/Ω	Feeder line style	Atom 1	Atom 2	Atom 3	Atom 4	Voting results	Fault line selection results
17	100	Overhead line S_1	0.9667 × 1	0.855 × 1	0.8358 × (−1)	0.7375 × 1	2.5592 > 0.8358	Vote S_1 is healthy line
		Overhead line S_2	0.9667 × 1	0.855 × 1	0.8358 × 1	0.7375 × (−1)	2.6575 > 0.7375	Vote S_2 is healthy line
		Hybrid line S_3	**0.9667 × (−1)**	**0.855 × 1**	**0.8358 × (−1)**	**0.7375 × (−1)**	**2.54 > 0.855**	**Vote S_3 is fault line**
		Cable line S_4	0.9667 × 1	0.855 × (−1)	0.8358 × 1	0.7375 × 1	2.54 > 0.855	Vote S_4 is healthy line
9	1000	Overhead line S_1	0.9667 × 1	0.855 × 1	0.8358 × 1	0.7375 × (−1)	2.6575 > 0.7375	Vote S_1 is healthy line
		Overhead line S_2	0.9667 × 1	0.855 × 1	0.8358 × 1	0.7375 × (−1)	2.6575 > 0.7375	Vote S_2 is healthy line
		Hybrid line S_3	**0.9667 × (−1)**	**0.855 × (−1)**	**0.8358 × (−1)**	**0.7375 × (−1)**	**3.395 > 0**	**Vote S_3 is fault line**
		Cable line S_4	0.9667 × 1	0.855 × 1	0.8358 × (−1)	0.7375 × 1	2.5592 > 0.8358	Vote S_4 is healthy line
17	1000	Overhead line S_1	0.9667 × 1	0.855 × 1	0.8358 × 1	0.7375 × (−1)	2.6575 > 0.7375	Vote S_1 is healthy line
		Overhead line S_2	0.9667 × 1	0.855 × 1	0.8358 × (−1)	0.7375 × 1	2.5592 > 0.8358	Vote S_2 is healthy line
		Hybrid line S_3	**0.9667 × (−1)**	**0.855 × (−1)**	**0.8358 × (−1)**	**0.7375 × (−1)**	**3.395 > 0**	**Vote S_3 is fault line**
		Cable line S_4	0.9667 × 1	0.855 × 1	0.8358 × 1	0.7375 × 1	3.395 > 0	Vote S_4 is healthy line
9	2000	Overhead line S_1	0.9667 × 1	0.855 × 1	0.8358 × 1	0.7375 × 1	3.395 > 0	Vote S_1 is healthy line
		Overhead line S_2	0.9667 × 1	0.855 × 1	0.8358 × 1	0.7375 × 1	3.395 > 0	Vote S_2 is healthy line
		Hybrid line S_3	**0.9667 × (−1)**	**0.855 × (−1)**	**0.8358 × (−1)**	**0.7375 × 1**	**2.6575 > 0.7375**	**Vote S_3 is fault line**
		Cable line S_4	0.9667 × 1	0.855 × 1	0.8358 × 1	0.7375 × 1	3.395 > 0	Vote S_4 is healthy line
17	2000	Overhead line S_1	0.9667 × 1	0.855 × 1	0.8358 × 1	0.7375 × 1	3.395 > 0	Vote S_1 is healthy line
		Overhead line S_2	0.9667 × 1	0.855 × 1	0.8358 × 1	0.7375 × 1	3.395 > 0	Vote S_2 is healthy line
		Hybrid line S_3	**0.9667 × (−1)**	**0.855 × (−1)**	**0.8358 × (−1)**	**0.7375 × (−1)**	**3.395 > 0**	**Vote S_3 is fault line**
		Cable line S_4	0.9667 × 1	0.855 × 1	0.8358 × 1	0.7375 × 1	3.395 > 0	Vote S_4 is healthy line

training samples with no deviation, and the calculation result is the best. Therefore, four ELM networks are used to train the fault atomic samples in four atomic libraries, respectively, of which the input layer neurons are 4000, the hidden layer neurons are 100, and the output layer neuron is 1.

According to the information entropy theory, the values of information entropy of main component atomic library, fundamental atomic library, and transient atomic libraries 1 and 2 are calculated as 0.9667, 0.95, 0.9833, and 0.9833 respectively. In addition, after the ELM networks train every atom library, the accuracy rate of the 4 test sets of ELM network is 100%, 90%, 85%, and 80%. Therefore, according to formula (27), the confidence degree of every atomic library is 0.9667, 0.855, 0.8358, and 0.7866. Table 7 shows the voting results of fault in overhead line S_1 when the initial phase angle is 0°. According to the fault voting mechanism, assuming that all the branch lines are healthy lines, if the line checked by the ELM network is a healthy line, then multiply the line selection credibility by "1" which shows "agree"; if the line checked is a fault line, then multiply the line selection credibility by "−1," which shows "against." Finally, FLS is achieved through the comparison between the votes of "agree" and of "against." As can be seen from Table 7, at different fault distance and different grounding resistance value, the fault in overhead line

TABLE 9: Fault voting result of cable line S_4 under 0°.

Fault location/km	Transition resistance/Ω	Feeder line style	Atom 1	Atom 2	Atom 3	Atom 4	Voting results	Fault line selection results
10	100	Overhead line S_1	0.9667 × 1	0.7125 × 1	0.9341 × (−1)	0.7375 × 1	2.4167 > 0.9341	Vote S_1 is healthy line
		Overhead line S_2	0.9667 × 1	0.7125 × 1	0.9341 × 1	0.7375 × 1	3.3508 > 0	Vote S_2 is healthy line
		Hybrid line S_3	0.9667 × 1	0.7125 × 1	0.9341 × 1	0.7375 × 1	3.3508 > 0	Vote S_3 is healthy line
		Cable line S_4	**0.9667 × (−1)**	**0.7125 × (−1)**	**0.9341 × (−1)**	**0.7375 × (−1)**	**3.3508 > 0**	**Vote S_4 is fault line**
6	1000	Overhead line S_1	0.9667 × 1	0.7125 × 1	0.9341 × 1	0.7375 × (−1)	2.6133 > 0.7375	Vote S_1 is healthy line
		Overhead line S_2	0.9667 × 1	0.7125 × 1	0.9341 × 1	0.7375 × 1	3.3508 > 0	Vote S_2 is healthy line
		Hybrid line S_3	0.9667 × 1	0.7125 × 1	0.9341 × 1	0.7375 × (−1)	2.6133 > 0.7375	Vote S_3 is healthy line
		Cable line S_4	**0.9667 × (−1)**	**0.7125 × (−1)**	**0.9341 × (−1)**	**0.7375 × (−1)**	**3.3508 > 0**	**Vote S_4 is fault line**
10	1000	Overhead line S_1	0.9667 × 1	0.7125 × (−1)	0.9341 × 1	0.7375 × 1	2.6383 > 0.7125	Vote S_1 is healthy line
		Overhead line S_2	0.9667 × 1	0.7125 × (−1)	0.9341 × 1	0.7375 × 1	2.6383 > 0.7125	Vote S_2 is healthy line
		Hybrid line S_3	0.9667 × 1	0.7125 × 1	0.9341 × 1	0.7375 × (−1)	2.6133 > 0.7375	Vote S_3 is healthy line
		Cable line S_4	**0.9667 × (−1)**	**0.7125 × (−1)**	**0.9341 × (−1)**	**0.7375 × (−1)**	**3.3508 > 0**	**Vote S_4 is fault line**
6	2000	Overhead line S_1	0.9667 × 1	0.7125 × 1	0.9341 × 1	0.7375 × 1	3.3508 > 0	Vote S_1 is healthy line
		Overhead line S_2	0.9667 × 1	0.7125 × (−1)	0.9341 × 1	0.7375 × 1	2.6383 > 0.7125	Vote S_2 is healthy line
		Hybrid line S_3	0.9667 × 1	0.7125 × 1	0.9341 × 1	0.7375 × (−1)	2.6133 > 0.7375	Vote S_3 is healthy line
		Cable line S_4	**0.9667 × (−1)**	**0.7125 × (−1)**	**0.9341 × (−1)**	**0.7375 × (−1)**	**3.3508 > 0**	**Vote S_4 is fault line**
10	2000	Overhead line S_1	0.9667 × 1	0.7125 × (−1)	0.9341 × 1	0.7375 × 1	2.6383 > 0.7125	Vote S_1 is healthy line
		Overhead line S_2	0.9667 × 1	0.7125 × (−1)	0.9341 × 1	0.7375 × 1	2.6383 > 0.7125	Vote S_2 is healthy line
		Hybrid line S_3	0.9667 × 1	0.7125 × 1	0.9341 × 1	0.7375 × 1	3.3508 > 0	Vote S_3 is healthy line
		Cable line S_4	**0.9667 × (−1)**	**0.7125 × (−1)**	**0.9341 × (−1)**	**0.7375 × 1**	**2.6133 > 0.7375**	**Vote S_4 is fault line**

S_1 is accurately checked out through the comparison of the values above, even if the grounding resistance value is as high as 1000 Ω.

Table 8 offers the selection result of hybrid line S_3 when the initial fault phase is 0°. An experiment of fault line under the end of the high resistance ground is made to further verify the accuracy of this method. Similarly, the entropy value of the main component atomic library, fundamental atomic library, and transient component atomic libraries 1 and 2 at this time is 0.9667, 0.95, 0.9833, and 0.9833 and the accuracy rate of the four test sets after being trained by the ELM network is 100%, 90%, 85%, and 75%; therefore the corresponding confidence degree of every atomic library is 0.9667, 0.855, 0.8358, and 0.7375. Table 8 shows that even when the fault occurs at the distance of 17 km under 2000 Ω, the voting result is 3.395 > 0, which indicates that fault occurs in line S_3.

Table 9 offers the selection result of cable line S_4 when the initial fault phase is 0°. The entropy value of every atomic library is 0.9667, 0.95, 0.9833, and 0.9833; the accuracy rate of the test sets after the atomic libraries are trained by the ELM network is 100%, 75%, 95%, and 75%; therefore the confidence degree of every atomic library is 0.9667, 0.7125, 0.9341, and 0.7375. The voting results prove that the method proposed

Table 10: Information entropy value of every atom library.

	Main component atom library	Fundamental atom library	Transient characteristic atom library number 1	Transient characteristic atom library number 2
Information entropy value	0.9792	0.9583	0.9861	0.9861

Table 11: Test result of every ELM network.

Samples style	ELM	
	Correct samples number/total number	Accuracy rate
Main component atom library	23/24	95.8333%
Fundamental atom library	20/24	83.3333%
Transient characteristic atom library number 1	21/24	87.5%
Transient characteristic atom library number 2	19/24	79.1667%
Total	83/96	86.4583%

in this paper can select fault line accurately when grounding fault of cable line occurs.

9. Applicability Analysis

Since the distribution network is exposed to the outdoor environment, when the fault occurs, the current signals collected contain large amounts of noise, which is a negative factor for FLS. In order to test the antinoise interference ability of the method proposed in this paper, a strong noise of 0.5 db is added to the zero sequence current signals. Figures 18(a) and 18(b) present the zero sequence current waveforms of overhead line S_1 when grounding fault occurs at 10 Ω and 2000 Ω. It shows that when the added noise is 0.5 db, compared with Figure 2, the zero sequence current signals of each line have changed greatly and there are lots of burrs on the waveforms due to noise interference, which makes the transient characteristics of fault line almost undistinguishable and which is harmful for FLS [32–34]. The zero sequence current signals of each line are much weaker in Figure 18(b) since it represents grounding fault under high resistance with noise interference. Therefore, whether or not accurate FLS of weak signals can be achieved with strong noise interference is important for testing the applicability of the proposed method. Table 10 shows the entropy values of each atomic library with noise interference, and Table 11 is the testing results of each ELM network.

It can be seen from Table 11 that, with the added 0.5 db noise, the overall accuracy of line selection of a single atomic library of ELM network can only reach 86.4583% without considering measuring instrument error, electromagnetic interference, and other factors, and the accuracy of the selection method based on one single fault characteristic is not successful in practice with all kinds of complicated conditions in consideration. So the method proposed in this paper tries to select fault line through fault voting of multiple atomic library fusion. Table 12 is the fault selection results of each line with strong noise interference when the initial phase angle is 0°.

Table 12 shows that, with the added 0.5 db noise, the method based on multiple atomic libraries of ELM model can still accurately select the fault line without the influence of transition resistance, fault distance, and other factors. Compared with the results based on one single atomic library shown in Table 11, effectively fusing with a variety of fault characteristics, this method improves the correct rate of FLS with its excellent fault tolerance and robustness.

10. Conclusions

A fault voting selection method based on the combination of atomic sparse decomposition and ELM is proposed in this paper. The following are the conclusions of the research.

(1) ASD breaks through the idea of fixed complete basis to decompose signal; it utilizes signal features to decompose signal by choosing adaptively appropriate base of atom library. Because the ASD has adaptive, analytical, and sparse characteristics, the algorithm has outstanding advantage of fault feature extraction of power system; the atoms extracted not only restore the main characteristic of initial signal well but also apply to judge the fault line with ELM network conveniently.

(2) It could get the unique optimum solution by set hidden layer neurons number of ELM network and does not need to adjust connection weight and hidden layer threshold. We construct four ELM networks and train and test each sample atom library to improve the accuracy of every sample test set, and it provided the base for FLS at next step. Through our research, we found that ELM network has fast learning speed, good generalization performance, and less adjustment parameters; it better applied to the fault diagnosis field of power system.

(3) Information entropy can measure the confidence degree of every sample library, combined the ELM network accuracy to establish FLS confidence degree, and then constructed fault vote selection mechanism by ELM network output and confidence degree value. As can be seen by voting, the accuracy of the method is 100%, and it is not affected by fault distance

Table 12: Fault voting result with strong noise.

(a) Fault voting result of overhead line S_1

Fault location/km	Transition resistance/Ω	Feeder line style	Atom 1	Atom 2	Atom 3	Atom 4	Voting results	Fault line selection results
5	5	**Overhead line S_1**	**0.9384 × (−1)**	**0.7986 × (−1)**	**0.8217 × (−1)**	**0.7807 × (−1)**	**3.3394 > 0**	**Vote S_1 is fault line**
		Overhead line S_2	0.9384 × 1	0.7986 × 1	0.8217 × (−1)	0.7807 × (−1)	1.737 > 1.6024	Vote S_2 is healthy line
		Hybrid line S_3	0.9384 × 1	0.7986 × 1	0.8217 × 1	0.7807 × 1	3.3394 > 0	Vote S_3 is healthy line
		Cable line S_4	0.9384 × 1	0.7986 × 1	0.8217 × 1	0.7807 × (−1)	2.5587 > 0.7807	Vote S_4 is healthy line
10	500	**Overhead line S_1**	**0.9384 × 1**	**0.7986 × (−1)**	**0.8217 × (−1)**	**0.7807 × (−1)**	**2.401 > 0.9384**	**Vote S_1 is fault line**
		Overhead line S_2	0.9384 × 1	0.7986 × 1	0.8217 × 1	0.7807 × 1	3.3394 > 0	Vote S_2 is healthy line
		Hybrid line S_3	0.9384 × 1	0.7986 × 1	0.8217 × 1	0.7807 × 1	3.3394 > 0	Vote S_3 is healthy line
		Cable line S_4	0.9384 × 1	0.7986 × 1	0.8217 × 1	0.7807 × 1	3.3394 > 0	Vote S_4 is healthy line

(b) Fault voting result of hybrid line S_3

Fault location/km	Transition resistance/Ω	Feeder line style	Atom 1	Atom 2	Atom 3	Atom 4	Voting results	Fault line selection results
5	500	Overhead line S_1	0.9384 × 1	0.7986 × 1	0.8217 × 1	0.7807 × (−1)	2.5587 > 0.7807	Vote S_1 is healthy line
		Overhead line S_2	0.9384 × 1	0.7986 × 1	0.8217 × 1	0.7807 × 1	3.3394 > 0	Vote S_2 is healthy line
		Hybrid line S_3	**0.9384 × (−1)**	**0.7986 × (−1)**	**0.8217 × 1**	**0.7807 × (−1)**	**2.5177 > 0.8217**	**Vote S_3 is fault line**
		Cable line S_4	0.9384 × 1	0.7986 × 1	0.8217 × (−1)	0.7807 × 1	2.5177 > 0.8217	Vote S_4 is healthy line
14	2000	Overhead line S_1	0.9384 × 1	0.7986 × 1	0.8217 × 1	0.7807 × (−1)	2.5587 > 0.7807	Vote S_1 is healthy line
		Overhead line S_2	0.9384 × 1	0.7986 × (−1)	0.8217 × 1	0.7807 × 1	2.5408 > 0.7986	Vote S_2 is healthy line
		Hybrid line S_3	**0.9384 × (−1)**	**0.7986 × (−1)**	**0.8217 × (−1)**	**0.7807 × (−1)**	**3.3394 > 0**	**Vote S_3 is fault line**
		Cable line S_4	0.9384 × 1	0.7986 × 1	0.8217 × 1	0.7807 × 1	3.3394 > 0	Vote S_4 is healthy line

(c) Fault voting result of cable line S_4

Fault location/km	Transition resistance/Ω	Feeder line style	Atom 1	Atom 2	Atom 3	Atom 4	Voting results	Fault line selection results
4	200	Overhead line S_1	0.9384 × 1	0.7986 × 1	0.8217 × 1	0.7807 × (−1)	2.5587 > 0.7807	Vote S_1 is healthy line
		Overhead line S_2	0.9384 × 1	0.7986 × 1	0.8217 × 1	0.7807 × 1	3.3394 > 0	Vote S_2 is healthy line
		Hybrid line S_3	0.9384 × 1	0.7986 × 1	0.8217 × 1	0.7807 × 1	3.3394 > 0	Vote S_3 is healthy line
		Cable line S_4	**0.9384 × (−1)**	**0.7986 × 1**	**0.8217 × (−1)**	**0.7807 × (−1)**	**2.5408 > 0.7986**	**Vote S_4 is fault line**
8	10	Overhead line S_1	0.9384 × 1	0.7986 × (−1)	0.8217 × 1	0.7807 × 1	2.5408 > 0.7986	Vote S_1 is healthy line
		Overhead line S_2	0.9384 × 1	0.7986 × 1	0.8217 × 1	0.7807 × 1	3.3394 > 0	Vote S_2 is healthy line
		Hybrid line S_3	0.9384 × 1	0.7986 × 1	0.8217 × 1	0.7807 × 1	3.3394 > 0	Vote S_3 is healthy line
		Cable line S_4	**0.9384 × (−1)**	**0.7986 × 1**	**0.8217 × (−1)**	**0.7807 × (−1)**	**2.5408 > 0.7986**	**Vote S_4 is fault line**

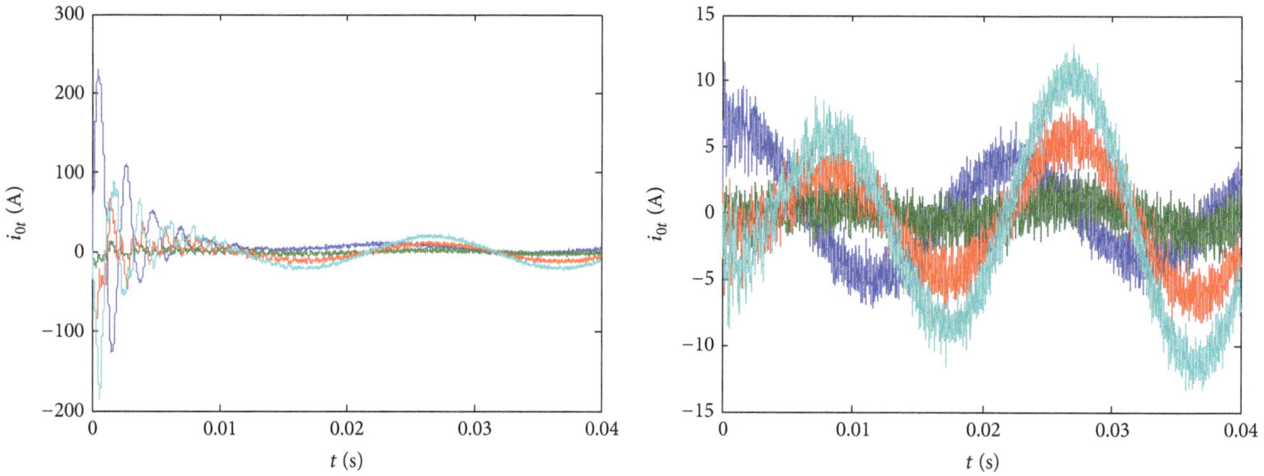

(a) 10 Ω to ground fault

(b) 2000 Ω to ground fault

FIGURE 18: Zero sequence current with strong noise.

and transition resistance value; besides, the method can accurately achieve FLS with 0.5 db strong noise interference.

(4) Because ASD algorithm adopts matching pursuit way to find the best atom in decomposition process, it needs a large number of inner product operations, so it needs to spend a long time. Therefore, the future work is how to reduce matching pursuit calculation time.

Notations

ASD: Atomic sparse decomposition
ELM: Extreme learning machine
FLS: Fault line selection
HHT: Hilbert-Huang transform.

Conflict of Interests

The authors declare that there is no conflict of interests regarding the publication of this paper.

Acknowledgments

This work was supported by National Natural Science Fund (61403127) of China, Science and Technology Research (12B470003, 14A470004, and 14A470001) of Henan Province, Control Engineering Lab Project (KG2011-15, KG2014-04) of Henan Province, China, and Doctoral Fund (B2014-023) of Henan Polytechnic University, China.

References

[1] X. Dong and S. Shi, "Identifying single-phase-to-ground fault feeder in neutral noneffectively grounded distribution system using wavelet transform," *IEEE Transactions on Power Delivery*, vol. 23, no. 4, pp. 1829–1837, 2008.

[2] J. Zhang, Z. He, and Y. Jia, "Fault line identification approach based on S-transform," *Proceedings of the CSEE*, vol. 31, no. 10, pp. 109–115, 2011.

[3] J. Ren, W. Sun, M. Zhou, and A. Cao, "Transient algorithm for single-line to ground fault selection in distribution networks based on mathematical morphology," *Automation of Electric Power Systems*, vol. 32, no. 1, pp. 70–75, 2008.

[4] T. Cui, X. Dong, Z. Bo, and A. Juszczyk, "Hilbert-transform-based transient/intermittent earth fault detection in noneffectively grounded distribution systems," *IEEE Transactions on Power Delivery*, vol. 26, no. 1, pp. 143–151, 2011.

[5] X. W. Wang, J. W. Wu, and R. Y. Li, "A novel method of fault line selection based on voting mechanism of prony relative entropy theory," *Electric Power*, vol. 46, no. 1, pp. 59–64, 2013.

[6] H. Shu, L. Gao, R. Duan, P. Cao, and B. Zhang, "A novel hough transform approach of fault line selection in distribution networks using total zero-sequence current," *Automation of Electric Power Systems*, vol. 37, no. 9, pp. 110–116, 2013.

[7] S. Lin, Z. He, T. Zang, and Q. Qian, "Novel approach of fault type classification in transmission lines based on rough membership neural networks," *Proceedings of the Chinese Society of Electrical Engineering*, vol. 30, no. 28, pp. 72–79, 2010.

[8] S. Zhang, Z.-Y. He, Q. Wang, and S. Lin, "Fault line selection of resonant grounding system based on the characteristics of charge-voltage in the transient zero sequence and support vector machine," *Power System Protection and Control*, vol. 41, no. 12, pp. 71–78, 2013.

[9] Q. Li and J. Z. Xu, "Power system fault diagnosis based on subjective Bayesian approach," *Automation of Electric Power Systems*, vol. 31, no. 15, pp. 46–50, 2007.

[10] X.-W. Wang, S. Tian, Y.-D. Li, T. Li, Y.-J. Zhang, and Q. Li, "A novel fault section location method for small current neutral grounding system based on characteristic frequency sequence of S-transform," *Power System Protection and Control*, vol. 40, no. 14, pp. 109–115, 2012.

[11] X. W. Wang, Y. X. Hou, S. Tian, Y.-D. Li, J. Gao, and X.-X. Wei, "A novel fault line selection method based on time-frequency atom decomposition and grey correlation analysis of small current to ground system," *Journal of China Coal Society*, 2014.

[12] H.-S. Zhao, L.-X. Yao, L.-F. Ke, and L. Wu, "A novel method for fault line selection and location in distribution system," *Power System Protection and Control*, vol. 38, no. 16, pp. 6–10, 2010.

[13] S. G. Mallat and Z. Zhang, "Matching pursuits with time-frequency dictionaries," *IEEE Transactions on Signal Processing*, vol. 41, no. 12, pp. 3397–3415, 1993.

[14] L. Lovisolo, E. A. B. da Silva, M. A. M. Rodrigues, and P. S. R. Diniz, "Efficient coherent adaptive representations of monitored electric signals in power systems using damped sinusoids," *IEEE Transactions on Signal Processing*, vol. 53, no. 10, part 1, pp. 3831–3846, 2005.

[15] L. Lovisolo, M. P. Tcheou, E. A. B. da Silva, M. A. M. Rodrigues, and P. S. R. Diniz, "Modeling of electric disturbance signals using damped sinusoids via atomic decompositions and its applications," *EURASIP Journal on Advances in Signal Processing*, vol. 2007, Article ID 29507, 2007.

[16] X. Zhang, J. Liang, F. Zhang, L. Zhang, and B. Xu, "Combined model for ultra short-term wind power prediction based on sample entropy and extreme learning machine," *Proceedings of the Chinese Society of Electrical Engineering*, vol. 33, no. 25, pp. 33–40, 2013.

[17] Q. Yuan, W. Zhou, S. Li, and D. Cai, "Approach of EEG detection based on ELM and approximate entropy," *Chinese Journal of Scientific Instrument*, vol. 33, no. 3, pp. 514–519, 2012.

[18] F. Y. Wu, X. F. Le, and D. L. Nan, "A short-term wind speed prediction model using phase-space reconstructed extreme learning machine," *Proceedings of the CSU-EPSA*, vol. 25, no. 1, pp. 136–141, 2013.

[19] G. B. Huang, Q. Y. Zhu, and C. K. Siew, "Extreme learning machine: theory and applications," *Neurocomputing*, vol. 70, no. 1–3, pp. 489–501, 2006.

[20] J. J. Jia, Q. W. Gong, X. Li, Q. Y. Guan, and S. L. Chen, "Single-phase adaptive recluse of shunt compensated transmission lines using atomic decomposition," *Automation of Electric Power Systems*, vol. 37, no. 5, pp. 117–123, 2013.

[21] Q. Jia, L. Shi, N. Wang, and H. Dong, "A fusion method for ground fault line detection in compensated power networks based on evidence theory and information entropy," *Transactions of China Electrotechnical Society*, vol. 27, no. 6, pp. 191–197, 2012.

[22] L. Fu, Z. Y. He, R. K. Mai, and Q. Q. Qian, "Application of approximate entropy to fault signal analysis in electric power system," *Proceedings of the Chinese Society of Electrical Engineering*, vol. 28, no. 28, pp. 68–73, 2008.

[23] Q.-Q. Jia, Q.-X. Yang, and Y.-H. Yang, "Fusion strategy for single phase to ground fault detection implemented through fault measures and evidence theory," *Proceedings of the CSEE*, vol. 23, no. 12, pp. 6–11, 2003.

[24] H. R. Karimi, P. J. Maralani, B. Lohmann, and B. Moshiri, "H_∞ control of parameter-dependent state-delayed systems using polynomial parameter-dependent quadratic functions," *International Journal of Control*, vol. 78, no. 4, pp. 254–263, 2005.

[25] S. Yin, S. X. Ding, A. Haghani, H. Hao, and P. Zhang, "A comparison study of basic data-driven fault diagnosis and process monitoring methods on the benchmark Tennessee Eastman process," *Journal of Process Control*, vol. 22, no. 9, pp. 1567–1581, 2012.

[26] H. M. Yang, L. Huang, C. F. He, and D. X. Yi, "Probabilistic prediction of transmission line fault resulted from disaster of ice storm," *Power System Technology*, vol. 36, no. 4, pp. 213–218, 2012.

[27] H. R. Karimi, "A computational method for optimal control problem of time-varying state-delayed systems by Haar wavelets," *International Journal of Computer Mathematics*, vol. 83, no. 2, pp. 235–246, 2006.

[28] H. Zhang, Z. He, and J. Zhang, "A fault line detection method for indirectly grounding power system based on quantum neural network and evidence fusion," *Transactions of China Electrotechnical Society*, vol. 24, no. 12, pp. 171–178, 2009.

[29] H. R. Karimi, "Robust H∞ filter design for uncertain linear systems over network with network-induced delays and output quantization," *Modeling, Identification and Control*, vol. 30, no. 1, pp. 27–37, 2009.

[30] Z. Kang, D. Li, and X. Liu, "Faulty line selection with non-power frequency transient components of distribution network," *Electric Power Automation Equipment*, vol. 31, no. 4, pp. 1–6, 2011.

[31] S. Yin, H. Luo, and S. X. Ding, "Real-time implementation of fault-tolerant control systems with performance optimization," *IEEE Transactions on Industrial Electronics*, vol. 61, no. 5, pp. 2402–2411, 2014.

[32] X. W. Wang, J. Gao, X. X. Wei, and Y. X. Hou, "A novel fault line selection method based on improved oscillator system of power distribution network," *Mathematical Problems in Engineering*, vol. 2014, Article ID 901810, 19 pages, 2014.

[33] H. R. Karimi, N. A. Duffie, and S. Dashkovskiy, "Local capacity H_∞ control for production networks of autonomous work systems with time-varying delays," *IEEE Transactions on Automation Science and Engineering*, vol. 7, no. 4, pp. 849–857, 2010.

[34] X. W. Wang, X. X. Wei, J. Gao, Y. X. Hou, and Y. F. Wei, "Stepped fault line selection method based on spectral kurtosis and relative energy entropy of small current to ground system," *Journal of Applied Mathematics*, vol. 2014, Article ID 726205, 18 pages, 2014.

Permissions

All chapters in this book were first published in JS, by Hindawi Publishing Corporation; hereby published with permission under the Creative Commons Attribution License or equivalent. Every chapter published in this book has been scrutinized by our experts. Their significance has been extensively debated. The topics covered herein carry significant findings which will fuel the growth of the discipline. They may even be implemented as practical applications or may be referred to as a beginning point for another development.

The contributors of this book come from diverse backgrounds, making this book a truly international effort. This book will bring forth new frontiers with its revolutionizing research information and detailed analysis of the nascent developments around the world.

We would like to thank all the contributing authors for lending their expertise to make the book truly unique. They have played a crucial role in the development of this book. Without their invaluable contributions this book wouldn't have been possible. They have made vital efforts to compile up to date information on the varied aspects of this subject to make this book a valuable addition to the collection of many professionals and students.

This book was conceptualized with the vision of imparting up-to-date information and advanced data in this field. To ensure the same, a matchless editorial board was set up. Every individual on the board went through rigorous rounds of assessment to prove their worth. After which they invested a large part of their time researching and compiling the most relevant data for our readers.

The editorial board has been involved in producing this book since its inception. They have spent rigorous hours researching and exploring the diverse topics which have resulted in the successful publishing of this book. They have passed on their knowledge of decades through this book. To expedite this challenging task, the publisher supported the team at every step. A small team of assistant editors was also appointed to further simplify the editing procedure and attain best results for the readers.

Apart from the editorial board, the designing team has also invested a significant amount of their time in understanding the subject and creating the most relevant covers. They scrutinized every image to scout for the most suitable representation of the subject and create an appropriate cover for the book.

The publishing team has been an ardent support to the editorial, designing and production team. Their endless efforts to recruit the best for this project, has resulted in the accomplishment of this book. They are a veteran in the field of academics and their pool of knowledge is as vast as their experience in printing. Their expertise and guidance has proved useful at every step. Their uncompromising quality standards have made this book an exceptional effort. Their encouragement from time to time has been an inspiration for everyone.

The publisher and the editorial board hope that this book will prove to be a valuable piece of knowledge for researchers, students, practitioners and scholars across the globe.

List of Contributors

Lorena Parra, Laura García and Jaime Lloret
Instituto de Investigación para la Gestión Integrada de Zonas Costeras, Universitat Politècnica de València, Spain

Sandra Sendra
Dep. de Teoría de la Señal, Telemática y Comunicaciones, ETS Ingenierías Informática y de Telecomunicación, Universidad de Granada, C/ Periodista Daniel Saucedo Aranda, s/n E-18071 Granada, Spain

Zhao-Ba Wang and Hai-yang Li
School of Information and Communication Engineering, North University of China, Taiyuan 030051, China

Hong-Ying Guo
School of Information and Communication Engineering, North University of China, Taiyuan 030051, China
Department of Electronics, Xinzhou Teachers University, Xinzhou 034000, China

Min Yao, Yanhua Jin, Min Zhao and Shaohua Xu
College of Automation Engineering, Nanjing University of Aeronautics and Astronautics, Nanjing, China

Hiroyuki Nakamoto Futoshi Kobayashi and Tokiya Yamaji
Graduate School of System Informatics, Kobe University, 1-1 Rokkodai-cho, Nada, Kobe 657-8501, Japan

Akio Yamamoto and Rei Ono
Graduate School of Health Science, Kobe University, 7-10-2 Tomogaoka, Suma, Kobe 654-0142, Japan

Hideo Ootaka and Yusuke Bessho
Bando Chemical Industries Ltd, 4-6-6 Minatojima Minamimachi, Chuo, Kobe 650-0047, Japan

Zhihe Zhao, Jiaheng Wang, Chenxu Fu, Zhenbang Liu, Dawei Liu and Bailiang Li
Xi'an Jiaotong-Liverpool University, Suzhou, China

Jaesung Lee, Wangduk Seo, Ho Han and Dae-Won Kim
School of Computer Science and Engineering, Chung-Ang University, 221, Heukseok-Dong, Dongjak-Gu, Seoul 06974, Republic of Korea

Zhiyuan Zhu, Kequan Xia and Zhiwei Xu
Ocean College, Zhejiang University, Zhoushan 316021, China

Hongze Zhang
Nanjing Electronic Devices Institute, Nanjing 210016, China

Sana Ullah Jan and Insoo Koo
School of Electrical Engineering, University of Ulsan, Ulsan 44610, Republic of Korea

Kyoo Nam Choi
Department of Information and Telecommunication Engineering, Incheon National University, Incheon 22012, Republic of Korea

Cihun-Siyong Alex Gong
Department of Electrical Engineering, School of Electrical and Computer Engineering, College of Engineering, Chang Gung University, Taoyuan, Taiwan
Portable Energy System Group of Green Technology Research Center, Chang Gung University, Taoyuan, Taiwan
Department of Ophthalmology, Chang Gung Memorial Hospital, Linkou, Taoyuan, Taiwan

Huang-Chang Lee
Department of Electrical Engineering, School of Electrical and Computer Engineering, College of Engineering, Chang Gung Taoyuan, Taiwan
Department of Neurosurgery, Chang Gung Memorial Hospital, Linkou, Taoyuan, Taiwan

Yu-Chieh Chuang
Department of Electrical Engineering, School of Electrical and Computer Engineering, College of Engineering, Chang Gung University, Taoyuan, Taiwan

Tien-Hua Li, Chih-Hui Simon Su, Lung-Hsien Huang and Chih-Hsiung Chang
AI-GTG Group, Taoyuan, Taiwan

Chih-Wei Hsu
Chiu Chau Enterprise Co. Ltd., Taoyuan, Taiwan

Yih-Shiou Hwang
Department of Ophthalmology, Chang Gung Memorial Hospital, Linkou, Taoyuan, Taiwan
Graduate Institute of Clinical Medical Sciences, College of Medicine, Chang Gung University, Taoyuan, Taiwan

Jiann-Der Lee
Department of Electrical Engineering, School of Electrical and Computer Engineering, College of Engineering, Chang Gung University, Taoyuan, Taiwan
Department of Neurosurgery, Chang Gung Memorial Hospital, Linkou, Taoyuan, Taiwan
Department of Electrical Engineering, Ming Chi University of Technology, New Taipei City, Taiwan

C. Fernandez-Blanco, C. Garcia-Cabezón and M. L. Rodriguez-Mendez
GroupUVaSens, Department of Inorganic Chemistry, Engineers School, Universidad de Valladolid, Valladolid, Spain

M. Mugica and M. A. Rodriguez-Perez
Cellular Materials Laboratory (CellMat), Condensed Matter Physics Department, University of Valladolid, Campus Miguel Delibes, Paseo de Belén n7, Valladolid, Spain

Nicola Testoni, Valentina Arditi, Federica Zonzini and Luca De Marchi
Department of Electrical, Electronic and Information Engineering, University of Bologna, Bologna, Italy

Cristiano Aguzzi and Tullio Salmon Cinotti
Department of Computer Science and Engineering, University of Bologna, Bologna, Italy

Alessandro Marzani
Department of Civil, Chemical, Environmental and Materials Engineering, University of Bologna, Bologna, Italy

Imran Ali Lakhiar, Gao Jianmin, Tabinda Naz Syed, Farman Ali Chandio and Noman Ali Buttar
Key Laboratory of Modern Agricultural Equipment and Technology, Ministry of Education, Institute of Agricultural Engineering, Jiangsu University, Zhenjiang, 212013 Jiangsu, China

Waqar Ahmed Qureshi
Research Centre of Fluid Machinery Engineering and Technology, Jiangsu University, Zhenjiang, 212013 Jiangsu, China

Hai-bo Zhu and Yong Zhang
School of Electronics and Information Engineering, University of Science and Technology Liaoning, Anshan 114051, China

Bo Yang, Xiaosu Xu, Tao Zhang, Yao Li and Jinwu Tong
Key Laboratory of Micro-Inertial Instrument and Advanced Navigation Technology, Ministry of Education, School of Instrument Science and Engineering, Southeast University, Nanjing 210096, China

Liyue Fu and Aiguo Song
School of Instrument Science and Engineering, Southeast University, Nanjing, Jiangsu 210096, China

Index